WITHDR
CHRIST CHURCH LIBRARY
OXFORD

D1145698

Random vibrations, spectral and wavelet analysis

The author's companion book *Mechanical Vibration Analysis and Computation* is also published by Longman

An introduction to

Random vibrations, spectral and wavelet analysis

Third edition

D. E. Newland

Professor of Engineering
University of Cambridge

Longman
Scientific &
Technical

Copublished in the United States with
John Wiley & Sons, Inc., New York

Longman Scientific & Technical,
Longman Group UK Limited,
Longman House, Burnt Mill, Harlow,
Essex CM20 2JE, England
and Associated Companies throughout the world.

Copublished in the United States with
John Wiley & Sons, Inc., 605 Third Avenue, New York, NY 10158

© D. E. Newland 1975, 1984, 1993

All rights reserved; no part of this publication may
be reproduced, stored in a retrieval system, or
transmitted in any form or by any means, electronic,
mechanical, photocopying, recording, or otherwise
without either the prior written permission of the
Publishers or a licence permitting restricted copying in
the United Kingdom issued by the Copyright Licensing
Agency Ltd, 90 Tottenham Court Road, London W1P 9HE.

First published 1975
Reprinted 1978, 1980, 1981, 1983
Second edition 1984
Reprinted 1986, 1987, 1989, 1991
Third edition 1993

British Library Cataloguing in Publication Data
A catalogue entry for this title is available from
the British Library

ISBN 0582 21584 6

Library of Congress Cataloging-in-Publication Data

ISBN 0-470-22153-4 (USA only)

Set in 10 on 12pt Times by 6

Printed in Malaysia by GPS

Contents

Preface to the first edition   ix

Preface to the second edition   xi

Preface to the third edition   xii

Acknowledgements   xiv

List of symbols   xvii

1   Introduction to probability distributions and averages   1

Probability density function   2
Gaussian distribution   7
Calculation of averages   7
Probability distribution function   10

2   Joint probability distributions, ensemble averages   12

Second-order probability functions   12
Second-order averages   14
Conditional probability   17
Second-order Gaussian distribution   18
Ensemble averaging   19

3   Correlation   21

Autocorrelation   25
Cross-correlation   29

4   Fourier analysis   33

Fourier integral   35
Complex form of the Fourier transform   38

5   Spectral density   41

Narrow band and broad band processes   44
Spectral density of a derived process   46
Cross-spectral density   48
Note on the units of spectral density   51

6 Excitation – response relations for linear systems   53

Classical approach   54
Frequency response method   55
Impulse response method   58
Relationship between the frequency response and impulse
response functions   60
Calculation of response to an arbitrary input   62

7 Transmission of random vibration   67

Mean level   68
Autocorrelation   69
Spectral density   71
Mean square response   73
Cross-correlation   77
Cross-spectral density   78
Probability distributions   79

8 Statistics of narrow band processes   82

Crossing analysis   85
Distribution of peaks   90
Frequency of maxima   92

9 Accuracy of measurements   95

Analogue spectrum analysis   96
Variance of the measurement   97
Analysis of finite length records   103
Confidence limits   108

10 Digital spectral analysis I: Discrete Fourier transforms   113

Discrete Fourier transforms   114
Fourier transforms of periodic functions   116
Aliasing   118
Calculation of spectral estimates   120

11 Digital spectral analysis II: Windows and smoothing   125

Relationship between linear and circular correlation   126
Fourier transform of a train of aperiodic functions   129
Basic lag and spectral windows   131
Smoothing spectral estimates   137
Extending record length by adding zeros   139
Summary   142
Practical considerations   146

12 The fast Fourier transform   150

Basic theory   150
Sample calculation   152

Programming flow charts   155
Practical value of FFT   162
Alternative algorithms   165

13 Pseudo random processes   167

Random binary process   168
Pseudo random binary signals   172
Random multi-level process   172
Spectrum of a multi-level process   179
Generation of random numbers   183
Synthesis of correlated noise sources   185

14 Application notes   189

Response of a resonant mode to broad band excitation   189
Fatigue and failure due to random vibration   191
Excitation by random surface irregularities   196
Simulation of random environments   201
Frequency response function and coherency measurements   206
Local spectral density calculations   218
Weibull distribution of peaks   219

15 Multi-dimensional spectral analysis   225

Two-dimensional Fourier series   225
Properties of the two-dimensional DFT   229
Spectral density of a multi-dimensional random process   233
Discrete spectral density and circular correlation
functions for a two-dimensional random process   239
Two-dimensional windows   245
Two-dimensional smoothing   254
Artificial generation of a two-dimensional random process   256
Generation of an isotropic surface   258
Cross-spectral density between parallel tracks across a
random surface   263

**16 Response of continuous linear systems to stationary
random excitation**   268

Response to excitation applied at a point   268
Response to distributed excitation   277
Normal mode analysis   280
Kinetic energy of a flat plate subjected to uncorrelated
random excitation   290
Single degree-of-freedom analogy   292

17 Discrete wavelet analysis   295
Basic ideas   295
Dilation equations   300

Dilation wavelets   303
Properties of the wavelet coefficients   307
Circular wavelet transforms   322
Discrete wavelet transforms   326
Properties of the DWT   339
Mean-square maps   348
Convolution by wavelets   353
Two-dimensional wavelet transforms   355
Harmonic wavelets   359
Discrete harmonic wavelet transform   364
Concluding comments   370

Appendices

Appendix 1   Table of integrals   371
Appendix 2   Computer programs   373
Appendix 3   The Gaussian probability integral   379
Appendix 4   Distribution of χ_κ^2 381
Appendix 5   Random numbers   383
Appendix 6   Distribution of reverse arrangements for a
             random series   385
Appendix 7   Wavelet programs   387

Problems   401

Answers to problems   457

References   465

Index   472

Preface to the first edition

The importance of random vibrations has increased in recent years. Whereas ten years ago undergraduate courses in engineering and applied science barely mentioned the subject, now a well-educated engineer needs at least some familiarity with the concepts and methods involved. Furthermore the applications of random process theory extend far beyond traditional engineering fields. The same methods of analysis which apply to the time-varying pressure in an aircraft's turbulent boundary layer apply also to the daily fluctuations of an economic index. Because of this growing importance, the theory of random vibrations is now being introduced to undergraduate students, and the methods of measurement and analysis are being increasingly used in research laboratories and in industry.

This book has two main objectives: first, to introduce the fundamental ideas of random vibrations, and, secondly, to deal in some depth with digital spectral analysis, which involves the measurement and analysis of random vibrations. Chapters 1 to 9, which take up about half the book, try to meet the first objective and provide the background for an introduction to the subject at undergraduate level. Chapters 10, 11 and 12 are then more specialized and cover digital spectral analysis and the fast Fourier transform. They deal in detail with the application of discrete Fourier transforms to the analysis of sampled data. The emphasis is on the so-called "direct method" of spectrum estimation based on the FFT (fast Fourier transform) and the approximations involved in this method are discussed at length. In the author's experience, many people who find themselves using the FFT for spectral analysis have an incomplete understanding of the nature of the approximations involved, and it is hoped that these chapters will provide the necessary explanation. The last two chapters of the book deal with the properties of pseudo-random processes and a variety of more specialized subjects including the measurement and application of coherence functions and a brief introduction to the analysis of non-stationary processes.

An important feature of the book is the set of carefully selected problems and answers; most of these have been specially prepared to illustrate points in the text. They serve the dual purpose of allowing readers to test their understanding while at the same time allowing the author to include alternative developments of the theory and different applications which could not otherwise have been discussed because of lack of space.

Although there is a good deal of mathematics, this has been kept as simple as a reasonably complete treatment of the subject allows. The knowledge assumed is limited to that normally possessed by final year undergraduates in the applied sciences, and no additional statistical background is required. This slightly restricts the mathematical scope of the book, but the text is supported by references to more specialized books and to the original literature where appropriate, and mathematically minded readers can pursue these sources if they wish.

The author will be pleased to hear from readers who spot errors and misprints, or who find points where amplification is desirable or see other ways of improving the book. All such suggestions will be gratefully received and most carefully studied.

Sheffield, 1974

D. E. NEWLAND

Disclaimer of warranty

Neither the author nor publisher makes any warranty, express or implied, that the calculation methods, procedures and programs included in this book are free from error. The application of these methods and procedures is at the user's own risk and the author and publisher disclaim all liability for damages, whether direct, incidental or consequential, arising from such application or from any other use of this book.

Preface to the second edition

For the second edition, two additional chapters have been added. Chapter 15 introduces multi-dimensional random processes and discusses multi-dimensional spectral analysis in detail. This is important when a random process depends on more than one independent variable. For example, the height of the sea depends on time and on position. The logic of the multi-dimensional discrete Fourier transform is explained and Appendix 2 now contains a two-dimensional DFT program as well as the original one-dimensional FFT program, which it uses. Chapter 16 extends the previous theory to the random vibration of a continuous linear system subjected to distributed random excitation. This introduces a specialized but growing field of random vibration analysis.

A new section on the Weibull distribution of peaks has been added to Chapter 14 and a number of other small additions made.

The computer programs in Appendix 2 are now listed in both Fortran and Basic languages. Readers with suitable mini-computers will be able to use the latter programs for some of the exercises. Because HP-enhanced Basic has been used, some modifications may be necessary for computers using other versions of Basic, but it is hoped that this will not present serious difficulties.

As before, the author will be glad to hear from readers and will be most grateful for the notification of errors and for suggestions for improvement.

Cambridge, 1984 D. E. NEWLAND

Preface to the third edition

Since completing the second edition in 1984, the most important development in signal analysis has been the wavelet transform. This overcomes a fundamental disadvantage of the Fourier transform. When applied to non-stationary processes, the Fourier transform gives frequency data averaged over the record length being analysed. The local position of particular features is lost. In contrast, *wavelet analysis* allows a general function of time to be decomposed into a series of orthogonal basis functions, called *wavelets*, of different lengths and different positions along the time axis. A particular feature can be located from the positions of the wavelets into which it is decomposed. Results of the wavelet transform can be presented in the form of a contour map in the frequency–time plane. This display, which has been compared to the way that notes are shown on a musical score, allows the changing spectral composition of non-stationary signals to be measured and compared.

Therefore it has seemed desirable to add a chapter on the wavelet transform with three objectives: (i) to include enough of the basic theory of wavelet analysis to establish the background for practical calculations; (ii) to explain how the discrete wavelet transform (DWT) works; (iii) to indicate some of the potential applications of the DWT. In addition it has been necessary to prepare computer programs for the DWT and its inverse, and for a two-dimensional DWT and inverse, and also for displaying their results. Considerable thought has been given to the language for these programs and MATLAB®* has been chosen as currently the most appropriate. To use the programs in Appendix 7 it will therefore be necessary to have MATLAB software. This is becoming increasingly popular in the teaching world as well as in industry and many readers will have access to it already. Others may not, and information is given at the end of the book about how MATLAB may be obtained. Although the program listings will not be usable by readers without MATLAB, the program logic is fully described in Chapter 17, and it is possible for readers who wish to recode the listings given in the language of their choice.

This edition includes all the material in the second edition (subject to minor editing and corrections) together with the following additions: a new section

*MATLAB is a registered trademark of The MathWorks Inc.

for Chapter 14 on local spectral density calculations; an extensive new Chapter 17, Discrete wavelet analysis; a new Appendix 7 with computer programs for wavelet analysis; and a set of supporting problems for Chapter 17. The objective of the new material is to put readers into the same position *vis-à-vis* wavelet analysis as readers of earlier editions have been in respect of spectral analysis. The program listings in Appendix 7 allow readers to make their own wavelet calculations, just as readers can use the programs in Appendix 2 to make their own spectral analysis calculations.

To recognize the importance of wavelet analysis in the overall scope of the new book, it has been decided to change the title from *Random Vibrations and Spectral Analysis* to *Random Vibrations, Spectral and Wavelet Analysis*. Although wavelets have been studied widely in the mathematical world, their application to vibration analysis is still new. The DWT provides an important procedure for analysing non-stationary vibration records, and it is hoped that this third edition will help vibration practitioners to understand and use the wavelet transform.

As for the earlier editions, the author would be glad to hear from readers with corrections and suggestions for improvement.

Cambridge, 1993   D. E. NEWLAND

Acknowledgements for the first edition

During the early 1960s I had the privilege of working with Professor J. P. Den Hartog and Professor Stephen H. Crandall at the Massachusetts Institute of Technology. My professional interest in vibrations dates from that period and it is a pleasure to acknowledge my debt to both these famous teachers. The textbook *Random Vibration in Mechanical Systems* by S. H. Crandall and W. D. Mark (Academic Press, 1963) has been a valuable reference during preparation of the first half of the present book.

I have been most fortunate to have had the untiring assistance of my father-in-law, Philip Mayne, M.A., formerly scholar of King's College, Cambridge. Beginning with the first handwitten draft, he has read every page, checked every formula and worked out every problem. Serious readers will appreciate the enthusiasm needed to undertake such a task and the resolution needed to pursue it to a conclusion. As a result of this help I have made many corrections and alterations to the original manuscript and I believe that many obscurities have been removed. I am more than grateful.

Completion of the manuscript coincided with a visit to Sheffield by Professor Philip G. Hill of Queen's University, Kingston, Ontario, who has been kind enough to read and comment helpfully on a major part of the book. Again I have been pleased to incorporate many of the changes suggested. I am also indebted to my colleague Dr H. K. Kohler, who provided information and advice on a wide range of topics, and to Professor R. E. D. Bishop, who reviewed the manuscript for Longman and gave encouragement at a critical stage. My publishers have always been enthusiastic and I would like to thank them for the patience and care with which this book has been prepared for publication. Lastly I must mention my secretary, Elaine Ibbotson, who toiled painstakingly with successive drafts, cheerfully accepting my determination to alter every section after, but never before, it had been typed.

Only an author who has written a book in his "spare time" knows the tremendous extra demands that this inevitably places on those nearest to him. Without the support and understanding of my wife and family this work could not have been brought to completion.

D.E.N.

Acknowledgements for the second edition

During the preparation of the second edition, I have been glad to have had the help of my doctoral research student David Cebon, formerly of Melbourne University. He has read and commented on the two new chapters and tested the Fortran version of the new computer program. A first draft of the new chapters was used as lecture notes at a short course at Monash University in 1983 and this allowed a number of errors and misprints to be identified. Professor J. D. C. Crisp organized this course and I am very glad to have had that opportunity to try out the new material for the second edition. I am grateful also for discussions at Monash with Professor W. H. Melbourne, as a result of which I decided to add a section on the Weibull distribution of peaks to Chapter 14.

I have been encouraged by the comments I have received from readers of the first edition. As a result of readers' suggestions, I have been able to make a number of corrections and improvements to the text, and I am most grateful to all those who have helped in this way.

D.E.N.

The author and publishers are indebted to the following for permission to reproduce copyright material: the McGraw-Hill Book Company for Fig. 12.7(a) which is based on Fig. 6.23 of Gold and Rader, *Digital Processing of Signals*, 1969; the Institute of Electrical and Electronics Engineers for the computer program in Appendix 2 which is based on that listed on p. 29 of the *IEEE Transactions on Education*, Vol. 12, No. 1, March 1969 (paper by Cooley, Lewis and Welch, "The Fast Fourier Transform and its Applications"); and to the authors and publishers of Fisher and Yates, *Statistical Tables for Biological, Agricultural and Medical Research* (6th edition), Longman, 1974 (previously published by Oliver and Boyd), from which Tables II(i), IV and XXXIII(i) are reproduced here as Appendices 3, 4 and 5 respectively.

Acknowledgements for the third edition

In preparing the new chapter on wavelets for this edition, I have studied many papers on the mathematics of wavelets. I am indebted to all their authors, but particularly to Professors Ingrid Daubechies and Stéphane Mallat whose pioneering papers have been immensely important in the subject's development. For me, a most valuable contribution has been the review paper by Professor Gilbert Strang [115]; this made clear many ideas that I had only gleaned roughly from other publications.

Many colleagues have helped in various ways, but I would like to mention particularly Dr Bill Fitzgerald who suggested key publications and commented on a first draft of Chapter 17, and Mrs Margaret Margereson who word-processed the text with care and good humour. My thanks are due to both of them.

There would not have been time to prepare this new edition without the opportunity afforded by sabbatical leave from the furious pace of university teaching, research and administration in the 1990s. I am grateful to Cambridge University for that opportunity.

D.E.N.

List of symbols

The principal symbols in chapters 1 to 14 are defined below with, where appropriate, reference to the chapter or equation in which each symbol first appears.

a (i) Amplitude of a periodic function.
(ii) Constant level $y = a$ during the analysis of crossings (Ch. 8).
(iii) Peak height for one cycle of a narrow band random process (Ch. 8).
(iv) Amplitude of a random binary signal (Ch. 13).

$a_0, a_1, a_2, \ldots, a_k, \ldots$ (i) Coefficients of Fourier series (4.1).
(ii) Coefficients of Fourier series (10.1) (which, except for a_0, differ from (4.1) by a factor 2).
(iii) Coefficients of differential equations of motion (6.1).

$\{a_r\}$ Re-ordered array of terms, $r = 1, 2, \ldots, N$, for entry to the FFT calculation, Fig. 12.4.

$\{a_r'\}$ Array of terms, $r = 1, 2, \ldots, N$, after completion of one row of butterfly calculations in the FFT algorithm, Fig. 12.4.

$a(t)$ Deterministic (non-random) function of time (14.84).

$b_0, b_1, b_2, \ldots, b_k, \ldots$ (i) Coefficients of Fourier series (4.1).
(ii) Coefficients of Fourier series (10.1) (which differ from (4.1) by a factor 2).
(iii) Coefficients of differential equations of motion (6.1).

c, c_1, c_2 Viscous damping coefficients.

d_r	Discrete value of the data weighting function $d(t)$, defined by $d_r = d(t = r\Delta)$ (Ch. 11).
$d(t)$	Data weighting function or "data window" (Ch. 11).
$e^{i\omega t}$	$\cos \omega t + i \sin \omega t$.
f	Frequency (Hz = c/s).
f_0	(i) Filter centre frequency (Hz) (Ch. 9). (ii) Maximum frequency of significant spectral components present in a signal (Hz) (Ch. 10). (iii) Clock frequency of a random binary process (Hz) (Ch. 13).
f_1	Lower cut-off frequency (Hz) of band-pass filter (Ch. 9).
$f(t)$	Mathematical function of t, usually representing force.
$f(x, y)$	Mathematical function of x and y.
$h(t)$	Impulse response function for a linear system subjected to an impulsive input of unit magnitude ((6.14) and (6.15)).
$h_1(t), h_2(t), \ldots$	Impulse response functions relating the output to a unit impulse at input 1, input 2, ..., respectively.
i	$\sqrt{-1}$.
j	Variable integer.
k	(i) Spring stiffness. (ii) Variable integer.
k_1, k_2	Spring stiffnesses.
l	Variable integer.
m	(i) Mean value of a random variable. (ii) Mass. (iii) Variable integer.
m_x, m_y	Mean value of the random variable x, y or of the stationary random process $x(t)$, $y(t)$.
n	(i) Variable integer. (ii) Number of sample values satisfying a certain condition.

$n_a^+(T)$ — Number of crossings of the level $y = a$ with positive slope in time T for a single member function of the random process $y(t)$ (Ch. 8).

$p_p(a)$ — Probability density function describing the distribution of the random variable a which represents the height of peaks of a narrow band random process (Ch. 8).

$p(T)$ — Probability density function for the distribution of the time T at which a first crossing occurs (Ch. 14).

$p(x)$ — (First order) probability density function describing the distribution of the random variable x (Ch. 1).

$p(x, y)$ — Second-order probability density function describing the joint distribution of the random variables x and y (Ch. 2).

$p(x|y)$ — Conditional probability density function for the distribution of x when y is specified (2.10).

$p(\chi_\kappa^2)$ — Probability density function describing the distribution of the random variable χ_κ^2 (defined in problem 9.3).

r — Variable integer.

s — Variable integer.

t — (i) Time.
(ii) Variable integer.

$t_0, t_1, t_2, \ldots$ — Instants of time.

u_r — Weighted version of $\hat{R}_r$ defined by

$$u_r = \frac{N - |r|}{N} \hat{R}_r \qquad 0 \leqslant |r| \leqslant N \quad (11.11)$$

and used with (11.10) to interpret the periodic series $\{R_r\}$ as values of a weighted estimate of the linear correlation function $R(\tau)$ (Ch. 11)

$u(\tau)$ — Continuous aperiodic function of τ from which the values u_r can be derived by putting

$$u_r = u(\tau = r\Delta) \quad 0 \leqslant |\tau| \leqslant T \quad \text{(Ch. 11)}.$$

$w(\tau)$ — Correlation weighting function (the "lag window") which satisfies (11.23) and is the inverse Fourier transform of the spectral window $W(\Omega)$.

x	Usually refers to a random variable obtained by sampling member functions of a random process $x(t)$.
$\lvert x \rvert$	Denotes the modulus (or absolute magnitude) of x.
x_0	Specific value of x.
$x_1, x_2, x_3, \ldots, x_r, \ldots$	(i) Different Gaussian random variables each with zero mean and unit variance (Ch. 9). (ii) Discrete values obtained by sampling a continuous function $x(t)$ at regular intervals (Ch. 10).
$\{x_r\}$	Series of discrete (real) values obtained by sampling a continuous function $x(t)$ which is usually the time history of a member function of the random process $\{x(t)\}$.
$x(t)$	(i) Random process composed of an ensemble of member functions each of which is a different function of time t. (ii) Input to a linear system which may be a deterministic (non-random) function of time or a random process.
$\{x(t)\}$	The brackets $\{\ \}$ are used to denote the ensemble of sample functions $x(t)$, definition (i) above, if this is not otherwise clear.
$x_1(t), x_2(t), \ldots, x_r(t),$ $x_s(t), \ldots$	(i) Member functions from the ensemble of functions which make up the random process $x(t)$. (ii) Different inputs to a linear system: these may be deterministic (non-random) functions of time or different random processes.
$\dot{x}(t)$	$\dfrac{\mathrm{d}x}{\mathrm{d}t}$
$\dot{x}(t + \tau)$	$\dfrac{\mathrm{d}}{\mathrm{d}(t + \tau)}\, x(t + \tau)$
$\ddot{x}(t)$	$\dfrac{\mathrm{d}^2 x}{\mathrm{d}t^2}$
$\ddot{x}(t + \tau)$	$\dfrac{\mathrm{d}^2}{\mathrm{d}^2(t + \tau)}\, x(t + \tau)$
y_0	Amplitude of a deterministic (non-random) function of time $y(t)$.

y_r	(i) rth sample value of $y(t)$.
	(ii) Alternate even values of $\{x_r\}$, $r = 0, 1, 2, \ldots,$ $(N - 1)$, defined by (12.1).
$y(t)$	(i) Random process composed of an ensemble of member functions each of which is a different function of t.
	(ii) Output from a linear system which may be a deterministic (non-random) function of time or a random process.
$y_1(t), y_2(t), \ldots,$ $y_r(t), \ldots$	(i) Member functions from the ensemble of functions which make up the random process $y(t)$.
	(ii) Different outputs from a linear system: when the inputs are deterministic they represent different deterministic functions of time, when the inputs are random they represent different random processes at the separate outputs.
z_r	Alternate odd values of $\{x_r\}$ $r = 0, 1, 2, \ldots, (N - 1)$, defined by (12.1).
$z(t)$	(i) Output of an analogue spectrum analyser (Fig. 9.1) which is itself a random process defined by (9.1).
	(ii) Stationary random process (14.84).
$A_0, A_1, A_2, \ldots$	Constant (real) coefficients of the complex frequency response function $H(\omega)$ defined in Appendix 1.
$A(\omega)$	Real part of Fourier transform.
B	Bandwidth of filter (Hz) (Ch. 9).
B_e	Effective bandwidth of spectral window (Hz) defined by (9.28).
$B_0, B_1, B_2, \ldots$	Constant (real) coefficients of the complex frequency response function $H(\omega)$ defined in Appendix 1.
$B(\omega)$	Imaginary part of Fourier transform.
$C(\omega)$	Real part of Fourier transform.
$D(\omega)$	Imaginary part of Fourier transform.
$E[\ ]$	Denotes the ensemble averaged value of the quantity in square brackets.

H_0	Peak magnitude of narrow band filter frequency response function (Fig. 9.1).
$H(\omega)$	Complex frequency response function for a linear system subjected to a harmonic input of unit amplitude ((6.10) and (6.11)). $H(\omega)$ is the Fourier transform of $h(t)$, see (6.21) and (6.22).
$H^*(\omega)$	Complex conjugate of $H(\omega)$.
$H_1(\omega), H_2(\omega), \dots,$ $H_r(\omega), H_s(\omega), \dots$	Complex frequency response functions relating the output to input 1, input 2, ..., input r, input s, ... respectively.
$H_{rs}(\omega)$	Complex frequency response function relating output r to input s (for the case when there is more than one output, see problem 7.5, equation (7.30), and Ch. 14).
I_1, I_2	Value of a definite integral.
L	Number of zero terms added to a discrete series of N terms to increase the total number of terms to a power of 2 for acceptance by the FFT (fast Fourier transform) (Ch. 11).
N	(i) Number of sample values. (ii) Number of separate inputs. (iii) Number of terms in a series. (iv) Number of terms in a pseudo random binary sequence. (v) Number of stress cycles to cause failure by fatigue.
$N(S)$	Number of cycles of stress level S which cause fatigue failure (Ch. 14).
$N_a^+(T)$	Average number of crossings of the level $y = a$ with positive slope in time T for a random process $y(t)$ (Ch. 8).
Prob	Abbreviation for "probability".
$P(x)$	(First-order) probability distribution function for the random variable x (1.21).
$P_0(T)$	Probability of there being no crossings (of the level a) in time T (Ch. 14).

R_r | Term of the periodic sequence $\{R_r\}$ defined by

$$R_r = \frac{1}{N} \sum_{s=0}^{N-1} x_s y_{s+r} \qquad (10.20)$$

$r = 0, 1, 2, \ldots, (N - 1)$.

Values of R_r are related to the continuous circular correlation function $R_c(\tau)$ by (10.21) and to the discrete linear correlation function estimate $\hat{R}_r$ by (11.10).

$\hat{R}_r$ | Term of the sequence $\{\hat{R}_r\}$ defined by

$$\hat{R}_r = \frac{1}{N - r} \sum_{s=0}^{N-1-r} x_s y_{s+r} \qquad (11.8)$$

$r = 0, 1, 2, \ldots, (N - 1)$, which is the discrete form of the linear correlation function estimate $\hat{R}(\tau)$.

$R(\tau)$ | Correlation function.

$R_x(\tau), R_{xx}(\tau)$ | Autocorrelation function for the stationary random process $x(t)$ (3.13).

$R'_x(\tau)$ | Autocorrelation function for a pseudo random binary signal $x(t)$ (Ch. 13).

$R_y(\tau), R_{yy}(\tau)$ | Autocorrelation function for the stationary random process $y(t)$.

$R_{xy}(\tau), R_{yx}(\tau)$ | Cross-correlation functions between the stationary random processes $x(t)$ and $y(t)$ (3.20).

$R_c(\tau)$ | Circular (cross-) correlation function defined so that values of $\{R_r\}$ can be derived by putting

$$R_r = R_c(\tau = r\Delta) \qquad (10.21)$$

$R_c(\tau)$ is periodic so that $R_c(\tau + T) = R_c(\tau)$ where T = record length. As defined here, it is a continuous function which is fitted to the values of $\{R_r\}$ and so is peculiar to the pair of records being analysed.

$\hat{R}(\tau), \hat{R}_{xy}(\tau)$ | Linear (cross-) correlation function estimate obtained by calculating a sample average from a continuous record of length T according to

$$\hat{R}_{xy}(\tau) = \frac{1}{T - \tau} \int_0^{T-\tau} x(t)\, y(t + \tau)\, dt \qquad (11.6)$$

$0 \leqslant \tau < T$.

S — Stress amplitude for calculation of fatigue life (Ch. 14).

S_0 — (i) (Constant) spectral density value.
(ii) Measured value of spectral density (Ch. 9).

$S(\omega)$ — Spectral density (strictly mean-square spectral density).

$\tilde{S}(\omega)$ — Smoothed spectral density obtained by averaging over a range of frequencies on either side of the centre frequency ω according to

$$\tilde{S}(\omega) = \int_{-\infty}^{\infty} W(\Omega - \omega)\, S(\Omega)\, d\Omega \qquad (9.26)$$

or the equivalent equations (11.27) and (11.28); also used more generally in Ch. 14 to denote a calculated (rather than a measured) value of spectral density (see, e.g. (14.61)).

$\hat{S}(\omega_k)$ — (Unweighted) average of spectral values in the region of ω_k defined by (11.39).

$\tilde{S}'(\omega_k)$ — Smoothed spectral density at centre frequency ω_k after correction for added zeros; defined by

$$\tilde{S}'(\omega_k) = \frac{N+L}{N}\tilde{S}(\omega_k) \qquad \text{(Ch. 11)}.$$

$S_c(\omega)$ — Fourier transform of the circular correlation function $R_c(\tau)$ (Ch. 10).

S_k, S_{xy_k} — Spectral coefficient obtained by calculating the DFT of $\{R_r\}$, $r = 0, 1, 2, \ldots, (N-1)$; an estimate for $(2\pi/T)\tilde{S}(\omega_k)$ where $\tilde{S}(\omega_k)$ is the weighted spectral density defined by (11.22) and T is the record length analysed. From (10.24)

$$S_k = S_{xy_k} = X_k^* Y_k$$

and, by definition, $S_{-k} = S_k^*$. Also $S_{xx_k} = X_k^* X_k$, $S_{yx_k} = Y_k^* X_k$ and $S_{yy_k} = Y_k^* Y_k$ (10.26).

$S_x(\omega), S_{xx}(\omega)$ — Auto-spectral density of the stationary random process $x(t)$ (5.1).

$S_{\dot{x}}(\omega)$ — Auto-spectral density of the $\dot{x}(t)$ process where $\dot{x}(t) = dx/dt$ and $x(t)$ is a stationary random process (5.17).

$S_{\ddot{x}}(\omega)$	Auto-spectral density of the $\ddot{x}(t)$ process where $\ddot{x}(t) = d^2x/dt^2$ and $x(t)$ is a stationary random process (5.19).
$S'_x(\omega)$	Spectral density for a pseudo random binary signal $x(t)$ (Ch. 13).
$S_x(\omega_1, \omega_2)$	Double frequency spectral density function for a non-stationary random process $x(t)$ (whose auto-correlation function depends on absolute time as well as on time lag) (14.79).
$S_{xy}(\omega), S_{yx}(\omega)$	Cross-spectral densities between the two stationary random processes $x(t)$ and $y(t)$ (5.20).
$S_y(\gamma)$	Spatial spectral density of surface irregularities y as a function of wavenumber γ.
T	(i) Time period. (ii) Averaging time (Ch. 9). (iii) Record length (Ch. 9) which is expressed as $T = N\Delta$ when Δ = sampling interval and N = number of sampled points (Ch. 10). (iv) Fatigue life (Ch. 14).
T_L	Length of record after adding L zero terms so that $$T_L = (N + L)\Delta \quad (\text{Ch. 11}).$$
$U(\omega)$	Fourier transform of $u(\tau)$ (Ch. 11).
W	Complex multiplying factor $e^{-i(2\pi/N)}$ (the "twiddle factor") arising in the FFT algorithm (12.8).
$W(\Omega)$	Spectral weighting function (the "spectral window") which satisfies (9.27) and is used to calculate the smoothed spectral density $\tilde{S}(\omega)$ according to (9.26). $W(\Omega)$ is the Fourier transform of the lag window $w(\tau)$ (11.25).
$W_a(\Omega)$	Spectral weighting function corrected for aliasing (the "aliased spectral window") – see equation (11.57), problem 11.7.
$W_x(f)$	Auto-spectral density of the stationary random process $x(t)$ expressed for positive frequencies only and with the units (mean square)/Hz (5.25).
X	Length parameter (14.32).

X_k	(i) Complex coefficient of the Fourier series (10.1) defined by

$$X_k = a_k - ib_k$$

	(ii) Term of the discrete Fourier transform (DFT) of the series $\{x_r\}$, $r = 0, 1, 2, \ldots, (N-1)$, defined by (10.8).
$X(\omega)$	Fourier transform of $x(t)$ (4.10).
$X^*(\omega)$	Complex conjugate of $X(\omega)$.
Y_k	Term of the discrete Fourier transform (DFT) of the series $\{y_r\}$, $r = 0, 1, 2, \ldots, (N-1)$, defined by (10.19). Also used for the DFT of a series of $N/2$ terms (12.2).
$Y(\omega)$	Fourier transform of $y(t)$.
Z_k	Term of the discrete Fourier transform (DFT) of the series $\{z_r\}$, $r = 0, 1, 2, \ldots, (N/2 - 1)$, defined by (12.2).
α	(i) Exponential decay rate (units of time^{-1}). (ii) Angle (rad). (iii) Constant (real) coefficient (13.18).
β	Constant (real) coefficient (13.18).
γ	Wavenumber of surface irregularities (rad/unit of length) (Ch. 14).
$\delta(\tau), \delta(x)$	Dirac's delta function (which is zero everywhere except when its argument τ, x is zero, when it is infinite in such a way that (5.9) is satisfied).
$\delta\left(\omega - \dfrac{2\pi k}{T}\right)$	Dirac's delta function, which is zero everywhere except when

$$\omega = \frac{2\pi k}{T}$$

	when it is infinite in such a way that (10.14) is satisfied.
ε	Quantization step for analogue-to-digital conversion, Fig. 11.9.
ζ	Damping ratio (14.4).
$\eta_{yx}^2(\omega)$	Single (or ordinary) coherence function between $y(t)$ and $x(t)$ (14.75).

$\eta^2_{y \cdot x}(\omega)$	Multiple coherence function between an output $y(t)$ and all the measured inputs $x_1(t)$, $x_2(t)$, etc. (14.62).
θ	(i) Phase angle (rad). (ii) Time variable (Ch. 6).
θ_1, θ_2	Different time variables.
κ	Number of statistical degrees-of-freedom in χ^2_κ (Ch. 9).
λ	Wavelength of surface irregularities (units of length) (Ch. 14).
μ_y	Average frequency of maxima of the random process $y(t)$.
v^+_a	Average frequency of positive slope crossings of the level $y = a$ for a random process $y(t)$ (Ch. 8).
v^+_0	Average frequency of positive slope crossings of the zero level (Ch. 8).
ξ	Dummy variable for x (1.23).
ρ_{xy}	Correlation coefficient (normalized covariance) for the random variables x and y (2.13).
σ	Standard deviation.
σ_x, σ_y	Standard deviation of the random variable x, y or of the stationary random process $x(t)$, $y(t)$.
$\sigma_{\dot{y}}$	Standard deviation of the stationary random process $$\dot{y}(t) = \frac{dy(t)}{dt} \quad \text{(Ch. 8)}.$$
$\sigma_{\ddot{y}}$	Standard deviation of the stationary random process $$\ddot{y}(t) = \frac{d^2 y(t)}{dt^2} \quad \text{(Ch. 8)}.$$
$\dfrac{\sigma}{m}$	Ratio of the standard deviation of a measurement (of spectral density) to its ensemble averaged mean value.
τ	(i) Time lag (time difference). (ii) Absolute time variable (Ch. 6).
τ'	Time lag (time difference).

xxviii   *List of symbols*

τ_0	Specific value of τ.
ϕ	(i) Phase angle (rad). (ii) Time variable (Ch. 7). (iii) Phase (time) of a sample function of a random binary process (Ch. 13).
χ_κ^2	Random variable which is defined as the sum of squares of κ independent Gaussian random variables each with zero mean and unit variance (9.30) and is used as a statistical model to calculate confidence limits for spectral measurements (χ is the Greek letter chi, pronounced "keye"; the subscript κ is Greek kappa).
ω	Angular frequency (rad/s).
ω_k	Angular frequency (rad/s) of the kth harmonic defined by

$$\omega_k = \frac{2\pi k}{T} \quad (4.3).$$

ω_N	Natural frequency of a resonant system (rad/s).
ω_0	(i) Specific angular frequency (rad/s). (ii) Filter centre frequency (rad/s) (Ch. 9). (iii) Maximum frequency of significant spectral components present in a signal (rad/s) (Ch. 10).
ω_1	(i) Specific angular frequency (rad/s). (ii) Dummy frequency variable (rad/s) describing frequency difference from a reference value (Ch. 11).
$\omega_2, \omega_3, \ldots$	Specific angular frequencies (rad/s).
Δ	(i) Deviation of a sample value of a random variable from its predicted value (Ch. 3). (ii) Sampling interval for analogue-to-digital conversion (Ch. 10).
Δt	Clock interval for a random binary process (Ch. 13).
$\Delta\omega$	(i) Frequency increment of magnitude $2\pi/T$ (4.4). (ii) Half bandwidth of resonance curve (Ch. 7). (iii) Bandwidth of spectral density curve (Ch. 8). (iv) Bandwidth of filter frequency response function (Ch. 9). (v) Spacing between spectral lines for a pseudo random binary signal (Ch. 13).

$\sum$	Denotes "the summation of".
Ω	Dummy frequency variable (rad/s) (9.26).
*	Denotes the complex conjugate of a quantity.

Chapter 1

Introduction to probability distributions and averages

A system is vibrating if it is shaking or trembling or moving backwards and forwards in some way. If this motion is unpredictable then the system is said to be in *random vibration*. For instance the motion of a leaf fluttering in the breeze is unpredictable. The leaf is subjected to random excitation as the wind's direction and strength change and as a result it moves backwards and forwards in random vibration. However the rate and amount of movement of the leaf are dependent not only on the severity of the wind excitation, but also on the mass, stiffness and inherent damping in the leaf system. The subject of random vibrations is concerned with finding out how the statistical (or average) characteristics of the motion of a randomly excited system, like a leaf, depend on the statistics of the excitation, in that case the wind, and the dynamic properties of the vibrating system, in that case the mass, stiffness and damping of the leaf system.

Figure 1.1 shows part of a possible time history for a system in random vibration. The displacement x from an arbitrary datum is plotted as a function

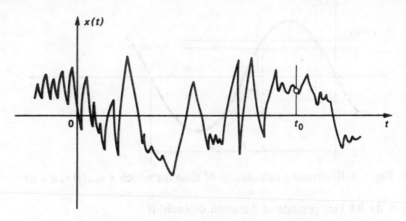

Fig. 1.1 Possible time history for a system in random vibration

of time t. Since motion is random, the precise value of x at any chosen time $t = t_0$ cannot be precisely predicted. The best we can do is to find the chance, or *probability*, that x at t_0 will lie within certain limits. The subject of probability is therefore at the heart of random vibration theory and we begin by considering some of the fundamental ideas of probability theory.

Probability density function

Suppose, first, that we are dealing with a time history which is non-random or *deterministic* and is, in fact, a sine wave, Fig. 1.2. In this case we can exactly

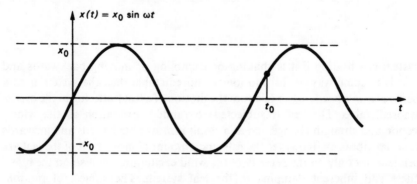

Fig. 1.2 Waveform for steady state deterministic vibration $x(t) = x_0 \sin \omega t$

predict the value of x for any given value of t. We can therefore calculate the proportion of time that the waveform spends between any two levels of x. With reference to Fig. 1.3, during one complete cycle, $x(t)$ lies in the band x

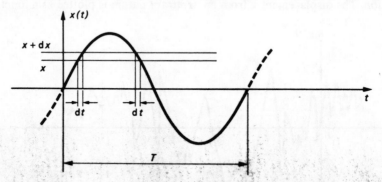

Fig. 1.3 Illustrating calculation of time for which $x \leqslant x(t) \leqslant x + \mathrm{d}x$

to $x + \mathrm{d}x$ for two periods of duration $\mathrm{d}t$ each. If

$$x = x_0 \sin \omega t \tag{1.1}$$

then

$$\mathrm{d}x = x_0 \omega \cos \omega t \, \mathrm{d}t$$

so that

$$dt = \frac{dx}{x_0 \omega \cos \omega t}. \tag{1.2}$$

Now substituting for

$$\cos \omega t = \sqrt{(1 - \sin^2 \omega t)}$$

from (1.1) into (1.2) gives

$$dt = \frac{dx}{x_0 \omega \sqrt{(1 - x^2/x_0^2)}}. \tag{1.3}$$

The *proportion* of time per cycle that $x(t)$ spends in the band x to $x + dx$ is therefore

$$\frac{2(dt)}{T} = \frac{2dx}{\omega T \sqrt{(x_0^2 - x^2)}}$$

which, putting

$$T = \frac{2\pi}{\omega}$$

for the period of the sine wave, gives

$$\frac{2(dt)}{T} = \frac{dx}{\pi \sqrt{(x_0^2 - x^2)}}. \tag{1.4}$$

For any complete number of cycles, equation (1.4) gives the proportion of the total elapsed time for which $x(t)$ lies within the x to $x + dx$ band.

Now consider a situation in which we have to choose an instant of time, $t = t_0$ (say), and find the value of x at this instant. Since $x(t)$ is a deterministic sine wave, as soon as we specify t_0, we immediately know $x(t_0)$. But suppose that t_0 is not precisely specified. Instead we are just told that it may lie anywhere along the time axis. In this case we cannot say what $x(t_0)$ will be; the best we can do is to predict what it *may* be. If t_0 is chosen perfectly arbitrarily, t_0 may lie anywhere during a complete cycle (the record is assumed to exist for ever with no beginning and no ending so that there is no question of t_0 falling in an unfinished cycle). The chance or *probability* that $x(t_0)$ lies in the band x to $x + dx$ will then depend only on how long per cycle $x(t)$ lies between x and $x + dx$. The probability that $x \leqslant x(t_0) \leqslant x + dx$ is therefore given by

Prob $(x \leqslant x(t_0) \leqslant x + dx) =$ Fraction of time per cycle for which $x(t)$ lies within the x to $x + dx$ band

$$= \frac{2(dt)}{T} \tag{1.5}$$

$$= \frac{dx}{\pi \sqrt{(x_0^2 - x^2)}} \quad \text{for } -x_0 \leqslant x \leqslant x_0$$

since x_0 is the amplitude of motion.

The *first-order probability density function* $p(x)$ is defined so that

$$\text{Prob}\,(x \leqslant x(t_0) \leqslant x + dx) = p(x)\,dx$$

(1.6)

therefore in this case, from (1.5),

$$p(x) = \frac{1}{\pi\sqrt{(x_0^2 - x^2)}} \quad \text{for } -x_0 \leqslant x \leqslant x_0.$$

(1.7)

This function is plotted in Fig. 1.4. Notice that this is a probability *density* curve so that it is the *area* under the curve which gives probability. Hence the

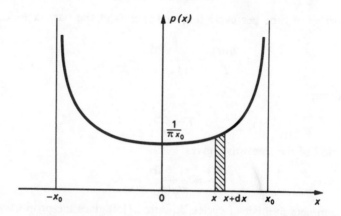

Fig. 1.4 Probability density function $p(x)$ for the sine wave of Fig. 1.2

probability that $x(t_0)$ lies in the band x to $x + dx$ is given by the shaded area shown in Fig. 1.4 and the probability that $x(t_0)$ has any value between $-x_0$ and $+x_0$ is the total area under the curve which is

$$\text{Prob}\,(-x_0 \leqslant x(t_0) \leqslant x_0) = \int_{-x_0}^{x_0} p(x)\,dx$$

$$= \int_{-x_0}^{x_0} \frac{1}{\pi\sqrt{(x_0^2 - x^2)}}\,dx.$$

(1.8)

Since, for the sine wave shown in Fig. 1.2, any value of x, chosen at random, must lie in the range $-x_0$ to $+x_0$, the integral on the r.h.s. of (1.8) must be unity, since the probability of a certainty is 1 or 100 per cent. This fact may easily be verified by reference to any standard list of integrals. The probability that $x(t_0)$ lies in any range of values outside the band $-x_0$ to x_0 is of course zero since a constant amplitude sine wave never strays outside its limiting values $-x_0$ and $+x_0$.

The probability density function $p(x)$ gives the density of the distribution of values of x and it may be helpful to see this as drawn in Fig. 1.5. Since a sine wave spends more of its time near its peak values than it does near its mean value, the probability density function increases towards the extremes of motion and is a minimum at the mean.

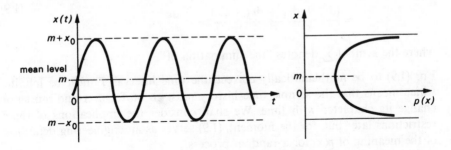

Fig. 1.5 Illustrating the probability density function $p(x)$ for a sine wave super-imposed on a constant mean level

Now consider the situation when $x(t)$ is no longer a sine wave, but instead represents a random process. When we say that $x(t)$ is random, we mean that the values of $x(t)$ cannot be precisely predicted *in advance*. As soon as they occur, these values can of course be accurately recorded and it is easy to imagine a situation in which we have an accurate record or time history of $x(t)$ for all time up to the present. Assuming that the statistical characteristics of $x(t)$ are not changing with time, then we can use this time history to calculate the proba-bility density function for $x(t)$ in exactly the same way as we have just done for a deterministic function. Figure 1.6 shows a sample time history for a

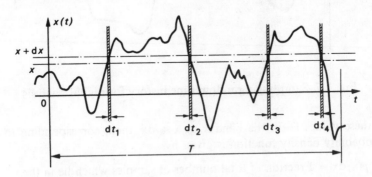

Fig. 1.6 Calculation of the probability density function $p(x)$ for a random process

random process with the times for which $x \leqslant x(t) \leqslant x + dx$ identified by the shaded strips. During the time interval T, $x(t)$ lies in the band of values x to $x + dx$ for a total time of $(dt_1 + dt_2 + dt_3 + dt_4)$. We can therefore say that, if T is long enough, the probability density function $p(x)$ is given by

$p(x)\,dx$ = Fraction of the total elapsed time for which $x(t)$ lies in the x to $x + dx$ band

$$= \frac{(dt_1 + dt_2 + dt_3 + \ldots)}{T} = \frac{\sum dt}{T} \tag{1.9}$$

where the symbol $\sum$ denotes "the summation of".

For (1.9) to be mathematically correct, the time interval T must be infinite, which means that the sample time history must go on for ever and must not change its character with time. We shall consider the implications of these restrictions later but, for the moment, (1.9) serves as an engineering definition of the meaning of $p(x)$ for a random process.

When $x(t)$ is a random function of t, we cannot use (1.9) to calculate a mathematical expression for $p(x)$. For any given sample time history the only thing to do is to measure $p(x)$ by laboriously dividing the sample record into many different levels, measuring the time spent at each band of values of x, and then using equation (1.9). Alternatively an instrument called a probability analyser will do the same thing much more quickly by sampling the time history at a series of closely spaced intervals. If there are N sample values, Fig. 1.7, and

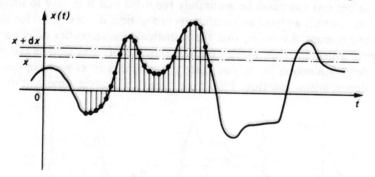

Fig. 1.7 Sampling a random time history for digital analysis

dn of these values lie in the band x to $x + dx$, then, corresponding to (1.9), the probability density function is given by

$p(x)\,dx$ = Fraction of total number of samples which lie in the x to $x + dx$ band

$$= \frac{dn}{N}. \tag{1.10}$$

Again the length of the record and the spacing between sampling points must be such that the true characteristics of the random process are identified, and we shall return to the requirements for accurate digital analysis later.

Gaussian distribution

It is an interesting fact of life that many naturally occurring random vibrations have the well-known "bell-shaped" probability distribution shown in Fig. 1.8.

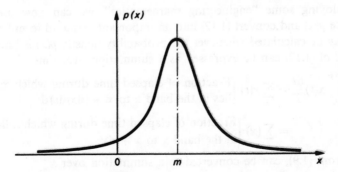

Fig. 1.8 First-order probability density for a normal (or Gaussian) process

When the shape of the bell is given by the equation

$$p(x) = \frac{1}{\sqrt{2\pi}\,\sigma} e^{-(x-m)^2/2\sigma^2} \tag{1.11}$$

where m and σ are constants which we shall identify shortly, the random variable x has a *normal* or *Gaussian* probability distribution, and this distribution is extensively used in random vibration theory to approximate the characteristics of random excitation.

Calculation of averages

Assuming that the probability density function $p(x)$ is available for a random process, it can be used to calculate certain statistics of the random process $x(t)$. To illustrate how this is done, we consider first the mean value of x which is usually denoted by $E[x]$ where the E stands for "the statistical expectation of".† With reference to Fig. 1.6, the mean value of the time history of x over the interval T is defined so that

$(E[x])T =$ Total area under the $x(t)$ curve during the interval T (areas below the zero line subtracting from the total area)

$$= \int_0^T x(t)\,dt$$

---

† Strictly the symbol E should be reserved for an ensemble average (see later) rather than a sample average, as here, but for the moment this distinction is ignored.

and hence

$$E[x] = \int_0^T x(t)\frac{dt}{T}. \tag{1.12}$$

By employing some "engineering mathematics", we can now use (1.9) to introduce $p(x)$ and convert (1.12) into an important standard form from which $E[x]$ may be calculated whenever the probability density $p(x)$ is known. First the r.h.s. of (1.12) can be expressed as a summation over time

$$\int_0^T x(t)\frac{dt}{T} = \sum_t x(t)\cdot\left(\begin{matrix}\text{Fraction of elapsed time during which } x(t) \\ \text{lies in the band } x \text{ to } x + (dx/dt)\,dt\end{matrix}\right)$$

$$= \sum_t (x)\cdot\left(\begin{matrix}\text{Fraction of elapsed time during which } x \text{ lies} \\ \text{in the band } x \text{ to } x + dx\end{matrix}\right)$$

which, from (1.9), can be converted to a summation over x

$$= \sum_x (x)\cdot\left(p(x)\,dx\right)$$

$$= \int_{-\infty}^\infty x\,p(x)\,dx \tag{1.13}$$

by a heuristic argument in which we replace an integral with respect to time by a summation, and then in turn replace the summation by an integral with respect of x. Combining (1.12) and (1.13) gives

$$E[x] = \int_{-\infty}^\infty x\,p(x)\,dx \tag{1.14}$$

which is really the fundamental definition of the mean value $E[x]$. Although this "derivation" of (1.14) is not mathematically rigorous, it does illustrate the relationship between (1.12) and (1.14), and, under certain conditions which we shall explore more fully later, both equations yield the same result for $E[x]$.

The mean square value of x, $E[x^2]$, is defined as the average value of x^2 which, corresponding to (1.12), is given by

$$E[x^2] = \int_0^T x^2(t)\frac{dt}{T} \tag{1.15}$$

and, corresponding to (1.14), by

$$E[x^2] = \int_{-\infty}^\infty x^2 p(x)\,dx \tag{1.16}$$

the latter equation being the rigorous definition.

Finally the standard deviation of x, usually denoted by σ, and the variance σ^2, are defined by the equation

$$\sigma^2 = E[(x - E[x])^2] \tag{1.17}$$

that is to say the variance is the mean of the square of the deviation of x from its mean level $E[x]$. Equation (1.17) may be simplified by multiplying out the terms which are squared to give

$$\sigma^2 = E[x^2 - 2xE[x] + (E[x])^2]$$
$$= E[x^2] - 2E[x]\cdot E[x] + (E[x])^2 \qquad (1.18)$$

since the average of a sum of terms is the same as the sum of the averages of each term separately, and the average of a constant is of course just the constant. Collecting terms, (1.18) gives

$$\sigma^2 = E[x^2] - (E[x])^2 \qquad (1.19)$$

or

$$(\text{variance}) = (\text{standard deviation})^2 = \{\text{Mean square} - (\text{Mean})^2\}.$$

Example

If we apply (1.14) and (1.16) to calculate the statistics of the Gaussian process defined by (1.11), we find that the mean level

$$E[x] = \int_{-\infty}^{\infty} x\, p(x)\, dx = \frac{1}{\sqrt{2\pi}\,\sigma} \int_{-\infty}^{\infty} x\, e^{-(x-m)^2/2\sigma^2}\, dx$$

and the mean square

$$E[x^2] = \int_{-\infty}^{\infty} x^2\, p(x)\, dx = \frac{1}{\sqrt{2\pi}\,\sigma} \int_{-\infty}^{\infty} x^2\, e^{-(x-m)^2/2\sigma^2}\, dx,$$

both of which involve difficult integrals. These may be evaluated by changing the variable of integration to $y = (x - m)$, so that

$$E[x] = \frac{1}{\sqrt{2\pi}\,\sigma} \int_{-\infty}^{\infty} (y + m)\, e^{-y^2/2\sigma^2}\, dy$$

$$E[x^2] = \frac{1}{\sqrt{2\pi}\,\sigma} \int_{-\infty}^{\infty} (y + m)^2\, e^{-y^2/2\sigma^2}\, dy$$

and then using the standard results

$$\int_{0}^{\infty} e^{-y^2/2\sigma^2}\, dy = \sqrt{\frac{\pi}{2}}\,\sigma$$

$$\int_{0}^{\infty} y\, e^{-y^2/2\sigma^2}\, dy = \sigma^2 \qquad (1.20)$$

$$\int_{0}^{\infty} y^2\, e^{-y^2/2\sigma^2}\, dy = \sqrt{\frac{\pi}{2}}\,\sigma^3$$

to obtain

$$E[x] = m$$

and

$$E[x^2] = \sigma^2 + m^2.$$

Hence m is the mean and it follows from (1.19) that σ is the standard deviation of the Gaussian process whose probability distribution is given by (1.11).

Probability distribution function

In addition to using the *probability density function* $p(x)$ to describe the distribution of values of a random variable, the closely related *probability distribution function* $P(x)$ is also often referred to and is usually computed by a probability analyser. $P(x)$ is defined by the equation

$$P(x) = \int_{-\infty}^{x} p(x)\,dx \qquad (1.21)$$

and may therefore be interpreted as the shaded area under the probability

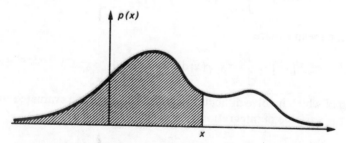

Fig. 1.9 The shaded area gives the value of the probability distribution function $P(x)$

density curve shown in Fig. 1.9. The value of $P(x)$ lies between zero and unity since

$$\text{Prob}\,(-\infty \leqslant x \leqslant \infty) = \int_{-\infty}^{\infty} p(x)\,dx = P(x = \infty) = 1 \qquad (1.22)$$

and gives the probability that a sample value of the random variable is *less than* x.

It will be seen that equation (1.21) might lead to some confusion because the variable of integration and the upper limit of the integral are the same. Strictly therefore (1.21) should appear as

$$P(x) = \int_{-\infty}^{x} p(\xi)\,d\xi \qquad (1.23)$$

where ξ is a dummy variable chosen so that

$$p(\xi = x) = p(x).$$

By differentiating (1.23) with respect to x, we can obtain another, final, relationship

$$\frac{\mathrm{d}P(x)}{\mathrm{d}x} = p(x) \tag{1.24}$$

which says that the slope of the probability distribution function is equal to the probability density function.

Chapter 2

Joint probability distributions, ensemble averages

Second-order probability density functions

The first-order probability density function $p(x)$ specifies the probability $p(x)\,dx$ that a random variable lies in the range of values x to $x + dx$. The *second-order probability density function* $p(x, y)$ is defined in the same way but extends the number of random variables from one to two, in this case x and y. The probability that the x random variable lies in the range x to $x + dx$ *and* that the y random variable lies in the range y to $y + dy$ is given by $p(x, y)dx\,dy$.

For instance, if $x(t)$ is one random function of time t and $y(t)$ is another random function of t, and if both functions are sampled at an arbitrary time t_0, then the joint probability that

$$\text{Prob}\,(x \leqslant x(t_0) \leqslant x + dx \quad and \quad y \leqslant y(t_0) \leqslant \dot{y} + dy) = p(x, y)\,dx\,dy \tag{2.1}$$

from the above definition of the second-order probability density function $p(x, y)$. In order to determine the joint probability that $x(t_0)$ and $y(t_0)$ lie within finite bands of values of x and y it is necessary to integrate the r.h.s. of (2.1) over the required bands of x and y so that

$$\text{Prob}\,(x_1 \leqslant x(t_0) \leqslant x_2 \ \text{and} \ y_1 \leqslant y(t_0) \leqslant y_2) = \int_{x_1}^{x_2} \int_{y_1}^{y_2} p(x, y)\,dx\,dy. \tag{2.2}$$

If both bands of values extend from $-\infty$ to $+\infty$, there is a 100 per cent certainty that both $x(t_0)$ and $y(t_0)$ must lie within these bounds, and so

$$\begin{array}{c} \text{Prob}\,(-\infty \leqslant x(t_0) \leqslant \infty \\ \text{and} \ -\infty \leqslant y(t_0) \leqslant \infty) \end{array} = \int_{-\infty}^{\infty} \int_{-\infty}^{\infty} p(x, y)\,dx\,dy = 1. \tag{2.3}$$

The joint probability density function $p(x, y)$ may therefore be represented as a two-dimensional surface, Fig. 2.1, for which the volume contained underneath the surface is unity (dimensionless). It follows that the dimensions of $p(x, y)$ must be the dimensions of $1/xy$.

Consider now the probability of finding the random variable $x(t_0)$ within the x to $x + dx$ band independently of the value of the random variable $y(t_0)$. From Chapter 1 this probability can be expressed in terms of the first-order

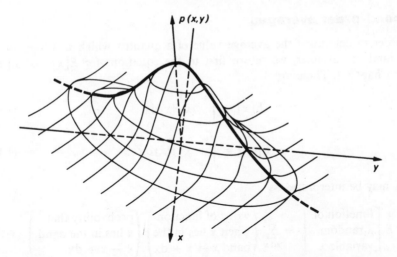

Fig. 2.1 Typical second-order or joint probability density function $p(x,y)$

probability density function $p(x)$ as

$$\text{Prob}\,(x \leqslant x(t_0) \leqslant x + \mathrm{d}x) = p(x)\,\mathrm{d}x. \tag{2.4}$$

But the same result must be obtained by using the second-order probability density function $p(x, y)$ and allowing the band of values for $y(t_0)$ to extend from $-\infty$ to $+\infty$. From (2.2)

$$\text{Prob}\,(x \leqslant x(t_0) \leqslant x + \mathrm{d}x \ \ and \ \ -\infty \leqslant y(t_0) \leqslant \infty) = \mathrm{d}x \int_{-\infty}^{\infty} p(x, y)\,\mathrm{d}y. \tag{2.5}$$

Since equations (2.4) and (2.5) are both expressing the same result, it follows that the r.h.s. of each equation is the same, so that

$$p(x)\,\mathrm{d}x = \mathrm{d}x \int_{-\infty}^{\infty} p(x, y)\,\mathrm{d}y$$

or

$$p(x) = \int_{-\infty}^{\infty} p(x, y)\,\mathrm{d}y. \tag{2.6}$$

The first-order probability density functions can therefore be obtained from the corresponding joint probability density function by integrating out the dependence on the unwanted random variable according to (2.6). By comparison with (2.6), the first-order probability density function for y is thus given by

$$p(y) = \int_{-\infty}^{\infty} p(x, y)\,\mathrm{d}x. \tag{2.7}$$

Second-order averages

In order to determine the average value of a quantity which is a function of two random variables, we return first to the equations for $E[x]$ and $E[x^2]$ from Chapter 1. These are

$$E[x] = \int_{-\infty}^{\infty} x\, p(x)\, dx \qquad (1.14)$$

and

$$E[x^2] = \int_{-\infty}^{\infty} x^2\, p(x)\, dx. \qquad (1.16)$$

They may be interpreted as

$$E\begin{bmatrix}\text{function of}\\ \text{a random}\\ \text{variable } x\end{bmatrix} = \sum_{\text{all } x}\begin{pmatrix}\text{value of function}\\ \text{when } x \text{ lies in the}\\ \text{band } x \to x + dx\end{pmatrix}\begin{pmatrix}\text{probability that}\\ x \text{ lies in the band}\\ x \to x + dx\end{pmatrix}$$

which may be extended to include the case of two random variables x and y, and then reads

$$E\begin{bmatrix}\text{function of}\\ \text{random}\\ \text{variables } x\\ \text{and } y\end{bmatrix} = \sum_{\substack{\text{all } x \\ \text{and} \\ \text{all } y}}\begin{pmatrix}\text{value of function}\\ \text{when } x \text{ lies in the}\\ \text{band } x \to x + dx\\ \textit{and } y \text{ lies in the}\\ \text{band } y \to y + dy\end{pmatrix}\begin{pmatrix}\text{probability that}\\ x \text{ lies in the}\\ \text{band } x \to x + dx\\ \textit{and } y \text{ lies in the}\\ \text{band } y \to y + dy\end{pmatrix}$$

In mathematical terms the latter becomes

$$E[f(x, y)] = \int_{-\infty}^{\infty}\int_{-\infty}^{\infty} f(x, y)\, p(x, y)\, dx\, dy \qquad (2.8)$$

and this is a general result for calculating the average value $E[f(x, y)]$ of a function $f(x, y)$ of two random variables x and y whose joint probability density function is $p(x, y)$.

Example

In order to try to illustrate these ideas we shall turn again to the analysis of sine wave time histories. In this case suppose that there are two sine waves of the same (constant) amplitude and frequency which are out-of-phase with each other by an unknown (random) phase angle ϕ so that, if

$$x(t) = x_0 \sin \omega t$$

then

$$y(t) = x_0 \sin(\omega t + \phi).$$

It is required to calculate the mean value of the product xy and of $(xy)^2$.

Suppose that both time histories are sampled at the same *arbitrary* instant of time t_0. In this case

$$x(t_0) = x_0 \sin (\omega t_0)$$

$$y(t_0) = x_0 \sin (\omega t_0 + \phi).$$

Since the time t_0 is arbitrary it may be anywhere along the time axis, but since the sine waves are periodic with period $T = 2\pi/\omega$ we need only consider the cases for which

$$0 \leqslant t_0 \leqslant \frac{2\pi}{\omega}.$$

If t_0 is chosen truly arbitrarily, i.e. it is a random variable which may fall with uniform probability anywhere on the time axis, then the first-order

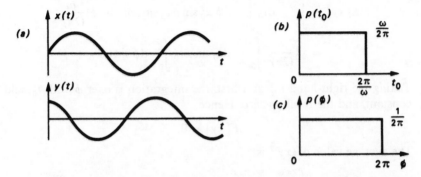

Fig. 2.2 Example of two sine waves with random phasing ϕ and sampled at random time t_0

probability density function $p(t_0)$ for t_0 will be as shown in Fig. 2.2(b). Similarly if the phase angle ϕ is a random variable which may fall with uniform probability anywhere between 0 and 2π, then the probability density function $p(\phi)$ for ϕ will be as shown in Fig. 2.2(c). Notice that, in both cases, the height of the probability density curve must be scaled so that, according to (1.22), the area under the curve is unity.

From the basic definitions we can now write

$$\text{Prob}\begin{pmatrix}\text{time of sampling lies} \\ \text{in the band } t_0 \text{ to } t_0 + dt_0\end{pmatrix} = p(t_0)\, dt_0 = \frac{\omega}{2\pi}\, dt_0$$

and

$$\text{Prob}\begin{pmatrix}\text{phase difference lies} \\ \text{in the band } \phi \text{ to } \phi + d\phi\end{pmatrix} = p(\phi)\, d\phi = \frac{1}{2\pi}\, d\phi.$$

To find the joint probability of the two "events" t_0 and ϕ we have to multiply together the probability of each event separately.† Hence the joint probability

$$\text{Prob}\begin{pmatrix}\text{sampling time in the} \\ \text{band } t_0 \to t_0 + dt_0 \\ \textit{and} \text{ phase difference} \\ \text{in the band } \phi \to \phi + d\phi\end{pmatrix} = \frac{\omega}{(2\pi)^2}\, dt_0\, d\phi$$

and the joint probability density function is

$$p(t_0, \phi) = \begin{cases} \dfrac{\omega}{(2\pi)^2} & \text{for} \quad \begin{cases} 0 \leqslant t_0 \leqslant 2\pi/\omega \\ 0 \leqslant \phi \leqslant 2\pi \end{cases} \\[2ex] 0 \text{ for all values of } t_0 \text{ and } \phi \text{ outside these} \\ \quad \text{limits.} \end{cases}$$

From (2.8), the average value of the product xy is

$$E[xy] = \int_0^{2\pi/\omega} dt_0 \int_0^{2\pi} d\phi\, x_0^2 \sin \omega t_0 \sin(\omega t_0 + \phi) \frac{\omega}{(2\pi)^2}$$

$$= x_0^2 \frac{\omega}{(2\pi)^2} \int_0^{2\pi/\omega} dt_0 \sin \omega t_0 \int_0^{2\pi} d\phi \sin(\omega t_0 + \phi).$$

Taking the right-hand integral first, the integration is over ϕ with t_0 held constant, and the result is zero. Hence

$$E[xy] = 0.$$

The average value of $(xy)^2$ is

$$E[x^2 y^2] = \int_0^{2\pi/\omega} dt_0 \int_0^{2\pi} d\phi\, x_0^4 \sin^2 \omega t_0 \sin^2(\omega t_0 + \phi) \frac{\omega}{(2\pi)^2}$$

$$= x_0^4 \frac{\omega}{(2\pi)^2} \int_0^{2\pi/\omega} dt_0 \sin^2 \omega t_0 \int_0^{2\pi} d\phi \sin^2(\omega t_0 + \phi).$$

The right-hand integral is again with respect to ϕ with t_0 held constant, and the result comes out to be π, independently of t_0. The left-hand integral may then be evaluated and has the value π/ω. Hence

$$E[x^2 y^2] = x_0^4 \cdot \frac{\omega}{(2\pi)^2} \cdot \frac{\pi}{\omega} \cdot \pi$$

$$= \tfrac{1}{4} x_0^4.$$

---

† This conclusion may easily be verified by thinking what happens when throwing a dice. The probability of a 4 (say) at the first throw is 1/6. The probability of a 4 at the second throw is also 1/6. The joint probability of two 4's in succession is however 1/36 since there are 36 possible combinations of numbers from throwing the dice twice and only one of these combinations gives two 4's.

Conditional probability

We have already shown how the first-order probability density functions can be obtained from the corresponding second-order probability density function, using equations (2.6) and (2.7). Equation (2.6), for instance, specifies the probability density for x to lie in the band x to $x + \mathrm{d}x$ independently of the value of y. Suppose now that, instead of allowing the random variable y to have any value, we restrict our attention only to those cases for which y lies in the band of values y to $y + \mathrm{d}y$. On the condition that y lies in this band, we want to find the distribution of x.

The meaning of this restriction may be seen by thinking again of the two random functions $x(t)$ and $y(t)$ which are sampled at arbitrary time t_0. By sampling the time histories many times (by changing t_0 arbitrarily) we obtain a population of sample values of x and y. From this population we select only those samples for which y lies in the range y to $y + \mathrm{d}y$, and by so doing obtain a much reduced population. Then for these selected samples only we analyse the distribution of x to obtain the so-called *conditional probability density function* for x given y. This is usually denoted $p(x|y)$.

We can in fact determine $p(x|y)$ from the second-order probability distribution. Figure 2.3 shows the joint probability density $p(x, y)$ as a function of

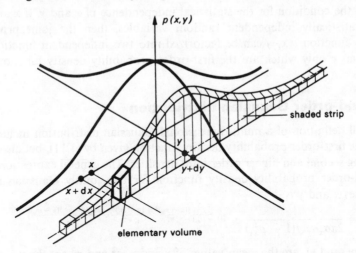

Fig. 2.3 Calculation of the conditional probability density function $p(x|y)$

x and y. The elementary volume shown represents the probability† that x lies in the x to $x + \mathrm{d}x$ band *and* that y lies in the y to $y + \mathrm{d}y$ band. The volume of the shaded strip shown represents the probability that $-\infty \leqslant x \leqslant \infty$ *and* that y lies in the y to $y + \mathrm{d}y$ band. Hence the conditional probability that,

---

†Strictly it is the *ratio* of the elementary volume to the total volume which represents probability, but since the total volume is normalized to unity, probability is equal to volume.

given y lies in the y to $y + dy$ band, x lies in the x to $x + dx$ band, is

$$p(x|y)\,dx = \frac{\text{elementary volume}}{\text{volume of shaded strip}}. \tag{2.9}$$

Substituting for the volumes on the r.h.s. of (2.9) gives

$$p(x|y)\,dx = \frac{p(x,y)\,dx\,dy}{dy\int_{-\infty}^{\infty} p(x,y)\,dx}$$

which, after simplifying and using (2.7), gives

$$p(x|y) = \frac{p(x,y)}{p(y)}. \tag{2.10}$$

This result is important because, if the conditional probability distribution for x is *independent* of y then we must have

$$p(x|y) = p(x)$$

in which case (2.10) gives

$$p(x,y) = p(x)p(y). \tag{2.11}$$

This is the condition for the statistical independence of x and y. If x and y are two statistically independent random variables, then the joint probability density function $p(x,y)$ can be factorized into two independent functions of x only and y only which are the first-order probability density functions for x and y.

Second-order Gaussian distribution

The full definition of a multi-dimensional Gaussian distribution includes, not only the first-order probability density function given by (1.11), but also corresponding second and higher order joint densities. The general expression for the second-order probability density function for two jointly Gaussian random variables x and y is

$$p(x,y) = \frac{1}{2\pi\sigma_x\sigma_y\sqrt{(1-\rho_{xy}^2)}}\,e^{-\frac{1}{2(1-\rho_{xy}^2)}\left\{\frac{(x-m_x)^2}{\sigma_x^2} + \frac{(y-m_y)^2}{\sigma_y^2} - \frac{2\rho_{xy}(x-m_x)(y-m_y)}{\sigma_x\sigma_y}\right\}} \tag{2.12}$$

where m_x and m_y are the mean values of x and y, σ_x^2 and σ_y^2 are the variances of x and y and ρ_{xy} is a correlation coefficient defined as

$$\rho_{xy} = \frac{E[(x-m_x)(y-m_y)]}{\sigma_x\sigma_y} \tag{2.13}$$

and called the normalized covariance. Notice that if $\rho_{xy} = 0$, x and y are statistically independent because then (2.12) may be factorized to give

$$p(x,y) = \left(\frac{1}{\sqrt{2\pi}\,\sigma_x}e^{-(x-m_x)^2/2\sigma_x^2}\right)\cdot\left(\frac{1}{\sqrt{2\pi}\,\sigma_y}e^{-(y-m_y)^2/2\sigma_y^2}\right) = p(x)\cdot p(y).$$

Ensemble averaging

So far we have been concerned with calculating averages for a single time history or, in this chapter, for two time histories of different events. Our mental picture is that of sampling the continuous time history at many points and so obtaining a representative population of sample values. This is what very often happens in practice and later we shall return to consider some of the pitfalls of digital sampling and the requirements for accurate digital analysis. However mathematically there is an alternative way of calculating averages. This is based on the concept of an ensemble or collection of sample functions $x_1(t)$, $x_2(t)$, $x_3(t)$, etc., which together make up the random process $x(t)$. The random process consists of a (theoretically) infinite number of sample functions each of which can be thought of as resulting from a separate experiment. For instance, suppose that the air pressure at a certain point on the wing of a jet airliner is the quantity being studied. It is necessary to know how the pressure varies during long-distance flying and so a large number of time histories are recorded for typical flights. It would obviously not be possible to obtain an infinite set of samples, but we assume that a large number is feasible, embracing the range of actual flying altitudes and weather conditions proportioned according to practical service experience. This collection of sample functions constitutes an engineer's approximation for the infinite ensemble of a random process.

Instead of being measured *along* a single sample, *ensemble averages* are measured *across* the ensemble, Fig. 2.4. By determining the values of enough

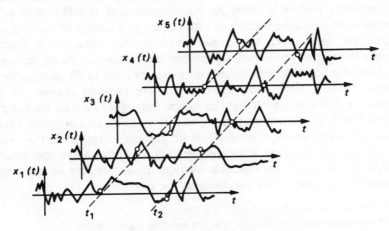

Fig. 2.4 Concept of ensemble averaging

sample functions at time t_1, the first-order probability distribution for x at t_1 can be calculated. If another series of measurements is made at time t_2, the second-order probability distribution for x at t_1 and x at t_2 can be found. Similarly, by making measurements at other times, higher order probability distributions for the ensemble can also be found.

For a *Gaussian random process*, the ensemble probability distributions must all be Gaussian. The first-order probability density functions for x at t_1 and for x at t_2 will then satisfy (1.11) (although the values of m and σ in the two cases may be different). The joint probability density for x at t_1 and x at t_2 will satisfy (2.12) when x becomes $x(t_1)$ and y becomes $x(t_2)$. Furthermore, for every finite set of times t_n, the random variables $x(t_n)$ will all have joint distributions described by the appropriate higher order Gaussian probability density functions.†

The random process is said to be *stationary* if the probability distributions obtained for the ensemble do not depend on absolute time (but only, for second and higher order probability distributions, on the time separation between measuring points). Of course the term "stationary" refers to the probability distributions and not to the samples themselves! This implies that all the averages are independent of absolute time and, specifically, that the mean, mean square, variance and standard deviation are independent of time altogether. Since all engineering random processes must have a beginning and and ending, they cannot be truly stationary, but for practical purposes it is very often adequate to assume that a process is stationary for the majority of its lifetime, or that it can be divided into several separate periods each of which is approximately stationary. The term *weakly stationary* is sometimes used to describe processes in which only the first- and second-order probability distributions are invariant with time; a *strictly stationary* process is one for which all probability distributions of the ensemble are invariant with time.

A stationary process is called an *ergodic process* if, in addition to all the ensemble averages being stationary with respect to a change of the time scale, the averages taken along any single sample are the same as the ensemble averages. In practical terms, each sample function is then completely representative of the ensemble that constitutes the random process. In discussing averages calculated from a single sample in this and the previous chapter, we were therefore implicitly assuming that this sample belonged to a stationary, ergodic process. Notice that if a process is ergodic it must also be stationary, because an average along a single sample will (in theory) extend from $t = -\infty$ to $t = +\infty$ and will therefore be independent of time. If the sample and ensemble averages are the same, the ensemble averages must therefore be independent of absolute time, and so the process must be stationary.

†See, for instance, Davenport *et al.* [22], p. 154.
Note: References, indicated by numbers in square brackets, begin on p. 366.

Chapter 3

Correlation

Consider a population of pairs of values of two random variables x and y. Suppose that each pair of values is represented by a point on a graph of y against x, Fig. 3.1. In Fig. 3.1(a) the values of x and y in each pair have no

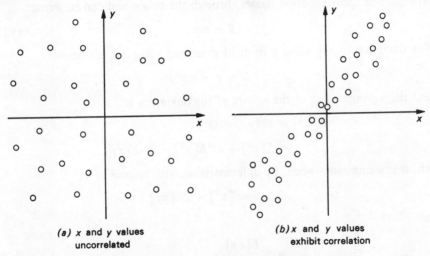

(a) x and y values uncorrelated

(b) x and y values exhibit correlation

Fig. 3.1 Illustrating correlation between two random variables x and y

apparent pattern whereas in Fig. 3.1(b) there is a definite pattern. For the latter case, in any pair, a large value of x is associated with a large value of y, a small value of x with a small value of y, and so on. The variables in Fig. 3.1(b) are accordingly said to be *correlated* whereas those in Fig. 3.1(a) are uncorrelated.

If, for Fig. 3.1(b), we wish to express an approximate functional relationship between x and y in the form of a straight line, one way of doing this is to minimize the square of the deviation of the actual values of y from their values predicted by the straight line approximation, Fig. 3.2(a). If the positions of the axes are adjusted so that

$$E[x] = E[y] = 0$$

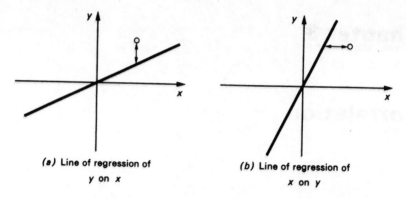

(a) Line of regression of
y on x

(b) Line of regression of
x on y

Fig. 3.2 Calculation of regression line for correlated data

i.e. so that the origin lies at the "centre of gravity" of the data points, then the straight line approximation passes through the origin and can be written

$$y = mx. \tag{3.1}$$

The deviation of any value y from its predicted value mx is then

$$\Delta = y - mx$$

and the average value of the square of the deviation is

$$E[\Delta^2] = E[(y - mx)^2]$$
$$= E[y^2] + m^2 E[x^2] - 2mE[xy]$$

which is a minimum when, by differentiating with respect of m,

$$0 = 2mE[x^2] - 2E[xy]$$

or

$$m = \frac{E[xy]}{E[x^2]}. \tag{3.2}$$

Substituting this optimum value of the slope m into (3.1) gives

$$y = \frac{E[xy]}{E[x^2]} \cdot x \tag{3.3}$$

or, since from (1.19) for zero means,

$$\sigma_x^2 = E[x^2] \qquad \text{and} \qquad \sigma_y^2 = E[y^2]$$

equation (3.3) may be written finally

$$\frac{y}{\sigma_y} = \left\{ \frac{E[xy]}{\sigma_x \sigma_y} \right\} \frac{x}{\sigma_x}. \tag{3.4}$$

This is the equation for the *line of regression of y on x*. Alternatively, had we calculated the deviation of x from its predicted value, Fig. 3.2(*b*), we should have obtained the *line of regression of x on y* given by

$$\frac{x}{\sigma_x} = \left\{\frac{E[xy]}{\sigma_x\sigma_y}\right\}\frac{y}{\sigma_y}. \tag{3.5}$$

In the case where x and y do not have zero means, as assumed above, the corresponding equations are (problem 3.1)

$$\frac{y - m_y}{\sigma_y} = \left\{\frac{E[(x - m_x)(y - m_y)]}{\sigma_x\sigma_y}\right\}\frac{x - m_x}{\sigma_x} \tag{3.6}$$

and

$$\frac{x - m_x}{\sigma_x} = \left\{\frac{E[(x - m_x)(y - m_y)]}{\sigma_x\sigma_y}\right\}\frac{y - m_y}{\sigma_y} \tag{3.7}$$

where m_x and m_y are the mean values of x and y respectively. The parameter

$$\rho_{xy} = \frac{E[(x - m_x)(y - m_y)]}{\sigma_x\sigma_y} \tag{3.8}$$

is called the *correlation coefficient* or *normalized covariance* and it will be clear that (3.6) and (3.7) only represent the same straight line if $\rho_{xy} = \pm 1$, in which case there is perfect correlation, Figs. 3.3(*a*) and (*b*). If $\rho_{xy} = 0$ there is no correlation and the regression lines are parallel to the x and y axes respectively, Fig. 3.3(*c*).

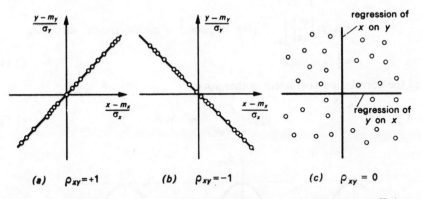

(*a*) $\rho_{xy} = +1$ (*b*) $\rho_{xy} = -1$ (*c*) $\rho_{xy} = 0$

Fig. 3.3 Regression lines for different values of the correlation coefficient ρ_{xy}

Consider now two sine waves of constant amplitude and frequency and with a fixed phase difference ϕ so that

$$x(t) = x_0 \sin \omega t$$
$$y(t) = y_0 \sin (\omega t + \phi). \tag{3.9}$$

Suppose that these two waves are sampled at an arbitrary time t_0, and we calculate the average value of the product $x(t_0)y(t_0)$. According to (2.8) the average value is given by

$$E[x(t_0)y(t_0)] = \int_{-\infty}^{\infty} x_0 y_0 \sin \omega t_0 \sin(\omega t_0 + \phi)p(t_0)\,dt_0 \qquad (3.10)$$

the double integral of (2.8) reducing to a single integral since t_0 is th only random variable. In this case we need only consider t_0 varying from zero to $2\pi/\omega$ to cover a single full cycle of the periodic motion, so that the probability

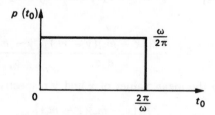

Fig. 3.4 Probability density function
for the random time of sampling t_0

density function for t_0 will be as shown in Fig. 3.4. Substituting this into (3.10) gives

$$E[x(t_0)y(t_0)] = x_0 y_0 \left(\frac{\omega}{2\pi}\right) \int_0^{2\pi/\omega} \sin \omega t_0 \sin(\omega t_0 + \phi)\,dt_0$$

$$= x_0 y_0 \left(\frac{\omega}{2\pi}\right) \int_0^{2\pi/\omega} \{\sin^2 \omega t_0 \cos \phi + \sin \omega t_0 \cos \omega t_0 \sin \phi\}\,dt_0$$

$$= \tfrac{1}{2} x_0 y_0 \cos \phi \qquad (3.11)$$

and the correlation coefficient is therefore

$$\rho_{xy} = \frac{E[xy]}{\sigma_x \sigma_y} = \cos \phi \qquad (3.12)$$

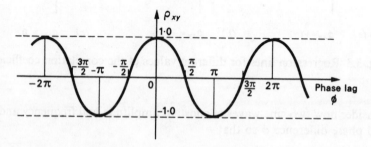

Fig. 3.5 Correlation coefficient for two sine waves with phase difference ϕ

since $\sigma_x = x_0/\sqrt{2}$ and $\sigma_y = y_0/\sqrt{2}$. The two sine waves may therefore be said to be perfectly correlated when their phase difference is 0 or 180° but uncorrelated when their phase difference is 90° or 270°, Fig. 3.5.

In general two harmonic functions of time will be correlated if they move in phase or anti-phase, and uncorrelated if they are in quadrature to each other. This conclusion is important in understanding the form of the autocorrelation function, to which we turn next.

Autocorrelation

The *autocorrelation function* for a random process $x(t)$ is defined as the average value of the product $x(t)x(t + \tau)$. The process is sampled at time t and then again at time $t + \tau$, Fig. 3.6, and the average value of the product, $E[x(t)x(t + \tau)]$, calculated for the ensemble.

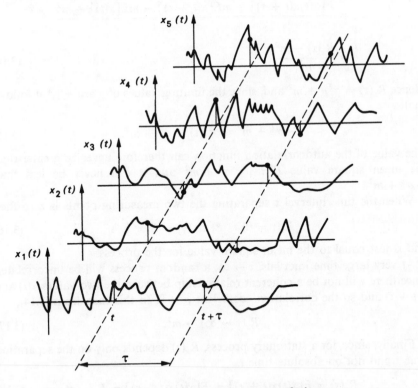

Fig. 3.6 Calculation of autocorrelation

Provided that the process is stationary, the value of $E[x(t)x(t + \tau)]$ will be independent of absolute time t and will depend only on the time separation τ

so that we may put

$$E[x(t)x(t + \tau)] = f(\tau) = R_x(\tau) \quad \text{(say)} \tag{3.13}$$

where $R_x(\tau)$ is the autocorrelation function for $x(t)$.

We can deduce at once some of the properties of $R_x(\tau)$. Firstly, if $x(t)$ is stationary, the mean and standard deviation will be independent of t, so that

$$E[x(t)] = E[x(t + \tau)] = m$$

and

$$\sigma_{x(t)} = \sigma_{x(t+\tau)} = \sigma.$$

The correlation coefficient for $x(t)$ and $x(t + \tau)$, defined by (3.8), is therefore given by

$$\rho = \frac{E[\{x(t) - m\}\{x(t + \tau) - m\}]}{\sigma^2}$$

$$= \frac{E[x(t)x(t + \tau)] - mE[x(t + \tau)] - mE[x(t)] + m^2}{\sigma^2}$$

$$= \frac{R_x(\tau) - m^2}{\sigma^2} \tag{3.14}$$

Hence $R_x(\tau) = \sigma^2 \rho + m^2$ and, since the limiting values of ρ are ± 1,† it follows that

$$-\sigma^2 + m^2 \leqslant R_x(\tau) \leqslant \sigma^2 + m^2. \tag{3.15}$$

The value of the autocorrelation function can therefore never be greater than the mean square value $E[x^2] = \sigma^2 + m^2$ and it can never be less than $-\sigma^2 + m^2$.

When the time interval τ separating the two measuring points is zero then

$$R_x(\tau = 0) = E[x(t)^2] = E[x^2] \tag{3.16}$$

and is just equal to the mean square value for the process.

At very large time intervals, $\tau \to \infty$, a random process will be uncorrelated since there will not be a coherent relationship between the two values $x(t)$ and $x(t + \tau)$ and so the correlation coefficient $\rho \to 0$. In this case, from (3.14),

$$R_x(\tau \to \infty) \to m^2. \tag{3.17}$$

Finally, since, for a stationary process, $R_x(\tau)$ depends only on the separation time τ and not on absolute time t,

$$R_x(\tau) = E[x(t)x(t + \tau)] = E[x(t)x(t - \tau)] = R_x(-\tau) \tag{3.18}$$

so that $R_x(\tau)$ is an even function of τ.

---

† For proof of this statement see, for instance, Meyer [48], p. 131.

All these properties are illustrated in a typical graph of an autocorrelation function $R_x(\tau)$ against separation time τ shown in Fig. 3.7.

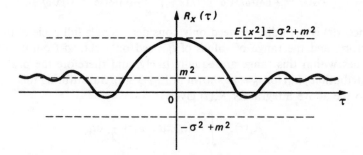

Fig. 3.7 Illustrating properties of the autocorrelation function $R_x(\tau)$ of a stationary random process $x(t)$

Example

Calculate the autocorrelation function for an ergodic random process $x(t)$ each of whose sample functions, Fig. 3.8(a), is a square wave of amplitude a and period T, but whose phase (that is the time of first switching after $t = 0$) is a random variable uniformly distributed between 0 and T.

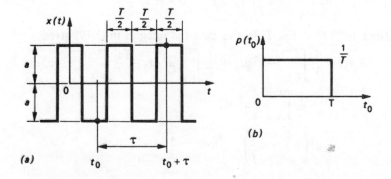

Fig. 3.8 Square wave sampled at arbitrary time t_0 and at $t_0 + \tau$ to calculate its autocorrelation function

There are two ways of making this calculation. The first is to calculate an ensemble average, by looking across the ensemble of sample functions at two fixed times t_0 and $t_0 + \tau$ (say) – see problem 3.2. The second way makes use of the fact that the process is ergodic, when, as we have seen, any one sample function is completely representative of the process as a whole. This method is to average along a single sample function, by thinking of the measuring time t_0 as a random variable uniformly distributed along the time axis. In this case, from (2.8),

$$R_x(\tau) = E[x(t_0)x(t_0 + \tau)] = \int_{-\infty}^{\infty} x(t_0)x(t_0 + \tau)p(t_0)\, dt_0.$$

Since $x(t)$ is periodic, we need only consider a single full cycle of the time history, and the range of values of t_0 need only extend from 0 to T. All values within this range are equally likely, and therefore the probability distribution $p(t_0)$ for t_0 is as shown in Fig. 3.8(b). Substituting this function into the above integral for $R_x(\tau)$ gives

$$R_x(\tau) = \int_0^T x(t_0)x(t_0 + \tau)\, \frac{1}{T}\, dt_0. \tag{3.19}$$

Because $x(t)$ is a discontinuous function, we must now proceed in steps. First, consider the case when $0 \leqslant \tau \leqslant T/2$. The integral in (3.19) then gives, after adjusting the time scale so that $t_0 = 0$ occurs at a switching point,

$$R_x(\tau) = \frac{1}{T}\int_0^{T/2 - \tau} a^2\, dt_0 + \frac{1}{T}\int_{T/2 - \tau}^{T/2} -a^2\, dt_0 + \frac{1}{T}\int_{T/2}^{T - \tau} a^2\, dt_0 +$$

$$+ \frac{1}{T}\int_{T - \tau}^T -a^2\, dt_0$$

$$= a^2\left(1 - 4\frac{\tau}{T}\right) \qquad \text{for} \qquad 0 \leqslant \tau \leqslant \frac{T}{2}.$$

Next, let $T/2 \leqslant \tau \leqslant T$. In this case the integral in (3.19) gives

$$R_x(\tau) = \frac{1}{T}\int_0^{T - \tau} -a^2\, dt_0 + \frac{1}{T}\int_{T - \tau}^{T/2} a^2\, dt_0 + \frac{1}{T}\int_{T/2}^{3T/2 - \tau} -a^2\, dt_0 +$$

$$+ \frac{1}{T}\int_{3T/2 - \tau}^T a^2\, dt_0$$

$$= a^2\left(-3 + 4\frac{\tau}{T}\right) \qquad \text{for} \qquad \frac{T}{2} \leqslant \tau \leqslant T.$$

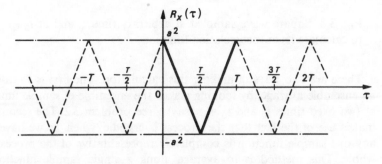

Fig. 3.9 Autocorrelation function for a square wave

If we plot these results they appear as the triangular wave shown in Fig. 3.9. By continuing the process of integration for successive stages the dotted sections can easily be obtained, and the autocorrelation function for a square wave is therefore a triangular wave of constant amplitude. Notice that, if $\tau = T, 2T, 3T$, etc., the sample values $x(t_0)$ and $x(t_0 + \tau)$ are always in phase with each other, so that there is perfect correlation and $R_x(\tau) = E[x^2] = a^2$. If $\tau = T/2, 3T/2, 5T/2$, etc., the sample values are always in antiphase and so $R_x(\tau) = -E[x^2] = -a^2$ with again perfect correlation. However for intermediate values of τ, the samples are sometimes in phase and sometimes in antiphase and so there is then incomplete correlation.

Cross-correlation

The cross-correlation functions between two different stationary random functions of time $x(t)$ and $y(t)$ are defined as

$$R_{xy}(\tau) = E[x(t)y(t + \tau)]$$

and

$$R_{yx}(\tau) = E[y(t)x(t + \tau)]. \tag{3.20}$$

Because the processes are stationary, it follows that

$$R_{xy}(\tau) = E[x(t - \tau)y(t)] = R_{yx}(-\tau)$$

and

$$R_{yx}(\tau) = E[y(t - \tau)x(t)] = R_{xy}(-\tau) \tag{3.21}$$

but in general $R_{xy}(\tau)$ and $R_{yx}(\tau)$ are not the same and, unlike the autocorrelation function, they are not even in τ.

From (3.8), each cross-correlation function can be expressed in terms of the corresponding normalized covariance ρ, to give

$$R_{xy}(\tau) = \sigma_x\sigma_y\rho_{xy}(\tau) + m_x m_y$$

and

$$R_{yx}(\tau) = \sigma_y\sigma_x\rho_{yx}(\tau) + m_y m_x \tag{3.22}$$

and since the limiting values of the ρ's are ± 1 for perfect in-phase or antiphase correlation, the limiting values of the cross-correlation functions must be

$$\pm\sigma_x\sigma_y + m_x m_y$$

so that, for instance,

$$-\sigma_x\sigma_y + m_x m_y \leqslant R_{xy}(\tau) \leqslant \sigma_x\sigma_y + m_x m_y. \tag{3.23}$$

For most random processes we expect that there will be no correlation between x and y when the time separation τ is very large, and therefore from (3.22),

$$R_{xy}(\tau \to \infty) \to m_x m_y$$

$$R_{yx}(\tau \to \infty) \to m_y m_x. \tag{3.24}$$

These properties are illustrated in Fig. 3.10 which shows the possible form of a graph of cross-correlation function $R_{xy}(\tau)$ against separation time τ. In this case it can be seen that the two random processes $x(t)$ and $y(t)$ show maximum correlation when $\tau = \tau_0$. It is shown in the next example that the phase of $y(t)$ will be lagging that of $x(t)$.

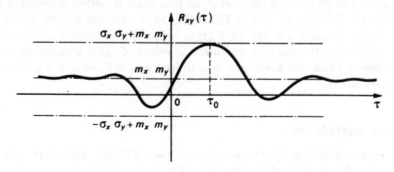

Fig. 3.10 Illustrating properties of the cross-correlation function $R_{xy}(\tau)$ of two stationary processes $x(t)$ and $y(t)$

Example

Consider two random processes $x(t)$ and $y(t)$ each of which consists of an ensemble of sample functions which are sine waves of the same constant amplitude and frequency. A typical sample of the $x(t)$ process is given by

$$x(t) = x_0 \sin(\omega t + \theta)$$

where θ is a (constant) phase angle. If θ were the same for all samples, $x(t)$ would not be a random process, but, as in the last example, we shall assume that the phase varies from one sample to the next and is itself a random variable. If we assume that all phase angles between 0 and 2π are equally likely for each new sample function, then we can write

$$p(\theta) = \begin{cases} 1/2\pi & \text{for } 0 \leqslant \theta \leqslant 2\pi \\ 0 & \text{elsewhere} \end{cases}$$

(we do not distinguish phase angles greater than 2π).

Now suppose that, although $y(t)$ is a random process, each sample is related to a corresponding sample of the $x(t)$ process so that, corresponding to the sample for which

$$x(t) = x_0 \sin(\omega t + \theta)$$

the corresponding sample of the $y(t)$ process is

$$y(t) = y_0 \sin(\omega t + \theta - \phi)$$

where ϕ is another constant phase angle which in this case we shall take to be the same for all samples. This could be the case if $y(t)$ is derived from the

$x(t)$ process by an electrical or mechanical system which introduces a phase lag in its response.

The cross-correlation function $R_{xy}(\tau)$ can now be calculated by finding the ensemble average and obtaining

$$R_{xy}(\tau) = E[x(t)y(t + \tau)] = E[x_0 y_0 \sin(\omega t + \theta)\sin(\omega t + \omega\tau + \theta - \phi)]$$

$$= \int_0^{2\pi} x_0 y_0 \sin(\omega t + \theta)\sin(\omega t + \omega\tau + \theta - \phi)\frac{1}{2\pi}\,d\theta$$

which, putting

$$\sin(\omega t + \omega\tau + \theta - \phi) = \sin(\omega t + \theta)\cos(\omega\tau - \phi) + \cos(\omega t + \theta)\sin(\omega\tau - \phi)$$

and evaluating the integrals, gives

$$R_{xy}(\tau) = \tfrac{1}{2}x_0 y_0 \cos(\omega\tau - \phi)$$

which is plotted in Fig. 3.11.

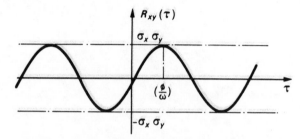

Fig. 3.11 Cross-correlation function for two sine waves with $y(t)$ lagging $x(t)$ by angle ϕ

The other cross-correlation function $R_{yx}(\tau)$ can be calculated in the same way, starting from

$$R_{yx}(\tau) = E[y(t)x(t + \tau)] = E[x_0 y_0 \sin(\omega t + \theta - \phi)\sin(\omega t + \omega\tau + \theta)]$$

and this leads to the result

$$R_{yx}(\tau) = \tfrac{1}{2}x_0 y_0 \cos(\omega\tau + \phi)$$

in agreement with (3.21).

Since we have assumed that θ is uniformly distributed between 0 and 2π, the processes $x(t)$ and $y(t)$ are both ergodic (notice that, if θ was not so distributed, the ensemble average would depend on absolute time, t – consider the case when θ is always 0). We therefore again have the option of calculating sample averages rather than ensemble averages. If we pursue this alternative approach, we can proceed as in the previous example by assuming that typical sample functions $x(t)$ and $y(t)$ are themselves sampled at an arbitrary time t_0 which may lie anywhere along the time axis. It is sufficient to let t_0 run from 0 to $2\pi/\omega$ (corresponding to one complete cycle) and we can imagine that t_0 is itself a random variable with distribution

$$p(t_0) = \begin{cases} \omega/2\pi & \text{for} \quad 0 \leqslant t_0 \leqslant 2\pi/\omega \\ 0 & \text{elsewhere.} \end{cases}$$

Then to find the sample average we have to integrate over t_0 and obtain

$$R_{xy}(\tau) = \int_0^{2\pi/\omega} x_0 y_0 \sin(\omega t_0 + \theta)\sin(\omega t_0 + \omega\tau + \theta - \phi)\frac{\omega}{2\pi} dt_0$$

which leads to the same result as before.

Chapter 4

Fourier analysis

Most engineers are familiar with the idea of frequency analysis by which a periodic function can be broken down into its harmonic components and readily accept that a periodic function may be synthesized by adding together its harmonic components. Incidentally, for many years this fact was not fully agreed by mathematicians and it is said that such famous people as Euler, d'Alembert and Lagrange held that arbitrary functions could not be represented by trigonometric series.† However it is certainly now agreed that if $x(t)$ is a periodic function of time t, with period T, as shown in Fig. 4.1, then we can

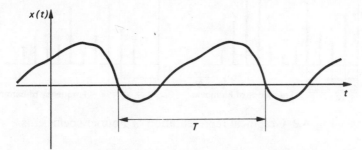

Fig. 4.1 Arbitrary periodic function of time

always express $x(t)$ as an infinite trigonometric series (a Fourier series) of the form

$$x(t) = a_0 + a_1 \cos\frac{2\pi t}{T} + a_2 \cos\frac{4\pi t}{T} + \cdots$$

$$+ b_1 \sin\frac{2\pi t}{T} + b_2 \sin\frac{4\pi t}{T} + \cdots$$

or, in more compact notation,

$$x(t) = a_0 + \sum_{k=1}^{\infty} \left(a_k \cos\frac{2\pi kt}{T} + b_k \sin\frac{2\pi kt}{T} \right) \tag{4.1}$$

† See, for instance, Kline [43] Ch. 20.

where a_0 and the a_k and b_k are constant Fourier coefficients given by

$$a_0 = \frac{1}{T} \int_{-T/2}^{T/2} x(t)\, dt$$

$$a_k = \frac{2}{T} \int_{-T/2}^{T/2} x(t) \cos \frac{2\pi k t}{T}\, dt \qquad (4.2)$$
$$\scriptstyle k \geqslant 1$$

$$b_k = \frac{2}{T} \int_{-T/2}^{T/2} x(t) \sin \frac{2\pi k t}{T}\, dt.$$
$$\scriptstyle k \geqslant 1$$

The mathematical conditions for the convergence of (4.1) are extremely general and cover practically every conceivable engineering situation.† The only important restriction is that, when $x(t)$ is discontinuous, the series gives the average value of $x(t)$ at the discontinuity.

Suppose that the position of the t axis in Fig. 4.1 is adjusted so that the mean value of $x(t)$ is zero. Then, according to the first of (4.2), the coefficient a_0 will be zero. The remaining coefficients a_k and b_k will in general all be different and their values may be illustrated graphically as shown in Fig. 4.2. The horizontal

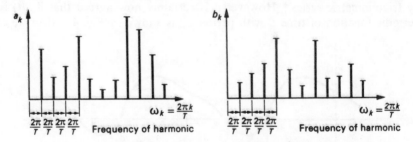

Fig. 4.2 Graphical representation of Fourier coefficients

axis in Fig. 4.2 is chosen to represent frequency and the location of the kth coefficient is at

$$\omega_k = \frac{2\pi k}{T} \qquad (4.3)$$

which is the frequency of the kth harmonic. The spacing between adjacent harmonics is

$$\Delta\omega = \frac{2\pi}{T} \qquad (4.4)$$

and it will be seen that, when the period T becomes large, the frequency spacing $\Delta\omega$ becomes small, and the Fourier coefficients become correspondingly tightly packed in Fig. 4.2. In the limit when $T \to \infty$, they will in fact actually merge together. Since in this case $x(t)$ no longer represents a periodic

†See, for instance, Churchill [8], Chapters IV and V.

phenomenon we can then no longer analyse it into discrete frequency components. Subject to certain conditions, we can however still follow the same line of thought except that the Fourier series (4.1) turns into a Fourier integral and the Fourier coefficients (4.2) turn into continuous functions of frequency called Fourier transforms. We now examine this extension of the Fourier series concept in detail.

Fourier integral

Substituting (4.2) into (4.1) gives, for $a_0 = 0$,

$$x(t) = \sum_{k=1}^{\infty} \left\{ \frac{2}{T} \int_{-T/2}^{T/2} x(t) \cos \frac{2\pi kt}{T} \, dt \right\} \cos \frac{2\pi kt}{T} +$$
$$+ \sum_{k=1}^{\infty} \left\{ \frac{2}{T} \int_{-T/2}^{T/2} x(t) \sin \frac{2\pi kt}{T} \, dt \right\} \sin \frac{2\pi kt}{T}.$$

Next, substituting for $2\pi k/T$ from (4.3) and $1/T$ from (4.4), gives

$$x(t) = \sum_{k=1}^{\infty} \left\{ \frac{\Delta\omega}{\pi} \int_{-T/2}^{T/2} x(t) \cos \omega_k t \, dt \right\} \cos \omega_k t +$$
$$+ \sum_{k=1}^{\infty} \left\{ \frac{\Delta\omega}{\pi} \int_{-T/2}^{T/2} x(t) \sin \omega_k t \, dt \right\} \sin \omega_k t.$$

When the period $T \to \infty$, $\Delta\omega \to d\omega$ and the $\sum$ becomes an integral with the limits $\omega = 0$ to $\omega = \infty$. In this case

$$x(t) = \int_{\omega=0}^{\infty} \frac{d\omega}{\pi} \left\{ \int_{-\infty}^{\infty} x(t) \cos \omega t \, dt \right\} \cos \omega t +$$
$$+ \int_{\omega=0}^{\infty} \frac{d\omega}{\pi} \left\{ \int_{-\infty}^{\infty} x(t) \sin \omega t \, dt \right\} \sin \omega t$$

or, putting

$$A(\omega) = \frac{1}{2\pi} \int_{-\infty}^{\infty} x(t) \cos \omega t \, dt$$
$$B(\omega) = \frac{1}{2\pi} \int_{-\infty}^{\infty} x(t) \sin \omega t \, dt$$

(4.5)

gives

$$x(t) = 2 \int_0^{\infty} A(\omega) \cos \omega t \, d\omega + 2 \int_0^{\infty} B(\omega) \sin \omega t \, d\omega. \quad (4.6)$$

The terms $A(\omega)$ and $B(\omega)$ defined by (4.5) are the components of the *Fourier transform of x(t)* and equation (4.6) is a representation of $x(t)$ by a *Fourier integral* or *inverse Fourier transform*.

This heuristic development is hardly rigorous, but it does indicate the logic which turns a discrete Fourier series representation into a Fourier integral. Classical Fourier analysis theory† considers the conditions that $x(t)$ must satisfy for (4.5) and (4.6) to be true. For engineering applications, the important condition is usually expressed in the form

$$\int_{-\infty}^{\infty} |x(t)| \, dt < \infty. \tag{4.7}$$

It means that classical theory applies only to functions which decay to zero when $|t| \to \infty$. We shall see later that this condition may be removed when impulse functions are introduced in the generalized theory of Fourier analysis, but we begin by considering only classical theory using ordinary functions which satisfy (4.7). As for discrete Fourier series, when there is a discontinuity in $x(t)$, equation (4.6) gives the average value of $x(t)$ at the discontinuity.

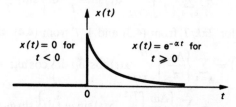

$x(t)$

$x(t) = 0$ for
$t < 0$

$x(t) = e^{-\alpha t}$ for
$t \geqslant 0$

0

t

Fig. 4.3 An aperiodic function for
Fourier transformation

Example
  Calculate the components $A(\omega)$ and $B(\omega)$ of the Fourier transform of the function shown in Fig. 4.3.

From (4.5)

$$A(\omega) = \frac{1}{2\pi} \int_0^{\infty} e^{-\alpha t} \cos \omega t \, dt$$

and

$$B(\omega) = \frac{1}{2\pi} \int_0^{\infty} e^{-\alpha t} \sin \omega t \, dt.$$

These integrals may be integrated by parts to give

$$\int_0^{\infty} e^{-\alpha t} \cos \omega t \, dt = \left[-\frac{1}{\alpha} e^{-\alpha t} \cos \omega t \right]_0^{\infty} - \int_0^{\infty} \frac{\omega}{\alpha} e^{-\alpha t} \sin \omega t \, dt$$

$$= \frac{1}{\alpha} - \frac{\omega}{\alpha} \int_0^{\infty} e^{-\alpha t} \sin \omega t \, dt.$$

†See previous footnote.

and

$$\int_0^\infty e^{-\alpha t} \sin \omega t \, dt = \left[-\frac{1}{\alpha} e^{-\alpha t} \sin \omega t \right]_0^\infty + \int_0^\infty \frac{\omega}{\alpha} e^{-\alpha t} \cos \omega t \, dt$$

$$= \frac{\omega}{\alpha} \int_0^\infty e^{-\alpha t} \cos \omega t \, dt.$$

Hence

$$\left(1 + \frac{\omega^2}{\alpha^2} \right) \int_0^\infty e^{-\alpha t} \cos \omega t \, dt = \frac{1}{\alpha}$$

and

$$\left(1 + \frac{\omega^2}{\alpha^2} \right) \int_0^\infty e^{-\alpha t} \sin \omega t \, dt = \frac{\omega}{\alpha^2}$$

so that

$$A(\omega) = \frac{1}{2\pi} \frac{\alpha}{\alpha^2 + \omega^2}$$

$$B(\omega) = \frac{1}{2\pi} \frac{\omega}{\alpha^2 + \omega^2}$$

and the Fourier integral representation is, from (4.6)

$$x(t) = \frac{1}{\pi} \int_0^\infty \frac{\alpha}{\alpha^2 + \omega^2} \cos \omega t \, d\omega + \frac{1}{\pi} \int_0^\infty \frac{\omega}{\alpha^2 + \omega^2} \sin \omega t \, d\omega.$$

The dependence of $A(\omega)$ and $B(\omega)$ on frequency ω may be plotted as shown in Fig. 4.4, which corresponds to Fig. 4.2 for the discrete coefficients of a Fourier series.

Notice that, when $t = 0$,

$$x(0) = \frac{1}{\pi} \int_0^\infty \frac{\alpha}{\alpha^2 + \omega^2} \, d\omega = \frac{1}{\pi} \left[\tan^{-1} \frac{\omega}{\alpha} \right]_0^\infty = \frac{1}{2}$$

which is the average value of $x(t)$ at the discontinuity.

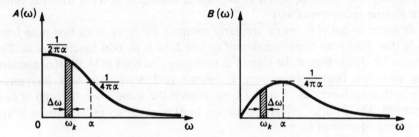

Fig. 4.4 Components of the Fourier transform of the function shown in Fig. 4.3

A Fourier integral may be regarded as the formal limit of a Fourier series as the period tends to infinity. The reason for introducing this concept is because Fourier integrals indicate the frequency composition of an aperiodic function; we shall see in the next chapter that in order to understand the frequency characteristics of random processes it is necessary to analyse the frequency composition of their correlation functions, which are generally aperiodic functions. It is helpful to compare the Fourier transform equations (4.5) with the second and third of (4.2) and the Fourier integral (4.6) with the infinite Fourier series (4.1). Note that the physical dimensions of the Fourier transform components $A(\omega)$ and $B(\omega)$ (which are those of x/ω) are different from those of the Fourier series coefficients a_k and b_k (which are those of x). With reference to Fig. 4.4, we can see that small elementary areas $A(\omega_k)\Delta\omega$ and $B(\omega_k)\Delta\omega$ have the same dimensions as a_k and b_k in Fig. 4.2, and may be compared with the half amplitudes of the harmonic components of a periodic function (note the factor 2 occurring in equation (4.6) which is not present in (4.1)).

Complex form of the Fourier transform

It has become customary in random vibration theory to write equations (4.5) and (4.6) in complex form, making use of the result that

$$e^{i\theta} = \cos\theta + i\sin\theta. \tag{4.8}$$

Defining $X(\omega)$ as

$$X(\omega) = A(\omega) - iB(\omega) \tag{4.9}$$

equations (4.5) may be combined to give

$$X(\omega) = \frac{1}{2\pi} \int_{-\infty}^{\infty} x(t)(\cos\omega t - i\sin\omega t)\,dt$$

$$= \frac{1}{2\pi} \int_{-\infty}^{\infty} x(t)\,e^{-i\omega t}\,dt \tag{4.10}$$

from (4.8). The latter equation is the formal definition of $X(\omega)$ which is called the *Fourier transform* of $x(t)$.

In order to put (4.6) into a similarly compact form, we must first note from (4.5) that $A(\omega)$ is an even function of ω and $B(\omega)$ is an odd function of ω. The reason for this is that, if the sign of ω is changed in both of (4.5), $A(\omega)$ remains the same but $B(\omega)$ changes sign. It follows that $A(\omega)\cos\omega t$ and $B(\omega)\sin\omega t$ are both even functions of ω and both remain the same when the sign of ω is changed. The Fourier integral equation (4.6) may therefore be written in the following equivalent form

$$x(t) = \int_{-\infty}^{\infty} A(\omega)\cos\omega t\,d\omega + \int_{-\infty}^{\infty} B(\omega)\sin\omega t\,d\omega \tag{4.11}$$

where the integrals now run from $-\infty$ to $+\infty$ instead of from 0 to ∞ and the factor 2 disappears. The idea of a "negative" frequency has been introduced but this is purely a mathematical artifice to simplify the equation (or perhaps confuse the issue, depending on your point of view). Next, since $A(\omega)$ is an even function and $\sin \omega t$ an odd function of ω, $A(\omega) \sin \omega t$ is an odd function and so

$$\int_{-\infty}^{\infty} A(\omega) \sin \omega t \, d\omega = 0$$

and, similarly, (4.12)

$$\int_{-\infty}^{\infty} B(\omega) \cos \omega t \, d\omega = 0.$$

We can therefore add the integrals in (4.12) to (4.11) without making any difference to the value of $x(t)$. This can then be written as

$$x(t) = \int_{-\infty}^{\infty} A(\omega) \cos \omega t \, d\omega + \int_{-\infty}^{\infty} B(\omega) \sin \omega t \, d\omega +$$

$$+ i \int_{-\infty}^{\infty} A(\omega) \sin \omega t \, d\omega - i \int_{-\infty}^{\infty} B(\omega) \cos \omega t \, d\omega$$

since both of the last two integrals are just zero, or, collecting terms, as

$$x(t) = \int_{-\infty}^{\infty} \{A(\omega) - iB(\omega)\} \{\cos \omega t + i \sin \omega t\} \, d\omega$$

$$= \int_{-\infty}^{\infty} X(\omega) e^{i\omega t} \, d\omega. \tag{4.13}$$

The two equations (4.10) and (4.13), which are repeated together below, are called a Fourier transform pair. $X(\omega)$ is the (complex) Fourier transform of $x(t)$,

$$X(\omega) = \frac{1}{2\pi} \int_{-\infty}^{\infty} x(t) e^{-i\omega t} \, dt \tag{4.10}$$

$$x(t) = \int_{-\infty}^{\infty} X(\omega) e^{i\omega t} \, d\omega \tag{4.13}$$

which may be regained from $X(\omega)$ by the inverse Fourier integral equation (4.13). Authors differ over the position of the factor $1/2\pi$ which appears in these equations. Some include the $1/2\pi$ in the inverse transform equation (4.13), while others include a factor $1/\sqrt{2\pi}$ in both (4.10) and (4.13). However the definition of $X(\omega)$ given here is the one popularly used in random vibration theory.

Example
    Use (4.10) to determine the (complex) Fourier transform of the function $x(t)$ shown in Fig. 4.3, and verify that the result agrees with that obtained in the previous example.

From (4.10)

$$X(\omega) = \frac{1}{2\pi} \int_0^\infty e^{-\alpha t} e^{-i\omega t} \, dt$$

$$= \frac{1}{2\pi} \int_0^\infty e^{-(\alpha + i\omega)t} \, dt$$

$$= \frac{1}{2\pi} \left[-\frac{1}{\alpha + i\omega} e^{-(\alpha + i\omega)t} \right]_0^\infty$$

$$= \frac{1}{2\pi} \left(\frac{1}{\alpha + i\omega} \right)$$

$$= \frac{1}{2\pi} \left(\frac{\alpha - i\omega}{\alpha^2 + \omega^2} \right)$$

$$= \frac{1}{2\pi} \left(\frac{\alpha}{\alpha^2 + \omega^2} \right) - i \frac{1}{2\pi} \left(\frac{\omega}{\alpha^2 + \omega^2} \right)$$

$$= A(\omega) - iB(\omega) \text{ from the previous example}$$

Chapter 5

Spectral density

We turn now to the frequency composition of a naturally occurring random process. Because the time history $x(t)$ of a sample function is not periodic, it cannot be represented by a discrete Fourier series. Also, for a stationary process, $x(t)$ goes on for ever and the condition

$$\int_{-\infty}^{\infty} |x(t)| \, dt < \infty$$

is not satisfied, so that the classical theory of Fourier analysis cannot be applied to a sample function. This difficulty can be overcome by analysing, not sample functions of the process itself, but its autocorrelation function $R_x(\tau)$.

The logic behind this approach is that the autocorrelation function gives information about the frequencies present in a random process indirectly. We saw in the examples in Chapter 3 that $R_x(\tau)$ was a maximum for values of τ for which $x(t)$ and $x(t + \tau)$ were in phase and a minimum for values of τ for which they were in antiphase. The frequencies present in a graph of $R_x(\tau)$ against τ therefore reflect the frequency content of sample functions of the random process $x(t)$. During the course of Chapters 10 and 11 we shall be examining closely how the frequency composition of sample functions of a random process and the frequency composition of its autocorrelation function compare, but for the present we shall concentrate only on the frequency analysis of $R_x(\tau)$ for the simple reason that we have no other choice if we are to use classical Fourier methods. If the zero value of the random process $x(t)$ is normalized (or adjusted) so that the mean value of the process $m = E[x]$ is zero, then, provided that $x(t)$ has no periodic components,

$$R_x(\tau \to \infty) = 0$$

and the condition

$$\int_{-\infty}^{\infty} |R_x(\tau)| \, d\tau < \infty$$

is satisfied. We can therefore apply the methods described in Chapter 4 to calculate the Fourier transform of $R_x(\tau)$.

From (4.10) and (4.13), the Fourier transform of $R_x(\tau)$, and its inverse, are given by

$$S_x(\omega) = \frac{1}{2\pi} \int_{-\infty}^{\infty} R_x(\tau) e^{-i\omega\tau} d\tau \tag{5.1}$$

and

$$R_x(\tau) = \int_{-\infty}^{\infty} S_x(\omega) e^{i\omega\tau} d\omega \tag{5.2}$$

where $S_x(\omega)$ is called the *spectral density* of the x process and is a function of angular frequency ω. The most important property of $S_x(\omega)$ becomes apparent when we put $\tau = 0$ in equation (5.2). In this case

$$R_x(\tau = 0) = \int_{-\infty}^{\infty} S_x(\omega) d\omega$$

which, from the fundamental definition of $R_x(\tau)$ in (3.13), gives

$$E[x^2] = \int_{-\infty}^{\infty} S_x(\omega) d\omega. \tag{5.3}$$

The mean square value of a stationary random process x is therefore given by the area under a graph of spectral density $S_x(\omega)$ against ω, Fig. 5.1. The units of $S_x(\omega)$ are accordingly those of (mean square)/(unit of frequency) and a more complete name for $S_x(\omega)$ is the *mean square spectral density*.

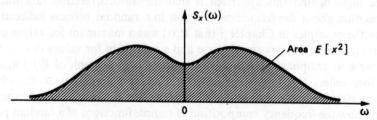

Fig. 5.1 The area under a spectral density curve is equal to $E[x^2]$

In Chapter 4 we showed that the complex Fourier transform can be expressed in terms of its real and imaginary parts by (4.9). In this case we have

$$S_x(\omega) = A(\omega) - iB(\omega) \tag{5.4}$$

where, from (4.5)

$$A(\omega) = \frac{1}{2\pi} \int_{-\infty}^{\infty} R_x(\tau) \cos \omega\tau \, d\tau \tag{5.5}$$

and

$$B(\omega) = \frac{1}{2\pi} \int_{-\infty}^{\infty} R_x(\tau) \sin \omega\tau \, d\tau. \tag{5.6}$$

Since $R_x(\tau)$ is an even function of τ while $\sin \omega\tau$ is an odd function, the product $R_x(\tau)\sin \omega\tau$ is an odd function and so the integral from $-\infty$ to 0 in (5.6) is exactly equal and opposite to the integral from 0 to ∞. $B(\omega)$ is therefore zero and

$$S_x(\omega) = A(\omega) \tag{5.7}$$

which, from (5.5), is a real even function of ω. In addition it can be shown† that $S_x(\omega)$ is never negative, a fact which is supported by considering (5.3). We would not expect the addition of frequency components $S_x(\omega)\,d\omega$ to reduce the mean square value $E[x^2]$. In summary, therefore, the mean square spectral density of a stationary random process $x(t)$ is a real, even and non-negative function of ω of the form illustrated in Fig. 5.1.

Example

Determine the mean square and autocorrelation function for the stationary random process $x(t)$ whose mean square spectral density is shown in Fig. 5.2.

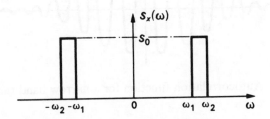

Fig. 5.2 Spectral density of a narrow band random process

From (5.3),

$$E[x^2] = \int_{-\infty}^{\infty} S_x(\omega)\,d\omega = 2S_0(\omega_2 - \omega_1)$$

From (5.2),

$$R_x(\tau) = \int_{-\infty}^{\infty} S_x(\omega)\,e^{i\omega\tau}\,d\omega$$

$$= \int_{-\infty}^{\infty} S_x(\omega)\cos \omega\tau\,d\omega$$

† See, for instance, Davenport, Jr., *et al.* [22] Ch. 6.

since $S_x(\omega)$ is an even function of ω. Hence

$$R_x(\tau) = 2\int_{\omega_1}^{\omega_2} S_0 \cos \omega\tau \, d\omega$$

$$= 2S_0\left[\frac{1}{\tau}\sin \omega\tau\right]_{\omega_1}^{\omega_2}$$

$$= \frac{2S_0}{\tau}(\sin \omega_2\tau - \sin \omega_1\tau)$$

$$= \frac{4S_0}{\tau}\cos\left(\frac{\omega_1 + \omega_2}{2}\right)\tau \cdot \sin\left(\frac{\omega_2 - \omega_1}{2}\right)\tau$$

which has the form shown in Fig. 5.3.

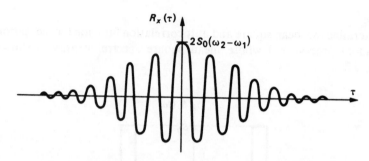

Fig. 5.3 Autocorrelation function for a narrow band random process

Narrow band and broad band processes

A process whose spectral density has the form shown in Fig. 5.2 is, not unnaturally, called a "narrow band" process because its spectral density occupies only a narrow band of frequencies. The autocorrelation function then has the form shown in Fig. 5.3 where the predominant frequency of $R_x(\tau)$ against τ is the average value $\frac{1}{2}(\omega_1 + \omega_2)$. Correlation is a maximum when

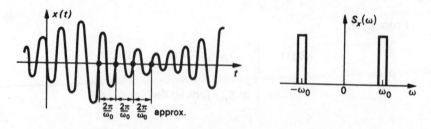

Fig. 5.4 Time history of a sample from a narrow band process

$\tau = 0$ and follows a cosine graph of decreasing amplitude as correlation at the in-phase values of τ is gradually lost with increasing time separation τ. The time history of a sample of a typical narrow band process is illustrated in Fig. 5.4.

A broad band process is one whose spectral density covers a broad band of frequencies and the time history is then made up of the superposition of the whole band of frequencies as shown in Fig. 5.5. In the limit when the frequency

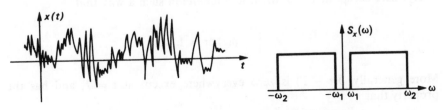

Fig. 5.5 Time history of a sample from a broad band process

band extends from $\omega_1 = 0$ to $\omega_2 = \infty$, the spectrum is called *white*. From (5.3) the mean square value of a white noise process† must be infinite, so white noise is only a theoretical concept, but in practical terms a spectrum is called white if it is broad band noise whose bandwidth extends well past all the frequencies of interest.

The form of the autocorrelation function corresponding to white noise can be derived by considering the result of the above example. When the lower frequency limit $\omega_1 = 0$, $R_x(\tau)$ becomes

$$R_x(\tau) = \frac{4S_0}{\tau} \cos\frac{\omega_2\tau}{2} \sin\frac{\omega_2\tau}{2} = 2S_0 \frac{\sin\omega_2\tau}{\tau} \tag{5.8}$$

which has the form shown in Fig. 5.6(a). When $\omega_2 \to \infty$, adjacent cycles pack

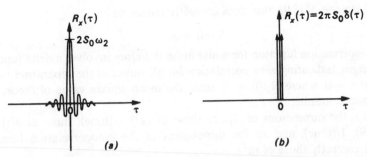

Fig. 5.6 Illustrating how the autocorrelation function becomes a delta function for white noise.

†The original analogy was with white light (whose spectrum is approximately constant over the range of visible frequencies) but the term "noise" is derived from the study of electrical noise. A broad band random process is now often described loosely as "noise".

together so tightly that they become indistinguishable from a single vertical spike of infinite height, zero width, and finite area which we shall show has magnitude $2\pi S_0$, Fig. 5.6(b). This behaviour may be represented mathematically by using Dirac's delta function $\delta(\tau)$, which is an example of a *generalized function*. As mentioned already, delta (or impulse) functions play an important role in generalized Fourier analysis and allow the scope of Fourier transforms to be greatly extended (Lighthill [46]). The delta function $\delta(\tau)$ is defined so that it is zero everywhere except at $\tau = 0$, when it is infinite in such a way that

$$\int_{-\infty}^{\infty} \delta(\tau)\,d\tau = 1.$$

More generally, $\delta(\tau - T)$ is zero everywhere except at $\tau = T$, and has the property that

$$\int_{-\infty}^{\infty} \delta(\tau - T)f(\tau)\,d\tau = f(\tau = T) \tag{5.9}$$

where $f(\tau)$ is any arbitrary continuous function of τ. Using this "delta function" notation, the autocorrelation function for a stationary random white noise process with spectral density S_0 may be written

$$R_x(\tau) = 2\pi S_0\,\delta(\tau). \tag{5.10}$$

The truth of this statement can be seen by noting that, according to (5.10), $R_x(\tau)$ is zero everywhere except at $\tau = 0$ where it is infinite, in agreement with Fig. 5.6(b). The area under $R_x(\tau)$ at $\tau = 0$ is $2\pi S_0$ and we can now verify that this is correct by calculating the Fourier transform of $R_x(\tau)$ to regain the spectral density $S_x(\omega)$, which, if all is well, should come out to be the constant value S_0 we originally specified. From (5.1),

$$S_x(\omega) = \frac{1}{2\pi}\int_{-\infty}^{\infty} 2\pi S_0\,\delta(\tau)\,e^{-i\omega\tau}\,d\tau$$

and making use of (5.9), this does correctly reduce to

$$S_x(\omega) = S_0.$$

The autocorrelation function for white noise therefore involves a delta function at the origin, indicating zero correlation for all values of the separation time τ except at $\tau = 0$, where $R_x(0) = \infty$ since the mean square value of theoretical white noise is infinite.

In (5.10) the dimensions of S_0 are those of $(x^2) \times$ (time), those of $\delta(\tau)$ are, from (5.9), 1/(time), and so the dimensions of the autocorrelation function $R_x(\tau)$ are correctly those of (x^2).

Spectral density of a derived process

If we know the spectral density $S_x(\omega)$ of a stationary random process $x(t)$, we can use this to calculate the mean square value $E[x^2]$ according to (5.3).

We can also use it to calculate the spectral density of processes which are obtained by differentiating x, for instance the velocity process $dx/dt = \dot{x}$ and the acceleration process $d^2x/dt^2 = \ddot{x}$.

The calculation begins with the autocorrelation function

$$R_x(\tau) = E[x(t)x(t + \tau)]$$

which may be written, for an ensemble average,

$$R_x(\tau) = \lim_{N \to \infty} \frac{1}{N} \sum_{r=1}^{N} x_r(t)x_r(t + \tau).$$

Consider differentiating $R_x(\tau)$ with respect to τ. In order to do so we have to differentiate each term of the form $x_r(t)x_r(t + \tau)$ in the summation, keeping t constant. This gives

$$\frac{d}{d\tau}\{x_r(t)x_r(t + \tau)\} = x_r(t)\frac{d}{d\tau}x_r(t + \tau)$$

$$= x_r(t)\frac{d}{d(t + \tau)}x_r(t + \tau)\cdot\frac{d(t + \tau)}{d\tau}$$

$$= x_r(t)\dot{x}_r(t + \tau)$$

and so we obtain

$$\frac{d}{d\tau}(R_x(\tau)) = E[x(t)\dot{x}(t + \tau)]. \tag{5.11}$$

For a stationary process, ensemble averages are independent of time t, so

$$E[x(t)\dot{x}(t + \tau)] = E[x(t - \tau)\dot{x}(t)] \tag{5.12}$$

giving

$$\frac{d}{d\tau}(R_x(\tau)) = E[x(t - \tau)\dot{x}(t)].$$

Differentiating again with respect to τ, gives

$$\frac{d^2}{d\tau^2}(R_x(\tau)) = -E[\dot{x}(t - \tau)\dot{x}(t)]$$

$$= -R_{\dot{x}}(\tau) \tag{5.13}$$

where $R_{\dot{x}}(\tau)$ is the autocorrelation function for the derived process $\dot{x}(t)$.

Also, from the Fourier integral (5.2),

$$R_x(\tau) = \int_{-\infty}^{\infty} S_x(\omega) e^{i\omega\tau} d\omega.$$

The r.h.s. of this equation is a definite integral with respect to ω, with τ held constant, and with the limits of integration independent of τ. It is therefore legitimate to differentiate with respect to τ under the integral sign to obtain

$$\frac{d}{d\tau}(R_x(\tau)) = \int_{-\infty}^{\infty} i\omega\, S_x(\omega)\, e^{i\omega\tau}\, d\omega$$

and

$$\frac{d^2}{d\tau^2}(R_x(\tau)) = -\int_{-\infty}^{\infty} \omega^2 S_x(\omega)\, e^{i\omega\tau}\, d\omega. \tag{5.14}$$

Combining (5.13) and (5.14), we now see that the autocorrelation function for the derived process can be expressed as

$$R_{\dot{x}}(\tau) = \int_{-\infty}^{\infty} \omega^2 S_x(\omega)\, e^{i\omega\tau}\, d\omega. \tag{5.15}$$

But since $R_{\dot{x}}(\tau)$ can also be written as the inverse transform of the spectral density $S_{\dot{x}}(\omega)$ according to (5.2), we also have

$$R_{\dot{x}}(\tau) = \int_{-\infty}^{\infty} S_{\dot{x}}(\omega)\, e^{i\omega\tau}\, d\omega \tag{5.16}$$

By comparing (5.15) and (5.16), it is clear that

$$S_{\dot{x}}(\omega) = \omega^2 S_x(\omega), \tag{5.17}$$

so that the spectral density of the derived process is just ω^2 times the spectral density of the original process. This is an important result because we can now calculate the mean square velocity $E[\dot{x}^2]$ from knowledge of $S_x(\omega)$, since

$$E[\dot{x}^2] = \int_{-\infty}^{\infty} S_{\dot{x}}(\omega)\, d\omega = \int_{-\infty}^{\infty} \omega^2 S_x(\omega)\, d\omega. \tag{5.18}$$

Similarly, the mean square acceleration $E[\ddot{x}^2]$ is given by

$$E[\ddot{x}^2] = \int_{-\infty}^{\infty} S_{\ddot{x}}(\omega)\, d\omega = \int_{-\infty}^{\infty} \omega^4 S_x(\omega)\, d\omega. \tag{5.19}$$

Cross-spectral density

We have seen how the spectral density of a random process is defined as the Fourier transform of its autocorrelation function. In the same way, the cross-spectral density of a pair of random processes is defined as the Fourier transform of the corresponding cross-correlation function for the two processes. Therefore if $R_{xy}(\tau)$ and $R_{yx}(\tau)$ are the two cross-correlation functions, we have

$$S_{xy}(\omega) = \frac{1}{2\pi} \int_{-\infty}^{\infty} R_{xy}(\tau)\, e^{-i\omega\tau}\, d\tau$$

and

$$ \tag{5.20}$$

$$S_{yx}(\omega) = \frac{1}{2\pi} \int_{-\infty}^{\infty} R_{yx}(\tau)\, e^{-i\omega\tau}\, d\tau$$

and their accompanying inverse transform relations which are

$$R_{xy}(\tau) = \int_{-\infty}^{\infty} S_{xy}(\omega) e^{i\omega\tau} d\omega$$

and

$$(5.21)$$

$$R_{yx}(\tau) = \int_{-\infty}^{\infty} S_{yx}(\omega) e^{i\omega\tau} d\omega.$$

According to classical Fourier transform theory we need

$$\int_{-\infty}^{\infty} |R_{xy}(\tau)| d\tau < \infty$$

for the integrals to exist, and this means that $x(t)$ and $y(t + \tau)$ must be uncorrelated when $\tau \to \infty$ and, from (3.24), that either m_x or m_y (i.e. one of the mean values) must be zero.

The cross-correlation functions are related by (3.21) and it follows that the two cross-spectra are also related. Putting

$$R_{xy}(\tau) = R_{yx}(-\tau)$$

in the first of (5.20) gives

$$S_{xy}(\omega) = \frac{1}{2\pi} \int_{-\infty}^{\infty} R_{yx}(-\tau) e^{-i\omega\tau} d\tau$$

and if we now substitute $\tau' = -\tau$ we obtain

$$S_{xy}(\omega) = \frac{1}{2\pi} \int_{\tau'=\infty}^{\tau'=-\infty} R_{yx}(\tau') e^{i\omega\tau'} (-d\tau')$$

$$= \frac{1}{2\pi} \int_{-\infty}^{\infty} R_{yx}(\tau') e^{i\omega\tau'} d\tau' \qquad (5.22)$$

which is the same as the second of (5.20) except that the sign of $i\omega$ has been changed. If, according to (4.8) and (4.9),

$$S_{xy}(\omega) = A(\omega) - iB(\omega)$$

and

$$S_{yx}(\omega) = C(\omega) - iD(\omega)$$

where $A(\omega)$, $B(\omega)$, $C(\omega)$ and $D(\omega)$ are real functions of ω, then by comparing (5.22) and the second of (5.20), we must have

$$C(\omega) = A(\omega)$$

and

$$D(\omega) = -B(\omega)$$

since the definite integrals are independent of the variables of integration, whether τ or τ'. Hence $S_{xy}(\omega)$ and $S_{yx}(\omega)$ are the same except that the sign of their imaginary parts is reversed. $S_{yx}(\omega)$ is therefore the complex conjugate of $S_{xy}(\omega)$, which is usually written

$$S_{yx}(\omega) = S_{xy}^*(\omega)$$

and, conversely, (5.23)

$$S_{xy}(\omega) = S_{yx}^*(\omega).$$

We shall see later how measurements of cross-spectra allow the characteristics of dynamic systems to be identified and how amplitude ratios and phase angles can be calculated as a function of frequency. In order to do this we must first review the dynamic properties of deterministic (non-random) systems and we turn to these in the next chapter.

Example

The white noise ergodic random process $y(t)$ is the result of delaying a similar process $x(t)$ by time T. (A sample of the $x(t)$ process may be thought of as being tape recorded and then replayed T seconds later as $y(t)$.) If the spectral density of $x(t)$ and $y(t)$ is S_0, determine the cross-correlation functions $R_{xy}(\tau)$ and $R_{yx}(\tau)$ and the cross-spectra $S_{xy}(\omega)$ and $S_{yx}(\omega)$.

From (3.20),

$$R_{xy}(\tau) = E[x(t)y(t + \tau)] = R_{yx}(-\tau).$$

In this case

$$y(t + T) = x(t)$$

so that

$$y(t) = x(t - T)$$

and

$$y(t + \tau) = x(t + \tau - T)$$

giving

$$R_{xy}(\tau) = E[x(t)x(t + \tau - T)].$$

The r.h.s. of this is just the autocorrelation function $R_x(\tau - T)$ which from (5.10) is given by $2\pi S_0 \delta(\tau - T)$, and so finally we have

$$R_{xy}(\tau) = 2\pi S_0 \delta(\tau - T) = R_{yx}(-\tau)$$

as shown in Fig. 5.7.

Since $x(t)$ and $y(t)$ (which is simply $x(t)$ delayed by T) are both white noise processes, their cross-correlation functions are zero everywhere except for the single value of τ for which $x = y$. Then

$$R_{xy}(\tau = T) = R_{yx}(\tau = -T) = R_x(\tau = 0)$$

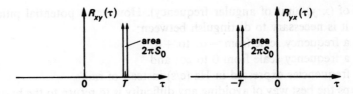

Fig. 5.7 Cross-correlation functions for delayed white noise

which is the delta function illustrated in Fig. 5.6(*b*). From the first of (5.20),

$$S_{xy}(\omega) = \frac{1}{2\pi} \int_{-\infty}^{\infty} 2\pi S_0\, \delta(\tau - T)\, e^{-i\omega\tau}\, d\tau$$

$$= S_0\, e^{-i\omega T}$$

after using (5.9).

From the second of (5.20)

$$S_{yx}(\omega) = \frac{1}{2\pi} \int_{-\infty}^{\infty} 2\pi S_0\, \delta(\tau + T)\, e^{-i\omega\tau}\, d\tau$$

$$= S_0\, e^{i\omega T}$$

thus confirming (5.23).

Putting $S_{xy}(\omega) = A(\omega) - iB(\omega)$ we see that

$$A(\omega) = S_0 \cos \omega t$$

$$B(\omega) = S_0 \sin \omega t$$

and if we plot the magnitude of $|S_{xy}(\omega)|$ and the angle $\theta = \tan^{-1} B(\omega)/A(\omega)$ these appear as shown in Fig. 5.8.

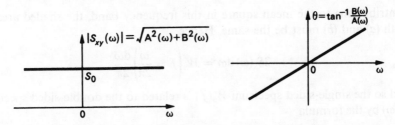

Fig. 5.8 Cross-spectral density function for delayed white noise

Note on the units of spectral density

We have already noted that the units of spectral density are those of $(x^2)/$(unit of angular frequency), while those of cross-spectral density are the units of $(x \cdot y)/$(unit of angular frequency). However a potential pitfall occurs when it is necessary to distinguish between:

(a) a frequency scale from $-\infty$ to $+\infty$;

(b) a frequency scale from 0 to ∞; and

(c) frequencies expressed in Hz (c/s) instead of rad/s.

Perhaps the best way of avoiding any difficulty is to return to the basic formula (5.3)

$$E[x^2] = \int_{-\infty}^{\infty} S_x(\omega)\,d\omega \tag{5.3}$$

and consider the practical form of this which is

$$E[x^2] = \int_{0}^{\infty} W_x(f)\,df \tag{5.24}$$

where f is frequency in Hz and $W_x(f)$ is the equivalent one-sided spectral density function, Fig. 5.9(b). The frequency band ω to $\omega + d\omega$ rad/s in Fig. 5.9(a) corresponds to $\omega/2\pi$ to $(\omega + d\omega)/2\pi$ Hz in Fig. 5.9(b), so that for equal

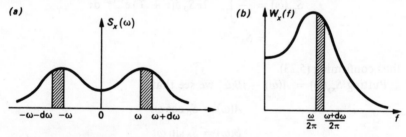

Fig. 5.9 Illustrating the relationship between alternative spectral density parameters

contributions to the mean square in this frequency band, the shaded areas in both (a) and (b) must be the same. Hence

$$2S_x(\omega)\,d\omega = W_x\!\left(f = \frac{\omega}{2\pi}\right)\frac{d\omega}{2\pi}$$

and so the single-sided spectrum $W_x(f)$ is related to the double-sided spectrum $S_x(\omega)$ by the formula

$$W_x\!\left(f = \frac{\omega}{2\pi}\right) = 4\pi S_x(\omega)$$

or by its equivalent   (5.25)

$$W_x(f) = 4\pi S_x(\omega = 2\pi f).$$

The same result also applies for the magnitude of the cross-spectral density function.

Chapter 6

Excitation – response relations for linear systems

We turn now to the response characteristics of physical systems. Before considering what happens when a system is subjected to random excitation, we must deal first with various different methods for describing the response of a general system to deterministic (non-random) excitation. The general system may be a vibrating structure or machine, or a complete building, or a small electric circuit. Whatever it is, we assume that there are a number of *inputs* $x_1(t)$, $x_2(t)$, $x_3(t)$, etc., which constitute the *excitation* and a number of *outputs* $y_1(t)$, $y_2(t)$, $y_3(t)$, etc., which constitute the *response*, Fig. 6.1. The $x(t)$ and $y(t)$ may be forces, pressures, displacements, velocities, accelerations, voltages, currents, etc., or a mixture of all of these. We shall restrict ourselves to *linear* systems for which each response variable $y(t)$ is related to the excitation by a linear differential equation of the form

$$
\begin{aligned}
a_n \frac{d^n y_1}{dt^n} + a_{n-1} \frac{d^{n-1} y_1}{dt^{n-1}} + \cdots + a_1 \frac{dy_1}{dt} + a_0 y_1 = \\
= \Bigg\{ b_r \frac{d^r x_1}{dt^r} + b_{r-1} \frac{d^{r-1} x_1}{dt^{r-1}} + \cdots + b_1 \frac{dx_1}{dt} + b_0 x_1 + \\
+ c_s \frac{d^s x_2}{dt^s} + c_{s-1} \frac{d^{s-1} x_2}{dt^{s-1}} + \cdots + c_1 \frac{dx_2}{dt} + c_0 x_2 + \\
+ d_t \frac{d^t x_3}{dt^t} + d_{t-1} \frac{d^{t-1} x_3}{dt^{t-1}} + \cdots + d_1 \frac{dx_3}{dt} + d_0 x_3 + \\
+ \ldots \ldots \ldots \ldots \ldots \ldots \ldots \ldots \Bigg\}.
\end{aligned}
\tag{6.1}
$$

The equations are linear because if $y_1'(t)$ is the response to an excitation $x_1'(t)$, $x_2'(t)$, $x_3'(t)$, etc., and $y_1''(t)$ is the response to a separate excitation $x_1''(t)$, $x_2''(t)$, $x_3''(t)$, etc., then the response to the combined excitation $x_1'(t) + x_1''(t)$, $x_2'(t) + x_2''(t)$, $x_3'(t) + x_3''(t)$, etc., is just $y_1'(t) + y_1''(t)$, i.e. the *principle of linear superposition* applies. The coefficients a, b, c, d, etc., may in general be functions of time, but we shall consider only cases when they are constant which means that the vibrating system does not change its characteristics with time.

On account of the principle of superposition, our problem is greatly simpli-

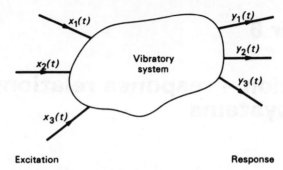

Fig. 6.1 Excitation and response parameters of a physical system

fied because we can consider how each output variable responds to a single input variable alone, and then just add together the separate responses to many input variables in order to obtain the response to the combined excitation at many points. Of course the assumption of linearity is a bold one, but since vibrations usually only involve small displacements from equilibrium, it is very often not too far from the truth. We can therefore simplify the system we should consider to that shown in Fig. 6.2 in which only one input variable and one output variable appears.

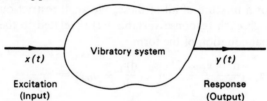

Fig. 6.2 For a linear system the response to each input variable may be considered separately

Classical approach

If the equations of motion for the constant parameter linear system in Fig. 6.2 can be determined, then there is a known linear differential equation relating $y(t)$ and $x(t)$ with the form

$$a_n \frac{d^n y}{dt^n} + a_{n-1} \frac{d^{n-1} y}{dt^{n-1}} + \cdots + a_1 \frac{dy}{dt} + a_0 y =$$

$$= b_r \frac{d^r x}{dt^r} + b_{r-1} \frac{d^{r-1} x}{dt^{r-1}} + \cdots + b_1 \frac{dx}{dt} + b_0 x. \tag{6.2}$$

For given excitation $x(t)$ and given initial conditions, this equation can be solved by classical methods to give a complete solution for $y(t)$. However such an approach is not usually helpful for random vibration problems for two reasons. First the differential equation (6.2) is seldom obtainable directly because there is inadequate data available and simple experimental methods for finding the coefficients a and b are not available. Secondly, even if the differential equation is known, a complete time history for $y(t)$ can only be calculated if we have a complete time history $x(t)$, and for random vibration problems this data is not of course available. In order to calculate average values of the output variables it is more convenient to concentrate on alternative ways of representing the relationship between $y(t)$ and $x(t)$.

Frequency response method

A completely different method of describing the dynamic characteristics of a linear system is to determine the response to a sine wave input. With reference to Fig. 6.2, if the input is a constant amplitude sine wave of fixed frequency

$$x(t) = x_0 \sin \omega t \qquad (6.3)$$

then, from equation (6.2), the steady state output must also be a sine wave of fixed amplitude, the same frequency ω and phase difference ϕ, so that

$$y(t) = y_0 \sin(\omega t - \phi). \qquad (6.4)$$

Incidentally we are assuming that, for no excitation, the system is dormant and there is no response, $y(t) = 0$. Unstable systems, in which self-excited oscillations can occur, are not considered.

Information about the amplitude ratio y_0/x_0 and the phase angle ϕ defines the transmission characteristics or transfer function of the system at the fixed frequency ω. By making measurements at a series of closely spaced frequencies, amplitude ratio and phase angle can be plotted as a function of frequency, and in theory if the frequency range extends from zero to infinity then the dynamic characteristics of the system are completely defined.

Example

Determine the amplitude ratio and phase angle for the transmission of sine wave excitation through the spring-damper system of Fig. 6.3(a). The excitation is the force $x(t)$ and the response is the displacement $y(t)$.

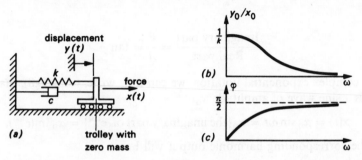

Fig. 6.3 Frequency response characteristics for a simple system

For a linear spring of stiffness k and a linear viscous damper of coefficient c, the equation of motion is

$$c\dot{y} + ky = x(t)$$

and so when $x(t)$ is a constant amplitude sine wave (6.3), and $y(t)$ is the response given by (6.4)

$$cy_0\omega\cos(\omega t - \phi) + ky_0\sin(\omega t - \phi) = x_0\sin\omega t$$

which gives, collecting terms

$$y_0\sin\omega t\left\{c\omega\sin\phi + k\cos\phi - \frac{x_0}{y_0}\right\} + y_0\cos\omega t\{c\omega\cos\phi - k\sin\phi\} = 0.$$

For this equation to be true, the terms in brackets must be separately zero, so that the amplitude ratio

$$\frac{y_0}{x_0} = \frac{1}{\sqrt{(c^2\omega^2 + k^2)}}$$

and the phase angle

$$\phi = \tan^{-1}\frac{c\omega}{k}.$$

Instead of thinking of amplitude ratio and phase angle as two separate quantities, it has become customary in vibration theory to use a single complex number to represent both quantities. This is called the (complex) frequency response function $H(\omega)$ which is defined so that its magnitude is equal to the amplitude ratio and the ratio of its imaginary part to its real part is equal to the tangent of the phase angle. If

$$H(\omega) = A(\omega) - iB(\omega) \tag{6.5}$$

where $A(\omega)$ and $B(\omega)$ are real functions of ω, then

$$|H(\omega)| = \sqrt{(A^2 + B^2)} = \frac{y_0}{x_0} \tag{6.6}$$

and

$$\frac{\text{Imaginary part}}{\text{Real part}} = \frac{B}{A} = \tan\phi. \tag{6.7}$$

Using complex exponential notation, we can now write, using (4.8), that if the input is a sine wave of amplitude x_0,

$$x(t) = x_0\sin\omega t = x_0\{\text{the imaginary part of } e^{i\omega t}\} = x_0\,\text{Im}(e^{i\omega t}), \tag{6.8}$$

then the corresponding harmonic output will be

$$y(t) = x_0\,\text{Im}\{H(\omega)e^{i\omega t}\}. \tag{6.9}$$

The proof that this is correct is easily obtained by substituting (6.5) into (6.9) to give

$$y(t) = x_0 \, \text{Im}\{A(\omega) - iB(\omega)\}(\cos\omega t + i\sin\omega t)$$

$$= x_0\{A(\omega)\sin\omega t - B(\omega)\cos\omega t\}$$

$$= x_0\sqrt{(A^2 + B^2)}\sin\left(\omega t - \tan^{-1}\frac{B}{A}\right)$$

$$= y_0\sin(\omega t - \phi)$$

after using (6.6) and (6.7).

In summary, therefore, if a constant amplitude harmonic input given by

$$x(t) = x_0 \, e^{i\omega t} \tag{6.10}$$

is applied to a linear system, the corresponding output $y(t)$ will be given by

$$y(t) = H(\omega)x_0 \, e^{i\omega t} \tag{6.11}$$

where $H(\omega)$ is the system's complex frequency response function evaluated at angular frequency ω. In interpreting (6.10) and (6.11) we mean either "the real part" or "the imaginary part" of the right-hand side of the equations (but not both) according to the convention agreed on. The magnitude of $H(\omega)$ gives the ratio of the amplitudes of $y(t)$ and $x(t)$ and its argument gives the phase angle between $y(t)$ and $x(t)$. Notice that the quantity x_0 in (6.10) and (6.11) need not necessarily be real; its magnitude gives the amplitude of $x(t)$ while the magnitude of $H(\omega)x_0$ gives the amplitude of $y(t)$.

Example

Calculate the complex frequency response function for the system of Fig. 6.3(a).

The equation of motion is

$$c\dot{y} + ky = x.$$

Putting $x = x_0 \, e^{i\omega t}$ (where either "the imaginary part of" or "the real part of" is understood) and

$$y = H(\omega)x_0 \, e^{i\omega t}$$

gives

$$(ci\omega + k)H(\omega)e^{i\omega t} = e^{i\omega t}$$

or

$$H(\omega) = \frac{1}{k + ic\omega} = A(\omega) - iB(\omega).$$

The amplitude ratio

$$\frac{y_0}{x_0} = |H(\omega)| = \frac{1}{\sqrt{(k^2 + c^2\omega^2)}} \tag{6.12}$$

and the phase angle ϕ is given by

$$\tan \phi = \frac{B(\omega)}{A(\omega)} = \frac{c\omega}{k} \qquad (6.13)$$

which checks with the previous example.

Notice that $H(\omega)$ has dimensions (in this case those of displacement/force) and that the output lags behind the input by angle ϕ.

Impulse response method

The frequency response function $H(\omega)$ gives the *steady state* response of a system to a sine wave input. By measuring $H(\omega)$ for all frequencies we completely define the dynamic characteristics of the system. An alternative approach is to measure *transient* response, initiated by a suitable disturbance. If we measure the transient response for all time, until static equilibrium has been regained after the disturbance, then this is another method of defining completely a system's dynamic characteristics.

It is usual to consider the result of exciting the system by an input disturbance which is short and sharp and is present for a very short (theoretically zero) time interval. The transient response is then not complicated by removal of the disturbance. Using the delta function notation described in Chapter 5, we can represent such a disturbance by the equation

$$x(t) = I\,\delta(t) \qquad (6.14)$$

where I is a constant parameter with the dimensions $(x) \times$ (time). For the case when $x(t)$ represents a force, (6.14) describes a hammer blow or *impulse* of magnitude

$$\int_{-\infty}^{\infty} x(t)\,\mathrm{d}t = I \int_{-\infty}^{\infty} \delta(t)\,\mathrm{d}t = I \text{ in (force)} \times \text{(time) units.} \qquad (6.15)$$

This terminology is carried over to the general case when $x(t)$ may represent any input parameter, whether a force or not, and the *impulse response* of a system is defined as the system's response to an "impulsive" input of the form (6.14) where I has the proper dimensions. The excitation is described as a *unit*

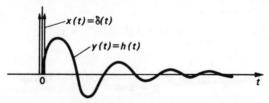

Fig. 6.4 Typical impulse response function

impulse when I is numerically unity in (6.14). In response to such an impulsive input, the initially dormant system suddenly springs to life and then gradually recovers its static equilibrium position as time passes. The response to a unit impulse at $t = 0$ is represented by the (unit) *impulse response function*, $h(t)$, Fig. 6.4. Notice that $h(t) = 0$ for $t < 0$ because $y(t) = 0$ before the impulse occurs.

Example

Determine the impulse response function for the system shown in Fig. 6.3(a).

Starting from the equation of motion

$$c\dot{y} + ky = x,$$

$h(t)$ is the solution $y(t)$ when $x(t) = \delta(t)$, i.e.

$$c\dot{h} + kh = \delta(t).$$

For $t > 0$, $\delta(t) = 0$ and so

$$c\dot{h} + kh = 0$$

whose solution is

$$h = C e^{-kt/c}$$

where C is a constant to be determined from the conditions at $t = 0$. In order to find what happens when the hammer blow falls, we can make use of the fact that $\delta(t)$ is zero everywhere except at $t = 0$. If $t = 0-$ is a point just to the left and $t = 0+$ a point just to the right of the origin, then

$$\int_{0-}^{0+} \delta(t)\,dt = 1$$

and integrating both sides of the equation of motion from $t = 0-$ to $t = 0+$ gives

$$c\int_{0-}^{0+} \dot{h}\,dt + k \int_{0-}^{0+} h\,dt = \int_{0-}^{0+} \delta(t)\,dt = 1.$$

To go further we need to employ some physical reasoning about what happens to the system when the impulse is applied. The massless trolley responds to the impulse with a sudden movement and its velocity will be infinite instantaneously. Because $h(t) = \dot{y}(t)$ is infinite at $t = 0$, the integral $\int_{0-}^{0+} \dot{h}\,dt$ across $t = 0$ will be finite, even though the integration time from $0-$ to $0+$ approaches zero in the limit. However, $h(t)$ is not infinite at $t = 0$, and so the value of the integral $\int_{0-}^{0+} h(t)\,dt$ must be zero in the limit. We therefore obtain

$$c\int_{0-}^{0+} \dot{h}\,dt = 1$$

so that

$$c(h(t = 0+) - 0) = 1$$

or

$$h(t = 0+) = \frac{1}{c}.$$

It follows that the constant C is given by

$$C = \frac{1}{c}$$

and the full solution for $h(t)$ is

$$h(t) = 0 \qquad \text{for} \qquad t < 0$$

$$h(t) = \frac{1}{c}e^{-kt/c} \qquad \text{for} \qquad t > 0$$

as sketched in Fig. 6.5. Notice that the units of $h(t)$ are those of (displacement)/(impulse) or, equivalently, of (velocity)/(force).

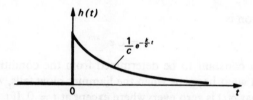

Fig. 6.5 Unit impulse response function for the system shown in Fig. 6.3(a)

Relationship between the frequency response and impulse response functions

Since complete information about either the frequency response function or the impulse response function fully defines the dynamic characteristics of a system, it follows that we should be able to derive one from the other and vice versa. The Fourier transform method of breaking an aperiodic function into its frequency spectrum provides the necessary link.

As we are dealing with stable systems which are dormant before they are excited and for which motion dies away after an impulse, we know that

$$\int_{-\infty}^{\infty} |h(t)| \, dt < \infty \tag{6.16}$$

and we may therefore take Fourier transforms of both the impulsive input $x(t) = \delta(t)$ and the transient output $y(t) = h(t)$. If we do this, we obtain

$$X(\omega) = \frac{1}{2\pi} \int_{-\infty}^{\infty} x(t)e^{-i\omega t} \, dt = \frac{1}{2\pi} \int_{-\infty}^{\infty} \delta(t)e^{-i\omega t} \, dt \tag{6.17}$$

and

$$Y(\omega) = \frac{1}{2\pi} \int_{-\infty}^{\infty} y(t)\,e^{-i\omega t}\,dt = \frac{1}{2\pi} \int_{-\infty}^{\infty} h(t)\,e^{-i\omega t}\,dt. \qquad (6.18)$$

The first of these equations may be simplified by expanding the complex exponential to obtain

$$X(\omega) = \frac{1}{2\pi} \int_{-\infty}^{\infty} \delta(t)\cos \omega t\,dt - i\frac{1}{2\pi} \int_{-\infty}^{\infty} \delta(t)\sin \omega t\,dt$$

and then using the property of a delta function (5.9) to show that the first integral is unity and the second zero (since $\sin \omega t = 0$ at $t = 0$) to obtain

$$X(\omega) = \frac{1}{2\pi}. \qquad (6.19$$

We have therefore obtained the Fourier transforms of an impulsive input $x(t) = \delta(t)$ in (6.19) and of the corresponding impulse response function $y(t) = h(t)$ in (6.18).

Furthermore, these Fourier transforms are related by the frequency response function $H(\omega)$. The nature of this relationship can be seen by the following argument.† We know that, when a linear system is subjected to steady state harmonic excitation at frequency ω, it responds with a steady harmonic output at the same frequency. It therefore seems reasonable to expect that, for an aperiodic input signal, frequency components $X(\omega)\,d\omega$ in the frequency band ω to $\omega + d\omega$ in the input will correspond with components $Y(\omega)\,d\omega$ in the same frequency band in the output. In this case, if we had a harmonic input of the form

$$x(t) = X(\omega)\,d\omega\,e^{i\omega t}$$

(where "the real part of" or "the imaginary part of" is understood) the corresponding harmonic output would be given by

$$y(t) = Y(\omega)\,d\omega\,e^{i\omega t}.$$

But, from (6.10) and (6.11), we also know that

$$y(t) = H(\omega)X(\omega)\,d\omega\,e^{i\omega t}$$

and so, by comparing these two expressions for $y(t)$, we obtain

$$Y(\omega) = H(\omega)X(\omega) \qquad (6.20)$$

which is a very important relation between the Fourier transforms of the input and output, $X(\omega)$ and $Y(\omega)$, and the frequency response function $H(\omega)$.

---

† The lines of a formal proof are indicated in problem 6.4.

Finally, substituting for $X(\omega)$ from (6.19) and for $Y(\omega)$ from (6.18) into (6.20), gives

$$\frac{1}{2\pi}\int_{-\infty}^{\infty} h(t)\,e^{-i\omega t}\,dt = H(\omega)\cdot\frac{1}{2\pi}$$

or

$$H(\omega) = \int_{-\infty}^{\infty} h(t)\,e^{-i\omega t}\,dt \qquad (6.21)$$

expressing the result that the frequency response function $H(\omega)$ is the Fourier transform of the impulse response function $h(t)$. Actually, by comparing (6.21) with the definition of a Fourier transform given in (4.10), it can be seen that there is a factor $1/2\pi$ missing from (6.21). However as already mentioned in Chapter 4, the position of this factor is optional so long as it appears in either the Fourier transform equation or the inverse Fourier transform equation. From (4.13), the inverse transform equation corresponding to (6.21) is

$$h(t) = \frac{1}{2\pi}\int_{-\infty}^{\infty} H(\omega)\,e^{i\omega t}\,d\omega \qquad (6.22)$$

and the impulse response function $h(t)$ is shown as a Fourier integral of the frequency response function $H(\omega)$.

Calculation of response to an arbitrary input

We turn now to how the frequency response and impulse response functions can be used to calculate how a system responds to a prescribed excitation. Suppose that a linear system has an arbitrary input $x(t)$ which is defined, which satisfies (4.7), and for which we wish to calculate the resulting output $y(t)$.

As already described in the previous section, the frequency response function can be used to relate the Fourier transforms of input and output, so that if

$$X(\omega) = \frac{1}{2\pi}\int_{-\infty}^{\infty} x(t)\,e^{-i\omega t}\,dt$$

then

$$Y(\omega) = H(\omega)X(\omega)$$

and taking the inverse transform to find $y(t)$ we obtain

$$y(t) = \int_{-\infty}^{\infty} H(\omega)\left\{\frac{1}{2\pi}\int_{-\infty}^{\infty} x(t)\,e^{-i\omega t}\,dt\right\}e^{i\omega t}\,d\omega \qquad (6.23)$$

which is a formal solution for the output $y(t)$. The integral with respect to ω is, in general, extremely difficult to evaluate, and so it is rarely possible to use (6.23) to obtain a neat closed expression for $y(t)$. We do not therefore pursue this approach for calculating $y(t)$ but turn to the impulse response method as a more promising alternative.

The impulse response function $h(t)$ gives the response at time t to a unit impulse applied at time $t = 0$, i.e. after a delay of duration t. It follows that $h(t - \tau)$ is the response at time t to a unit impulse or "hammer blow" at time τ, i.e. after a delay of duration $t - \tau$. Now think of an arbitrary input function $x(t)$ as being made up of a continuous series of small impulses. The "impulse" corresponding to the input $x(t)$ between the time limits τ and $\tau + d\tau$, Fig. 6.6,

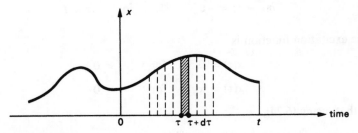

Fig. 6.6 Breaking down an arbitrary input $x(t)$ into a series of "impulses"

has the magnitude $x(\tau)\,d\tau$ shown shaded. The response at time t to this "impulse" alone is just the fraction

$$\frac{x(\tau)\,d\tau}{1}$$

of the response to a unit impulse at $t = \tau$, which is $h(t - \tau)$. The shaded area in Fig. 6.6 therefore contributes an amount

$$h(t - \tau)x(\tau)\,d\tau$$

to the total response at time t. Furthermore, since the principle of super-position applies for a linear system, we may obtain the total response $y(t)$ at t by adding together all the separate responses to all the small "impulses" which make up the total time history of $x(t)$ back to $t = -\infty$. We therefore integrate the response to $x(\tau)\,d\tau$ back to $\tau = -\infty$ to give

$$y(t) = \int_{-\infty}^{t} h(t - \tau)x(\tau)\,d\tau \tag{6.24}$$

which is another formal expression for the response $y(t)$ at time t as a result of an excitation whose value at time τ is $x(\tau)$ and which may exist from $\tau = -\infty$ to the present time $\tau = t$.

Example

Calculate the response at time $t > 0$ of the system shown in Fig. 6.3(a) when it is subjected to a step input $x(t) = x_0$ at $t = 0$.

From the previous example we already have

$$h(t) = 0 \qquad \text{for} \quad t < 0$$

$$h(t) = \frac{1}{c} e^{-kt/c} \qquad \text{for} \quad t > 0$$

so that

$$h(t - \tau) = 0 \qquad \text{for} \quad \tau > t$$

$$h(t - \tau) = \frac{1}{c} e^{-k(t - \tau)/c} \qquad \text{for} \quad \tau < t.$$

The excitation function is

$$x(\tau) = 0 \qquad \text{for} \quad \tau < 0$$

$$x(\tau) = x_0 \qquad \text{for} \quad \tau > 0$$

so that, using (6.24),

$$y(t) = \int_{-\infty}^{t} h(t - \tau)x(\tau)\,d\tau = \int_{0}^{t} \frac{1}{c} e^{-k(t - \tau)/c} x_0\,d\tau \qquad \text{for} \quad t > 0$$

the lower limit of the integral being changed to 0 since $x(\tau) = 0$ for $\tau < 0$. Evaluating the integral gives

$$y(t) = \frac{x_0}{k}(1 - e^{-kt/c}) \qquad \text{for} \quad t > 0.$$

The response $y(t)$ is obviously zero for $t < 0$ before the step input has occurred.

The *superposition* or *convolution* integral (6.24) is a most important input–output relationship for a linear system. Provided that the system is passive so that it only responds to past inputs and $h(t)$ decays eventually to static equilibrium, so that

$$\int_{-\infty}^{\infty} |h(t)|\,dt < \infty,$$

then (6.24) applies for any input $x(t)$ whose magnitude $|x(t)|$ is in general bounded by a finite level. Notice that $x(t)$ does not need to satisfy (4.7) which is necessary in the classical Fourier transform (frequency response) approach.

There are three alternative versions of (6.24) which can easily be derived by physical reasoning. First we must recall that $h(t - \tau)$ is the response to a unit impulse at

$$(t - \tau) = 0$$

that is at $t = \tau$. For $(t - \tau) < 0$, there is no response, as there has been no impulse applied. Hence, for $\tau > t$

$$h(t - \tau) = 0.$$

We may therefore extend the upper limit of the integral in (6.24) from $\tau = t$ to $\tau = \infty$ without changing the result, since $h(t - \tau)$ is zero in this interval. The first alternative form of (6.24) is therefore

$$y(t) = \int_{-\infty}^{\infty} h(t - \tau)x(\tau)\,d\tau. \tag{6.25}$$

Next consider changing the variable in (6.24) by putting

$$\theta = t - \tau$$

where θ may be interpreted as the time delay between the occurrence of an impulse and the instant when its result is being calculated. The limits of integration $\tau = -\infty$ and $\tau = t$ now become $\theta = \infty$ (notice that time t is a constant) and $\theta = 0$ and $d\tau$ becomes $-d\theta$, so that substituting in (6.24) gives

$$y(t) = \int_{\infty}^{0} h(\theta)x(t - \theta)(-d\theta)$$

or changing over the limits of integration to dispense with the minus sign in front of $d\theta$

$$y(t) = \int_{0}^{\infty} h(\theta)x(t - \theta)\,d\theta. \tag{6.26}$$

Finally, the third alternative form can be obtained either by putting $\theta = (t - \tau)$ in (6.25) or by noting that $h(\theta) = 0$ for $\theta < 0$, since there is no response before the impulse occurs, and using (6.26), to obtain

$$y(t) = \int_{-\infty}^{\infty} h(\theta)x(t - \theta)\,d\theta. \tag{6.27}$$

Collecting together these important alternative results, the response $y(t)$ of a passive linear system to an arbitrary input $x(t)$ can be calculated by evaluating one of the superposition integrals

$$y(t) = \int_{-\infty}^{t} h(t - \tau)x(\tau)\,d\tau \tag{6.24}$$

or

$$y(t) = \int_{-\infty}^{\infty} h(t - \tau)x(\tau)\,d\tau \tag{6.25}$$

or

$$y(t) = \int_{0}^{\infty} h(\theta)x(t - \theta)\,d\theta \tag{6.26}$$

or

$$y(t) = \int_{-\infty}^{\infty} h(\theta)x(t - \theta)\,d\theta \tag{6.27}$$

where $h(t)$ is the response at time t to a unit impulse $x(t) = \delta(t)$ applied at $t = 0$.

Example

Use equation (6.26) to calculate the response of the system shown in Fig. 6.3(a) to a step input $x(t) = x_0$ at $t = 0$.

From the earlier example

$$h(\theta) = \frac{1}{c} e^{-k\theta/c} \qquad \text{for} \qquad \theta > 0$$

so that, if we want the response at time $t = t_1$ (say), then, from (6.26)

$$y(t_1) = \int_0^\infty \frac{1}{c} e^{-k\theta/c} x(t_1 - \theta) \, d\theta. \tag{6.28}$$

In order to evaluate this integral we have to determine $x(t_1 - \theta)$ as a function of θ for t_1 constant. This must be calculated from the known input function $x(t)$ shown in Fig. 6.7(a).

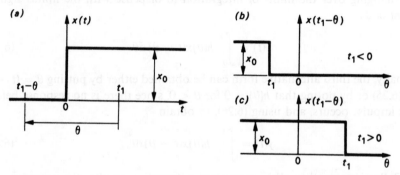

Fig. 6.7 Result of changing the variable of integration from t to $(t - \theta)$

First we select the *constant* value $t = t_1$. Then, since θ is the delay between the occurrence of an "impulse" and time t_1, the variable θ is measured *backwards* along the time axis, as shown in Fig. 6.7(a). We can now read off values of $x(t_1 - \theta)$ and plot two separate Figs. 6.7(b) and (c) for the two cases $t_1 < 0$ and $t_1 > 0$.

The function $x(t_1 - \theta)$ that we need to substitute in (6.28) is therefore different for the two cases and so we obtain two different analytical results: one for the case $t_1 < 0$ and the other for the case $t_1 > 0$. These results are

$$y(t_1) = 0 \qquad \text{for} \qquad t_1 < 0$$

$$y(t_1) = \int_0^{t_1} \frac{1}{c} e^{-k\theta/c} x_0 \, d\theta = \frac{x_0}{k}(1 - e^{-kt_1/c}) \qquad \text{for} \qquad t_1 > 0$$

in agreement with the previous example in which we applied (6.24) rather than (6.26).

Chapter 7

Transmission of random vibration

We are now ready to consider how the characteristics of random signals are changed by transmission through stable linear systems. We shall consider the response $y(t)$ to two separate random inputs $x_1(t)$ and $x_2(t)$. The response to a single input can then be obtained directly by putting either x_1 or x_2 zero, and the response to more than two inputs can be inferred from the form of the solution for two inputs.

The mental picture we need is of an infinity of experiments, all proceeding simultaneously, and each with an identical linear system for which the impulse response functions are $h_1(t)$ and $h_2(t)$ and the corresponding frequency response functions are $H_1(\omega)$ and $H_2(\omega)$, Fig. 7.1. Each experiment is excited by sample functions from the $x_1(t)$ and $x_2(t)$ random processes and the response is a sample

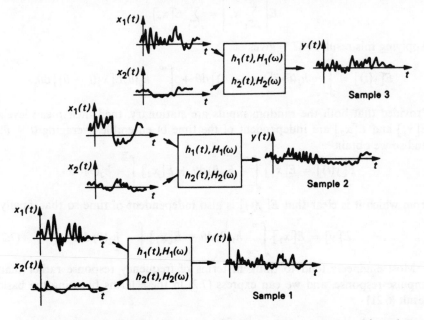

Fig. 7.1 Concept of ensemble averaging for a linear system subjected to random excitation

function from the $y(t)$ random process. We want to know how the character-istics of the output process $y(t)$ depend on the characteristics of the two input processes $x_1(t)$ and $x_2(t)$ and on the input–output characteristics of the system. The functions $h_1(t)$ and $H_1(\omega)$ give the response $y(t)$ due to an input $x_1(t)$ and the functions $h_2(t)$ and $H_2(\omega)$ give the response $y(t)$ due to an input $x_2(t)$.

Mean level

According to (6.27), the response $y(t)$ of a typical sample experiment to the inputs $x_1(t)$ and $x_2(t)$ may be expressed as

$$y(t) = \int_{-\infty}^{\infty} h_1(\theta)x_1(t-\theta)\,d\theta + \int_{-\infty}^{\infty} h_2(\theta)x_2(t-\theta)\,d\theta. \qquad (7.1)$$

If we are now to calculate the ensemble average $E[y(t)]$ we have to determine the average values of both of the integrals on the r.h.s. of (7.1). To do this we need to remember that an integral is just the limiting case of a summation and that the average of a sum of numbers is the same as the sum of the average numbers, for instance

$$E[x_1 + x_2 + x_3 + \cdots] = E[x_1] + E[x_2] + E[x_3] + \cdots$$

or using the summation sign

$$E\left[\sum_{r=1}^{N} x_r\right] = \sum_{r=1}^{N} E[x_r].$$

Applying this result to (7.1) gives

$$E[y(t)] = \int_{-\infty}^{\infty} h_1(\theta)E[x_1(t-\theta)]\,d\theta + \int_{-\infty}^{\infty} h_2(\theta)E[x_2(t-\theta)]\,d\theta.$$

Provided that both the random inputs are stationary, then their mean levels $E[x_1]$ and $E[x_2]$ are independent of the time of ensemble averaging $(t-\theta)$, and so we obtain

$$E[y(t)] = E[x_1]\int_{-\infty}^{\infty} h_1(\theta)\,d\theta + E[x_2]\int_{-\infty}^{\infty} h_2(\theta)\,d\theta$$

from which it is clear that $E[y(t)]$ is also independent of time so that, finally,

$$E[y] = E[x_1]\int_{-\infty}^{\infty} h_1(\theta)\,d\theta + E[x_2]\int_{-\infty}^{\infty} h_2(\theta)\,d\theta. \qquad (7.2)$$

Most engineers tend to think in terms of frequency response rather than impulse response, and we can express (7.2) in these terms by using the basic result (6.21)

$$H(\omega) = \int_{-\infty}^{\infty} h(t)e^{-i\omega t}\,dt \qquad (6.21)$$

which, for $\omega = 0$, becomes

$$H(\omega = 0) = \int_{-\infty}^{\infty} h(t)\,dt \tag{7.3}$$

and on substitution into (7.2) gives

$$E[y] = E[x_1]H_1(0) + E[x_2]H_2(0) \tag{7.4}$$

where

$$H_1(0) = \frac{\text{constant level of } y}{\text{constant level of } x_1}$$

$$H_2(0) = \frac{\text{constant level of } y}{\text{constant level of } x_2}$$

or, in electrical engineering terms,

$$H_1(0) = \frac{\text{d.c. (direct current) level of } y}{\text{d.c. level of } x_1}$$

$$H_2(0) = \frac{\text{d.c. level of } y}{\text{d.c. level of } x_2}.$$

The mean levels of stationary random vibration are therefore transmitted just as though they are constant non-random signals and the superimposed random excursions have no effect on the relationships between mean levels.

Autocorrelation

The autocorrelation function for the output process $y(t)$ is

$$E[y(t)y(t + \tau)]. \tag{7.5}$$

According to (7.1) we can write formal solutions for $y(t)$ and $y(t + \tau)$ and, putting θ_1 and θ_2 instead of θ to avoid confusion, these are

$$y(t) = \int_{-\infty}^{\infty} h_1(\theta_1)x_1(t - \theta_1)\,d\theta_1 + \int_{-\infty}^{\infty} h_2(\theta_1)x_2(t - \theta_1)\,d\theta_1 \tag{7.6}$$

and

$$y(t + \tau) = \int_{-\infty}^{\infty} h_1(\theta_2)x_1(t + \tau - \theta_2)\,d\theta_2 + \int_{-\infty}^{\infty} h_2(\theta_2)x_2(t + \tau - \theta_2)\,d\theta_2. \tag{7.7}$$

Substituting (7.6) and (7.7) into (7.5) gives

$$E[y(t)y(t + \tau)] = E\left[\int_{-\infty}^{\infty} h_1(\theta_1)x_1(t - \theta_1)\,d\theta_1 \int_{-\infty}^{\infty} h_1(\theta_2)x_1(t + \tau - \theta_2)\,d\theta_2 +\right.$$

$$+ \int_{-\infty}^{\infty} h_1(\theta_1)x_1(t - \theta_1)\,d\theta_1 \int_{-\infty}^{\infty} h_2(\theta_2)x_2(t + \tau - \theta_2)\,d\theta_2 +$$

$$+ \int_{-\infty}^{\infty} h_2(\theta_1)x_2(t - \theta_1)\,d\theta_1 \int_{-\infty}^{\infty} h_1(\theta_2)x_1(t + \tau - \theta_2)\,d\theta_2 +$$

$$+ \left. \int_{-\infty}^{\infty} h_2(\theta_1)x_2(t - \theta_1)\,d\theta_1 \int_{-\infty}^{\infty} h_2(\theta_2)x_2(t + \tau - \theta_2)\,d\theta_2 \right].$$

$$(7.8)$$

Since we are dealing with a stable system, the integrals in (7.8) converge and it is legitimate to write each product of two integrals as a single double integral,† so that, for instance,

$$\int_{-\infty}^{\infty} h_1(\theta_1)x_1(t - \theta_1)\,d\theta_1 \int_{-\infty}^{\infty} h_1(\theta_2)x_1(t + \tau - \theta_2)\,d\theta_2 =$$

$$= \int_{-\infty}^{\infty}\int_{-\infty}^{\infty} h_1(\theta_1)h_1(\theta_2)x_1(t - \theta_1)x_1(t + \tau - \theta_2)\,d\theta_1\,d\theta_2. \qquad (7.9)$$

When we come to average the expression (7.8), we have then to average these double integrals to find, for example,

$$E\left[\int_{-\infty}^{\infty}\int_{-\infty}^{\infty} h_1(\theta_1)h_1(\theta_2)x_1(t - \theta_1)x_1(t + \tau - \theta_2)\,d\theta_1\,d\theta_2 \right]$$

which is equal to

$$\int_{-\infty}^{\infty}\int_{-\infty}^{\infty} h_1(\theta_1)h_1(\theta_2)E[x_1(t - \theta_1)x_1(t + \tau - \theta_2)]\,d\theta_1\,d\theta_2.$$

Provided that the input process $x_1(t)$ is stationary, its autocorrelation function is independent of absolute time t, and so

$$E[x_1(t - \theta_1)x_1(t + \tau - \theta_2)] = R_{x_1}(\tau - \theta_2 + \theta_1).$$

The first term on the r.h.s. of (7.8) may therefore be written

$$\int_{-\infty}^{\infty}\int_{-\infty}^{\infty} h_1(\theta_1)h_1(\theta_2)R_{x_1}(\tau - \theta_2 + \theta_1)\,d\theta_1\,d\theta_2$$

and is independent of time t. The same reasoning can be applied to the other three terms in (7.8), and so, for stationary excitation, the output auto-correlation function is independent of absolute time t and can be expressed by

---

†For a brief review of the properties of multiple integrals see, for instance, Kreyszig [45] Ch. 6. The order of integration of a multiple integral can be changed provided that convergence is retained when the integrand is replaced by its modulus. It can be shown that this condition is always met for a stable system subjected to random inputs provided that the mean square values of the inputs are finite (statistically the condition is met with probability one) (Davenport *et al.* [22] pp. 65–6 and 181–5).

the following rather lengthy expression

$$R_y(\tau) = \int_{-\infty}^{\infty} \int_{-\infty}^{\infty} h_1(\theta_1)h_1(\theta_2)R_{x_1}(\tau - \theta_2 + \theta_1)\,d\theta_1\,d\theta_2 +$$

$$+ \int_{-\infty}^{\infty} \int_{-\infty}^{\infty} h_1(\theta_1)h_2(\theta_2)R_{x_1x_2}(\tau - \theta_2 + \theta_1)\,d\theta_1\,d\theta_2 +$$

$$+ \int_{-\infty}^{\infty} \int_{-\infty}^{\infty} h_2(\theta_1)h_1(\theta_2)R_{x_2x_1}(\tau - \theta_2 + \theta_1)\,d\theta_1\,d\theta_2 +$$

$$+ \int_{-\infty}^{\infty} \int_{-\infty}^{\infty} h_2(\theta_1)h_2(\theta_2)R_{x_2}(\tau - \theta_2 + \theta_1)\,d\theta_1\,d\theta_2 \qquad (7.10)$$

involving double convolutions of the input autocorrelation and cross-correlation functions

The autocorrelation function for the output process $R_y(\tau)$ is independent of absolute time t for stationary excitation and this is a general result for the response of any constant parameter linear system. It turns out that *all* averages of the output process are time invariant for stationary excitation, and the output process is therefore itself stationary.

Spectral density

Although (7.10) is such a complicated expression, fortunately considerable simplifications emerge if we take Fourier transforms of both sides to find $S_y(\omega)$, the spectral density of the output process. From (5.1) the Fourier transform of the first double integral on the r.h.s. of (7.10) is

$$I_1 = \frac{1}{2\pi} \int_{-\infty}^{\infty} d\tau\,e^{-i\omega\tau} \left\{ \int_{-\infty}^{\infty} d\theta_1 \int_{-\infty}^{\infty} d\theta_2\, h_1(\theta_1)h_1(\theta_2)R_{x_1}(\tau - \theta_2 + \theta_1) \right\}. \qquad (7.11)$$

Changing the order of integration this may be written

$$I_1 = \frac{1}{2\pi} \int_{-\infty}^{\infty} d\theta_1\, h_1(\theta_1) \int_{-\infty}^{\infty} d\theta_2\, h_1(\theta_2) \int_{-\infty}^{\infty} d\tau\,e^{-i\omega\tau} R_{x_1}(\tau - \theta_2 + \theta_1).$$

The last integral is with respect to τ with θ_1 and θ_2 constant. We can therefore legitimately write this as

$$e^{i\omega(\theta_1 - \theta_2)} \int_{-\infty}^{\infty} d(\tau - \theta_2 + \theta_1)e^{-i\omega(\tau - \theta_2 + \theta_1)} R_{x_1}(\tau - \theta_2 + \theta_1)$$

which, from (5.1), is equal to

$$e^{i\omega(\theta_1 - \theta_2)} \cdot 2\pi S_{x_1}(\omega)$$

so that (7.11) may be written

$$I_1 = \int_{-\infty}^{\infty} d\theta_1 \, h_1(\theta_1) \int_{-\infty}^{\infty} d\theta_2 \, h_1(\theta_2) \, e^{i\omega(\theta_1 - \theta_2)} S_{x_1}(\omega)$$

$$= \int_{-\infty}^{\infty} d\theta_1 h_1(\theta_1) e^{i\omega\theta_1} \int_{-\infty}^{\infty} d\theta_2 h_1(\theta_2) e^{-i\omega\theta_2} \, S_{x_1}(\omega). \tag{7.12}$$

The two remaining integrals in (7.12) can now be related to the frequency response function $H(\omega)$ since, from (6.21),

$$H_1(\omega) = \int_{-\infty}^{\infty} h_1(\theta_2) e^{-i\omega\theta_2} \, d\theta_2$$

and the complex conjugate of $H_1(\omega)$

$$H_1^*(\omega) = \int_{-\infty}^{\infty} h_1(\theta_1) e^{i\omega\theta_1} \, d\theta_1.$$

Hence we obtain finally

$$I_1 = H_1^*(\omega) H_1(\omega) \, S_{x_1}(\omega)$$

for the Fourier transform of the first double integral on the r.h.s. of (7.10).

The same procedure may be applied to the rest of (7.10) and the final result of taking the Fourier transforms of both sides of the equation is the following expression for the spectral density of the output process

$$S_y(\omega) = H_1^*(\omega)H_1(\omega)S_{x_1}(\omega) + H_1^*(\omega)H_2(\omega)S_{x_1x_2}(\omega) +$$
$$+ H_2^*(\omega)H_1(\omega)S_{x_2x_1}(\omega) + H_2^*(\omega)H_2(\omega)S_{x_2}(\omega). \tag{7.13}$$

This is a most important result. By considering more than two inputs, it is not difficult to show that, for N inputs, the corresponding expression is

$$S_y(\omega) = \sum_{r=1}^{N} \sum_{s=1}^{N} H_r^*(\omega)H_s(\omega)S_{x_rx_s}(\omega) \tag{7.14}$$

when we define

$$S_{x_rx_r} = S_{x_r}$$

for the spectral density of the rth input.

Equation (7.14) is the central result of random vibration theory and its simplicity justifies our faith in the Fourier transform and frequency response approach.

In the case of response to a single input, (7.14) gives

$$S_y(\omega) = H^*(\omega)H(\omega)S_x(\omega) \tag{7.15}$$

or, since the product of a complex number and its complex conjugate is equal to the magnitude of the number squared,

$$S_y(\omega) = |H(\omega)|^2 \, S_x(\omega). \tag{7.16}$$

For uncorrelated inputs for which the cross-spectral density terms are all zero, (7.16) can be generalized to

$$S_y(\omega) = \sum_{r=1}^{N} |H_r(\omega)|^2 S_{x_r}(\omega). \tag{7.17}$$

Mean square response

Once the response spectral density has been determined, the mean square response can be calculated directly from (5.3)

$$E[y^2] = \int_{-\infty}^{\infty} S_y(\omega)\, d\omega \tag{5.3}$$

which, for a single input, becomes

$$E[y^2] = \int_{-\infty}^{\infty} |H(\omega)|^2 S_x(\omega)\, d\omega \tag{7.18}$$

and, for many *uncorrelated* inputs, is

$$E[y^2] = \sum_{r=1}^{N} \int_{-\infty}^{\infty} |H_r(\omega)|^2 S_{x_r}(\omega)\, d\omega. \tag{7.19}$$

For uncorrelated inputs, the mean square response is therefore the sum of the mean square responses due to each input separately. However in general this is not the case and, for correlated inputs, the mean square response is *not* just the sum of the separate mean square responses. In these cases, (7.14) must be used to find the response spectral density $S_y(\omega)$ and then the integral in (5.3) evaluated.

Example 1

Determine the output spectral density $S_y(\omega)$ for the single degree-of-freedom oscillator shown in Fig. 7.2 when it is excited by a forcing function $x(t)$ whose spectral density $S_x(\omega) = S_0$.

From (7.16),

$$S_y(\omega) = |H(\omega)|^2 S_0$$

where $H(\omega)$ is the complex frequency response function. To find $H(\omega)$, put $x(t) = e^{i\omega t}$ and $y(t) = H(\omega)e^{i\omega t}$ in the equation of motion

$$m\ddot{y} + c\dot{y} + ky = x(t)$$

to obtain

$$(-m\omega^2 + ci\omega + k)H(\omega) = 1$$

and

$$H(\omega) = \frac{1}{-m\omega^2 + ic\omega + k}.$$

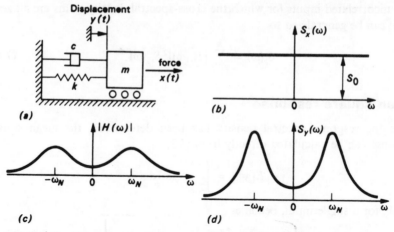

Fig. 7.2 Response spectral density $S_y(\omega)$ for a single degree-of-freedom oscillator subjected to a white noise force input $S_x(\omega) = S_0$.

Hence the output spectral density is

$$S_y(\omega) = \frac{S_0}{(k - m\omega^2)^2 + c^2\omega^2}$$

as sketched in Fig. 7.2(d). The area under the spectral density curve is equal to the mean square, which may be written

$$E[y^2] = \int_{-\infty}^{\infty} \left| \frac{1}{-m\omega^2 + ic\omega + k} \right|^2 S_0 \, d\omega.$$

A list of definite integrals of this form can be found in the literature† and the result is

$$E[y^2] = \frac{\pi S_0}{kc}$$

which is independent of the magnitude of the mass *m*. This is a surprising result as we would naturally expect *m* to affect the overall mean square level of the output $E[y^2]$. The explanation can be seen by considering the height and width of the spectral peak in the right-hand half of Fig. 7.2(d). (Remember that the peak in the left-hand half of the figure is just the mirror image of the peak in the right-hand half, since $S_y(\omega)$ is an even function of ω.)

The peak value occurs, for small damping, when

$$\omega \simeq \sqrt{\frac{k}{m}} = \omega_N$$

---

†See, for instance, James *et al.* [38] pp. 333–9, from which a selection of results is given in Appendix 1.

and its height is therefore

$$S_y(\omega_N) = \frac{S_0}{c^2\omega_N^2} = \frac{S_0 m}{c^2 k}$$

which is proportional to m. The "width" of the spectral peak needs definition, but suppose that we arbitrarily define this as the difference in frequency $2\Delta\omega$ between the two points on either side of ω_N whose height is half the peak height (the so-called "half-power" bandwidth). For small damping

$$\Delta\omega \ll \omega_N$$

and so

$$2\Delta\omega \simeq \frac{c}{m}$$

which is inversely proportional to m. Hence we can see that increasing the mass m increases the height of the spectral peak, but at the same time reduces its width. As we have seen, it turns out that these two opposite effects exactly cancel out and the total area under the spectral density curve, and therefore the mean square value, is independent of m.

Example 2
The massless trolley shown in Fig. 7.3 is connected to two abutments by springs and viscous dashpots as shown. If the abutments move distances

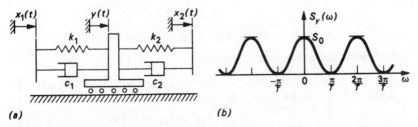

(a)   (b)

Fig. 7.3 Response spectral density for a system with two correlated inputs

$x_1(t)$ and $x_2(t)$ with spectral densities $S_{x_1}(\omega) = S_{x_2}(\omega) = S_0$ (constant) but $x_2(t + T) = x_1(t)$, so that the cross-spectra are

$$S_{x_1 x_2}(\omega) = S_0 e^{-i\omega T}$$

and

$$S_{x_2 x_1}(\omega) = S_0 e^{i\omega T}$$

determine the response spectral density $S_y(\omega)$ and the mean square response $E[y^2]$.

For equilibrium of the trolley

$$k_1(x_1 - y) + c_1(\dot{x}_1 - \dot{y}) = k_2(y - x_2) + c_2(\dot{y} - \dot{x}_2)$$

so that the equation of motion is

$$(c_1 + c_2)\dot{y} + (k_1 + k_2)y = k_1 x_1 + c_1 \dot{x}_1 + k_2 x_2 + c_2 \dot{x}_2.$$

To determine the frequency response functions, first put $x_1 = e^{i\omega t}$, $x_2 = 0$, $y = H_1(\omega)e^{i\omega t}$ to find

$$H_1(\omega) = \frac{k_1 + ic_1\omega}{k_1 + k_2 + i(c_1 + c_2)\omega}$$

and then put $x_1 = 0$, $x_2 = e^{i\omega t}$, $y = H_2(\omega)e^{i\omega t}$ to find

$$H_2(\omega) = \frac{k_2 + ic_2\omega}{k_1 + k_2 + i(c_1 + c_2)\omega}.$$

Hence, according to (7.13),

$$S_y(\omega) = \frac{k_1^2 + c_1^2\omega^2}{(k_1 + k_2)^2 + (c_1 + c_2)^2\omega^2} S_0 +$$

$$+ \frac{(k_1 - ic_1\omega)(k_2 + ic_2\omega)}{(k_1 + k_2)^2 + (c_1 + c_2)^2\omega^2} S_0 e^{-i\omega T}$$

$$+ \frac{(k_1 + ic_1\omega)(k_2 - ic_2\omega)}{(k_1 + k_2)^2 + (c_1 + c_2)^2\omega^2} S_0 e^{i\omega T}$$

$$+ \frac{k_2^2 + c_2^2\omega^2}{(k_1 + k_2)^2 + (c_1 + c_2)^2\omega^2} S_0$$

which, after collecting terms, becomes

$$S_y(\omega) = S_0 \left\{ \frac{k_1^2 + k_2^2 + c_1^2\omega^2 + c_2^2\omega^2 + 2(k_1 k_2 + c_1 c_2\omega^2)\cos\omega T + {} + 2(k_1 c_2\omega - k_2 c_1\omega)\sin\omega T}{(k_1 + k_2)^2 + (c_1 + c_2)^2\omega^2} \right\}.$$

When the delay time T is zero, so that $x_1(t) = x_2(t)$, then we see that $S_y(\omega) = S_0$ and motion of the trolley is always equal to motion of the abutments.

For large ω (high frequencies)

$$S_y(\omega) \to S_0 \left\{ \frac{c_1^2 + c_2^2 + 2c_1 c_2 \cos\omega T}{(c_1 + c_2)^2} \right\}$$

which is finite and so

$$E[y^2] = \int_{-\infty}^{\infty} S_y(\omega)\, d\omega \to \infty$$

on account of the characteristics of white noise excitation which has an infinite mean square value.

For the case when the two spring stiffnesses and the two damper coefficients are the same,

$$S_y(\omega) = \frac{S_0}{2}(1 + \cos \omega T)$$

which has the form shown in Fig. 7.3(b).

Cross-correlation

In the case of a system excited by many inputs, it is sometimes helpful to determine the cross-correlation between the output and one of the inputs. Beginning with the definition (3.20)

$$R_{x_1 y}(\tau) = E[x_1(t)y(t + \tau)] \tag{3.20}$$

for the case of two inputs, using (7.7)

$$R_{x_1 y}(\tau) = E\left[x_1(t) \int_{-\infty}^{\infty} h_1(\theta)x_1(t + \tau - \theta)\,d\theta + \right.$$
$$\left. + x_1(t) \int_{-\infty}^{\infty} h_2(\theta)x_2(t + \tau - \theta)\,d\theta \right].$$

Since $x_1(t)$ is not a function of θ, it may be moved under the integral signs and the averaging process carried out to give

$$R_{x_1 y}(\tau) = \int_{-\infty}^{\infty} h_1(\theta)E[x_1(t)x_1(t + \tau - \theta)]\,d\theta +$$
$$+ \int_{-\infty}^{\infty} h_2(\theta)E[x_1(t)x_2(t + \tau - \theta)]\,d\theta$$

or, in terms of the input autocorrelation and cross-correlation functions,

$$R_{x_1 y}(\tau) = \int_{-\infty}^{\infty} h_1(\theta)R_{x_1}(\tau - \theta)\,d\theta + \int_{-\infty}^{\infty} h_2(\theta)R_{x_1 x_2}(\tau - \theta)\,d\theta \tag{7.20}$$

which expresses the cross-correlation between input $x_1(t)$ and output $y(t)$ in terms of the autocorrelation of $x_1(t)$, the cross-correlation between $x_1(t)$ and the other input $x_2(t)$, and the impulse response functions between x_1 and y and x_2 and y.

For the special case when x_1 is white noise so that, from (5.10)

$$R_{x_1}(\tau - \theta) = 2\pi S_0 \,\delta(\tau - \theta)$$

and x_1 and x_2 are uncorrelated so that

$$R_{x_1 x_2}(\tau - \theta) = 0$$

then (7.20) gives

$$R_{x_1 y}(\tau) = \int_{-\infty}^{\infty} h_1(\theta) 2\pi S_0\, \delta(\tau - \theta)\, d\theta$$

$$= h_1(\tau) 2\pi S_0 \quad \text{from (5.9).} \tag{7.21}$$

The cross-correlation function between a white noise input $x_1(t)$ and the output $y(t)$ is therefore the same as the impulse response at y for a unit impulse at x_1 multiplied by the factor $2\pi S_0$. This is an interesting result which is sometimes used to obtain the impulse response function experimentally.

Cross-spectral density

Taking the Fourier transform of both sides of (7.20) gives

$$S_{x_1 y}(\omega) = \frac{1}{2\pi} \int_{-\infty}^{\infty} d\tau\, e^{-i\omega\tau} \left\{ \int_{-\infty}^{\infty} h_1(\theta) R_{x_1}(\tau - \theta)\, d\theta + \right.$$

$$\left. + \int_{-\infty}^{\infty} h_2(\theta) R_{x_1 x_2}(\tau - \theta)\, d\theta \right\}.$$

Rearranging terms in the same way as for the previous calculation of spectral density gives

$$S_{x_1 y}(\omega) = \frac{1}{2\pi} \int_{-\infty}^{\infty} d\theta\, h_1(\theta) e^{-i\omega\theta} \int_{-\infty}^{\infty} d\tau\, R_{x_1}(\tau - \theta) e^{-i\omega(\tau - \theta)} +$$

$$+ \frac{1}{2\pi} \int_{-\infty}^{\infty} d\theta\, h_2(\theta) e^{-i\omega\theta} \int_{-\infty}^{\infty} d\tau\, R_{x_1 x_2}(\tau - \theta) e^{-i\omega(\tau - \theta)}.$$

The integrals with respect to τ are with θ constant, and so if $(\tau - \theta)$ is replaced by ϕ (say) then $d\tau$ becomes $d\phi$. Using (5.1) and (5.20) to evaluate the integrals with respect to ϕ, and (6.21) to evaluate the integrals with respect to θ which then remain, we obtain

$$S_{x_1 y}(\omega) = H_1(\omega) S_{x_1}(\omega) + H_2(\omega) S_{x_1 x_2}(\omega). \tag{7.22}$$

When there are N separate inputs, of which $x_r(t)$ is a typical one, (7.22) becomes the summation

$$S_{x_r y}(\omega) = \sum_{s=1}^{N} H_s(\omega) S_{x_r x_s}(\omega) \tag{7.23}$$

where, as for equation (7.14),

$$S_{x_r x_r} = S_{x_r}.$$

For uncorrelated inputs, we can see that

$$S_{xy}(\omega) = H(\omega) S_x(\omega) \tag{7.24}$$

where $H(\omega)$ is the complex frequency response function relating the input

$x(t)$ to the output $y(t)$, $S_x(\omega)$ is the spectral density of the input process, and $S_{xy}(\omega)$ is the cross-spectral density between the input and output. From the properties of cross-spectra (5.23), it follows that

$$S_{yx}(\omega) = S_{xy}^*(\omega) = H^*(\omega)S_x(\omega) \tag{7.25}$$

since $S_x(\omega)$ is of course always a real quantity.

The extension of these results to the case of more than one output variable is left as an exercise for the reader – see problem 7.5 where the cross-spectral density $S_{y_r y_s}(\omega)$ between two output processes $y_r(t)$ and $y_s(t)$ is expressed in terms of the input spectra and cross-spectra and the relevant frequency response functions.

Probability distributions

Lastly in this chapter we consider how the probability distribution for the response of a linear system depends on the probability distribution of the excitation. We can say at once that there is no simple relationship. There is no general method for obtaining the output probability distributions for a linear system except for the special case when the input probability distributions are Gaussian.

The fact that we can calculate the output probability distributions for a linear system subjected to Gaussian excitation arises from the special properties of Gaussian processes. Firstly there is a general theorem† which says that if y_1 and y_2 are a pair of (jointly) Gaussian random variables, then if y is defined so that

$$y = y_1 + y_2 \tag{7.26}$$

the new random variable y will also be Gaussian. Secondly, this result may be applied to show that the response $y(t)$ of a linear system will be a Gaussian process if the excitation $x(t)$ is Gaussian. Using the convolution integral (6.24) we know that

$$y(t) = \int_{-\infty}^{t} h(t - \tau)x(\tau)\,d\tau \tag{7.27}$$

where $h(t)$ is the impulse response function. It can be shown mathematically† that the integral in (7.27) can be thought of as the limiting case of a linear sum of random variables, of which (7.26) is the simplest example. Hence if $x(t)$ is a Gaussian process, then $y(t)$ must be the same. Finally these results can be extended to the case of more than one (jointly) Gaussian input to show that the output processes after transmission through a linear system will also be (jointly) Gaussian. The output of a linear system subjected to Gaussian inputs is therefore Gaussian and the output probability distributions can be calculated if the respective mean values, variances and covariances (2.13) are known. Also, since a derivative can be expressed as the limiting case of the difference $\{y(t + \Delta t) - y(t)\}/\Delta t$ between two random variables $y(t + \Delta t)$ and $y(t)$, if the

† See, for instance, Davenport, Jr., *et al.* [22] sect. 8.3 and 8.4.

process $y(t)$ is Gaussian then so is its derivative $\dot{y}(t)$ (and so are higher derivatives such as the acceleration $\ddot{y}(t)$).

Example

A linear system is subjected to stationary Gaussian excitation as a result of which the response $y(t)$ has a mean level m_y, standard deviation σ_y, and an autocorrelation function $R_y(\tau)$. Determine the probability density function $p(y_1, y_2)$ for the joint distribution of y at t_1 and y at t_2 where $t_2 = t_1 + \tau$.

Referring to the definition of the second-order Gaussian probability density in Chapter 2, from (2.13) the normalized covariance $\rho_{y_1 y_2}$ is given by

$$\rho_{y_1 y_2} = \frac{E[(y_1 - m_y)(y_2 - m_y)]}{\sigma_y^2} = \frac{R_y(\tau) - m_y^2}{\sigma_y^2} = \rho \quad \text{(say)} \quad (7.28)$$

and (2.12) gives

$$p(y_1, y_2) = \frac{1}{2\pi\sigma_y^2 \sqrt{(1-\rho^2)}} e^{-\frac{1}{2\sigma_y^2(1-\rho^2)}\{(y_1-m_y)^2 + (y_2-m_y)^2 - 2\rho(y_1-m_y)(y_2-m_y)\}}. \quad (7.29)$$

When the time interval $\tau \to \infty$, $R_y(\tau) \to m_y^2$ and, from (7.28), $\rho \to 0$. In this case (7.29) becomes

$$p(y_1, y_2) = \frac{1}{\sqrt{2\pi}\,\sigma_y} e^{-\frac{(y_1-m_y)^2}{2\sigma_y^2}} \cdot \frac{1}{\sqrt{2\pi}\,\sigma_y} e^{-\frac{(y_2-m_y)^2}{2\sigma_y^2}} = p(y_1) \cdot (p(y_2)$$

and $y_1 = y(t_1)$ and $y_2 = y(t_2)$ are then statistically independent.

For non-Gaussian random processes, the theory becomes much more complicated. It has been shown that the output probability distribution for the response of an arbitrary linear system can be expressed in terms of a series expansion whose terms involve the input statistics and the system's characteristics† but the calculations needed are difficult and time-consuming. Fortunately, in many practical problems of random vibration it turns out the excitation is Gaussian or nearly so and this becomes another "standard" assumption – for engineering calculations the world is often assumed to be linear, stationary, ergodic and Gaussian. Actually there is an important theorem of probability theory, called the *central limit theorem*, which helps to explain why the probability distributions of many naturally occurring processes should be Gaussian.‡ Roughly speaking, the central limit theorem says that when a random process results from the summation of infinitely many random elementary events, then this process will tend to have Gaussian probability distributions. We therefore have good reason to expect that, for instance, the noise generated by falling rain, or by a turbulent fluid boundary layer, or by the random emission of electrons in a thermionic device, will all have probability distributions which

---

†See Ohta *et al.* [90].
‡See, for instance, Meyer [48] Ch. 12.

approximate to Gaussian. Furthermore even when the excitation is non-Gaussian, a system's response may approximate to Gaussian if it is a narrow band response derived from broad band excitation because the convolution integral (7.27) may again be thought of as the limiting case of a linear sum of approximately independent random variables. However, there should be one word of caution. An assumed Gaussian probability distribution may give a poor approximation at the tails of the distribution and predictions of maximum excursion of a random process, based on such an assumed distribution, should be treated with suspicion. The Gaussian distribution forecasts an occasional very large excursion which for obvious engineering reasons may just not be possible in practice.

Chapter 8

Statistics of narrow band processes

In the previous chapter we considered general relations between the input and the output of a linear system subjected to random excitation. The characteristics of the excitation are modified by the response of the system which, in electrical engineering terms, acts as a filter. In most vibration problems, the system has at least one resonant frequency at which large amplitudes can be generated by small inputs. At other frequencies transmission is reduced and, at very high frequencies, the effective mass may be so high that the output is not measurable. A typical frequency response function $H(\omega)$ for such a resonant system is shown in Fig. 8.1, which also shows how the characteristics of broad band noise are changed by transmission through this system. Because the output spectrum is confined to a narrow band of frequencies in the vicinity of the resonant frequency, the response $y(t)$ is a *narrow band random process* and the typical time history of $y(t)$ resembles a sine wave of varying amplitude and phase as shown in Fig. 8.1.

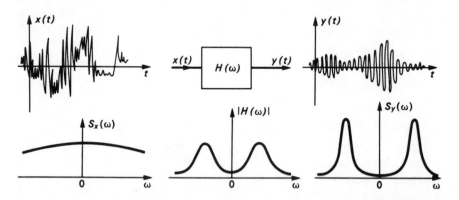

Fig. 8.1 Narrow band response of a resonant system excited by broad band noise

In order to make some simple calculations, suppose that the frequency response function has very sharp cut-offs above and below the resonant frequency so that the response spectral density $S_y(\omega)$ has the idealized form shown in Fig. 8.2(*a*). We have already worked out the corresponding auto-correlation function (in the first example in Chapter 5) and this is

$$R_y(\tau) = 4S_0 \frac{\sin(\Delta\omega \ \tau/2)}{\tau} \cos \omega_0 \tau \qquad (8.1)$$

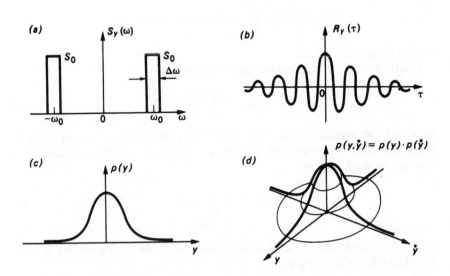

Fig. 8.2 Characteristics of a stationary, Gaussian, narrow band process

as shown in Fig. 8.2(*b*). If the excitation is Gaussian, we can find the probability distributions for *y* and, for later reference, we shall need the first-order probability density function $p(y)$ and the second-order probability density function $p(y, \dot{y})$ for the joint probability of *y* and its derivative $\dot{y}$. Both of these functions are given by the standard expressions (1.11) and (2.12) and are known provided that the statistics

$$m_y, m_{\dot{y}}, \sigma_y, \sigma_{\dot{y}} \text{ and } \rho_{y\dot{y}}$$

are known.

First, consider the mean level of *y*, m_y. If this were to be other than zero, the spectral density for *y* would have to show a delta function at $\omega = 0$, because otherwise there cannot be a finite contribution to $E[y^2]$ at $\omega = 0$. Since the spectral density $S_y(\omega)$ shown in Fig. 8.2(*a*) does not have this delta function, the mean level is zero, $m_y = 0$.

Next, since the spectral density of the $\dot{y}$ process is given according to (5.17) by

$$S_{\dot{y}}(\omega) = \omega^2 S_y(\omega), \qquad (8.2)$$

the $S_{\dot{y}}(\omega)$ function also cannot have a delta function at $\omega = 0$, so the mean level of $\dot{y}$ is also zero, $m_{\dot{y}} = 0$.

In order to calculate the variances σ_y^2 and $\sigma_{\dot{y}}^2$ we use (1.19) and (5.3) to find

$$\sigma_y^2 = E[y^2] = \int_{-\infty}^{\infty} S_y(\omega)\,d\omega = 2S_0\Delta\omega \qquad (8.3)$$

and

$$\sigma_{\dot{y}}^2 = E[\dot{y}^2] = \int_{-\infty}^{\infty} \omega^2 S_y(\omega)\,d\omega \simeq 2S_0\omega_0^2\Delta\omega \qquad (8.4)$$

for $\Delta\omega \ll \omega_0$.

Lastly, from (2.13), the normalized covariance

$$\rho_{y\dot{y}} = \frac{E[y\dot{y}]}{\sigma_y\sigma_{\dot{y}}} \qquad (8.5)$$

for which the correlation function $E[y\dot{y}]$ must be calculated. From (5.11) we know that this can be expressed in terms of the derivative of the autocorrelation function, in the form

$$E[y\dot{y}] = \frac{d}{d\tau}R_y(\tau) \Big|_{\tau=0} \qquad (8.6)$$

and, if $R_y(\tau)$ is expressed as the Fourier integral of the corresponding spectral density, using (5.2), this gives

$$E[y\dot{y}] = i\int_{-\infty}^{\infty} \omega\, S_y(\omega)\,d\omega. \qquad (8.7)$$

Now, since $S_y(\omega)$ is a real even function of frequency ω, the integrand $\omega S_y(\omega)$ appearing in (8.7) is a real odd function of ω. When integrated over the range $-\infty$ to ∞, the contribution from $-\infty$ to 0 is exactly equal but opposite in sign to the contribution from 0 to ∞. Hence the integral in (8.7) must be zero and we obtain

$$E[y\dot{y}] = 0. \qquad (8.8)$$

It is therefore a property of any stationary random process $y(t)$ that y and its derivative $\dot{y}$ are uncorrelated and so the normalized covariance $\rho_{y\dot{y}}$ is always zero.

We now have all the parameters we need and can substitute in (1.11) and (2.12) to obtain the probability density functions

$$p(y) = \frac{1}{\sqrt{2\pi}\,\sigma_y}\, e^{\frac{-y^2}{2\sigma_y^2}} \qquad (8.9)$$

and

$$p(y,\dot{y}) = \frac{1}{2\pi\,\sigma_y\sigma_{\dot{y}}}\, e^{-\frac{1}{2}\left(\frac{y^2}{\sigma_y^2}+\frac{\dot{y}^2}{\sigma_{\dot{y}}^2}\right)} = p(y)p(\dot{y}) \qquad (8.10)$$

where σ_y and $\sigma_{\dot{y}}$ are given by (8.3) and (8.4). These functions are sketched in Fig. 8.2 alongside the corresponding spectral density and autocorrelation curves.

Crossing analysis

Although the data in Fig. 8.2 says a lot about the narrow band process $y(t)$ by describing its frequency composition and its amplitude and velocity distributions, it is possible to go further and obtain important information about the distribution of peak values, that is to say information about the amplitude of the fluctuating sine wave that makes up the process. Suppose that we enquire how many "cycles" of $y(t)$ have amplitudes greater than the level $y = a$ during the time period T, Fig. 8.3. For the sample shown there are three cycles. Another

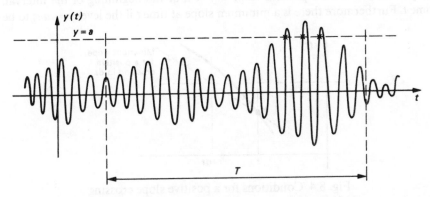

Fig. 8.3 Typical sample of a narrow band process

way of saying this is that there are three crossings with positive slope of the level $y = a$ in time T. Each of these positive slope crossings is marked with a cross in the figure.

Now consider Fig. 8.3 as one sample function of an ensemble of functions which make up the stationary random process $y(t)$. Let $n_a^+(T)$ denote the number of positive slope crossings† of $y = a$ in time T for a typical sample and let the mean value for all the samples be $N_a^+(T)$ where

$$N_a^+(T) = E[n_a^+(T)]. \tag{8.11}$$

Since the process is stationary, if we take a second interval of duration T immediately following the first we shall obtain the same result, and for the two intervals together (total time $2T$) we shall therefore obtain

$$N_a^+(2T) = 2N_a^+(T) \tag{8.12}$$

---

†The notation used in this chapter follows Crandall *et al.* [18].

from which it follows that, for a stationary process, the average number of crossings is proportional to the time interval T. Hence

$$N_a^+(T) \propto T$$

or

$$N_a^+(T) = v_a^+ T \tag{8.13}$$

where v_a^+ is the average frequency of positive slope crossings of the level $y = a$. We now consider how the frequency parameter v_a^+ can be deduced from the underlying probability distributions for $y(t)$.

Consider a small length of duration dt of a typical sample function, Fig. 8.4. Since we are assuming that the narrow band process $y(t)$ is a smooth function of time, with no sudden ups and downs, if dt is small enough, the sample can only cross $y = a$ with positive slope if $y < a$ at the beginning of the interval, time t. Furthermore there is a minimum slope at time t if the level $y = a$ is to be

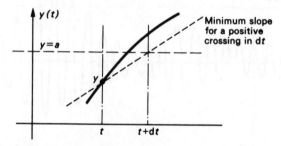

Fig. 8.4 Conditions for a positive slope crossing of $y = a$ in time interval dt

crossed in time dt depending on the value of y at time t. From Fig. 8.4 this is

$$\frac{a - y}{dt}$$

and so there will be a positive slope crossing of $y = a$ in the next time interval dt if, at time t,

$$y < a \quad \text{and} \quad \frac{dy}{dt} > \frac{a - y}{dt}. \tag{8.14}$$

Actually what we really mean is that there will be a high probability of a crossing in time dt if these conditions are satisfied, but we will overlook this fine statistical point by referring the reader to the original and classic papers on the subject (Rice [58]).

In order to determine whether the conditions (8.14) are satisfied at any arbitrary time t, we must find how the values of y and $\dot{y}$ are distributed by considering their joint probability density $p(y, \dot{y})$. Suppose that the level $y = a$ and time interval dt are specified. Then we are only interested in values of $y < a$ and values of $\dot{y} = (dy/dt) > (a - y)/dt$, which means the shaded wedge of values of y and $\dot{y}$ shown in Fig. 8.5(a). The wedge angle α is chosen so that

$$\tan \alpha = \frac{a - y}{\dot{y}} = dt \qquad (8.15)$$

in order to satisfy (8.14). If the values of y and $\dot{y}$ lie within this shaded wedge, then there will be a positive slope crossing of $y = a$ in time dt. If they do not

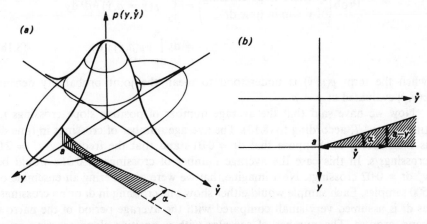

Fig. 8.5 Calculation of the probability that there will be a positive slope crossing of $y = a$ in time interval dt

lie in the shaded wedge, then there will not be a crossing. The probability that they do lie in the shaded wedge can be calculated from the joint probability density function $p(y,\dot{y})$ and is just the shaded volume shown in Fig. 8.5(a), i.e. the volume under the probability surface above the shaded wedge of acceptable values of y and $\dot{y}$. Hence

$$\text{Prob}\begin{pmatrix}\text{Positive slope crossing} \\ \text{of } y = a \text{ in time d}t\end{pmatrix} = \iint p(y,\dot{y})\,dy\,d\dot{y}$$

over the shaded wedge in Fig. 8.5(a)

$$= \int_0^\infty d\dot{y} \int_{a - \dot{y}\tan\alpha}^a dy\, p(y,\dot{y}). \qquad (8.16)$$

When d$t \to 0$, the angle of the wedge $\alpha = dt \to 0$ and in this case it is legitimate to put

$$p(y,\dot{y}) = p(y = a,\dot{y})$$

since at large values of y and $\dot{y}$ the probability density function approaches zero fast enough. Hence (8.16) may be written

$$\text{Prob}\begin{pmatrix}\text{Positive slope crossing} \\ \text{of } y = a \text{ in time d}t\end{pmatrix} = \int_0^\infty d\dot{y} \int_{a - \dot{y}\tan\alpha}^a dy\, p(y = a,\dot{y}) \quad (8.17)$$

in which the integrand is no longer a function of y so that the first integral is just

$$\int_{a-\dot{y}\tan\alpha}^{a} \mathrm{d}y\, p(y=a, \dot{y}) = p(y=a, \dot{y})\dot{y}\tan\alpha$$

and hence, with $\tan\alpha = \mathrm{d}t$,

$$\mathrm{Prob}\begin{pmatrix}\text{Positive slope crossing}\\ \text{of } y=a \text{ in time } \mathrm{d}t\end{pmatrix} = \int_{0}^{\infty} p(y=a, \dot{y})\dot{y}\,\mathrm{d}t\,\mathrm{d}\dot{y}$$

$$= \mathrm{d}t \int_{0}^{\infty} p(a, \dot{y})\dot{y}\,\mathrm{d}\dot{y} \qquad (8.18)$$

when the term $p(a, \dot{y})$ is understood to mean the joint probability density $p(y, \dot{y})$ evaluated at $y = a$.

Now we have said that the average number of positive slope crossings in time T is $v_a^+ T$, according to (8.13). The average number of crossings in time $\mathrm{d}t$ is therefore $v_a^+ \,\mathrm{d}t$. Suppose that $\mathrm{d}t = 0{\cdot}01\,\mathrm{s}$ and that the frequency $v_a^+ = 2{\cdot}0$ crossings/s. In this case the average number of crossings in $0{\cdot}01\,\mathrm{s}$ would be $v_a^+ \,\mathrm{d}t = 0{\cdot}02$ crossings. Next imagine that we were considering an ensemble of 500 samples. Each sample would either show one crossing in $\mathrm{d}t$ or no crossings, as $\mathrm{d}t$ is assumed very small compared with the average period of the narrow band process. The number of samples with a positive slope crossing must therefore be 10 since $10/500 = 0{\cdot}02$. But $10/500$ is also the probability that any one sample chosen at random has a crossing in time $\mathrm{d}t$. So we arrive at the following result

$$\begin{pmatrix}\text{Average no. of positive crossings}\\ \text{of } y=a \text{ in time } \mathrm{d}t\end{pmatrix} = \mathrm{Prob}\begin{pmatrix}\text{Positive slope crossing}\\ \text{of } y=a \text{ in time } \mathrm{d}t\end{pmatrix} \quad (8.19)$$

which is only true because $\mathrm{d}t$ is small and the process $y(t)$ is smooth so that there cannot be more than one crossing of $y = a$ in time $\mathrm{d}t$. Accepting (8.19) and substituting from (8.13) and (8.18) gives

$$v_a^+ \,\mathrm{d}t = \mathrm{d}t \int_{0}^{\infty} p(a, \dot{y})\dot{y}\,\mathrm{d}\dot{y}$$

from which $\mathrm{d}t$ cancels to give the following result for the frequency parameter v_a^+ in terms of the joint probability density function $p(y\,\dot{y})$

$$v_a^+ = \int_{0}^{\infty} p(a, \dot{y})\dot{y}\,\mathrm{d}\dot{y}. \qquad (8.20)$$

This is a general result which applies for any probability distribution, but for the special case of a Gaussian process we know, from (8.10), that

$$p(a, \dot{y}) = \frac{1}{\sqrt{2\pi}\,\sigma_y} e^{-a^2/2\sigma_y^2} \frac{1}{\sqrt{2\pi}\,\sigma_{\dot{y}}} e^{-\dot{y}^2/2\sigma_{\dot{y}}^2}$$

which, on substitution into (8.20), gives

$$v_a^+ = \frac{1}{\sqrt{2\pi}\,\sigma_y}\,e^{-a^2/2\sigma_y^2} \int_0^\infty \frac{1}{\sqrt{2\pi}\,\sigma_{\dot{y}}}\,e^{-\dot{y}^2/2\sigma_{\dot{y}}^2}\,\dot{y}\,d\dot{y}.$$

The integral is one of the standard results (1.20) and its value is $\sigma_{\dot{y}}/\sqrt{2\pi}$, so that the final result is, for a Gaussian process,

$$v_a^+ = \frac{1}{2\pi}\frac{\sigma_{\dot{y}}}{\sigma_y}e^{-a^2/2\sigma_y^2}. \tag{8.21}$$

A special case occurs if we take the level $a = 0$ because this gives a statistical average frequency for crossing the level $y = 0$, Fig. 8.6, which may be thought

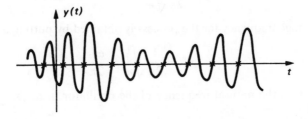

Fig. 8.6 Positive slope crossings of the level $y = 0$

of as a statistical average frequency for the process. Notice that v_0^+ is obtained by averaging across the ensemble and so it is not the same as the average frequency along the time axis unless the process is ergodic.

Example
   Calculate the frequency of positive crossings of the level $y = a$ for the single degree-of-freedom oscillator shown in Fig. 7.2(a) when it is subjected to Gaussian white noise of spectral density S_0, Fig. 7.2(b).
   In example 1 of Chapter 7 we have worked out that the frequency response function is

$$H(\omega) = \frac{1}{-m\omega^2 + ic\omega + k}$$

so that

$$\sigma_y^2 = \int_{-\infty}^\infty \left|\frac{1}{-m\omega^2 + ic\omega + k}\right|^2 S_0\,d\omega = \frac{\pi S_0}{kc}$$

as already calculated. The frequency response function relating $\dot{y}(t)$ to the excitation $x(t)$ is obtained by multiplying $H(\omega)$ by $i\omega$ to obtain

$$H'(\omega) = \frac{i\omega}{-m\omega^2 + ic\omega + k}$$

so that

$$\sigma_{\dot{y}^2} = \int_{-\alpha}^{\alpha} \left| \frac{i\omega}{-m\omega^2 + ic\omega + k} \right|^2 S_0 \, d\omega.$$

To evaluate this integral refer to the list of standard integrals in Appendix 1, to obtain

$$\sigma_{\dot{y}^2} = \frac{\pi S_0}{mc}.$$

Finally substituting in (8.21) gives

$$v_a^+ = \frac{1}{2\pi} \sqrt{\frac{k}{m}} e^{-a^2/(2\pi S_0/kc)}.$$

The average frequency for the process is obtained by putting $a = 0$ to give

$$v_0^+ = \frac{1}{2\pi} \sqrt{\frac{k}{m}} = \frac{\omega_N}{2\pi}$$

where ω_N is the natural frequency of the oscillator in rad/s.

Distribution of peaks

Having obtained the frequency of crossings of $y = a$, it is not difficult to extend this calculation to determine the probability distribution of peaks (Powell [56]). Let $p_p(a) \, da$ be the probability that the magnitude of a peak, chosen at random,

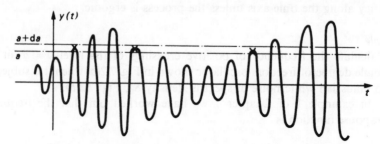

Fig. 8.7 Identification of peaks in the band $y = a$ to $y = a+da$

lies in the range a to $a + da$, Fig. 8.7. The probability that any peak is greater than a is therefore

$$\text{Prob}\left(\text{Peak value exceeds } y = a\right) = \int_a^\infty p_p(a) \, da. \tag{8.22}$$

Now in time T we know that, on average, there will be $v_0^+ T$ cycles (since one positive crossing of $y = 0$ occurs for each full cycle of the narrow band process) of which only $v_a^+ T$ will have peak values exceeding $y = a$. The proportion of cycles whose peak value exceeds $y = a$ is therefore

$$\frac{v_a^+}{v_0^+}$$

and this must be the probability that any peak value, chosen at random, exceeds $y = a$. Hence we obtain

$$\int_a^\infty p_p(a)\,\mathrm{d}a = \frac{v_a^+}{v_0^+} \tag{8.23}$$

and this equation may be differentiated with respect to a to give

$$-p_p(a) = \frac{1}{v_0^+}\frac{\mathrm{d}}{\mathrm{d}a}(v_a^+) \tag{8.24}$$

which is a general result for the probability density function for the occurrence of peaks. It applies for any narrow band process provided that this is a smooth process with each cycle crossing the mean level $y = 0$ so that all the maxima occur above $y = 0$ and all the minima occur below $y = 0$. Equation (8.24) applies for any probability distribution, but if $y(t)$ is Gaussian then there is a simple and important result for $p_p(a)$. Substituting (8.21) into (8.24) gives

$$-p_p(a) = \frac{\mathrm{d}}{\mathrm{d}a}(e^{-a^2/2\sigma_y^2}) = -\frac{a}{\sigma_y^2}e^{-a^2/2\sigma_y^2}$$

or

$$p_p(a) = \frac{a}{\sigma_y^2}e^{-a^2/2\sigma_y^2} \qquad 0 \leqslant a \leqslant \infty \tag{8.25}$$

which is the well-known Rayleigh distribution, Fig. 8.8. The function $p_p(a)$ has its maximum value at $a = \sigma_y$, the standard deviation of the y process, and

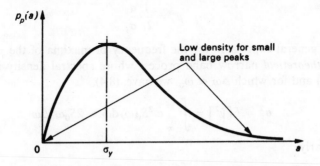

Fig. 8.8 Rayleigh distribution of peaks for a Gaussian narrow band process

it is clear from Fig. 8.8 that the majority of peaks have about this magnitude. The probability of finding very small or very large peaks is small and the probability that any peak, chosen at random, exceeds a is, from (8.23),

$$\text{Prob(Peak value exceeds } a) = e^{-a^2/2\sigma_y^2}. \tag{8.26}$$

Example

Calculate the probability that any peak value of a Gaussian narrow band process $y(t)$ exceeds $3\sigma_y$ where σ_y is its standard deviation.

From (8.26), the required probability is

$$e^{-a^2/2\sigma_y^2} = e^{-4.5} = 0{\cdot}011$$

so on average only about 1 peak in a 100 exceeds the $3\sigma_y$ level.

Frequency of maxima

Our analysis of peaks, leading to the Rayleigh distribution for a Gaussian process, is based on the assumption that the narrow band process $y(t)$ resembles a sine wave of varying amplitude and phase. We can investigate the validity of this assumption by calculating the distribution of local maxima of $y(t)$ by another approach. We know that y is at an extremum when $dy/dt = 0$ and that this extremum is a maximum if, at the same time, $d^2y/dt^2 = -\text{ve}$. Therefore the frequency of maxima of $y(t)$ must be the frequency of negative zero crossings of the derived process $\dot{y}(t)$ and, since there is one negative crossing for each positive crossing, this is the same as the frequency of positive zero crossings of $\dot{y}(t)$. Hence if μ_y is the frequency of maxima of $y(t)$, and $v_{\dot{y}=0}^+$ is the frequency of zero crossings of $\dot{y}(t)$, we have

$$\mu_y = v_{\dot{y}=0}^+ \tag{8.27}$$

where $v_{\dot{y}=0}^+$ can be calculated from (8.21) by substituting $\sigma_{\dot{y}}$ for σ_y and $\sigma_{\ddot{y}}$ for $\sigma_{\dot{y}}$ and putting the level $\dot{y} = a = 0$ to obtain

$$\mu_y = \frac{1}{2\pi} \frac{\sigma_{\ddot{y}}}{\sigma_{\dot{y}}}. \tag{8.28}$$

This is a general expression for the frequency of maxima of the process $y(t)$.

For a *theoretical* narrow band process whose spectral density is shown in Fig. 8.2(a) and for which $\Delta\omega \ll \omega_0$, we have, (8.4),

$$\sigma_{\dot{y}}^2 = E[\dot{y}^2] = \int_{-\infty}^{\infty} \omega^2 S_y(\omega)\,d\omega \simeq 2S_0\omega_0^2\Delta\omega$$

and similarly

$$\sigma_{\ddot{y}}^2 = E[\ddot{y}^2] = \int_{-\infty}^{\infty} \omega^4 S_y(\omega)\,d\omega \simeq 2S_0\omega_0^4\Delta\omega$$

in which case (8.28) gives

$$\mu_y \simeq \frac{\omega_0}{2\pi} \tag{8.29}$$

for the frequency of maxima, compared with, from (8.21),

$$v^+_{y=0} = \frac{1}{2\pi} \frac{\sigma_{\dot{y}}}{\sigma_y} \simeq \frac{\omega_0}{2\pi} \qquad (8.30)$$

for the frequency of zero crossings of $y(t)$. The frequency of maxima and the frequency of zero crossings are therefore the same and our assumption that there is only one peak for each zero crossing is justified.

If the bandwidth of the narrow band process is not narrow enough to assume $\Delta\omega \ll \omega_0$, this conclusion is however modified. Suppose that the spectral

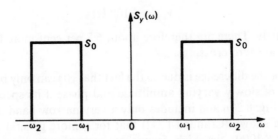

Fig. 8.9 Spectral density of a theoretical band
limited process

density of $y(t)$ has the form shown in Fig. 8.9. In this case the variances of y, $\dot{y}$ and $\ddot{y}$ are

$$\sigma_y^2 = E[y^2] = \int_{-\infty}^{\infty} S_y(\omega)\,d\omega \quad = 2S_0(\omega_2 - \omega_1)$$

$$\sigma_{\dot{y}}^2 = E[\dot{y}^2] = \int_{-\infty}^{\infty} \omega^2 S_y(\omega)\,d\omega = \tfrac{2}{3}S_0(\omega_2^3 - \omega_1^3)$$

$$\sigma_{\ddot{y}}^2 = E[\ddot{y}^2] = \int_{-\infty}^{\infty} \omega^4 S_y(\omega)\,d\omega = \tfrac{2}{5}S_0(\omega_2^5 - \omega_1^5)$$

and the frequency of maxima is now

$$\mu_y = \frac{1}{2\pi} \frac{\sigma_{\ddot{y}}}{\sigma_{\dot{y}}} = \frac{1}{2\pi} \sqrt{\left\{ \frac{3(\omega_2^5 - \omega_1^5)}{5(\omega_2^3 - \omega_1^3)} \right\}} \qquad (8.31)$$

compared with the frequency of zero crossings which is

$$v^+_{y=0} = \frac{1}{2\pi} \frac{\sigma_{\dot{y}}}{\sigma_y} = \frac{1}{2\pi} \sqrt{\left\{ \frac{\omega_2^3 - \omega_1^3}{3(\omega_2 - \omega_1)} \right\}} \qquad (8.32)$$

and these are clearly not the same.

Example

Calculate the frequency of maxima and the frequency of zero crossings for a Gaussian process whose spectrum is flat and covers an octave bandwidth from $\omega_1/2\pi = 70\cdot7$ Hz (c/s) to $\omega_2/2\pi = 141\cdot4$ Hz, Fig. 8.9, with a centre frequency of 100 Hz.

Note that for an octave bandwidth the upper cut-off frequency is twice the lower cut-off frequency.

Substituting numbers into (8.31) gives

$$\mu_y = 115 \text{ Hz}$$

and into (8.32) gives

$$v_{y=0}^+ = 108 \text{ Hz}$$

approximately. There are therefore about 6·5 per cent more local maxima than there are zero crossings.

The reason for the difference is due to the fact that $y(t)$ can only be represented by a sine wave of slowly varying amplitude and phase if its spectrum has the form shown in Fig. 8.2(a) and includes only a very narrow band of frequencies $\Delta\omega$. It has been shown (Crandall [21]) that for a more general narrow-band spectrum, Fig. 8.10(a), the high frequency components present introduce irregularities in the smooth form of the sine wave approximation and it is these irregularities which cause the additional maxima. In many applications we are concerned only with the large amplitude excursions of a narrow band process and the Rayleigh distribution of peaks, which assumes only one maximum for each zero crossing, is then a valuable guide to the probability of occurrence of large peak values. The more general Weibull distribution of peaks, which is not based on a Gaussian assumption for the underlying narrow-band process, is described in Chapter 14.

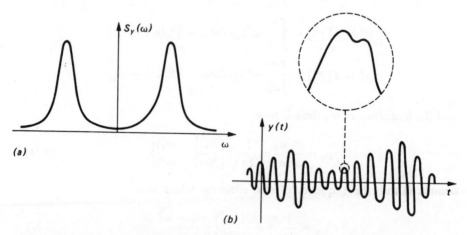

Fig. 8.10 Illustrating how local irregularities in a narrow band process give more than one maximum per zero crossing.

Chapter 9

Accuracy of measurements

So far we have been concerned with the basic mathematical theory of random process analysis. We turn now to a more practical aspect of the subject: experimental measurements. We shall concentrate almost entirely on measuring the spectral density of a random process or the cross-spectral density between two random processes. This emphasis on spectral measurements is justified by the central role which spectra occupy in the theory of random vibrations. Their importance comes from the simple form of the input–output relations for spectral density for a linear system subjected to random excitation. In this chapter we shall describe the operation of an analogue spectrum analyser and discuss at length the factors which affect the accuracy of any measurement of spectral density. In Chapters 10, 11 and 12 we shall then turn to digital methods of determining spectral density and cross-spectral density by analysing recorded time histories of random processes by computer.

The fundamental experimental problem is that our measurements must be made on one sample function of a theoretically infinite ensemble, or, at the most, on only a few sample functions from the infinite ensemble. Furthermore, we are only able to analyse a limited length of any given sample function because we cannot go on taking measurements for ever. The fact that we are limited to a single sample and that we can only analyse a finite length of it means that, automatically, we shall introduce errors into the measured spectrum. Even assuming that the random process we are studying is ergodic, in which any one sample function completely represents the infinity of functions which make up the ensemble, we are still introducing errors when we only deal with a finite length of a sample function. Remember that a sample average for an ergodic process is only the same as an ensemble average when the sample averaging time is infinite. Obviously this is not a practical proposition.

If we follow the mathematical definition of spectral density, we must begin by measuring the autocorrelation for the process being studied and then devise a way of calculating the Fourier transform of the autocorrelation function. However in practice this is not the best way to proceed. It turns out that it is easier to measure spectra by a procedure which does not involve first calculating correlation functions. Although the experimental procedure follows a route which is not mathematically rigorous, and therefore cannot be used to define the spectral density functions, we shall see that we can obtain approximations for the true spectra which are correct to any stated accuracy; furthermore the measurements are considerably simpler and quicker than they would otherwise be.

Analogue spectrum analysis

Most engineers will be familiar with an instrument called a frequency analyser. The output of an accelerometer, or other vibration transducer, is fed into the instrument which is essentially a variable frequency narrow-band filter with an r.m.s. meter to display the filter output. Usually the filter centre frequency is continuously variable and the experimenter adjusts this as he searches for the predominant frequencies present in a vibration signal. An analogue *spectrum analyser* is a similar instrument except that it has more accurate filters and precisely calibrated filter bandwidths, Fig. 9.1(*a*). Suppose that the input $x(t)$ is

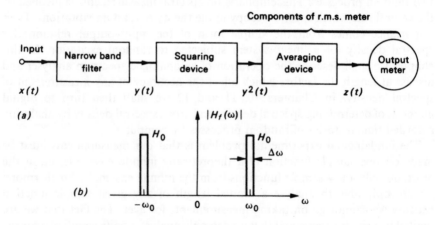

Fig. 9.1 Schematic of a spectrum analyser showing the theoretical filter frequency response

a sample function of an ergodic (and therefore stationary) random process. This signal is filtered by a filter whose theoretical frequency response is shown in Fig. 9.1(*b*). The filter output $y(t)$ is squared and then the time average $z(t)$ calculated where

$$z(t) = \frac{1}{T} \int_0^T y^2(t)\,dt. \tag{9.1}$$

Since the averaging time T cannot be infinite, $z(t)$ is itself a function of time, and fluctuates about its true mean value (the ensemble average). However if T is long enough, the fluctuations are small and the mean level $E[z]$ can be approximately determined from the analyser's output meter. From (9.1) we know that

$$E[z] = \frac{1}{T} \int_0^T E[y^2] \, dt = E[y^2] \tag{9.2}$$

since $y(t)$ is stationary, and, from (7.18),

$$E[y^2] = \int_{-\infty}^{\infty} |H(\omega)|^2 \, S_x(\omega) \, d\omega$$

which for the filter frequency response shown in Fig. 9.1(*b*), with $\Delta\omega \ll \omega_0$, can be approximated by

$$E[y^2] \simeq 2H_0^2 \, \Delta\omega \, S_x(\omega_0). \tag{9.3}$$

Combining (9.2) and (9.3), the average output of the spectrum analyser is proportional to the spectral density of the input process at the filter centre frequency ω_0, or turning the formula round,

$$S_x(\omega_0) \simeq \frac{E[z]}{2H_0^2 \, \Delta\omega}. \tag{9.4}$$

The mean output level $E[z]$ is therefore a direct measure of the input spectral density.

We shall now investigate the accuracy of this measurement. Before plunging into the details, which are quite complicated, it is probably already clear that we are likely to improve the accuracy if we use an instrument with a long averaging time, because then the output depends on an average (9.1) integrated over a long period of time. Also, if we want to distinguish sharp peaks in the curve of spectral density against frequency, we should use a sharp filter with a very narrow bandwidth $\Delta\omega$. In the next section we shall find an expression which relates the accuracy of measurement to both these quantities.

Variance of the measurement

From (9.4), the spectral density $S_x(\omega)$ can be determined if H_0, $\Delta\omega$ and $E[z]$ are all known. We can determine the first two to any desired accuracy by using precisely calibrated narrow-band filters, Fig. 9.1(*b*). However $E[z]$ cannot be precisely determined since it is an ensemble average and therefore not obtainable from measurements of finite length on a single sample. All we can do is to try to make sure that $z(t)$ never differs very much from its mean value $E[z]$ so that a spot value of $z(t)$ is likely to be a close approximation for $E[z]$. The variance of $z(t)$ is a measure of the magnitude of its fluctuations about the mean, and we define

$$\sigma^2 = E[z^2] - (E[z])^2 \tag{9.5}$$

as the variance of the measurement, according to (1.19). We shall now seek to determine σ^2. Clearly this will depend on the characteristics of the $y(t)$ random process, which is the output of the spectrum analyser's narrow-band filter, Fig. 9.1(a), since $z(t)$ is a function of $y(t)$, equation (9.1). We begin by substituting for z in terms of y in (9.5). Using (9.2), the $E[z]$ term can be replaced by $E[y^2]$ to obtain

$$\sigma^2 = E[z^2] - (E[y^2])^2 \tag{9.6}$$

but the $E[z^2]$ term is more difficult. Returning to the definition of $z(t)$, (9.1), we can write

$$z^2 = \left\{ \frac{1}{T} \int_0^T y^2(t)\,dt \right\}^2 = \left\{ \frac{1}{T} \int_0^T y^2(t_1)\,dt_1 \right\} \left\{ \frac{1}{T} \int_0^T y^2(t_2)\,dt_2 \right\} \tag{9.7}$$

where the two different time variables t_1 and t_2 are introduced so that this product of two integrals may be written as the equivalent double integral

$$z^2 = \frac{1}{T^2} \int_0^T dt_1 \int_0^T dt_2\, y^2(t_1) y^2(t_2). \tag{9.8}$$

Averaging equation (9.8) for the ensemble then gives

$$E[z^2] = \frac{1}{T^2} \int_0^T dt_1 \int_0^T dt_2\, E[y^2(t_1)y^2(t_2)] \tag{9.9}$$

which, substituting into (9.6), gives the following expression for the variance of the measurement, σ^2,

$$\sigma^2 = \left\{ \frac{1}{T^2} \int_0^T dt_1 \int_0^T dt_2\, E[y^2(t_1)y^2(t_2)] \right\} - (E[y^2])^2. \tag{9.10}$$

Although at present in an unpalatable form, this is a most important result and it is worth spending some time to reduce it to simpler terms.

In order to use (9.10), we first relate the fourth order average† $E[y^2(t_1)y^2(t_2)]$ to the autocorrelation function for the y process $R_y(\tau)$. This is easy to do if $y(t)$ is a Gaussian process. Fortunately this is likely to be a fair approximation since $y(t)$ is the output from a narrow band filter and so, as explained in Chapter 7, we may expect its probability distribution to approach a Gaussian distribution when the input to the filter is broad band noise. For a Gaussian process with zero mean, the fourth order average $E[y_1 y_2 y_3 y_4]$ can be expressed in terms of second order averages by the following equation‡

$$E[y_1 y_2 y_3 y_4] = E[y_1 y_2] \cdot E[y_3 y_4] + E[y_2 y_3] \cdot E[y_4 y_1] + E[y_1 y_3] \cdot E[y_2 y_4] \tag{9.11}$$

which, for $y_3 = y_1$ and $y_4 = y_2$, simplifies to

$$E[y_1^2 y_2^2] = 2(E[y_1 y_2])^2 + (E[y^2])^2 \tag{9.12}$$

---

† The terminology is an extension of that introduced in Ch. 2. A fourth order average is the statistical average value of the product of four random variables.

‡ See, for example, Middleton [49] p. 343.

if $E[y_1^2] = E[y_2^2] = E[y^2]$. If now we put $y_1 = y(t_1)$ and $y_2 = y(t_2)$ we obtain

$$E[y^2(t_1)y^2(t_2)] = 2(E[y(t_1)y(t_2)])^2 + (E[y^2])^2 \qquad (9.13)$$

in which, since $y(t)$ is a stationary process,

$$E[y(t_1)y(t_2)] = R_y(t_2 - t_1)$$

and so, finally, for a Gaussian process,

$$E[y^2(t_1)y^2(t_2)] = 2R_y^2(t_2 - t_1) + (E[y^2])^2. \qquad (9.14)$$

Substituting (9.14) into (9.10) gives the following expression for the variance of the measurement, σ^2:

$$\sigma^2 = \frac{2}{T^2} \int_0^T \mathrm{d}t_1 \int_0^T \mathrm{d}t_2 \, R_y^2(t_2 - t_1). \qquad (9.15)$$

For a stationary input, the autocorrelation function depends only on the time difference $\tau = t_2 - t_1$, so changing one of the variables t_2 to $\tau + t_1$ (where t_1 is a constant for the integration with respect to τ) we obtain

$$\sigma^2 = \frac{2}{T^2} \int_0^T \mathrm{d}t_1 \int_{-t_1}^{T-t_1} \mathrm{d}\tau \, R_y^2(\tau). \qquad (9.16)$$

Since the integrand $R_y^2(\tau)$ is a function of only one of the two variables of integration, we can integrate immediately with respect to the other variable. However the limits of integration require some thought. The range of values of τ and t_1 covered by the double integral is shown in Fig. 9.2(a). Integrating with respect to t_1 with τ constant, i.e. along the shaded strip in Fig. 9.2(a), gives

$$R_y^2(\tau)(T - |\tau|) \qquad \text{for} \qquad -T \leqslant \tau \leqslant T$$

and then, integrating with respect to the other variable over its full range from $-T$ to $+T$, we obtain

$$\sigma^2 = \frac{2}{T} \int_{-T}^T R_y^2(\tau)\left(1 - \frac{|\tau|}{T}\right)\mathrm{d}\tau. \qquad (9.17)$$

In order to evaluate this integral, we must introduce an expression for the autocorrelation function $R_y(\tau)$. Since $y(t)$ is the output of a narrow band filter, Fig. 9.1(a), its autocorrelation function will have the form (8.1), which may be written

$$R_y(\tau) = R_y(0)\left\{\frac{\sin(\Delta\omega \, \tau/2)}{(\Delta\omega \, \tau/2)}\right\} \cos \omega_0 \tau. \qquad (9.18)$$

We now want to substitute (9.18) into (9.17) and then evaluate the integral to obtain the variance σ^2. Fortunately we are only looking for the order of magnitude of σ^2, rather than an exact value, and it is sufficient to integrate only approximately. First we note that (9.18) includes the quotient

$$\left\{\frac{\sin(\Delta\omega \, \tau/2)}{(\Delta\omega \, \tau/2)}\right\}$$

which, when squared and plotted as a function of τ, has the form shown in Fig. 9.2(b). This may be approximated quite closely by the straight line function

$$\left\{1 - \frac{|\tau|}{(2\pi/\Delta\omega)}\right\} \qquad \text{for} \qquad |\tau| \leqslant \frac{2\pi}{\Delta\omega}$$

$$0 \qquad \text{for} \qquad |\tau| > \frac{2\pi}{\Delta\omega}$$

shown in Fig. 9.2 (c). With this assumption, (9.17) becomes, after substituting for $R_y(\tau)$,

$$\sigma^2 \simeq \frac{2}{T}\int_{-2\pi/\Delta\omega}^{2\pi/\Delta\omega} R_y^2(0)\left\{1 - \frac{|\tau|}{(2\pi/\Delta\omega)}\right\}\cos^2\omega_0\tau\left(1 - \frac{|\tau|}{T}\right)d\tau.$$

Secondly, we shall assume that the averaging time T of the r.m.s. meter in Fig. 9.1 is long, so that

$$T \gg \frac{2\pi}{\Delta\omega} \tag{9.19}$$

in which case we can make a further simplification of the integrand to obtain

$$\sigma^2 \simeq \frac{2}{T}\int_{-2\pi/\Delta\omega}^{2\pi/\Delta\omega} R_y^2(0)\left\{1 - \frac{|\tau|}{(2\pi/\Delta\omega)}\right\}\cos^2\omega_0\tau\, d\tau.$$

This integral is represented by the shaded area shown in Fig. 9.2(d). Provided that the bandwidth $\Delta\omega$ is small compared with the centre frequency of the filter ω_0, then there will be many cycles of $\cos^2\omega_0\tau$ inside the triangular envelope of the figure, and, in this case, the shaded area is equal to half the total area enclosed by the dotted triangle in Fig. 9.2(d). Hence the integral is given approximately by

$$R_y^2(0)\cdot\left(\frac{2\pi}{\Delta\omega}\right)\cdot\frac{1}{2}$$

and so, finally, we arrive at the approximate result

$$\sigma^2 \simeq \frac{2\pi}{T\Delta\omega}R_y^2(0)$$

or, putting the symbol B to denote bandwidth in Hz, so that

$$B = \frac{\Delta\omega}{2\pi} \tag{9.20}$$

then

$$\sigma^2 \simeq \frac{1}{BT}R_y^2(0). \tag{9.21}$$

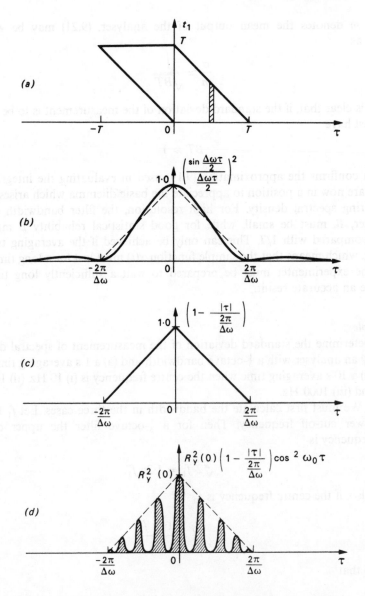

Fig. 9.2 Illustrating the steps in an approximate calculation of variance for the output of a spectrum analyser

This is the simplified version of (9.10) we have been looking for. It gives the variance of the output of a spectrum analyser. Since the correct output should be, from (9.5),

$$E[z] = E[y^2] = R_y(0) = m \quad \text{(say)}$$

where m denotes the mean output of the analyser, (9.21) may be written finally as

$$\frac{\sigma}{m} \simeq \frac{1}{\sqrt{BT}} \qquad (9.22)$$

and it is clear that, if the standard deviation of the measurement is to be small, we must have

$$BT \gg 1 \qquad (9.23)$$

(which confirms the approximation (9.19) used in evaluating the integrals).

We are now in a position to appreciate the basic dilemma which arises when measuring spectral density. For high resolution, the filter bandwidth of the analyser, B, must be small, while for good statistical reliability B must be large compared with $1/T$. This can only be achieved if the averaging time T is long, which means that the sample function $x(t)$ must last for a long time and that the experimenter must be prepared to wait a sufficiently long time to achieve an accurate result.

Example

Determine the standard deviation of the measurement of spectral density by an analyser with a $\frac{1}{3}$-octave bandwidth and (a) a 1 s averaging time and (b) a 10 s averaging time when the centre frequency is (i) 10 Hz, (ii) 100 Hz and (iii) 1000 Hz.

We must first calculate the bandwidth in the three cases. Let f_1 be the lower cut-off frequency. Then for a $\frac{1}{3}$-octave filter the upper cut-off frequency is

$$\sqrt[3]{2}\, f_1 = 1 \cdot 26\, f_1.$$

Also, if the centre frequency is f_0

$$\frac{f_0}{f_1} = \frac{1 \cdot 26\, f_1}{f_0}$$

so that

$$f_0 = 1 \cdot 12\, f_1$$

and the bandwidth is

$$0 \cdot 26\, f_1 = 0 \cdot 23\, f_0.$$

Hence the filter bandwidths are (i) 2·3 Hz, (ii) 23 Hz and (iii) 230 Hz approximately.

The ratio of standard deviation to mean output, σ/m, can now be tabulated by putting numbers into (9.22).

	(a) Averaging time $T = 1$ s	(b) Averaging time $T = 10$ s
Case (i) $f_0 = 10$ Hz	$\dfrac{\sigma}{m} = 0{\cdot}66$	$\dfrac{\sigma}{m} = 0{\cdot}21$
Case (ii) $f_0 = 100$ Hz	$\dfrac{\sigma}{m} = 0{\cdot}21$	$\dfrac{\sigma}{m} = 0{\cdot}07$
Case (iii) $f_0 = 1000$ Hz	$\dfrac{\sigma}{m} = 0{\cdot}07$	$\dfrac{\sigma}{m} = 0{\cdot}02$

This means that, when an experimenter makes a spot measurement of spectral density, his result is subject to an error on account of the fluctuations of the output meter. The standard deviation of the measurement (and therefore of the error) is expressed above as a ratio of the mean output level m.

Before leaving this subject it should be mentioned that the output of a spectral analyser will also be susceptible to a steady state or *bias error* when the filter bandwidth covers a range of frequencies in which the spectral density is changing rapidly with frequency. The mean output of the analyser is, from (7.18),

$$E[y^2] = 2 \int_{\omega_0 - \Delta\omega/2}^{\omega_0 + \Delta\omega/2} H_0^2 S_x(\omega)\, d\omega$$

and our assumption in (9.3) that this may be written

$$E[y^2] \simeq 2H_0^2 S_x(\omega_0) \Delta\omega$$

is of course not accurate if $S_x(\omega)$ is changing fast with frequency in the band $\Delta\omega$. The only practical solution to this problem is to achieve greater resolution by employing an analyser with narrower band filters (provided that it is possible to use a long enough averaging time to keep the variance of the measurement acceptable).

Analysis of finite length records

We have seen how the standard deviation of a single measurement of spectral density is affected by the bandwidth B and averaging time T of an analogue spectrum analyser and, to a good approximation, the ratio of the standard deviation σ to its mean value m is independent of the variance of the input signal and depends only on B and T, according to the fundamental result (9.22). If the instrument's averaging time is T, its output at any instant is based only on the input values for the immediately preceding time interval T, equation (9.1). The basic problem is that it is just not possible to calculate a precise estimate of spectral density when only a limited length of data is available for analysis.

Now, with an analogue spectrum analyser, operating on a continuous random process, we can go on taking instantaneous measurements of spectral density and, by watching the movement of the instrument's output meter, quickly appreciate the variability of the output. Usually an analogue instrument has two or more alternative averaging times and several different filter bandwidths, and, by altering the settings of the instrument, it is soon possible to judge how the instrument has to be adjusted to give reasonable accuracy with an acceptable bandwidth, and therefore adequate resolution of close spectral peaks. However analogue spectrum analysis takes time, since many separate readings have to be taken to cover a wide range of frequencies; also the maximum resolution obtainable from analogue filters is limited. Digital data analysis methods are therefore extensively used and virtually all random data analysis except for so-called "quick-look" spectrum analysis is now carried out digitally. The sample function $x(t)$ is first digitized by an analogue-to-digital converter and then a digital computer is used to make calculations on the digitized data. In the following chapters we shall be concerned with the computer algorithms (or logical procedures) which are used to calculate spectral density values from the data fed into a computer. However, even for a computer calculation which involves no approximations or inaccuracies, there is the same fundamental problem that we have just met for analogue spectrum analysis. Since there is a limit to the number of data points that can be fed into a computer, there is a limit to the length of the sample function that can be analysed. This restricted length of sample causes the same loss of precision as that occurring in an analogue instrument with a finite averaging time. It turns out that accuracy still depends on record length T and bandwidth B, except that the bandwidth is no longer that of an analogue filter, Fig. 9.1, but has to be interpreted in a different way. We shall now consider the general problem of analysing a record of finite length and introduce the concept of a *spectral window* which is used to define the equivalent bandwidth B_e of a digital calculation.

Suppose that $\{x(t)\}$ is a stationary random process consisting of an ensemble of sample functions.† Since the process is stationary, each sample function theoretically goes on for ever. However suppose that records are only available for the period $t = 0$ to $t = T$, Fig. 9.3. In this case the autocorrelation function

$$R_x(\tau) = E[x(t)x(t + \tau)]$$

can only be determined for $|\tau| \leqslant T$. We cannot therefore calculate the corresponding spectral density $S_x(\omega)$ from its fundamental definition (5.1)

$$S_x(\omega) = \frac{1}{2\pi} \int_{-\infty}^{\infty} R_x(\tau) e^{-i\omega\tau} d\tau \qquad (5.1)$$

since we do not know $R_x(\tau)$ for $|\tau| > T$. The best we can do is to approximate

---

† The brackets $\{\ \}$ are used to designate the ensemble of sample functions $x(t)$ if this is not otherwise clear.

$S_x(\omega)$ by truncating the integral in (5.1) to give the approximation

$$S_x(\omega) \simeq \frac{1}{2\pi} \int_{-T}^{T} R_x(\tau)\, e^{-i\omega\tau}\, d\tau. \tag{9.24}$$

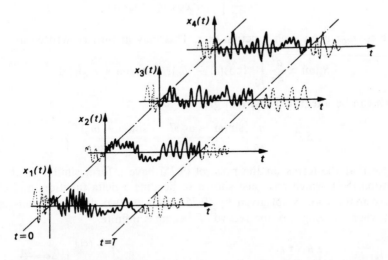

Fig. 9.3 Finite length records of duration T from the stationary
random process $\{x(t)\}$

Although, as already mentioned, the calculation procedure carried out in a
computer may not actually involve finding the autocorrelation function $R_x(\tau)$,
nevertheless the basic difficulty which arises is the inherent loss of accuracy
resulting from an approximation equivalent to (9.24).

To illustrate how the approximate spectrum (9.24) is likely to differ from
its true value calculated from (5.1), suppose that

$$x(t) = a\sin(\omega_0 t + \phi)$$

where the amplitude a and the frequency ω_0 are constant, and the phase angle ϕ
is constant for each sample function, but varies randomly from sample to
sample with all values between 0 and 2π being equally likely. In this case

$$R_x(\tau) = E[x(t)x(t+\tau)] = \int_{0}^{2\pi} a^2\sin(\omega_0 t + \phi)\sin(\omega_0 t + \omega_0\tau + \phi)p(\phi)\,d\phi$$

which, putting $p(\phi) = 1/2\pi$, gives

$$R_x(\tau) = \frac{a^2}{2}\cos\omega_0\tau.$$

Now suppose that the records $x(t)$ are only defined for $t = 0$ to $t = T$, so that
we only know $R_x(\tau)$ for $|\tau| \leqslant T$, and consider calculating the approximation

for $S_x(\omega)$ given by (9.24). Substituting for $R_x(\tau)$ in (9.24) gives

$$S_x(\omega) \simeq \frac{1}{2\pi} \int_{-T}^{T} \frac{a^2}{2} \cos \omega_0 \tau\, e^{-i\omega\tau}\, d\tau$$

$$= \frac{1}{2\pi} \int_{0}^{T} a^2 \cos \omega_0 \tau \cos \omega \tau\, d\tau$$

since $\cos \omega_0 \tau$ is an even function of τ. This may in turn be written as

$$S_x(\omega) \simeq \frac{a^2}{4\pi} \int_{0}^{T} \{\cos(\omega - \omega_0)\tau + \cos(\omega + \omega_0)\tau\}\, d\tau$$

which can be integrated easily to give

$$S_x(\omega) \simeq \frac{a^2}{4\pi} \left\{ \frac{\sin(\omega - \omega_0)T}{\omega - \omega_0} + \frac{\sin(\omega + \omega_0)T}{\omega + \omega_0} \right\}. \qquad (9.25)$$

Notice that the terms on the r.h.s. of (9.25) have a form similar to those in equation (5.8), which has been shown to become a delta function, and so the approximation for $S_x(\omega)$ given by (9.25) tends to two delta functions in the limit when the length of the record, T, becomes infinite.

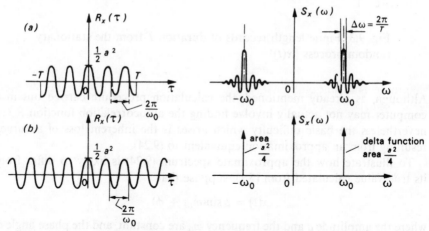

Fig. 9.4 Fourier transform of a finite length of a cosine wave compared with the transform of an infinite length

In Fig. 9.4(a) the approximate result from (9.25) for T finite is shown compared with, in Fig. 9.4(b), the exact result when $T \to \infty$. The conclusion we can draw from this is that the result of analysing only a finite length of record is to "smear" out a sharp spectral line over a band of frequencies of width $\Delta\omega = 2\pi/T$ approximately. In order to resolve two nearby spectral peaks, the length of record, T, must be long enough for their frequency difference to be large compared with $1/T$ Hz.† However accurate the analogue-to-digital con-

†Units of time are assumed to be seconds unless otherwise stated.

version and however large the computer, close spectral peaks can only be distinguished if the record length T is long enough.

A further limitation on the accuracy of numerical calculations can be seen by considering the fluctuations of $S(\omega)$ on either side of $\omega = \omega_0$ in Fig. 9.4(a). These fluctuations remain as the length of the record increases although their frequency increases as shown in Fig. 9.5(a) (which is drawn for positive frequencies only). One way of removing them, or at least of reducing their

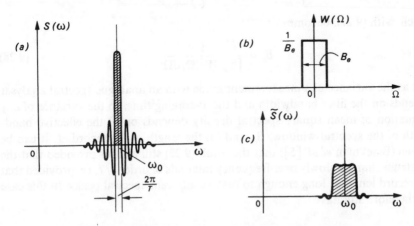

Fig. 9.5 Smoothing with a rectangular spectral window

magnitude, is to smooth the spectrum so that instead of plotting $S(\omega)$ we plot a *smoothed spectrum* $\tilde{S}(\omega)$ given by

$$\tilde{S}(\omega) = \int_{-\infty}^{\infty} W(\Omega - \omega)S(\Omega)\,d\Omega \qquad (9.26)$$

where Ω is a dummy frequency variable and $W(\Omega)$ is a weighting function which satisfies

$$\int_{-\infty}^{\infty} W(\Omega)\,d\Omega = 1. \qquad (9.27)$$

Suppose, for example, that $W(\Omega)$ is the rectangular function shown in Fig. 9.5(b). In this case the smoothed form of the spectrum defined by (9.25) is sketched in Fig. 9.5(c). Since positive and negative half cycles of $S(\omega)$ tend to cancel each other out, the smoothed spectrum has the approximate form shown and, on account of (9.27), the shaded areas in Figs. 9.5(a) and (c) are approximately equal.

The function $W(\Omega)$ is called a *spectral window* function† and the shape of a graph of $W(\Omega)$ against Ω is said to be the "shape of the window". Many different shapes have been suggested, but the effective width of the window is

†The term "window" was introduced by Blackman and Tukey [5] and is now accepted jargon.

what matters because this defines the band of frequencies over which averaging occurs, equation (9.26). For a rectangular window, Fig. 9.5(b), the bandwidth B_e is just the full width of the function. However, for other functions which may rise gradually to a peak and then fall off gradually, it is necessary to calculate an effective width and this is usually defined by the effective bandwidth B_e where

$$B_e \int_{-\infty}^{\infty} W^2(\Omega)\,d\Omega = \left\{ \int_{-\infty}^{\infty} W(\Omega)\,d\Omega \right\}^2$$

which, with (9.27), becomes

$$B_e = \frac{1}{\int_{-\infty}^{\infty} W^2(\Omega)\,d\Omega}. \tag{9.28}$$

Just as the variance of a measurement made with an analogue spectral analyser depends on the filter bandwidth and the averaging time, so the variance of *any* estimation of mean square spectral density depends on (i) the effective bandwidth of the spectral window, B_e, and (ii) the length of the record, T. It can be shown (Blackman *et al.* [5]) that the result (9.22) still applies provided that the spectrum changes slowly over frequency intervals of order $1/T$, i.e. provided that the record length is long enough to resolve adjacent spectral peaks. In this case we therefore still have

$$\frac{\sigma}{m} \simeq \frac{1}{\sqrt{(B_e T)}} \tag{9.29}$$

where σ is the standard deviation of a measurement whose mean value is m, B_e is the effective bandwidth of the spectral window and T is the record length. Furthermore (9.29) also applies approximately to measurements of the auto-correlation function.† In this case the effective bandwidth B_e is the bandwidth of the entire input process so an individual measurement of a correlation function has greater statistical reliability than an individual measurement of spectral density from the same length of record. However in transforming from the time domain to the frequency domain we are forced to employ selective filters which automatically reduce the effective bandwidth and therefore reduce the statistical reliability of the results obtained.

Confidence limits

We have been able to find an expression for the ratio of the standard deviation σ to the true mean m of a measurement of spectral density. If σ/m is small, there is a high confidence that a sample measurement lies close to the true mean value; if σ/m is large, then confidence that a sample measurement is near to the true value will be low. In this section we shall make an approximate

---

†Crandall [17] Ch. 2. See also Bendat *et al.* [3] and Jenkins *et al.* [39].

calculation of the *level of confidence* that can be attached to a measurement of (auto) spectral density when σ/m is known. Our analysis will be based on a sweeping assumption about the distribution of measured values of the spectral density. We have seen that the measured value z (see Fig. 9.1) is itself a random variable whose mean value is m (we neglect any bias error) and whose variance is σ. We shall assume in what follows that z can be expressed as the sum of squares of a number of statistically independent Gaussian random variables each with zero mean and the same variance. The plausibility of this assumption can be seen from (9.1) which can be written in the form

$$z(t) = \frac{1}{T}\left\{ \int_0^{T/n} y^2(t)\,dt + \int_{T/n}^{2T/n} y^2(t)\,dt + \cdots + \int_{T(1-1/n)}^T y^2(t)\,dt \right\}$$

so it is possible to think of z as the sum of a number of squared terms, but we cannot prove that separate terms carry equal weight or that they are statistically independent. The best we can do is to say that it has been found experimentally that measured values of z have approximately the probability distribution which would occur if our assumption held good.

Consider the random variable χ_κ^2 (written chi-square and pronounced "keye-square" rhyming with "eye") which is the sum of squares of κ independent Gaussian random variables x_1, x_2, etc., so that

$$\chi_\kappa^2 = x_1^2 + x_2^2 + x_3^2 + \cdots + x_\kappa^2. \tag{9.30}$$

If the mean value of each of the x variables is zero and each of their variances is 1, then the mean value of χ_κ^2 is just

$$m = E[\chi_\kappa^2] = \kappa E[x^2] = \kappa. \tag{9.31}$$

The variance of χ_κ^2 is

$$\sigma^2 = E[(\chi_\kappa^2)^2] - (E[\chi_\kappa^2])^2 \tag{9.32}$$

$$= E[(x_1^2 + x_2^2 + \cdots + x_\kappa^2)^2] - \kappa^2$$

$$= E[(x_1^4 + x_2^4 + \cdots + x_\kappa^4) + 2(x_1^2 x_2^2 + x_1^2 x_3^2 + \cdots + x_{\kappa-1}^2 x_\kappa^2)] - \kappa^2. \tag{9.33}$$

This can be simplified by making use of equation (9.11) to expand the fourth order averages and obtain

$$E[x^4] = 3(E[x^2])^2 = 3 \tag{9.34}$$

and

$$E[x_1^2 x_2^2] = 2(E[x_1 x_2])^2 + E[x^2] = 1 \tag{9.35}$$

since there is no correlation between statistically independent variables. Substituting (9.34) and (9.35) into (9.33), and noting that there are $\frac{1}{2}\kappa(\kappa - 1)$ separate terms of the form $x_1 x_2$, we obtain

$$\sigma^2 = (3\kappa) + \kappa(\kappa - 1) - \kappa^2 = 2\kappa. \tag{9.36}$$

The ratio of the standard deviation σ to the mean m of a χ^2_κ random process is therefore

$$\frac{\sigma}{m} = \sqrt{\frac{2}{\kappa}} \qquad (9.37)$$

where κ is the number of statistical degrees-of-freedom in χ^2_κ (not to be confused with the kinematic degrees-of-freedom of a dynamic system: although the terminology is the same, the meaning is completely different). We see that if there is a large number of statistical degrees-of-freedom the ratio σ/m is small so that there will then be a high confidence that a sample measurement lies close to the true mean. If κ is small, confidence will be correspondingly low.

The probability density function $p(\chi^2_\kappa)$ for χ^2_κ is given in books on probability theory† and its form depends on the number of degrees-of-freedom κ. In Fig. 9.6, $p(\chi^2_\kappa)$ is sketched for values of $\kappa = 1, 2, 4$ and 10. When κ becomes very large

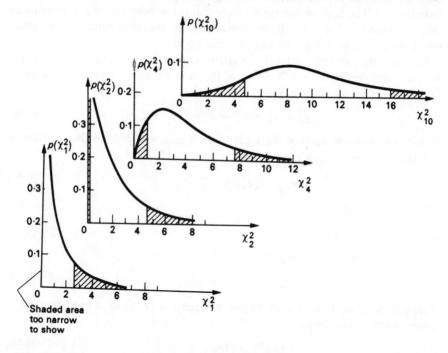

Fig. 9.6 Chi-square probability density functions for 1, 2, 4 and 10 degrees-of freedom

the probability density function becomes indistinguishable from a Gaussian function, as predicted by the Central Limit Theorem. From statistical tables‡ of χ^2_κ we can determine the "tails" of each distribution in which there is a 10 per

†See, for instance, Parzan [54] and also problem 9.3.
‡See Appendix 4.

cent (say) probability of finding χ_κ^2 and these areas are shown shaded in Fig. 9.6. This information about the width of the distributions may then all be combined on a single graph of χ_κ^2 against κ, Fig. 9.7. According to (9.31) the mean value m is equal to k and so the mean of the distribution is represented by the straight line shown. Since, from (9.37), σ/m diminishes as κ increases, the limits of the distribution converge towards the mean value as κ increases. Two limits are shown in Fig. 9.7 for the middle 80 per cent and 99 per cent of the distribution;

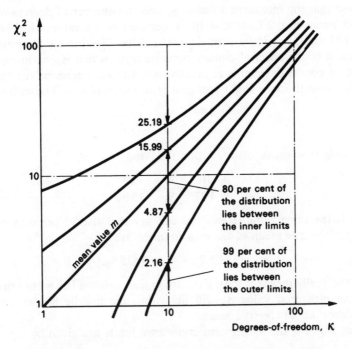

Fig. 9.7 Dependence of the χ_κ^2 distribution on the number of statistical degrees-of-freedom κ (see Appendix 4).

for the latter, $\frac{1}{2}$ per cent of samples will lie above the upper 99 per cent limit and $\frac{1}{2}$ per cent of samples below the lower 99 per cent limit.

We are now ready to apply this information to obtain approximate *confidence limits* for a measurement of spectral density, z. As already explained, we shall assume that the distribution of z follows a chi-square law. This distribution is completely defined by specifying its mean value m and standard deviation σ. The equivalent number of degrees of freedom κ is then determined by (9.37). According to (9.29), if B_e is the effective bandwidth and T is the record length or averaging time of the spectral measurement, then

$$\frac{\sigma}{m} \simeq \frac{1}{\sqrt{(B_e T)}}$$

and so the approximate number of degrees of freedom of χ^2_κ can be obtained from (9.37), by putting

$$\sqrt{\frac{2}{\kappa}} = \frac{1}{\sqrt{(B_e T)}}$$

to give

$$\kappa = 2 B_e T. \tag{9.38}$$

Suppose that the measured value is S_0 and the number of degrees-of-freedom $\kappa = 10$. From Fig. 9.7, for $\kappa = 10$, 80 per cent of the values of χ^2_κ lie between $\chi^2_\kappa = 4\cdot87$ and $\chi^2_\kappa = 15\cdot99$ with the mean value $E[\chi^2_\kappa] = 10$. Suppose that the true mean of the spectral density measurement, which is unknown, is m. We cannot of course determine m exactly, but we can determine the 80 per cent confidence limits for m. Eighty per cent of all values of S_0 will lie within the band

$$\frac{4\cdot87}{10} < \frac{S_0}{m} < \frac{15\cdot99}{10}$$

or, turning this upside down, within the band

$$\frac{10}{4\cdot87} > \frac{m}{S_0} > \frac{10}{15\cdot99}$$

which is the same thing. Hence we can say that the 80 per cent confidence limits for the true mean of the measurement are

$$2\cdot05 S_0 > m > 0\cdot625 S_0.$$

What this really means is that, if the true mean value m lies within these limits, then the measured value S_0 will lie within the middle 80 per cent of the distribution about the true mean.

From Fig. 9.7, the 99 per cent confidence limits are given by

$$\frac{2\cdot16}{10} < \frac{S_0}{m} < \frac{25\cdot19}{10}$$

that is by

$$4\cdot63 S_0 > m > 0\cdot397 S_0$$

from which it will be clear that, in order to achieve really reliable results, the number of statistical degrees-of-freedom of a measurement must be much greater than 10.

Chapter 10

Digital spectral analysis I: Discrete Fourier transforms

We have already mentioned that most experimental measurements of random processes are carried out digitally. A typical function $x(t)$ of the process to be measured is fed through an analogue-to-digital converter. This samples $x(t)$ at a series of regularly spaced times, Fig. 10.1. If the sampling interval is Δ

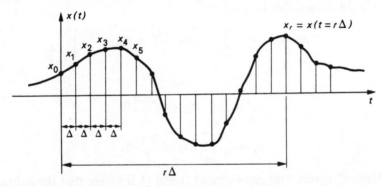

Fig. 10.1 Sampling a continuous function of time at regular intervals

(constant) then the discrete value of $x(t)$ at time $t = r\Delta$ is written x_r, and the sequence $\{x_r\}$, $r = \ldots, -1, 0, 1, 2, 3, \ldots$, is called a *discrete time series*. Since the discrete series has been derived from a continuous function of time, it has become customary to refer to the latter as a *continuous time series*. The words "time series" are therefore used either to refer to a sequence of discrete numbers x_r ordered in time, or to refer to the original continuous time sample $x(t)$ from which the discrete series has been obtained. The objectives of *time series analysis* are to determine the statistical characteristics of the original function $x(t)$ by manipulating the series of discrete numbers x_r. The main interest is the frequency composition of $x(t)$ and in this chapter we are concerned with estimating the spectrum of a random process $x(t)$ by analysing the discrete time series obtained by sampling a finite length of a sample function.

In order to estimate spectra from measured data, the obvious method is to estimate the appropriate correlation function first and then to Fourier transform this function to obtain the required spectrum. Until the late 1960s, this approach was the basis of practical calculation procedures which followed the formal mathematical route by which spectra are defined as Fourier transforms of correlation functions. The assumptions and approximations involved were studied in detail and there is an extensive literature† on the "classical" method. However the position was changed by the advent of the *fast Fourier transform* (or FFT for short). This is a remarkably efficient way of calculating the Fourier transform of a time series. Instead of estimating spectra by first determining correlation functions and then calculating their Fourier transforms, it is now quicker and more accurate to calculate spectral estimates directly from the original time series by a method which we shall describe in detail. We begin by returning to the Fourier analysis of a periodic (non-random) function of time $x(t)$.

Discrete Fourier transforms

We saw in Chapter 4 that, if $x(t)$ is a periodic function with period T, Fig. 4.1, then it is always possible to write

$$x(t) = a_0 + 2 \sum_{k=1}^{\infty} \left(a_k \cos \frac{2\pi kt}{T} + b_k \sin \frac{2\pi kt}{T} \right) \tag{10.1}$$

where

$$\begin{aligned} a_k &= \frac{1}{T} \int_0^T x(t) \cos \frac{2\pi kt}{T} \, dt \\ {\scriptstyle k \geqslant 0} \\ b_k &= \frac{1}{T} \int_0^T x(t) \sin \frac{2\pi kt}{T} \, dt. \\ {\scriptstyle k \geqslant 1} \end{aligned} \tag{10.2}$$

This "recipe" agrees with equations (4.1) and (4.2) except that the definitions of the Fourier coefficients have been slightly altered and the range of the integrals in (10.2) runs from 0 to T instead of from $-T/2$ to $T/2$ (which makes no difference since $x(t)$ has period T). If we use complex notation, equations (10.2) can be combined into a single equation by defining

$$X_k = a_k - ib_k \tag{10.3}$$

and putting

$$e^{-i(2\pi kt/T)} = \cos \frac{2\pi kt}{T} - i \sin \frac{2\pi kt}{T}$$

to give

$$X_k = \frac{1}{T} \int_0^T x(t) e^{-i(2\pi kt/T)} \, dt. \tag{10.4}$$
$${\scriptstyle k \geqslant 0}$$

Now consider what happens if the continuous time series $x(t)$ is not known and only equally spaced samples are available. Suppose that these are

---

† See, for instance, Blackman *et al.* [5]; Bendat *et al.* [2] Jenkins *et al.* [39].

represented by the discrete series $\{x_r\}$, $r = 0, 1, 2, \ldots, (N - 1)$, where $t = r\Delta$, Fig. 10.1, and $\Delta = T/N$. In this case, the integral in (10.4) may be replaced approximately by the summation

$$X_k = \frac{1}{T} \sum_{r=0}^{N-1} x_r e^{-i(2\pi k/T)(r\Delta)} \, \Delta. \tag{10.5}$$

This amounts to assuming that the total area under the curve shown in Fig. 10.2 is given by the sum of all the shaded strips. Substituting $T = N\Delta$ into (10.5) then gives

$$X_k = \frac{1}{N} \sum_{r=0}^{N-1} x_r e^{-i(2\pi kr/N)} \tag{10.6}$$

which may be regarded as an approximate formula for calculating the coefficients of the Fourier series (10.1).

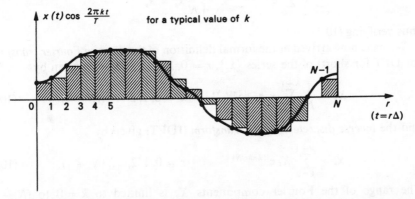

Fig. 10.2 Approximation involved in calculating Fourier coefficients from a discrete rather than a continuous series

Although (10.6) does not provide enough information to allow the continuous time series $x(t)$ to be obtained, it is a most important fact that it does allow all the discrete values of the series $\{x_r\}$ to be regained *exactly*. Any typical value x_r of the series $\{x_r\}$ is given by the inverse formula

$$x_r = \sum_{k=0}^{N-1} X_k e^{i(2\pi kr/N)}. \tag{10.7}$$

The truth of this statement may be verified by putting $r = s$ in (10.6) (to avoid muddling terms later) and then substituting this into the r.h.s. of (10.7), to obtain

$$\sum_{k=0}^{N-1} X_k e^{i(2\pi kr/N)} = \sum_{k=0}^{N-1} \left\{ \frac{1}{N} \sum_{s=0}^{N-1} x_s e^{-i(2\pi ks/N)} \right\} e^{i(2\pi kr/N)}$$

$$= \sum_{k=0}^{N-1} \sum_{s=0}^{N-1} \frac{1}{N} x_s e^{-i(2\pi k/N)(s-r)}$$

Interchanging the order of summation makes this

$$= \sum_{s=0}^{N-1} \left\{ \sum_{k=0}^{N-1} e^{-i(2\pi k/N)(s-r)} \right\} \frac{1}{N} x_s$$

and, since k, s, r and N are all integers, the exponentials all sum to zero unless $s = r$, so that the term in { } is given by either

$$\left\{ \sum_{k=0}^{N-1} e^{-i(2\pi k/N)(s-r)} \right\} = 0 \qquad \text{for} \qquad s \neq r$$

or

$$\left\{ \sum_{k=0}^{N-1} e^{-i(2\pi k/N)(s-r)} \right\} = N \qquad \text{for} \qquad s = r$$

and hence

$$\sum_{s=0}^{N-1} \left\{ \sum_{k=0}^{N-1} e^{-i(2\pi k/N)(s-r)} \right\} \frac{1}{N} x_s = x_r$$

thus verifying (10.7).

We have now arrived at the formal definition of the *discrete Fourier transform* (or DFT for short) of the series $\{x_r\}$, $r = 0, 1, 2, \ldots, (N-1)$, given by

$$X_k = \frac{1}{N} \sum_{r=0}^{N-1} x_r e^{-i(2\pi kr/N)} \qquad k = 0, 1, 2, \ldots, (N-1) \qquad (10.8)$$

and the *inverse discrete Fourier transform* (IDFT) given by

$$x_r = \sum_{k=0}^{N-1} X_k e^{i(2\pi kr/N)} \qquad r = 0, 1, 2, \ldots, (N-1). \qquad (10.9)$$

The range of the Fourier components X_k is limited to $k = 0$ to $(N-1)$ (corresponding to harmonics of frequency $\omega_k = 2\pi k/T = 2\pi k/N\Delta$) in order to maintain the symmetry of the transform pair (10.8) and (10.9). We shall see shortly that harmonic components with frequencies in excess of a certain limiting value fixed by the sampling interval Δ, Fig. 10.1, can in any case not be identified by analysing sampled data, so no information is lost by restricting the allowable range of the integer k. Although we have introduced the DFT by considering the properties of continuous Fourier series, it is important to realize that the discrete Fourier transform (10.8) has the exact inverse defined by (10.9) and that the properties of DFT's are exact properties rather than approximate properties based on the corresponding results for continuous Fourier transforms.

Fourier transforms of periodic functions

In Chapter 4 we discussed the continuous Fourier transform pair

$$X(\omega) = \frac{1}{2\pi} \int_{-\infty}^{\infty} x(t) e^{-i\omega t} \, dt \qquad (10.10)$$

$$x(t) = \int_{-\infty}^{\infty} X(\omega) e^{i\omega t} \, d\omega \qquad (10.11)$$

and explained that classical Fourier transform theory is restricted to a function $x(t)$ which satisfies (4.7), which in this case becomes

$$\int_{-\infty}^{\infty} |x(t)| \, dt < \infty.$$

However by allowing generalized functions to be used, this restriction may be lifted. Suppose that $x(t)$ is a periodic function of time represented by the Fourier series (10.1), and, for simplicity, consider only the kth harmonic,

$$x(t) = 2\left(a_k \cos \frac{2\pi k}{T} t + b_k \sin \frac{2\pi k}{T} t\right). \qquad (10.12)$$

The Fourier transform of this $x(t)$ can be shown to be

$$X(\omega) = (a_k - ib_k)\delta\left(\omega - \frac{2\pi k}{T}\right) + (a_k + ib_k)\delta\left(\omega + \frac{2\pi k}{T}\right) \qquad (10.13)$$

where $\delta(\omega)$ is the Dirac delta function, first introduced in Chapter 5. $\delta(\omega)$ has the property of being zero everywhere except at $\omega = 0$ where it is infinite in such a way that

$$\int_{-\infty}^{\infty} \delta(\omega) \, d\omega = 1.$$

Similarly $\delta(\omega - (2\pi k/T))$ is zero everywhere except at $\omega = 2\pi k/T$ where it satisfies

$$\int_{-\infty}^{\infty} \delta\left(\omega - \frac{2\pi k}{T}\right) d\omega = 1. \qquad (10.14)$$

The best way to verify (10.13) is to use the inverse transform (10.11). If we substitute the assumed transform (10.13) into (10.11) and make use of (10.14) when evaluating the integral, then it is easy to show that (10.12) is regained.

The real and imaginary parts of the Fourier transform of a harmonic wave (10.12) can therefore be represented by a pair of delta functions (10.13) and, by extending the analysis to many harmonics, we can see that the real and imaginary parts of the Fourier transforms of a periodic Fourier series $x(t)$, (10.1), can be represented by the infinite "combs" of delta functions shown in Fig. 10.3. If the period of $x(t)$ is T, then the spectral lines (delta functions) are spaced $2\pi/T$ rad/s apart.†

The same result can also be expressed mathematically by extending (10.13) to a summation covering all harmonics. If for this purpose we put, from (10.3),

$$X_k = a_k - ib_k \qquad (10.3)$$

and define

$$X_{-k} = X_k^* = a_k + ib_k \qquad (10.15)$$

†Units of time are again assumed to be seconds.

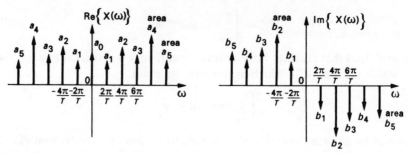

Fig. 10.3 Real and imaginary parts of the Fourier transform $X(\omega)$ of a periodic function

$$x(t) = a_0 + 2 \sum_{k=1}^{\infty} \left(a_k \cos \frac{2\pi kt}{T} + b_k \sin \frac{2\pi kt}{T} \right)$$

then (10.13) may be generalized to

$$X(\omega) = \sum_{k=-\infty}^{\infty} X_k \delta \left(\omega - \frac{2\pi k}{T} \right) \tag{10.16}$$

which is an important equation that we shall need again later.

In summary, we have seen that the DFT of a discrete time series $\{x_r\}$, $r = 0, 1, 2, \ldots, (N-1)$, yields a set of complex coefficients

$$X_k = a_k - ib_k$$

which can be interpreted in two ways. If $\{x_r\}$ is thought of as having been derived by sampling a *single cycle* of a continuous *periodic function* $x(t)$, the components of $\{X_k\}$ are an approximation for:

(i) the coefficients of a Fourier series expansion of $x(t)$; and

(ii) the areas of the teeth of combs of delta functions which are the real and imaginary parts of the Fourier transform of $x(t)$ (and which are, as we have seen, the same thing as (i)).

Provided that $x(t)$ is a continuous function (no sudden discontinuities) the approximation may be made as close as we like by making the sampling interval Δ smaller and smaller. We can therefore relate the DFT of a discrete time series to the Fourier transform of the underlying continuous function, and this is important in understanding how the DFT is used to calculate spectral density estimates from sampled data. We shall return to this subject shortly, but first we consider some fundamental properties of DFT's.

Aliasing

We have seen that the DFT of the series $\{x_r\}$, $r = 0, 1, 2, \ldots, (N-1)$, is defined by

$$X_k = \frac{1}{N} \sum_{r=0}^{N-1} x_r e^{-i(2\pi kr/N)} \tag{10.8}$$

for $k = 0, 1, 2, \ldots, (N - 1)$. Suppose that we try calculating values of X_k for the case when k is greater than $(N - 1)$. Let

$$k = N + l \quad \text{(say)}.$$

Then

$$X_{N+l} = \frac{1}{N} \sum_{r=0}^{N-1} x_r e^{-i(2\pi r/N)(N+l)}$$

$$= \frac{1}{N} \sum_{r=0}^{N-1} x_r e^{-i(2\pi rl/N)} e^{-i2\pi r}$$

which, since $e^{-i2\pi r}$ is always equal to 1 whatever the value of r, gives

$$X_{N+l} = X_l. \tag{10.17}$$

The coefficients X_k therefore just repeat themselves for $k > (N - 1)$, so that if we plot the magnitudes $|X_k|$ along a frequency axis $\omega_k = 2\pi k/N\Delta$, the graph

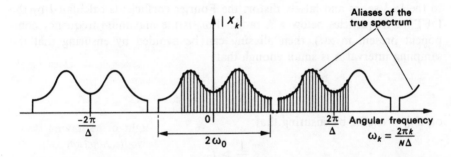

Fig. 10.4 Periodicity of Fourier coefficients calculated by the DFT

repeats itself periodically as shown in Fig. 10.4. Furthermore, it is also easy to see from (10.8) that, provided terms in the $\{x_r\}$ series are real,

$$X_{-l} = X_l^* \quad \text{(the complex conjugate of } X_l) \tag{10.18}$$

in agreement with (10.15). Hence

$$|X_{-l}| = |X_l|$$

and Fig. 10.4 is therefore symmetrical about the zero frequency position. The unique part of the graph occupies the frequency range $|\omega| \leqslant \pi/\Delta$ rad/s. Higher frequencies just show spurious Fourier coefficients which are repetitions of those which apply at frequencies below π/Δ rad/s. We can therefore see that the coefficients X_k calculated by the DFT are only correct Fourier coefficients for frequencies up to

$$\omega_k = \frac{2\pi k}{N\Delta} = \frac{\pi}{\Delta}$$

that is for k in the range $k = 0, 1, 2, \ldots, N/2$. Moreover if there are frequencies above π/Δ rad/s present in the original signal, these introduce a distortion of the graph called *aliasing*, Fig. 10.5. The high frequency components contribute

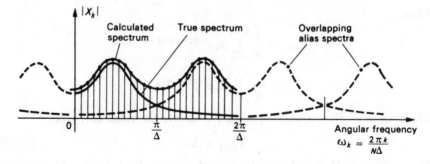

Fig. 10.5 Aliasing distortion when signal bandwidth exceeds π/Δ rad/s.

to the $\{x_r\}$ series and falsely distort the Fourier coefficients calculated by the DFT for frequencies below π/Δ rad/s. If ω_0 is the maximum frequency component present in $x(t)$, then aliasing can be avoided by ensuring that the sampling interval Δ is small enough that

$$\frac{\pi}{\Delta} > \omega_0$$

or, if $f_0 = \omega_0/2\pi$, by ensuring that

$$\frac{1}{2\Delta} > f_0.$$

The frequency $1/2\Delta$ Hz is called the *Nyquist frequency* (or sometimes the *folding frequency*) and is the maximum frequency that can be detected from data sampled at time spacing Δ (seconds).

The phenomenon of aliasing is most important when analysing practical data. The sampling frequency $1/2\Delta$ must be high enough to cover the full frequency range of the continuous time series. Otherwise the spectrum from equally spaced samples will differ from the true spectrum because of aliasing. In some cases the only way to be certain that this condition is met may be to filter the time series to remove intentionally all frequency components higher than $1/2\Delta$ before beginning the analysis.

Calculation of spectral estimates

We are now ready to see how the DFT may be used to calculate an estimate for the spectral density of a random process. In this chapter we shall be concerned only with the basic "recipe" for the calculation. Chapter 11 deals with how to interpret the recipe and the assumptions and approximations inherent in it, and then, in Chapter 12, we shall come eventually to the FFT algorithm, by which the recipe can be implemented efficiently on a computer.

Since spectra are defined as Fourier transforms of their corresponding correlation functions, we begin with correlation functions. However, once the calculation recipe has been derived, it is not necessary to calculate correlation functions in order to determine spectra and spectral estimates can be obtained directly from the original time series. We want to find the (auto) spectral density $S_{xx}(\omega)$ defined by

$$S_{xx}(\omega) = \frac{1}{2\pi} \int_{-\infty}^{\infty} R_{xx}(\tau) e^{-i\omega\tau} \, d\tau \tag{5.1}$$

and the cross-spectral density $S_{xy}(\omega)$ defined by

$$S_{xy}(\omega) = \frac{1}{2\pi} \int_{-\infty}^{\infty} R_{xy}(\tau) e^{-i\omega\tau} \, d\tau. \tag{5.20}$$

If we consider only the cross-spectral density in the following, then the auto-spectral density can be easily obtained by just putting $y = x$ in the resulting formulae.

Suppose that we have sample functions $x(t)$ and $y(t)$ from two random processes, each in the form of a continuous record which lasts from $t = 0$ to $t = T$. In what follows it is necessary to assume that we are analysing deterministic (non-random) *periodic* records of which $x(t)$ and $y(t)$ constitute a single cycle. Of course this is not really the case, but we shall leave a discussion of the significance of the assumption until Chapter 11. From these records we can generate the discrete time series $\{x_r\}$ and $\{y_r\}$, $r = 0, 1, 2, 3, \ldots, (N-1)$, by using an analogue-to-digital converter with a sampling time $\Delta = T/N$. We can then calculate the DFT's $\{X_k\}$ and $\{Y_k\}$ of the discrete time series $\{x_r\}$ and $\{y_r\}$ given, according to (10.8), by

$$X_k = \frac{1}{N} \sum_{r=0}^{N-1} x_r e^{-i(2\pi kr/N)}$$

$$Y_k = \frac{1}{N} \sum_{r=0}^{N-1} y_r e^{-i(2\pi kr/N)} \qquad k = 0, 1, 2, \ldots, (N-1). \tag{10.19}$$

Now consider calculating the cross-correlation function between $x(t)$ and $y(t)$ from the sampled data $\{x_r\}$ and $\{y_r\}$. Clearly we can only estimate this for time lags τ which are integral multiples of the sampling interval Δ and, if we write R_r as an estimate for the correlation function when $\tau = r\Delta$, then we can define

$$R_r = \frac{1}{N} \sum_{s=0}^{N-1} x_s y_{s+r} \qquad r = 0, 1, 2, \ldots, (N-1) \tag{10.20}$$

where, since $y(t)$ is assumed to be periodic,

$$y_{s+r} = y_{s+r-N} \qquad \text{when} \qquad s + r \geqslant N.$$

Since the sequence $\{R_r\}$ is periodic, it cannot faithfully represent the (ensemble averaged) correlation function for the random processes of which $x(t)$ and $y(t)$ are sample functions. In any case, $\{R_r\}$ is calculated from a single pair of sample functions and there will be statistical differences between these and the other functions in the ensembles. To deal with this problem, we define a periodic or *circular correlation function* $R_c(\tau)$, which is peculiar to the pair of records being analysed, and is chosen so as to fit the R_r values calculated from (10.20) so that

$$R_r = R_c(\tau = r\Delta). \tag{10.21}$$

In Chapter 11 we shall pursue the differences between $R_c(\tau)$ and the true ensemble averaged correlation function, but for the moment we shall just assume that we can work with the circular function $R_c(\tau)$. From (10.16), we know that the Fourier transform of this periodic function may be written

$$S_c(\omega) = \sum_k S_k \delta\left(\omega - \frac{2\pi k}{T}\right) \quad \text{for} \quad -\frac{\pi}{\Delta} < \omega < \frac{\pi}{\Delta} \tag{10.22}$$

where S_k is the DFT of the discrete series $\{R_r\}$ defined by

$$S_k = \frac{1}{N} \sum_{r=0}^{N-1} R_r e^{-i(2\pi kr/N)} \tag{10.23}$$

and we understand from (10.18) that $S_{-k} = S_k^*$, the complex conjugate of S_k. Notice that we have introduced the subscript c to $S_c(\omega)$ in (10.22) in order to remember that this represents the spectrum of a circular correlation function rather than of the true linear correlation function. The frequency range of (10.22) is limited by the Nyquist frequency and it is assumed that the sampling interval is small enough that aliasing does not occur.

We shall now show that the terms S_k in (10.23) are related to X_k and Y_k by the simple equation

$$S_k = X_k^* Y_k. \tag{10.24}$$

To prove (10.24), we start with the definition of S_k as the DFT of the discrete series $\{R_r\}$, equation (10.23), and substitute for R_r from (10.20), to obtain

$$S_k = \frac{1}{N} \sum_{r=0}^{N-1} \left\{ \frac{1}{N} \sum_{s=0}^{N-1} x_s y_{s+r} \right\} e^{-i(2\pi kr/N)}$$

which, rearranging terms, may be written

$$S_k = \frac{1}{N^2} \sum_{r=0}^{N-1} \sum_{s=0}^{N-1} x_s e^{i(2\pi ks/N)} y_{s+r} e^{-i(2\pi k(s+r)/N)}.$$

Consider those terms with the variable integer r which are involved in the summation over r. These may be grouped together as shown below

$$S_k = \frac{1}{N} \sum_{s=0}^{N-1} x_s e^{i(2\pi ks/N)} \left\{ \frac{1}{N} \sum_{r=0}^{N-1} y_{s+r} e^{-i(2\pi k(s+r)/N)} \right\}. \tag{10.25}$$

If we define a new variable $t = (s + r)$ then the terms in brackets become

$$\frac{1}{N} \sum_{t=s}^{(N-1)+s} y_t e^{-i(2\pi kt/N)}$$

which, since we are assuming that the sequence $\{y_r\}$ is periodic with period N so that $y_{N+s} = y_s$, is just Y_k. The remaining terms outside the $\{\}$ bracket in (10.25) are those in the first equation (10.19) except that the sign of the exponent is reversed. They are therefore equal to the complex conjugate of X_k, denoted by X_k^*. The final form

$$S_k = X_k^* Y_k$$

then follows from (10.25) to prove the result already stated in (10.24).

We can now summarize progress so far. The calculation procedure involves:

(i) Generating discrete time series $\{x_r\}$ and $\{y_r\}$, $r = 0, 1, 2, \ldots, (N-1)$, by sampling the continuous records at time interval $\Delta = T/N$.
(ii) Calculating the discrete Fourier transforms (DFT's) of $\{x_r\}$ and $\{y_r\}$ according to the formulae

$$X_k = \frac{1}{N} \sum_{r=0}^{N-1} x_r e^{-i(2\pi kr/N)}$$

$$Y_k = \frac{1}{N} \sum_{r=0}^{N-1} y_r e^{-i(2\pi kr/N)}$$

(10.19)

where $k = 0, 1, 2, \ldots, (N-1)$.
(iii) Calculating the discrete series $\{S_{xx_k}\}$, $\{S_{xy_k}\}$, $\{S_{yx_k}\}$ and $\{S_{yy_k}\}$ from $\{X_k\}$ and $\{Y_k\}$ by forming the products

$$S_{xx_k} = X_k^* X_k \qquad\qquad S_{xy_k} = X_k^* Y_k$$

$$S_{yx_k} = Y_k^* X_k \qquad\qquad S_{yy_k} = Y_k^* Y_k.$$

(10.26)

The appropriate spectrum is then given by

$$S_c(\omega) = \sum_k S_k \delta\left(\omega - \frac{2\pi k}{T}\right) \qquad \text{for} \qquad -\frac{\pi}{\Delta} < \omega < \frac{\pi}{\Delta}. \qquad (10.22)$$

Since this is the spectrum corresponding to a circular correlation function represented by the series $\{R_r\}$, it is *not* the spectrum we are looking for, but, as we shall see shortly, it is closely related to the true spectrum.

(iv) If, in addition to calculating spectra, the corresponding (circular) correlation functions are required, these can be determined by calculating the inverse discrete Fourier transforms (IDFT's) of the spectra to generate the discrete series $\{R_{xx_r}\}$, $\{R_{xy_r}\}$, $\{R_{yx_r}\}$ and $\{R_{yy_r}\}$ according to (10.9)

$$R_r = \sum_{k=0}^{N-1} S_k e^{i(2\pi kr/N)}$$

(10.27)

where $r = 0, 1, 2, \ldots, (N - 1)$, and

$$R_r = R_c(\tau = r\Delta) \tag{10.21}$$

where Δ is the sampling interval.

In the next chapter we shall relate the coefficients S_k to the true spectrum $S_{xy}(\omega)$, investigate the assumptions and errors present in the calculation, and discuss techniques of smoothing the spectral estimates to improve their statistical reliability.

Chapter 11

Digital spectral analysis II: Windows and smoothing

The procedure described in Chapter 10 for calculating spectral estimates from discrete time series is based on the definition (10.20) of the correlation function R_r given by

$$R_r = \frac{1}{N} \sum_{s=0}^{N-1} x_s y_{s+r} \qquad r = 0, 1, 2, \ldots, (N-1) \qquad (10.20)$$

where it is implicitly assumed (in the definition of the DFT) that the sequences $\{x_s\}$ and $\{y_s\}$ are periodic so that

$$x_{s+N} = x_s$$

and (11.1)

$$y_{s+N} = y_s.$$

Because of this implied periodicity, R_r differs from the true correlation function it is meant to represent. The reason for the difference is illustrated in Fig. 11.1

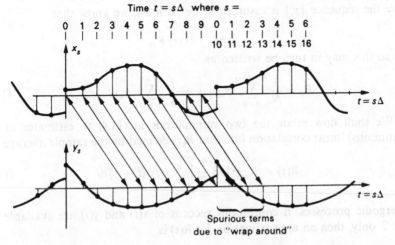

Fig. 11.1 Effect of "wrap around" on the calculation of correlation functions

for the case $N = 10$, $r = 4$. In calculating the summation in equation (10.20), the first six terms $x_s y_{s+r}$ are correct data values for calculating $R(\tau = 4\Delta)$ but the remaining four terms only arise because the $\{y_s\}$ sequence repeats itself. If we write out the summation required by (10.20) we obtain, from Fig. 11.1

$$R_4 = \frac{1}{10}\left\{x_0 y_4 + x_1 y_5 + x_2 y_6 + x_3 y_7 + x_4 y_8 + x_5 y_9\right\} +$$

$$+ \frac{1}{10}\left\{x_6 y_0 + x_7 y_1 + x_8 y_2 + x_9 y_3\right\}. \tag{11.2}$$

The first term in $\{\ \}$ brackets is an estimate for $R(\tau = 4\Delta)$ but the second term in $\{\ \}$ brackets has nothing to do with $R(\tau = 4\Delta)$ at all and is instead an estimate for $R(\tau = -6\Delta)$. Both terms are *biased* estimates for $R(\tau)$ since the denominator would obviously have to be 6 (not 10) in the first case and 4 (not 10) in the second case if we were to find correct mean values in each case. It is therefore apparent that there are some subtle approximations involved in the calculations described in the last chapter, and we shall now investigate the nature of these in some detail.

Relationship between linear and circular correlation

We can begin by considering the general form of equation (11.2), by writing (10.20) in two parts as

$$R_r = \frac{1}{N}\sum_{s=0}^{N-1-r} x_s y_{s+r} + \frac{1}{N}\sum_{s=N-r}^{N-1} x_s y_{s+r}. \tag{11.3}$$

Since the sequence $\{y_r\}$ is assumed to be periodic, we know that

$$y_{s+r} = y_{s+r-N}$$

and so this may in turn be written as

$$R_r = \frac{1}{N}\sum_{s=0}^{N-1-r} x_s y_{s+r} + \frac{1}{N}\sum_{s=N-r}^{N-1} x_s y_{s-(N-r)}. \tag{11.4}$$

We shall now relate the two summations in (11.4) to estimates of the (continuous) linear correlation function $R(\tau)$ defined by the sample average

$$R(\tau) = R_{xy}(\tau) = \frac{1}{T}\int_{0}^{T} x(t)y(t + \tau)\,dt \tag{11.5}$$
$$\lim T \to \infty$$

for ergodic processes. If continuous records of $x(t)$ and $y(t)$ are available for time T only, then an approximation for $R(\tau)$ is

$$\hat{R}(\tau) = \frac{1}{T-\tau}\int_{0}^{T-\tau} x(t)y(t + \tau)\,dt \tag{11.6}$$

for $0 \leqslant \tau < T$ since the available integration length depends on the lag τ and clearly we can make no calculations at all for τ greater than the record length T. For negative lag times, the expression corresponding to (11.6) is

$$\hat{R}(-\tau) = \frac{1}{T - \tau} \int_{\tau}^{T} x(t)y(t - \tau)\,dt \qquad (11.7)$$

for $0 \leqslant \tau < T$.

Now consider the case when the continuous signals $x(t)$ and $y(t)$ are not available and we only have finite sequences $\{x_s\}$ and $\{y_s\}$, $s = 0, 1, 2, \ldots, (N - 1)$, obtained by sampling the continuous records at regular sampling intervals Δ, where $\Delta N = T$. The discrete forms of (11.6) and (11.7) are then, where $\tau = r\Delta$,

$$\hat{R}_r = \frac{1}{N - r} \sum_{s=0}^{N-1-r} x_s y_{s+r} \qquad \text{for} \qquad r = 0, 1, 2, \ldots, (N - 1) \quad (11.8)$$

and

$$\hat{R}_{-r} = \frac{1}{N - r} \sum_{s=r}^{N-1} x_s y_{s-r} \qquad \text{for} \qquad r = 0, 1, 2, \ldots, (N - 1). \quad (11.9)$$

These are *consistent* estimates of the true correlation functions, since by taking the ensemble average

$$E[\hat{R}_r] = \frac{1}{N - r} \sum_{s=0}^{N-1-r} E[x_s y_{s+r}]$$

$$= \frac{1}{N - r} \sum_{s=0}^{N-1-r} R(\tau = r\Delta)$$

$$= R(\tau = r\Delta)$$

and similarly

$$E[\hat{R}_{-r}] = R(\tau = -r\Delta)$$

so that the ensemble average of each estimate converges to the true value of $R(\tau)$ (for the values of τ for which $\hat{R}(\tau)$ is defined).

Next consider what happens when we substitute from (11.8) and (11.9) into the definition of the periodic or circular correlation function R_r defined by (11.4). We obtain the result

$$R_r = \left(\frac{N - r}{N}\right)\hat{R}_r + \left(\frac{N - (N - r)}{N}\right)\hat{R}_{-(N-r)} \qquad r = 0, 1, 2, \ldots, (N - 1) \quad (11.10)$$

so that R_r can be expressed as the sum of two terms which turn out to be biased (or weighted) estimates of the true or linear correlation function $R(\tau = r\Delta)$. If we plot the two parts of R_r against an abscissa of lag time $r\Delta$, then, for the case $N = 10$, these might appear as shown in Figs. 11.2(a) and (b). Since the discrete series $\{x_s\}$ and $\{y_s\}$ are periodic in s according to (11.1), if we extend the abscissa in Fig. 11.2 outside the range $s = 0$ to $s = (N - 1) = 9$, the two functions

$$\left(\frac{N-r}{N}\right)\hat{R}_r \quad \text{and} \quad \left(\frac{N-(N-r)}{N}\right)\hat{R}_{-(N-r)}$$

will repeat themselves periodically as shown. If we now join the function

$$\left(\frac{N-(N-r)}{N}\right)\hat{R}_{-(N-r)}$$

from the period with r negative to the function

$$\left(\frac{N-r}{N}\right)\hat{R}_r$$

from the next period with r positive, we can obtain the function represented by the solid heavy line in Fig. 11.2(c).

Time delay $r\Delta$ where $r =$

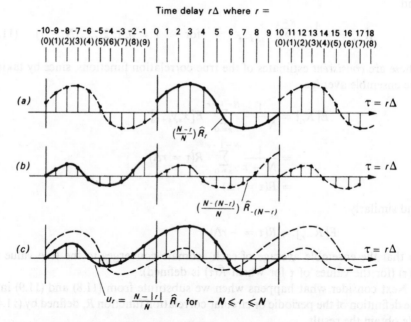

Fig. 11.2 Interpretation of the circular correlation function R_r defined by (10.20) as the superposition of biased estimates $\dfrac{N-|r|}{N}\hat{R}_r$ where $\hat{R}_r$ is defined by (11.8).

By substituting specific values of r, a little thought will show that the equation for points on the solid line, which we will call u_r, can be written as

$$u_r = \frac{N-|r|}{N}\hat{R}_r \quad \text{for} \quad -N \leqslant r \leqslant N. \tag{11.11}$$

In the same way, the two parts of R_r from every pair of successive periods can be joined together, as shown by the dotted curves in Fig. 11.2(c), and we therefore conclude that the circular correlation function, represented by the periodic series $\{R_r\}$, may be interpreted as the superposition of infinitely many biased estimates

$$u_r = \left(\frac{N - |r|}{N}\right)\hat{R}_r, \qquad 0 \leqslant |r| \leqslant N$$

of the true linear correlation function $R(\tau = r\Delta)$, each spaced equal distances $T = N\Delta$ apart along the lag time axis, Fig. 11.2(c).†

We shall now assume that the discrete time series $\{u_r\}$ has been derived from a continuous time series $u(\tau)$, where

$$u_r = u(\tau = r\Delta) \qquad 0 \leqslant |\tau| \leqslant T.$$

Our next step is then to investigate how the DFT of the periodic series $\{R_r\}$ is related to the (continuous) Fourier transform of $u(\tau)$.

Fourier transform of a train of aperiodic functions

Consider the general problem of Fourier analysing a train of overlapping identical functions of time $u(\tau)$, each of which satisfies (4.7), and each of which is

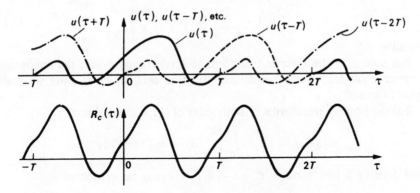

Fig. 11.3 Generation of a periodic function $R_c(\tau)$ by the superposition of a train of aperiodic functions $u(\tau)$, $u(\tau - T)$, etc., at spacing T.

spaced time T apart, Fig. 11.3. Let $R_c(\tau)$ represent the summation of these functions at any particular instant of time τ (the subscript c indicating that this is the circular correlation function and should be distinguished from the

†An alternative derivation of this result, which may be helpful, is indicated in the first part of problem 11.5.

true (linear) correlation function $R(\tau)$). Then we can write

$$R_c(\tau) = u(\tau) + u(\tau - T) + u(\tau + T) + u(\tau - 2T) + u(\tau + 2T) + \cdots$$

$$= \sum_{m=-\infty}^{\infty} u(\tau - mT).$$ (11.12)

Now if $R_c(\tau)$ is sampled N times in the interval 0 to T to generate the sequence $\{R_r\}$, the DFT of $\{R_r\}$ is

$$S_k = \frac{1}{N} \sum_{k=0}^{N-1} R_r e^{-i(2\pi rk/N)} \qquad r = 0, 1, 2, \ldots, (N-1)$$ (10.6)

and we know from Chapter 10 that the Fourier transform of $R_c(\tau)$ can be expressed as

$$S_c(\omega) = \sum_k S_k \delta\left(\omega - \frac{2\pi k}{T}\right) \qquad -\frac{\pi}{\Delta} < \omega < \frac{\pi}{\Delta}.$$ (10.22)

Hence, provided that the sampling interval Δ is small enough that the Nyquist frequency π/Δ rad/s exceeds the maximum frequency of significant Fourier components, then the continuous Fourier transform of $R_c(\tau)$ can be accurately represented by "combs" of delta functions, the area of each tooth of which is given by the appropriate component of the coefficients of the DFT of $\{R_r\}$. More generally, if we define $S_k = a_k - ib_k$ and $S_{-k} = S_k^*$, where a_k and b_k are the coefficients of an infinite Fourier series expansion of (11.12), then

$$S_c(\omega) = \sum_{k=-\infty}^{\infty} S_k \delta\left(\omega - \frac{2\pi k}{T}\right)$$ (11.13)

for all ω.

But how does the Fourier transform of $R_c(\tau)$ compare with the continuous Fourier transform of the constituent function $u(\tau)$, which is what we really want to know?

Taking Fourier transforms of both sides of (11.12), we obtain

$$S_c(\omega) = \sum_{m=-\infty}^{\infty} \frac{1}{2\pi} \int_{-\infty}^{\infty} u(\tau - mT) e^{-i\omega\tau} \, d\tau$$

and defining a new variable $\tau' = \tau - mT$, this may be written as

$$S_c(\omega) = \sum_{m=-\infty}^{\infty} e^{-i\omega mT} \frac{1}{2\pi} \int_{-\infty}^{\infty} u(\tau') e^{-i\omega\tau'} \, d\tau'$$

$$= \sum_{m=-\infty}^{\infty} e^{-i\omega mT} U(\omega)$$ (11.14)

where $U(\omega)$ is the Fourier transform of $u(\tau)$, which is the one we want. At first sight (11.14) looks rather awkward, but fortunately it can be expressed in the following alternative form

$$S_c(\omega) = \frac{2\pi}{T} U(\omega) \sum_{k=-\infty}^{\infty} \delta\left(\omega - \frac{2\pi k}{T}\right).$$ (11.15)

In order to prove this result, consider

$$\sum_{k=-\infty}^{\infty} \delta\left(\omega - \frac{2\pi k}{T}\right)$$

as a periodic function of ω which can be represented by an infinite trigonometric series of the form (10.1) with ω replacing t. If we calculate the appropriate Fourier coefficients a_m and b_m according to (10.2) and substitute into (10.1), then it is not difficult to show that†

$$\sum_{k=-\infty}^{\infty} \delta\left(\omega - \frac{2\pi k}{T}\right) = \frac{T}{2\pi} \sum_{m=-\infty}^{\infty} e^{-i\omega mT} \tag{11.16}$$

which is the result needed to go from (11.14) to (11.15).

By comparing (11.15) with (11.13), we have now determined how the continuous spectrum $U(\omega)$ we require is related to the coefficients S_k, since

$$\frac{2\pi}{T}U(\omega) \sum_{k=-\infty}^{\infty} \delta\left(\omega - \frac{2\pi k}{T}\right) = \sum_{k=-\infty}^{\infty} S_k \delta\left(\omega - \frac{2\pi k}{T}\right) \tag{11.17}$$

and so, by integrating over a small frequency band near $\omega_k = 2\pi k/T$, we obtain

$$\frac{2\pi}{T}U(\omega_k) = S_k. \tag{11.18}$$

Equation (11.18) is an important result, expressing the continuous Fourier transform $U(\omega)$ of the *single aperiodic function* $u(\tau)$ in terms of the coefficients S_k. Strictly these are the coefficients of an infinite Fourier series expansion of $R_c(\tau)$, but, as we have seen, they are approximated by the terms of the DFT of the periodic sequence $\{R_r\}$.

Basic lag and spectral windows

We are now ready to combine the results of the last two sections to interpret the meaning of the spectral estimates S_k calculated by the procedure described in Chapter 10.

First, we have found that, when we calculate S_k according to the recipe (10.24)

$$S_{xy_k} = X_k^* Y_k, \tag{10.24}$$

we are actually calculating the DFT of a periodic sequence

$$R_r = \left(\frac{N-r}{N}\right)\hat{R}_r + \left(\frac{N-(N-r)}{N}\right)\hat{R}_{-(N-r)} \qquad r = 0, 1, 2, \ldots, (N-1) \tag{11.10}$$

where $\hat{R}_r$ is an estimate of the (linear) correlation function defined by (11.8).

†A rigorous proof is given in Lighthill [46] sect. 5.4.

Second, the $\{R_r\}$ sequence may be interpreted as an infinite train of overlapping sequences

$$u_r = \frac{N - |r|}{N} \hat{R}_r \qquad 0 \leqslant |r| \leqslant N \qquad (11.11)$$

at regular spacing N. Third, the coefficients S_k obtained by taking the DFT of $\{R_r\}$ give discrete values of the Fourier transform $U(\omega)$ of the continuous aperiodic function $u(\tau)$ (from which the $\{u_r\}$ sequence is derived) according to the formula

$$\frac{2\pi}{T} U(\omega_k) = S_k \qquad (11.18)$$

where $\omega_k = 2\pi k/T$.

Now imagine that we have an infinite ensemble of experimental records to analyse. We can then carry out the analysis many times for different records, and average our results. Suppose that we begin by averaging both sides of (11.11). Since the equation is true for each record separately, the average value of the l.h.s. must remain equal to the average value of the r.h.s., and therefore

$$E[u_r] = \frac{N - |r|}{N} E[\hat{R}_r] \qquad 0 \leqslant |r| \leqslant N.$$

But we have already shown that $\hat{R}_r$ is a consistent estimate for $R(\tau = r\Delta)$, so that

$$E[\hat{R}_r] = R(\tau = r\Delta)$$

and therefore

$$E[u_r] = \frac{N - |r|}{N} R(\tau = r\Delta) \qquad 0 \leqslant |r| \leqslant N.$$

Also we have assumed that the discrete series $\{u_r\}$ has been derived from a continuous function of time $u(\tau)$, so that

$$u_r = u(\tau = r\Delta)$$

and hence we can put

$$E[u(\tau = r\Delta)] = \frac{N - |r|}{N} R(\tau = r\Delta) \qquad 0 \leqslant |r| \leqslant N.$$

This result is true for all values of r within the range 0 to $\pm N$, and holds however small (or large) the sampling interval Δ. If we put $r = \tau/\Delta$ and $N = T/\Delta$ we can therefore conclude that

$$E[u(\tau)] = \frac{T - |\tau|}{T} R(\tau) \qquad 0 \leqslant |\tau| \leqslant T. \qquad (11.19)$$

Next consider the result of averaging equation (11.18). This gives, after reversing the order of the two sides,

$$E[S_k] = \frac{2\pi}{T} E[U(\omega_k)] \tag{11.20}$$

where $E[S_k]$ represents the average value of the coefficient S_k (which would be obtained if we carried out the analysis many times and then averaged the results). $E[U(\omega_k)]$ represents the average value of $U(\omega_k)$ which we know is the Fourier transform of $u(\tau)$ so that

$$U(\omega_k) = \frac{1}{2\pi} \int_{-\infty}^{\infty} u(\tau) e^{-i\omega_k \tau} d\tau$$

and, after taking the ensemble average,

$$E[U(\omega_k)] = \frac{1}{2\pi} \int_{-\infty}^{\infty} E[u(\tau)] e^{-i\omega_k \tau} d\tau.$$

Combining this result with (11.20), we see that

$$E[S_k] = \frac{2\pi}{T} \left\{ \frac{1}{2\pi} \int_{-\infty}^{\infty} E[u(\tau)] e^{-i\omega_k \tau} d\tau \right\}$$

which, using (11.19), gives

$$E[S_k] = \frac{2\pi}{T} \left\{ \frac{1}{2\pi} \int_{-T}^{T} \frac{T - |\tau|}{T} R(\tau) e^{-i\omega_k \tau} d\tau \right\}.$$

This is the final result we need. By introducing the symbol $\tilde{S}(\omega)$ to denote the term in brackets, we can express it more conveniently in the following standard form:

$$E[S_k] = \frac{2\pi}{T} \tilde{S}(\omega_k) \tag{11.21}$$

where

$$\tilde{S}(\omega_k) = \frac{1}{2\pi} \int_{-T}^{T} \frac{T - |\tau|}{T} R(\tau) e^{-i\omega_k \tau} d\tau. \tag{11.22}$$

From these two important equations we conclude that the calculation procedure described in Chapter 10 yields a series of coefficients S_k which are *estimates* of $(2\pi/T)\tilde{S}(\omega_k)$ where $\tilde{S}(\omega_k)$ is the weighted spectral density defined by (11.22). The weighting arises on account of the factor $(T - |\tau|)/T$ in (11.22) which is described as the basic *lag window* through which $R(\tau)$ is "viewed" by the calculation procedure.

We shall now consider how the weighted spectrum $\tilde{S}(\omega)$ compares with the true spectrum $S(\omega)$. To keep this part of the analysis general, we consider any symmetrical lag window $w(\tau)$, restricted only in that we shall specify

$$w(\tau) = w(-\tau)$$

$$w(\tau = 0) = 1 \tag{11.23}$$

and

$$\int_{-\infty}^{\infty} |w(\tau)|\, d\tau < \infty.$$

For this lag window the corresponding weighted spectrum $\tilde{S}(\omega)$ is defined by

$$\tilde{S}(\omega) = \frac{1}{2\pi} \int_{-\infty}^{\infty} w(\tau)R(\tau)e^{-i\omega\tau}\, d\tau \tag{11.24}$$

compared with the true spectrum which is defined by

$$S(\omega) = \frac{1}{2\pi} \int_{-\infty}^{\infty} R(\tau)e^{-i\omega\tau}\, d\tau. \tag{5.1}$$

If we introduce a new quantity $W(\omega)$ which is the Fourier transform of $w(\tau)$ defined by

$$W(\omega) = \frac{1}{2\pi} \int_{-\infty}^{\infty} w(\tau)e^{-i\omega\tau}\, d\tau, \tag{11.25}$$

then we can express $w(\tau)$ as the inverse transform of $W(\omega)$ according to

$$w(\tau) = \int_{-\infty}^{\infty} W(\omega)e^{i\omega\tau}\, d\omega. \tag{11.26}$$

We shall now use the latter to substitute for $w(\tau)$ in (11.24). To avoid confusing the two ω's, we first put $\omega = \omega_1$ in (11.26), and then obtain

$$\tilde{S}(\omega) = \frac{1}{2\pi} \int_{-\infty}^{\infty} \left\{ \int_{-\infty}^{\infty} W(\omega_1)e^{i\omega_1\tau}\, d\omega_1 \right\} R(\tau)e^{-i\omega\tau}\, d\tau.$$

Since the separate integrals converge they can be rewritten as

$$\tilde{S}(\omega) = \int_{-\infty}^{\infty} d\omega_1\, W(\omega_1) \left\{ \frac{1}{2\pi} \int_{-\infty}^{\infty} d\tau\, R(\tau)e^{-i(\omega-\omega_1)\tau} \right\}$$

which, using the definition of $S(\omega)$ given by (5.1), can in turn be written as

$$\tilde{S}(\omega) = \int_{-\infty}^{\infty} d\omega_1\, W(\omega_1)S(\omega - \omega_1). \tag{11.27}$$

Finally, if we change the variable of integration to $\Omega = (\omega - \omega_1)$, we obtain

$$\tilde{S}(\omega) = \int_{\infty}^{-\infty} (-d\Omega)W(\omega - \Omega)S(\Omega)$$

or

$$\tilde{S}(\omega) = \int_{-\infty}^{\infty} W(\omega - \Omega)S(\Omega)\, d\Omega \tag{11.28}$$

so that the estimated spectrum $\tilde{S}(\omega)$ turns out to be a smoothed version of the true spectrum $S(\omega)$. Smoothing has the same form as that already discussed in Chapter 9 (see equation (9.26)). Since $w(\tau)$ has been chosen to be an even function of τ, then $W(\omega)$ will be an even function of ω (see Chapter 5) and so an alternative version of (11.28) is

$$\tilde{S}(\omega) = \int_{-\infty}^{\infty} W(\Omega - \omega)S(\Omega)\,d\Omega \qquad (11.29)$$

which is identical with (9.26). This is a most important result. It is true for any even function $w(\tau)$ whose Fourier transform is $W(\omega)$. The only extra restriction is that, if the spectral weighting function $W(\omega)$ is to be normalized so that

$$\int_{-\infty}^{\infty} W(\omega)\,d\omega = 1 \qquad (11.30)$$

in agreement with (9.27), then from (11.26) we must have $w(\tau = 0) = 1$ as already specified in (11.23).

For the calculation procedure described in Chapter 10 we have seen that a certain lag window is automatically introduced and is inherent in the procedure adopted. From (11.21), this is given by

$$w(\tau) = \begin{cases} \left\{1 - \dfrac{|\tau|}{T}\right\} & \text{for} \quad 0 \leqslant |\tau| \leqslant T \\[2mm] 0 \text{ elsewhere} \end{cases} \qquad (11.31)$$

where T is the length of the records being analysed. It is called a triangular lag window for obvious reasons, Fig. 11.4(a). The basic *spectral window* $W(\omega)$ corresponding to this $w(\tau)$ is given by

$$\begin{aligned} W(\omega) &= \frac{1}{2\pi}\int_{-T}^{T}\left\{1 - \frac{|\tau|}{T}\right\}e^{-i\omega\tau}\,d\tau \\[2mm] &= \frac{1}{\pi}\int_{0}^{T}\left\{1 - \frac{\tau}{T}\right\}\cos\omega\tau\,d\tau \\[2mm] &= \frac{T}{2\pi}\left|\frac{\sin(\omega T/2)}{\omega T/2}\right|^{2} \end{aligned} \qquad (11.32)$$

and the shape of this function is shown, approximately to scale, in Fig. 11.4(b).

The position is therefore now as follows. We have set up a procedure in Chapter 10 for using the DFT to calculate a series of spectral coefficients S_k (10.19) by the so-called "direct method". Estimates of the smoothed spectrum $\tilde{S}(\omega)$ are given at frequencies $\omega_k = 2\pi k/T$ by, from (11.21),

$$\tilde{S}(\omega_k) \simeq \frac{T}{2\pi}S_k \qquad (11.33)$$

and the smoothed spectrum $\tilde{S}(\omega)$ is related to the true spectrum $S(\omega)$ by

$$\tilde{S}(\omega) = \int_{-\infty}^{\infty} W(\Omega - \omega)S(\Omega)\,d\Omega \qquad (11.29)$$

where the window function $W(\omega)$ is given by (11.32) and shown in Fig. 11.4(*b*).

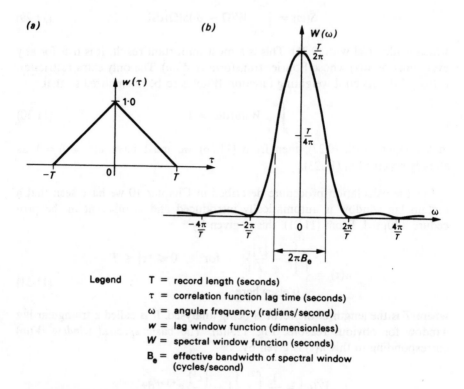

Fig. 11.4 Basic lag and spectral windows for spectral analysis by the direct method (see chapter 10)

Ideally $W(\omega)$ would be a rectangular window of the form shown in Fig. 9.5(*b*) because we would then know that $\tilde{S}(\omega)$ was a simple average value of the true spectrum $S(\omega)$ in the immediate vicinity of ω. In fact the triangular lag window $w(\tau)$ inherent in our calculation procedure yields a spectral window which weights adjacent values differently. However spectral components at frequencies further than $2\pi/T$ rad/s from the centre frequency contribute very little to the smoothed value $\tilde{S}(\omega)$, (see Fig. 11.4(*b*)) and the width of the window can be made narrower by increasing the record length T, so that the non-uniform weighting implied by (11.32) is not usually a problem. A much more serious problem is the accuracy of the spectral estimates we have obtained, and we turn to this question next.

Smoothing spectral estimates

In Chapter 9 the accuracy of a spectral measurement was shown to depend on the effective bandwidth of the measurement B_e (Hz) and the record length T (s). If σ is the standard deviation of a measurement of (auto) spectral density whose mean value is m, then we know from Chapter 9 that

$$\frac{\sigma}{m} \simeq \frac{1}{\sqrt{(B_e T)}} \tag{9.29}$$

provided that the spectrum is changing slowly over frequency intervals of order $1/T$. In the case of a measurement of the cross-spectral density between two processes $x(t)$ and $y(t)$, the same formula can still be used and now indicates an approximate upper bound on σ.† However in this case m is no longer the mean value of the measurement (which may be zero) but instead is given by

$$m = \sqrt{(m_x m_y)}$$

where m_x and m_y denote the mean values of the auto-spectral densities of the $x(t)$ and $y(t)$ processes. Also, since cross-spectral density is in general a complex quantity, the standard deviation σ now applies to measured values of the real and imaginary parts taken separately. Notice that, although the cross-spectral density between two uncorrelated processes is zero, *estimates* of this quantity will be scattered about zero with variance depending on the auto-spectral densities of $x(t)$ and $y(t)$.

In order to apply (9.29) we need the bandwidth of the spectral window $W(\omega)$ since we are forced to calculate spot values of $\tilde{S}(\omega)$ which are the result of smoothing the true spectrum $S(\omega)$ by $W(\omega)$. From (9.28), the effective bandwidth of a spectral window is defined by

$$B_e = \frac{1}{\int_{-\infty}^{\infty} W^2(\omega)\, d\omega} \tag{11.34}$$

and in the literature values of B_e have been exactly calculated for a variety of spectral windows including (11.32). It turns out ‡ that

$$B_e \simeq \frac{1}{T} \text{ Hz} \tag{11.35}$$

is a satisfactory approximation for most practical calculations (see Fig. 11.4(b)). In this case, from (11.35)

$$B_e T \simeq 1 \tag{11.36}$$

so that, from (9.29),

$$\frac{\sigma}{m} \simeq 1 \tag{11.37}$$

† Bendat *et al.* [3] p. 188. See also Jenkins *et al.* [39] p. 378.
‡ See, for instance, Blackman *et al.* [5]

and the standard deviation of the measurement is equal to the mean value, which means that accuracy is very poor. If the distribution of measured values of $\tilde{S}(\omega)$ can be represented approximately by a chi-square distribution, as suggested in Chapter 9, then, from (9.37), the number of statistical degrees-of-freedom κ in χ_κ^2 will be only 2. Furthermore increasing the record length T, so that more data is analysed, does not make any difference. This is a surprising result, because it is natural to expect that if we analyse more data we shall achieve greater accuracy. However the reason is simple. As T becomes longer, the spectral bandwidth decreases, and the product (bandwidth) × (record length) remains the same. Although we analyse more data, the width of the spectral window decreases proportionately, so we achieve better resolution but only the same accuracy.

In order to improve the accuracy of our results the only thing to do is to average adjacent estimates of the smoothed spectrum $\tilde{S}(\omega_k)$. According to (11.33), we use $(T/2\pi)S_k$ as our first estimate for $\tilde{S}(\omega_k)$ at frequency $\omega_k = 2\pi k/T$. A convenient and simple recipe for improving this estimate is to calculate the arithmetic average of several adjacent estimates. For instance if $\hat{S}(\omega_k)$ is the result of averaging three adjacent values of $\tilde{S}(\omega_k)$ then

$$\hat{S}(\omega_k) = \tfrac{1}{3}\{\tilde{S}(\omega_{k-1}) + \tilde{S}(\omega_k) + \tilde{S}(\omega_{k+1})\} \tag{11.38}$$

and, in the general case, if $2n + 1$ adjacent values are averaged, we obtain

$$\hat{S}(\omega_k) = \frac{1}{2n+1} \sum_{m=-n}^{n} \tilde{S}(\omega_{k+m}). \tag{11.39}$$

The spectral window through which $\hat{S}(\omega_k)$ "views" the true spectrum $S(\omega)$ can be obtained quite simply by substituting from (11.29) into (11.39) to obtain

$$\hat{S}(\omega_k) = \frac{1}{2n+1} \sum_{m=-n}^{n} \int_{-\infty}^{\infty} W(\Omega - \omega_{k+m})S(\Omega)\,d\Omega$$

$$= \int_{-\infty}^{\infty} \left\{ \frac{1}{2n+1} \sum_{m=-n}^{n} W(\Omega - \omega_{k+m}) \right\} S(\Omega)\,d\Omega. \tag{11.40}$$

The weighting function in {} brackets is just the linear superposition of $(2n + 1)$ basic spectral windows $W(\Omega)$ spaced $\Delta\omega_k = 2\pi/T$ apart along the frequency axis, and divided by $(2n + 1)$ to keep the total window area normalized to unity. For the case $(2n + 1) = 3$, equation (11.38), the shape of the resulting spectral window is sketched in Fig. 11.5.

The equivalent bandwidth is now $(2n + 1)\,2\pi/T$ instead of $2\pi/T$ so

$$\frac{\sigma}{m} \simeq \sqrt{\left(\frac{1}{2n+1}\right)}$$

corresponding to $(4n + 2)$ statistical degrees-of-freedom for a chi-square distribution. Once again statistical accuracy can only be obtained at the expense of frequency resolution.

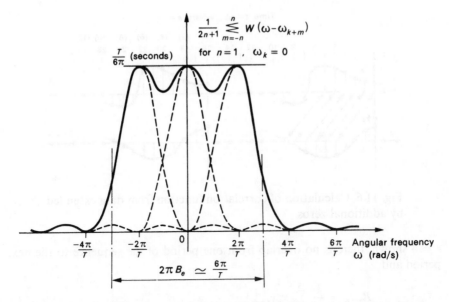

Fig. 11.5 Equivalent spectral window after averaging three adjacent spectral estimates according to equation (11.38)

Extending record length by adding zeros

The raison d'être behind the calculation procedure set out in Chapter 10 and analysed in this chapter, is the fast Fourier transform (FFT). As already mentioned, this is a highly efficient computer algorithm for calculating discrete Fourier transforms, and we shall discuss how it works in the next chapter. One feature of most FFT algorithms is that the length N of the sequence of data points to be analysed (for instance the length N of the sequence $\{x_r\}$, $r = 0, 1, 2, ..., (N - 1)$) must be a power of 2, i.e. $N = 2^n$ where n is a positive integer. In many practical applications, where N is large, this will not be the case because the length of available records and the choice of a suitable sampling interval Δ will not work out to give $N = 2^n$. Rather than abandoning part of the data to reduce N to the nearest lower power of 2, it is better to add zeros to the data to increase N to at least the next higher power of 2. As well as permitting computer library programs for the FFT to be used, this also brings certain other advantages which we shall now investigate briefly.

Suppose that each sequence to be analysed has N data points and L additional zeros, so the length of the sequence is now $(N + L)$. Consider calculating the cross-correlation function

$$R_r = \frac{1}{N + L} \sum_{s=0}^{N+L-1} x_s y_{s+r} \qquad (10.20)$$

for the example illustrated in Fig. 11.6 where $N = 10$ and $L = 6$. For the case

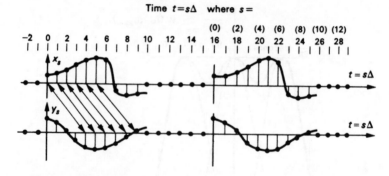

Fig. 11.6 Calculation of correlation function from data extended by additional zeros

$r = 4$ shown, there is no overlap from one period of the sequence to the next period and

$$R_{r=4} = \frac{1}{16} \sum_{s=0}^{5} x_s y_{s+4}$$

$$= \left(\frac{N-r}{N+L} \right) \hat{R}_r \qquad (r = 4, N = 10, L + N = 16) \qquad (11.41)$$

where, from (11.8),

$$\hat{R}_r = \frac{1}{N-r} \sum_{s=0}^{N-1-r} x_s y_{s+r}. \qquad (11.42)$$

There will be no overlap for other values of r up to $r = 6$, but for $r = 7, 8, 9$ there will be overlap and

$$R_r = \left(\frac{N-r}{N+L} \right) \hat{R}_r + \left(\frac{N-(N+L-r)}{N+L} \right) \hat{R}_{-(N+L-r)} \qquad (r = 7, 8, 9) \qquad (11.43)$$

corresponding to (11.10). For $r = 10$ to 15 there will again be no overlap, so that

$$R_r = \left(\frac{N-(N+L-r)}{N+L} \right) \hat{R}_{-(N+L-r)} \qquad (r = 10, 11, 12, 13, 14, 15). \qquad (11.44)$$

The circular correlation function R_r may therefore be interpreted, following Fig. 11.2, as a train of overlapping correlation functions

$$\left(\frac{N-|r|}{N+L} \right) \hat{R}_r, \qquad -N \leqslant r \leqslant N,$$

each spaced $(N + L)$ points apart, Fig. 11.7. If we now follow the same line of reasoning as already described for the case of no additional zeros, we conclude that the coefficients S_k calculated by determining the DFT of R_r give

Time delay $r\Delta$ where $r =$

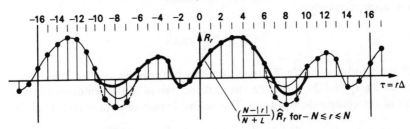

$$\left(\frac{N-|r|}{N+L}\right)\hat{R}_r, \text{ for } -N \leqslant r \leqslant N$$

Fig. 11.7 Graph of circular correlation function R_r plotted against lag time $r\Delta$ for the case $N = 10$ data points, $L = 6$ added zeros (see Fig. 11.6). R_r can be interpreted as the superposition of biased estimates $\frac{N-|r|}{N+L}\hat{R}_r$ (equation (11.8)) each running from $-N \leqslant r \leqslant N$ and spaced $(L + N)$ points apart

approximate values of the continuous Fourier transform $\tilde{S}(\omega)$ according to the formula

$$\tilde{S}(\omega_k) \simeq \frac{T_L}{2\pi} S_k \tag{11.45}$$

where the record length $T_L = (N + L)\Delta$. However in this case $\tilde{S}(\omega)$ is related to the true spectrum $S(\omega)$ by, corresponding to (11.29),

$$\tilde{S}(\omega) = \frac{N}{N + L} \int_{-\infty}^{\infty} W(\Omega - \omega)S(\Omega)\,d\Omega \tag{11.46}$$

where, if $T = N\Delta$, then from (11.32),

$$W(\omega) = \frac{T}{2\pi}\left|\frac{\sin(\omega T/2)}{\omega T/2}\right|^2 \tag{11.47}$$

so that the effective bandwidth of the basic spectral window remains approximately $1/T$ Hz.

The frequency spacing between adjacent spectral estimates S_k is now closer than before $(2\pi/T_L = 2\pi/(L + N)\Delta$ rad/s apart) although the bandwidth of the basic spectral window $(2\pi/T = 2\pi/N\Delta$ rad/s) is not affected by the number L of additional zeros. When adjacent spectral estimates are averaged to improve statistical reliability according to (11.39), the resulting equivalent spectral window is now the result of superimposing the required number of basic spectral windows spaced $2\pi/T_L$ apart and therefore overlapping more than shown in Fig. 11.5 (which are $2\pi/T$ apart). Since the overlap is more, the "ripple" along the top edge of the window is reduced and the "fall-off" at each side of the window is slightly improved.†

†See, for instance, Cooley *et al.* [10].

The total effective bandwidth of the $(2n + 1)$ overlapping spectral windows is now approximately

$$(2n + 1)\frac{2\pi}{T_L} \text{ rad/s}$$

since the windows are $2\pi/T_L$ apart (for this purpose we neglect the difference between $2\pi/T_L$ and $2\pi/T$ for the two outside halves of the windows at the two edges of the composite window). The record length of data is still T and so

$$\frac{\sigma}{m} \simeq \frac{1}{\sqrt{(B_e T)}} = \frac{1}{\sqrt{((2n + 1)T/T_L)}} \tag{11.48}$$

which, from (9.37), implies that the number of degrees-of-freedom for a chi-square distribution would be

$$\kappa = (4n + 2)\frac{T}{T_L}.$$

In order to separate completely the two overlapping parts of the linear correlation function in Fig. 11.7, the number of additional zeros L must be sufficient that

$$L \geqslant N - 1. \tag{11.49}$$

If the correlation functions are required, these can be computed as described in Chapter 10, by taking the IDFT of the corresponding S_k sequences and, provided enough zeros have been added to satisfy (11.49), we then know that the discrete values R_r so obtained are the result of carrying out the summations

$$R_r = \frac{1}{(N + L)} \sum_{s=0}^{N-1-r} x_s y_{s+r} \qquad r = 0, 1, 2, \ldots, (N - 1). \tag{11.50}$$

Summary

At this point the reader will hardly be surprised to hear that digital spectral analysis has become a highly developed field and that there are many specialist publications in the literature† dealing with different variations of the basic procedure. However for many practical applications the procedure outlined in Chapter 10 will prove adequate and we can now summarize all the essential steps in determining spectra by digital analysis.

1. Estimate the frequency range of interest and the maximum frequency of significant spectral components in the signals to be analysed. If necessary filter the signals to remove excessive high frequency components.

---

†In addition to [10] there are many important papers by Cooley, Lewis and Welch. These include [11] and [14]. See also other references quoted in the next section and those on the fast Fourier transform in Ch. 12.

2. Choose an appropriate sampling interval Δ (s) so that the Nyquist frequency $1/2\Delta$ (Hz) exceeds the maximum frequency present and is at least (say) twice the frequency of interest.

3. Decide the required accuracy σ/m.

4. Estimate the required frequency resolution and hence specify the maximum effective bandwidth of the calculation B_e (Hz).†

5. Calculate the required record length T (*excluding* any added zeros) from the formula

$$\frac{\sigma}{m} \simeq \frac{1}{\sqrt{(B_e T)}}.$$

6. Determine the number of data points $N = T/\Delta$ which must be obtained by sampling each record.

7. Find the number of added zeros L needed to increase the length $(N + L)$ of each sequence of data points to the nearest power of 2.

8. Determine the number $(2n + 1)$ of adjacent spectral estimates which must be averaged (with equal weighting) to give the required bandwidth according to the formula (from (11.48))

$$(2n + 1)\left(\frac{N}{N + L}\right) = B_e T.$$

9. Execute the calculation procedure set out in the last section of Chapter 10 including an extra step to adjust the data sequences to zero mean (to avoid difficulties with the spectral density becoming very large at zero frequency) as follows:

   (i) Generate the discrete time series $\{u_r\}$ and $\{v_r\}$, $r = 0, 1, 2, ..., (N - 1)$, by sampling the records at time interval $\Delta = T/N$.

   (ii) Calculate the mean values

$$\hat{u} = \frac{1}{N} \sum_{r=0}^{N-1} u_r \quad \text{and} \quad \hat{v} = \frac{1}{N} \sum_{r=0}^{N-1} v_r.$$

   (iii) Generate the new sequences $\{x_r\}$ and $\{y_r\}$, $r = 0, 1, 2, ..., (N + L - 1)$, where

$$\begin{aligned} x_r &= u_r - \hat{u} \quad &\text{for} \quad &r = 0, 1, 2, ..., (N - 1) \\ x_r &= 0 \quad &\text{for} \quad &r = N, (N + 1), ..., (N + L - 1) \end{aligned}$$

   and

$$\begin{aligned} y_r &= v_r - \hat{v} \quad &\text{for} \quad &r = 0, 1, 2, ..., (N - 1) \\ y_r &= 0 \quad &\text{for} \quad &r = N, (N + 1), ..., (N + L - 1). \end{aligned}$$

   Both $\{x_r\}$ and $\{y_r\}$ now have zero mean values and have been extended by L additional zeros.

---

† See the example calculation following this section.

(iv) Calculate the DFT's of the series $\{x_r\}$ and $\{y_r\}$ according to the formulae

$$X_k = \frac{1}{N+L} \sum_{r=0}^{N+L-1} x_r e^{-i(2\pi kr/N+L)}$$

$$Y_k = \frac{1}{N+L} \sum_{r=0}^{N+L-1} y_r e^{-i(2\pi kr/N+L)}$$

where $k = 0, 1, 2, \ldots, (N + L - 1)$.

(v) Calculate the required series of spectral coefficients $\{S_k\}$ by forming the appropriate products

$$S_{xx_k} = X_k^* X_k \qquad\qquad S_{xy_k} = X_k^* Y_k$$
$$S_{yx_k} = Y_k^* X_k \qquad\qquad S_{yy_k} = Y_k^* Y_k ,$$

$k = 0, 1, 2, \ldots, (N + L - 1)$.

(vi) If needed, the corresponding (circular) correlation functions R_r can be obtained by taking the IDFT of $\{S_k\}$ according to

$$R_r = \sum_{k=0}^{N+L-1} S_k e^{i(2\pi kr/N+L)},$$

$r = 0, 1, 2, \ldots, (N + L - 1)$.

10. Calculate estimates of the continuous spectrum from the formula (11.45)

$$\tilde{S}(\omega_k) \simeq \frac{T_L}{2\pi} S_k$$

(where $\omega_k = 2\pi k/T_L$ rad/s and $T_L = (N + L)\Delta$). The range of k is now adjusted to run from

$$-\frac{(N+L)}{2} \qquad \text{to} \qquad \frac{(N+L)}{2}$$

where $S_{-k} = S_k^*$. The reason for this is to restrict the range of ω to $-\pi/\Delta < \omega < \pi/\Delta$ since π/Δ is the Nyquist frequency and $\omega_k = \pi/\Delta$ when $k = (N + L)/2$. This change does not involve losing any information since values of S_k in the range $(N + L)/2 < k \leqslant (N + L)$ just repeat values in the range $0 \leqslant k < (N + L)/2$ (see equations (10.17) and (10.18)).

11. Modify these estimates to correct for the added zeros by multiplying by a correction factor $(N + L)/N$, equation (11.46), to give

$$\tilde{S}'(\omega_k) \simeq \left(\frac{N+L}{N}\right)\left(\frac{T_L}{2\pi}\right) S_k$$

where $k = 0, \pm 1, \pm 2, \ldots, \pm(N + L)/2$.

12. Carry out final smoothing by calculating the average of adjacent spectral

estimates according to (11.39), which becomes

$$\hat{S}(\omega_k) = \frac{1}{2n + 1} \sum_{m=-n}^{n} \tilde{S}'(\omega_{k+m})$$

where

$$\omega_k = \frac{2\pi k}{T_L} = \frac{2\pi k}{(L + N)\Delta}$$

$k = 0, \pm 1, \pm 2, \ldots, \pm(N + L)/2$. Values of $\tilde{S}'(\omega_{k+m})$ for $|k + m| > (N + L)/2$ are assumed to be zero. The result is a set of ordinates for a smoothed graph of $S(\omega)$ against ω over the frequency range $-\pi/\Delta < \omega < \pi/\Delta$ (rad/s).

Example

Choose a suitable record length and sampling interval for digital (auto) spectral analysis of a continuous time series in which the maximum significant frequency component is 500 Hz (c/s) and it is necessary to achieve a resolution of 1 Hz in the frequency range of interest which is 100 to 200 Hz.

The Nyquist frequency $1/2\Delta = 500$ Hz, so that

$$\Delta = 0.001 \text{ s.}$$

Suppose that we want to achieve sufficient accuracy that

$$\frac{\sigma}{m} \simeq \frac{1}{3}.$$

Then, from (9.29),

$$B_e T = \frac{1}{(\sigma/m)^2} = 9$$

where B_e is the effective bandwidth and T is the record length. In order to resolve a frequency difference of 1 Hz, the effective bandwidth should certainly not exceed 1 Hz and preferably should be less. Suppose that we put

$$B_e \simeq \tfrac{1}{2} \text{ Hz.}$$

Then the required record length is

$$T = \frac{9}{\frac{1}{2}} = 18 \text{ s.}$$

We shall therefore need $N = T/\Delta = 18\,000$ data points and have to add $L = 14\,768$ zeros to make

$$(N + L) = 2^{15}$$

as required by most FFT algorithms.

Spectral estimates will be spaced $1/T_L$ Hz apart where

$$T_L = \frac{N + L}{N} T = \frac{32\,768}{18\,000}(18) = 33 \text{ s (approx.)}.$$

If we average $(2n + 1)$ adjacent estimates, the bandwidth of the averaged estimate is approximately

$$(2n + 1)\frac{1}{33} \text{ Hz}$$

and if this is to equal 0·5 Hz approx. we obtain $(2n + 1) = 16$ approx., and so $n = 8$ (nearest higher integer value).

Hence we conclude that 18 s of data requires analysis at a sampling interval of 0·001 s, that the total record length must be increased to approximately 33 s by adding zeros to give 2^{15} data points, and that final smoothing should involve averaging every 17 adjacent spectral estimates. The graph of the required spectrum will then have ordinates spaced $1/33 = 0·03$ Hz approx. apart, each value will be the result of smoothing over 0·5 Hz, and the frequency range will run from -500 Hz to $+500$ Hz.

The units of the resulting spectral estimates will be those of (mean square)/(rad/s) and, to convert to (mean square)/(Hz) for positive frequencies only, must be multiplied by 4π as explained in Chapter 5.

Practical considerations

Before leaving this subject, we should mention some practical "tricks-of-the-trade" for improving the accuracy of spectral estimates. The first involves using a *tapered data window* to smooth the data at each end of the record before it is analysed (which has the effect of sharpening the spectral window); the second is a method of *correcting for slow trends* in the recorded data; and the third is a method for shaping the spectrum before analysis called *prewhitening*.

Before carrying out DFT calculations it may be desirable to modify the data first by weighting the middle of the record more heavily than the two ends. The procedure is to multiply the series $\{x_r\}$ by a weighting function $\{d_r\}$ which is

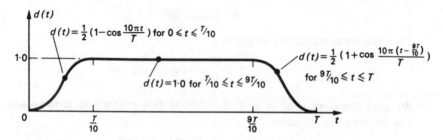

Fig. 11.8 Cosine taper data window after Bingham *et al.* [4], where $d_r = d(t = r\varDelta)$

zero at $r = 0$ and $r = N$ and rises smoothly to a maximum in between. A *cosine data taper* function is often used and an example of this is shown in Fig. 11.8.

We can think of this operation as smoothing the discontinuity at the two ends of the record. Remember that, when using the DFT, we are effectively calculating the Fourier coefficients of a periodic function of which $x(t)$, $0 \leqslant t \leqslant T$, is a single cycle. The *data window* smooths discontinuities at the "joints" of the periodic function.

A considerable amount has been written about data windows, and there are contrary views about whether to use a tapered window or not, and, if so, what shape this should have (Sloane [65]; Otnes *et al.* [52]). As shown in problem 11.5, the basic spectral window of Fig. 11.4 is modified by non-uniform data weighting before analysis, and the application of a tapered data window slightly improves the shape of the resulting spectral window. It does so at the expense of a slightly increased effective bandwidth and of a slight reduction in the statistical accuracy of the resulting spectral estimates. However the spectral window $W(\omega)$ can be significantly improved in the sense that, when ω is large, $W(\omega) \to 0$ faster when a data window has been used than without one (Cooley *et al.* [10]). Although the basic spectral window, Fig. 11.4, can always be "sharpened" and made to decay towards zero faster by increasing the record length T, a judiciously chosen data window may be valuable when only a limited length of record is available for analysis.

The advantages of different window shapes have been discussed in the literature (Durrani *et al.* [25], Harris [83]) but it appears that, if the purpose is to obtain a spectral window whose "tails" decay rapidly to zero, then the Bingham window shown in Fig. 11.8 may be perfectly adequate. It is shown in problem 11.5 that, if the input data x_r is modified by a weighting function d_r before analysis, the resulting spectral estimates must be divided by

$$\frac{1}{N} \sum_{r=0}^{N-1} d_r^2$$

to correct for the loss of amplitude caused by the data taper.

We have seen that it is necessary to normalize the time series being analysed so that it has zero mean in order to avoid difficulties with a large spike in the spectrum at $\omega \simeq 0$. On account of the blurring effect of the finite spectral window, this spike would falsely distort the nearby spectral estimates and so lead to errors. If there is a slow drift on the sample time series being analysed, due perhaps to instrumentation errors which have nothing to do with the process being studied, this also distorts the spectrum falsely in the region of zero frequency. Methods for compensating for such slow trends in the data involve calculating averages over a part length of the sample record at its beginning and over the same length part at its end (usually $\frac{1}{3}$ of the record is suggested) and then assuming that the average varies linearly from one end of the record to the other end, and normalizing the raw data accordingly to subtract out this linearly varying mean. The reader will find more details in specialist texts on spectral analysis, but this is essentially a semi-empirical correction and perhaps the best advice for someone needing good accuracy at very low frequencies is to remove all sources of spurious drift *before* taking measurements rather than hoping that drift will be linear and relying on a correction to the data after measurements have been taken.

The process of "prewhitening" involves operating on raw data before spectral analysis in such a way that the processed data has an approximately flat spectrum. Normalizing to zero mean and correcting for slow trends are two examples of this. Such initial processing of data before analysis is carried out for two reasons. One is that, because large peaks in the spectrum are reduced, spectral "leakage" due to the imperfect shape of the spectral window is reduced. The other arises because the standard error of a measurement, σ, is proportional to the mean value of this measurement, m (see equation (9.29)). The variance of a large spectral estimate is therefore proportionally more than the variance of a small estimate. By whitening the spectrum before analysis, the estimated spectrum can be given uniform accuracy at all frequencies.

In order to carry out prewhitening, it is necessary to estimate first the spectrum of the unprocessed data, and then, having determined the location and magnitude of the spectral peaks, to devise digital filters to "whiten" the raw data and then "recolour" the estimated spectrum. This is itself a specialist subject and the reader is again referred to the specialist texts.† Again we should perhaps note that a more straightforward way of improving accuracy is to analyse a longer record length, provided of course that this is available or can be obtained.

A source of error in digital calculations that we have not previously considered is the *quantization* error which occurs in the process of analogue to

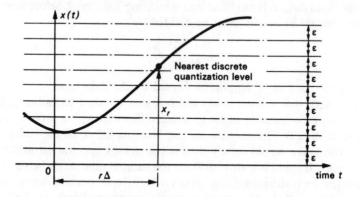

Fig. 11.9 Quantization of an analogue signal

digital conversion of a time series. This arises when a discrete value x_r is rounded off to the nearest quantization level, of which there can only be a finite number, Fig. 11.9. Typically an A to D converter might have a digital

† See, for instance, Gold *et al.* [29].

output with a word length of 9 binary digits (bits). One bit would be used to indicate the sign of x_r, whether positive or negative, and 8 bits to represent its quantitative value. The maximum discrete value available is then

$$(1\ 1\ 1\ 1\ 1\ 1\ 1\ 1)_2 = 2^7 + 2^6 + 2^5 + 2^4 + 2^3 + 2^2 + 2^1 + 2^0 = 255$$

and, if the spacing between discrete levels is ε, Fig. 11.9, the full range of the signal is 255ε. The maximum error due to rounding is $\pm\frac{1}{2}\varepsilon$ and the quantizing error of a full scale reading is hence limited to $\pm(0.5/255)100 = \pm 0.2$ per cent.

A complete error analysis for A to D conversion would consider equipment errors (involving the stability of the quantization levels and the transient response of the equipment to rapid changes of input) as well as quantization errors arising due to round-off. However with modern equipment, errors introduced in the conversion from analogue to digital form are generally negligible compared with the experimental accuracy of the original analogue signal.†

†See, for instance, Hoeschele, Jr. [36] and Watts [67].

Chapter 12

The fast Fourier transform

Basic theory

The fast Fourier transform (or FFT for short) is a computer algorithm for calculating discrete Fourier transforms (DFT's). We have seen that the DFT of a finite sequence $\{x_r\}$, $r = 0, 1, 2, ..., (N - 1)$, is a new finite sequence $\{X_k\}$ defined as

$$X_k = \frac{1}{N} \sum_{r=0}^{N-1} x_r e^{-i(2\pi kr/N)} \qquad k = 0, 1, 2, ..., (N - 1). \qquad (10.8)$$

If we were working out values of X_k by a direct approach we should have to make N multiplications of the form $(x_r) \times (e^{-i(2\pi kr/N)})$ for each of N values of X_k and so the total work of calculating the full sequence X_k would require N^2 multiplications. As we shall see shortly, the FFT reduces this work to a number of operations of the order $N \log_2 N$. If $N = 2^{15}$, as required in the example calculation in Chapter 11, $N^2 \simeq 1 \cdot 1 \times 10^9$ whereas $N \log_2 N = 4 \cdot 9 \times 10^5$ which is only about 1/2000th of the number of operations. The FFT therefore offers an enormous reduction in computer processing time. Moreover there is the added bonus of an increase in accuracy. Since fewer operations have to be performed by the computer, round-off errors due to the truncation of products by the limited word size (limited number of available digits) of the computer are reduced, and accuracy is accordingly increased.

The FFT works by partitioning the full sequence $\{x_r\}$ into a number of shorter sequences. Instead of calculating the DFT of the original sequence, only the DFT's of the shorter sequences are worked out. The FFT then combines these together in an ingenious way to yield the full DFT of $\{x_r\}$. This sounds complicated but actually the logic is surprisingly simple.

Suppose that $\{x_r\}$, $r = 0, 1, 2, ..., (N - 1)$ is the sequence shown in Fig. 12.1(a) where N is an even number and that this is partitioned into two shorter sequences $\{y_r\}$ and $\{z_r\}$ as shown in Fig. 12.1(b) where

$$y_r = x_{2r}$$

$$r = 0, 1, 2, ..., (N/2 - 1). \qquad (12.1)$$

$$z_r = x_{2r+1}$$

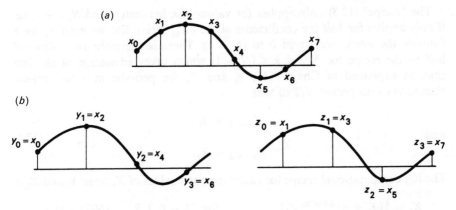

Fig. 12.1 Partitioning the sequences $\{x_r\}$ into two half sequences $\{y_r\}$ and $\{z_r\}$

The DFT's of these two short sequences are Y_k and Z_k where, from (10.8),

$$Y_k = \frac{1}{(N/2)} \sum_{r=0}^{N/2-1} y_r\, e^{-i\frac{2\pi kr}{(N/2)}}$$

$$Z_k = \frac{1}{(N/2)} \sum_{r=0}^{N/2-1} z_r\, e^{-i\frac{2\pi kr}{(N/2)}} \qquad k = 0, 1, 2, \ldots, (N/2 - 1). \qquad (12.2)$$

Now return to the DFT of the original sequence $\{x_r\}$ and consider rearranging the summation into two separate sums similar to those occurring in (12.2). We first separate the odd and the even terms in the $\{x_r\}$ sequence to obtain

$$X_k = \frac{1}{N} \sum_{r=0}^{N-1} x_r\, e^{-i\frac{2\pi rk}{N}}$$

$$= \frac{1}{N} \left\{ \sum_{r=0}^{N/2-1} x_{2r}\, e^{-i\frac{2\pi(2r)k}{N}} + \sum_{r=0}^{N/2-1} x_{2r+1}\, e^{-i\frac{2\pi(2r+1)k}{N}} \right\}.$$

Then, substituting from (12.1),

$$X_k = \frac{1}{N} \left\{ \sum_{r=0}^{N/2-1} y_r\, e^{-i\frac{2\pi rk}{(N/2)}} + e^{-i\frac{2\pi k}{N}} \sum_{r=0}^{N/2-1} z_r\, e^{-i\frac{2\pi rk}{(N/2)}} \right\}$$

from which we can see by comparison with (12.2) that

$$X_k = \tfrac{1}{2}\{Y_k + e^{-i(2\pi k/N)} Z_k\} \qquad (12.3)$$

for $k = 0, 1, 2, \ldots, (N/2 - 1)$.

The DFT of the original sequence can therefore be obtained directly from the DFT's of the two half-sequences Y_k and Z_k according to the recipe (12.3). This equation is the heart of the FFT method. If the original number of samples N in the sequence $\{x_r\}$ is a power of 2, then the half-sequences $\{y_r\}$ and $\{z_r\}$ may themselves be partitioned into quarter-sequences, and so on, until eventually the last sub-sequences have only one term each.

The "recipe" (12.3) only applies for values of k between 0 and $N/2 - 1$, i.e. it only applies for half the coefficients of the $\{X_k\}$ series. But we need X_k for k running the whole way from 0 to $(N - 1)$. There is therefore an additional half to the recipe for $N/2 \leqslant k \leqslant (N - 1)$ which takes advantage of the fact that, as explained in Chapter 10, Y_k and Z_k are periodic in k and repeat themselves with period $N/2$ so that

$$Y_{k-N/2} = Y_k$$

and $\hspace{10cm}$ (12.4)

$$Z_{k-N/2} = Z_k.$$

The full computational recipe for calculating the values of X_k from Y_k and Z_k is

$$X_k = \tfrac{1}{2}\{Y_k + e^{-i(2\pi k/N)} Z_k\} \qquad \text{for} \quad k = 0, 1, 2, \ldots, (N/2 - 1)$$

(12.5)

$$X_k = \tfrac{1}{2}\{Y_{k-N/2} + e^{-i(2\pi k/N)} Z_{k-N/2}\} \quad \text{for} \quad k = N/2, (N/2 + 1), \ldots, (N - 1)$$

or, if we only permit k to run from 0 to $N/2$, then an alternative equivalent recipe is

$$X_k = \tfrac{1}{2}\{Y_k + e^{-i(2\pi k/N)} Z_k\}$$
$$X_{k+N/2} = \tfrac{1}{2}\{Y_k + e^{-i(2\pi(k+N/2)/N)} Z_k\} \qquad k = 0, 1, 2, \ldots, (N/2 - 1) \quad (12.6)$$

which, making use of the fact that $e^{-i\pi} = -1$, can be simplified to

$$X_k = \tfrac{1}{2}\{Y_k + e^{-i(2\pi k/N)} Z_k\}$$
$$X_{k+N/2} = \tfrac{1}{2}\{Y_k - e^{-i(2\pi k/N)} Z_k\} \qquad k = 0, 1, 2, \ldots, (N/2 - 1). \quad (12.7)$$

Finally, if we define a new complex variable W where

$$W = e^{-i(2\pi/N)}$$

(12.8)

we can obtain the so-called computational "butterfly"

$$X_k = \tfrac{1}{2}\{Y_k + W^k Z_k\}$$
$$X_{k+N/2} = \tfrac{1}{2}\{Y_k - W^k Z_k\} \qquad k = 0, 1, 2, \ldots, (N/2 - 1) \quad (12.9)$$

which occurs in most FFT computer programs.

Sample calculation

In order to illustrate the FFT "in action" we can consider a simple case in which the original sequence x_r has only a few terms, and work through the whole calculation. For instance, suppose that $\{x_r\}$ has only four terms. This can be partitioned into sub-sequences as shown in Fig. 12.2 until the "one-quarter" sequences have only a single term each. If we consider the basic

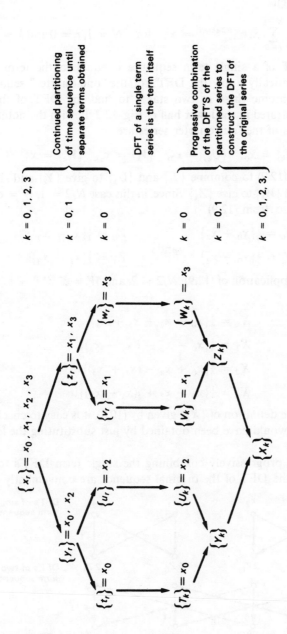

Fig. 12.2 Logical steps during operation of the FFT algorithm on a four-term sequence $\{x_r\}$

definition (10.8) of a DFT, it is evident that, if there is only one term in the sequence $\{x_r\}$, then

$$X_k = \frac{1}{N}\sum_{r=0}^{N-1} x_r\,\mathrm{e}^{-i(2\pi kr/N)} = x_0 \quad \text{for} \quad N = 1, r = 0 \text{ and } k = 0 \quad (12.10)$$

so that the DFT of a single term sequence is equal to the term itself. We therefore automatically know the DFT's of the "one-quarter" sequences and have only to combine these in two stages to find the DFT of the original sequence, as illustrated in the lower half of Fig. 12.2. If, with the notation in the figure, the DFT's of the four quarter series are

$$\{T_k\} = x_0, \quad \{U_k\} = x_2, \quad \{V_k\} = x_1, \quad \{W_k\} = x_3$$

then we can use (12.9) to combine $\{T_k\}$ and $\{U_k\}$ to give $\{Y_k\}$ and similarly to combine $\{V_k\}$ and $\{W_k\}$ to give $\{Z_k\}$. Since, in this case $N/2 = 1$, $W = \mathrm{e}^{-i(2\pi/N)} = \mathrm{e}^{-i\pi} = -1$, and so, from (12.9)

$$\begin{aligned} Y_0 &= \tfrac{1}{2}\{x_0 + x_2\} & & & Z_0 &= \tfrac{1}{2}\{x_1 + x_3\} \\ Y_1 &= \tfrac{1}{2}\{x_0 - x_2\} & &\text{and} & Z_1 &= \tfrac{1}{2}\{x_1 - x_3\}. \end{aligned} \quad (12.11)$$

For the second application of (12.9), $N/2 = 2$, and $W = \mathrm{e}^{-i2\pi/N} = \mathrm{e}^{-i\pi/2} = -i$, so that

$$\begin{aligned} X_0 &= \tfrac{1}{4}\{x_0 + x_2 + x_1 + x_3\} \\ X_1 &= \tfrac{1}{4}\{x_0 - x_2 - i(x_1 - x_3)\} \\ X_2 &= \tfrac{1}{4}\{x_0 + x_2 - (x_1 + x_3)\} \\ X_3 &= \tfrac{1}{4}\{x_0 - x_2 + i(x_1 - x_3)\}. \end{aligned} \quad (12.12)$$

By referring to the definition of $\{X_k\}$ given by (10.8), it is easy to check that this is the result that would have been obtained by just substituting the four values of $\{x_r\}$ into (10.8).

These steps of progressively combining the single term DFT's to generate all the terms of the DFT of the original sequence are conveniently shown by

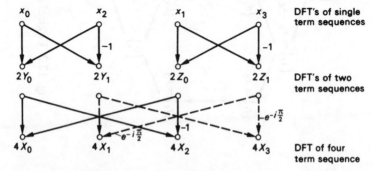

Fig. 12.3 Butterfly signal flow graphs illustrate successive steps in the FFT algorithm

"butterfly" diagrams, Fig. 12.3. The upper two "butterflies" represent equations (12.11) and show how $\{Y_k\}$ and $\{Z_k\}$ are generated from the four single-term DFT's x_0, x_2, x_1 and x_3. A dot on the butterfly represents a variable and the arrows terminating at this dot indicate which other variables contribute to its value. The contributions are all additive unless a multiplier (such as -1 to indicate subtraction) appears close to the appropriate arrowhead. The lower two overlapping butterflies in Fig. 12.3 show how the DFT $\{X_k\}$ of the original sequence is generated from $\{Y_k\}$ and $\{Z_k\}$ according to (12.12).

This application of signal flow graphs, as they are called, is most helpful in indicating the logical steps in an FFT algorithm written to accommodate an initial data sequence of arbitrary length and we shall therefore now pursue this approach in more detail.

Programming flow charts

Consider the case when $N = 2^n$ and the initial data sequence has been successively partitioned into sub-sequences to give N single term sequences. To save dividing by 2 at every stage of the calculation, as required by the recipe (12.9), we can divide by N at the beginning. Assume that this has been done and let the resulting single terms (in their new order) be called $a_1, a_2, a_3, \ldots, a_N$. Incidentally, we begin with a_1 rather than a_0 because it is not customary to use zero subscripts in computer programming languages. In Fig 12.4, the top row of separate butterflies indicates how these terms are combined in pairs, according to (12.9), to give the DFT's of corresponding two-term sequences. Rather than calling these DFT's $\{Y_k\}$, $\{Z_k\}$, etc., as in Fig. 12.3, we really do not need to invent any new names and can just write these values into the same registers (or computer pigeon holes) from which the original values were taken, since the original values are no longer needed for further computation. We thereby economize on computer storage space. As long as we understand that the new values $a_1, a_2, a_3, \ldots, a_N$ are now components of the two-term DFT's, the meaning of the top line of Fig. 12.4 should still be clear.

In the second line of the figure, each butterfly involves alternate terms in the $a_1, a_2, a_3, \ldots, a_N$ sequence and the multiplier $W_2 = e^{-i\pi/2}$ appears on appropriate arrows as required by (12.9). In the third line, the butterflies are 2^2 terms wide and the multiplier $W_3 = e^{-i\pi/4}$ appears in ascending powers up to W_3^3. In the fourth line, the butterflies are 2^3 terms wide and $W_4 = e^{-i\pi/8}$ appears in powers up to W_4^7, and so on.

A logical path through the calculation is to proceed sequentially from one row of butterflies to the next. For row m, data points are taken in groups of 2^m and each butterfly has a width of $k = \frac{1}{2}(2^m) = 2^{m-1}$. The multiplier (or "twiddle factor" as it is sometimes called) is now $W_m = e^{-i\pi/k}$ and it occurs in powers up to $W_m^{(k-1)}$. One approach is to calculate values of

$$W_m^r = e^{-i\pi r/k} = \cos\frac{\pi r}{k} - i\sin\frac{\pi r}{k}$$

and then to store these values in the computer memory for use when needed. However this involves a large number of sine and cosine evaluations and uses a considerable amount of storage space. An alternative approach is to calculate only one value of the twiddle factor (for $r = 0$) at each stage of the calculation and then generate the others by continuously multiplying the initial value W_m by itself to yield W_m^2, W_m^3, etc. This means that the initial value must be accurate enough that, when raised to its highest power, there is still adequate accuracy. Typical programming systems permit function evaluations (including sine and cosine) to at least eight significant figures. For the last step of a calculation on a sequence of length 2^n, the highest power of $W_m = W_n$ is $(k - 1) = (2^{n-1} - 1)$. For $n = 15$ this is approximately $1 \cdot 6 \times 10^4$ and so $W_n^{(k-1)}$ is still likely to have adequate accuracy for sequences of medium length.

A programming flow chart to carry out the calculation described above is shown in Fig. 12.5. The flow convention is from top to bottom of the chart unless otherwise indicated. The logical operations shown are followed in order, and an arrow pointing to the left means that the expression on the right is computed and its value is allotted to the quantity on the left. The calculation begins with step $m = 1$ corresponding to the top row of butterflies in Fig. 12.4. In this case $U = 1, j = 1, k = 1, W = -1$ and $l = 1$, so that the computer allots the value of a_2 to t, and the butterfly calculation gives

$$\text{new value of } a_2 = \text{old } a_1 - \text{old } a_2$$
$$\text{new value of } a_1 = \text{old } a_1 + \text{old } a_2$$

or, if we use a prime ' to denote the new values,

$$a_2' = a_1 - a_2$$
$$a_1' = a_1 + a_2$$

in agreement with the first line of Fig. 12.4. The dummy parameter t in the flow chart has to be introduced, because otherwise the old value of a_2 would be lost when the first line of the butterfly is calculated and would not be available to compute the second line. Continuing down the chart, l is increased from 1 to 3, and we go round the inside loop (1) to find $t = a_4$ and so

$$a_4' = a_3 - a_4$$
$$a_3' = a_3 + a_4$$

and so on, along the top line of butterflies of Fig. 12.4. When the end of the row is reached, the value of l exceeds 2^n. We then stop going round loop (1) and move downwards again. The variable j is incremented from 1 to 2, which exceeds $2^{m-1} = 2^0 = 1$ and so m is incremented to 2, to move round loop (3) to the second row of butterflies. Here U and j are reset to 1, k becomes 2 and $W = e^{-i\pi/2}$. Proceeding down the chart again, $t = a_3$ and the first butterfly of the second row is

$$a_3' = a_1 - a_3$$
$$a_1' = a_1 + a_3$$

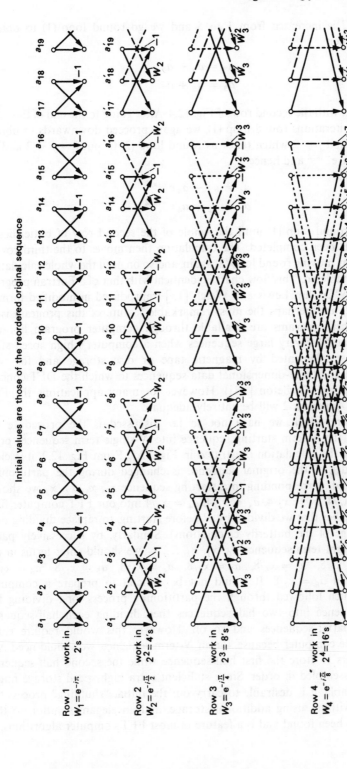

Fig. 12.4 Butterfly signal flow graphs to illustrate FFT programming logic

whereupon l is increased from 1 to 5 and we go round loop (1) to obtain $t = a_7$ and then

$$a'_7 = a_5 - a_7$$
$$a'_5 = a_5 + a_7$$

in agreement with the second row of Fig. 12.4. After going to the end of this row, by repeated iterations round loop (1), we again proceed downwards to obtain $U = e^{-i\pi/2}$ and $j = 2$ which takes us round loop (2) to put $l = j = 2$ and so obtain $t = a_4 e^{-i\pi/2}$ and hence

$$a'_4 = a_2 - a_4 e^{-i\pi/2}$$
$$a'_2 = a_2 + a_4 e^{-i\pi/2}$$

and so on round loop (1) until the whole of the second row of butterflies in Fig. 12.4 has been completed. The calculation then moves to the third row of butterflies, by moving round loop (3) again, and so on until the whole calculation has been completed. This flow chart is identical with that of a Fortran program published by Cooley, Lewis and Welch [13] (reproduced in modified form in Appendix 2) and perhaps the most remarkable feature of this program is its brevity. Other programs are available through computer program libraries and these cater for very large sequences when a computer's high speed store has to be supplemented by magnetic tape or disc storage and for two-dimensional and three-dimensional data sequences to which the DFT concept can be extended (Singleton [64]). However for many applications the FFT program in Appendix 2 will be entirely adequate.

One feature which we have not so far discussed is the procedure for partitioning the N-term starting sequence into N single term sequences prior to beginning the calculation illustrated in Fig. 12.5. From Fig. 12.2, it is clear that the order of the original sequence is changed during the partitioning procedure and, corresponding to a starting sequence x_0, x_1, x_2, x_3 we should enter $a_1 = x_0/4$, $a_2 = x_2/4$, $a_3 = x_1/4$, $a_4 = x_3/4$ into our FFT computer flow chart (remember that we divided by N before starting, to replace dividing by 2 at every stage of the butterfly calculations). Similarly, by successively partitioning an eight-term sequence x_0, x_1, x_2, ..., x_7 we should enter terms in the order $a_1 = x_0/8$, $a_2 = x_4/8$, $a_3 = x_2/8$, $a_4 = x_6/8$, $a_5 = x_1/8$, $a_6 = x_5/8$, $a_7 = x_3/8$ and $a_8 = x_7/8$. It would not be difficult to prepare a computer program which followed through the partitioning process, first dividing the original sequence into two half-sequences then dividing each half-sequence into two quarter-sequences and so on. However this would require extra storage in the computer because, for an N-term sequence, we should need $N/2$ extra registers to store the first half-sequence while the second half-sequence was being assembled in order. Since sufficient extra high-speed storage might not be available, it is desirable to carry out the initial "shuffling" process "in place", i.e. without using additional storage, and an elegant solution to this problem has been found and is a feature of most FFT computer algorithms.

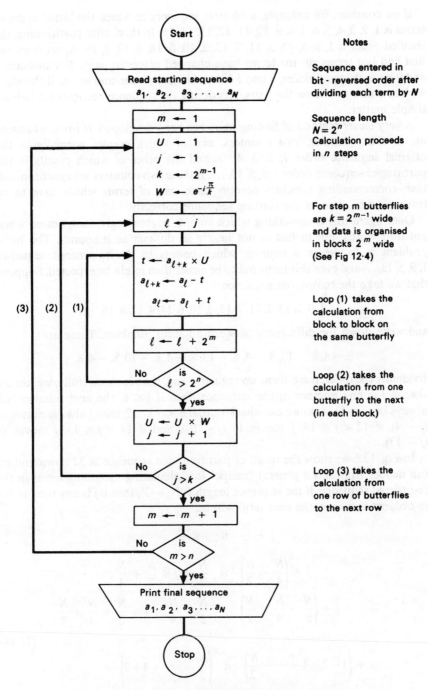

Fig. 12.5 Programming flow chart for a simple FFT algorithm
(After Cooley *et al.* [13])

If we consider, for example, a 16-term sequence in which the initial order of terms is 1, 2, 3, 4, 5, 6, 7, 8, 9, 10, 11, 12, 13, 14, 15, 16 then, after partitioning, the shuffled order is 1, 9, 5, 13, 3, 11, 7, 15, 2, 10, 6, 14, 4, 12, 8, 16. Apart from the first and last terms, all the terms have changed places in pairs. For instance 2 and 9 have changed places, 3 and 5 have changed places, and so on. If therefore we know how to choose the pairs, reordering the sequence becomes a relatively simple matter.

A very clever method of finding pairs has been developed. It involves setting up, in the computer, two counters, one of which counts normally in the original sequence order 1, 2, 3, 4, ... and the other of which counts in the partitioned sequence order 1, 9, 5, 13, ... If these two counters are synchronized, then corresponding numbers identify each pair of terms which have to be transposed to re-order the starting sequence correctly.

Once again this is something which sounds frighteningly complicated when put into words, but in fact is not nearly as difficult as it sounds. The main problem is to devise a counter which counts in the re-ordered sequence 1, 9, 5, 13, ... and even this turns out to be easier than might be expected. Suppose that we take the re-ordered sequence

$$1, 9, 5, 13, 3, 11, 7, 15, 2, 10, 6, 14, 4, 12, 8, 16$$

and write down the differences between adjacent numbers. These are:

$$8, -4, 8, -10, 8, -4, 8, -13, 8, -4, 8, -10, 8, -4, 8.$$

Evidently the difference 8 turns up regularly and, if we look carefully, we can see that, if j is any number in the sequence, then if $j \leqslant 8$, the next number will always be $(j + 8)$. Also we can check that, if $8 < j \leqslant 12$, then j always moves to $(j - 4)$, if $12 < j \leqslant 14$, j moves to $(j - 10)$, and if $14 < j \leqslant 15$, j moves to $(j - 13)$.

In Fig. 12.6 we show the result of partitioning a sequence of 32 terms and we can deduce from this a general "recipe" for calculating ascending terms in the re-ordered sequence. If the sequence length is $N = 2^n$, then if j is any term in the re-ordered sequence, the next term will be

$$j + \frac{N}{2} \quad \text{if} \quad j \leqslant \frac{N}{2}$$

$$j + \left(\frac{N}{4} - \frac{N}{2}\right) \quad \text{if} \quad \frac{N}{2} < j \leqslant \frac{N}{2} + \frac{N}{4}$$

$$j + \left(\frac{N}{8} - \frac{N}{4} - \frac{N}{2}\right) \quad \text{if} \quad \frac{N}{2} + \frac{N}{4} < j \leqslant \frac{N}{2} + \frac{N}{4} + \frac{N}{8}$$

$$\cdots\cdots\cdots \qquad\qquad\qquad \cdots\cdots\cdots \qquad (12.13)$$

$$j + \left(1 - 2 - 4 - \cdots - \frac{N}{2}\right) \quad \text{if} \quad \left(\frac{N}{2} + \cdots + 4 + 2\right)$$

$$< j \leqslant \left(\frac{N}{2} + \cdots + 4 + 2 + 1\right) = (N - 1).$$

N = 32

Original sequence in order:
1 2 3 4 5 6 7 8 9 10 11 12 13 14 15 16 17 18 19 20 21 22 23 24 25 26 27 28 29 30 31 32

after partitioning into ½ series:
1 3 5 7 9 11 13 15 17 19 21 23 25 27 29 31 | 2 4 6 8 10 12 14 16 18 20 22 24 26 28 30 32

after partitioning into ¼ series:
1 5 9 13 17 21 25 29 | 3 7 11 15 19 23 27 31 | 2 6 10 14 18 22 26 30 | 4 8 12 16 20 24 28 32

after partitioning into ⅛ series:
1 9 17 25 | 5 13 21 29 | 3 11 19 27 | 7 15 23 31 | 2 10 18 26 | 6 14 22 30 | 4 12 20 28 | 8 16 24 32

after partitioning into 1/16 series:
1 17 | 9 25 | 5 21 | 13 29 | 3 19 | 11 27 | 7 23 | 15 31 | 2 18 | 10 26 | 6 22 | 14 30 | 4 20 | 12 28 | 8 24 | 16 32

Table of differences between adjacent terms in fully partitioned sequence

	$+\frac{N}{2}$	$-\frac{N}{4}=(\frac{N}{4}-\frac{N}{2})$	$(\frac{N}{8}-\frac{N}{4}-\frac{N}{2})$	$(\frac{N}{16}-\frac{N}{8}-\frac{N}{4}-\frac{N}{2})$	$(\frac{N}{32}-\frac{N}{16}-\frac{N}{8}-\frac{N}{4}-\frac{N}{2})$
16					
	−8				
16					
		−20			
16					
	−8				
16					
			−26		
16					
	−8				
16					
		−20			
16					
	−8				
16					
				−29	
16					
	−8				
16					
		−20			
16					
	−8				
16					
			−26		
16					
	−8				
16					
		−20			
16					
	−8				
16					

Fig. 12.6 Table of differences between the adjacent terms of a re-ordered sequence of 32 points

It is not difficult to incorporate this logic into a computer flow chart as shown in Fig. 12.7(*a*) and from it to draw the flow chart of a counter which will count in the re-ordered sequence, Fig. 12.7(*b*). In the latter, a single decision point (1) (represented by a diamond-shaped box) included in loop (1) does the job of all the separate decision points (i), (ii), (iii), etc., of Fig. 12.7(*a*). When decision point (1) is reached from the box above, it asks the question is $j \leqslant N/2$, corresponding to decision (i) of the left-hand figure. If the answer to this question is no, loop (1) is traversed, and decision point (1) then asks the question is $j - N/2 \leqslant N/4$ which corresponds to decision point (ii) of the other figure, and so on. The programming flow chart of Fig. 12.7(*b*) has an additional loop (2) to re-enter the program with each new number and so generate all the numbers in the re-ordered sequence in turn. Each number is fed out from the program as it is generated at the exit point shown.

This system is usually called a *bit-reversed counter* because, if they are expressed as binary numbers, terms in the re-ordered sequence can be obtained from corresponding terms in the original sequence (if this begins with zero) by just reversing the order of bits (or binary digits). For instance, for an eight-term sequence the numbers 0, 1, 2, 3, 4, 5, 6, 7 become 0, 4, 2, 6, 1, 5, 3, 7. Number 1 = (001) in binary notation has become 4 = (100). Number 2 = (010) remains 2. Number 3 = (011) becomes 6 = (110), and so on. A counter running in the order of the re-ordered sequence is therefore running in reversed binary digit order.

We now have a counter which generates the re-ordered (or bit-reversed) sequence $j = 1, 9, 5, 13, \ldots$ (for a 16-point sequence) and this can be combined with a normal counter counting $l = 1, 2, 3, 4, \ldots$ in order to identify corresponding terms in each pair which must change places. We have then only to make the transposition

$$a_l \rightleftarrows a_j \tag{12.14}$$

in order to correctly re-order the starting sequence $a_1, a_2, a_3, a_4, \ldots$ prior to executing the butterfly calculations already described. A flow chart of the complete calculation is given in Fig. 12.8, which can be seen to carry out the transposition of terms required by (12.14). The only feature that this includes which we have not already mentioned is a by-pass loop which by-passes (12.14) when $l > j$. If this were omitted the calculation would transpose the same terms a second time and so undo the re-ordering already completed. For instance, for a 16-point sequence, at the second term a_2 is exchanged with a_9 in the re-ordering procedure. When we reach the ninth term, it would clearly be wrong to exchange a_9 with a_2 again as this would just put the sequence back in its original order. The by-pass loop for $l > j$ in Fig. 12.8 prevents this happening.

Practical value of the FFT

We have spent a considerable amount of time discussing the theory of the FFT and how it is implemented in computer programs and the reason for going into

(a) Basic flow chart to implement equations (12.13)

(After Gold et al. [29] Fig. 6.23, Courtesy of McGraw-Hill, Inc.)

(b) Programming flow chart following the same logical path as (a)

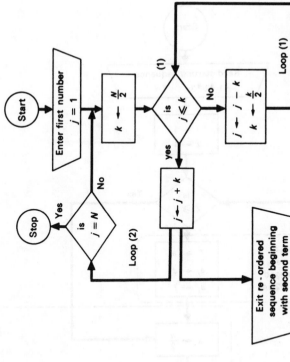

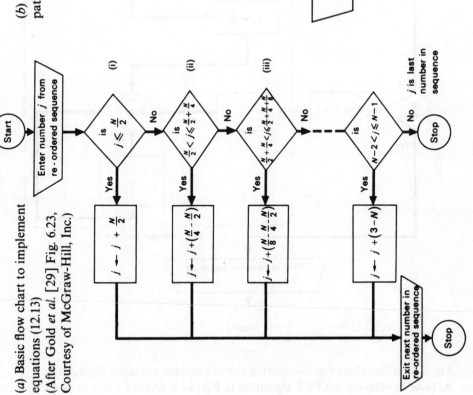

Fig. 12.7 Flow charts for bit-reversed counter

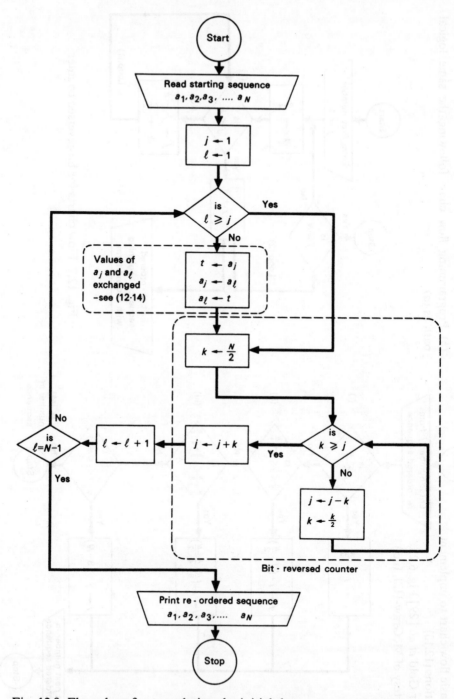

Fig. 12.8 Flow chart for re-ordering the initial data sequence $a_1, a_2, a_3, \ldots,$ a_N prior to entering the FFT algorithm in Fig. 12.5 (After Cooley *et al.* [13])

such detail is the important role which the FFT plays in spectral analysis. Since 1965, when the FFT was brought onto the scene,† it has profoundly changed our approach to digital spectral analysis. Instead of calculating spectra by first calculating discrete estimates of the appropriate correlation function and then finding the DFT of the correlation function, it is now quicker to calculate the DFT of the original discrete time series directly and then to manipulate this transformed sequence to yield the required spectral estimates as explained in Chapters 10 and 11. The reason for this different approach is just that it is quicker to calculate the DFT of a discrete series by the FFT and follow the procedure of Chapter 10 than it is to calculate the sums of products needed to estimate correlation functions.

A precise statement about how much faster the FFT is than other calculations depends on the specific FFT computer algorithm used. More complicated logic than that shown in Fig. 12.5 can effect additional time savings by taking advantage of the symmetries of the calculation. For instance multiplications by W^0 and $W^m = -1$ can be replaced by simple additions and subtractions. Similarly multiplications by $W^{m/2} = -i$ and $W^{3m/2} = i$ can be removed from the main loop of the program, and so on. However, neglecting these possible additional savings, we can see from Fig. 12.4 that, if we follow the flow chart of Fig. 12.5, there will have to be $N/2$ complex multiplications in each line of butterflies and N complex additions (including subtractions). Since there are $n = \log_2 N$ rows of butterflies, there will therefore have to be $\frac{1}{2}N \log_2 N$ complex multiplications and $N \log_2 N$ complex additions to compute the DFT of a sequence of N terms. To compute the same DFT by the direct method (of substituting into (10.8)) would require, by comparison, N^2 complex multiplications and $N(N - 1)$ complex additions. Similarly, to compute the correlation function

$$R_r = \frac{1}{N} \sum_{s=0}^{N-1} x_s y_{s+r} \qquad r = 0, 1, 2, \dots, N - 1 \qquad (10.13)$$

also requires a number of operations of the order of N^2. If we compare N^2 and $N \log_2 N$ for large values of N we can begin to appreciate the real impact of the FFT on digital spectral analysis. Table 12.1 shows the ratio $N^2/N \log_2 N$ for a range of values of the sequence length N and it is clear that simply enormous reductions in computer time can be achieved by using the FFT to analyse long data sequences.

Alternative algorithms

In this chapter we have considered only the so-called radix-2 FFT for which the sequence length N must be a power of 2 so that the original sequence can be partitioned into two half-sequences of equal length, and so on. When N is not a power of 2, but has other factors, the same basic ideas still apply with suitable

†Cooley *et al.* [9]. See also Cooley *et al.* [12].

Sequence length N	$\log_2 N$	Ratio$\left\{\dfrac{N^2}{N\log_2 N}\right\}$
4	2	2
16	4	4
64	6	10·7
256	8	32
1 024	10	102·4
4 096	12	341·3
16 384	14	1 170·3
65 536	16	4 096
262 144	18	14 563·6
1 048 576	20	52 428·8

Table 12.1 Computational efficiency of the FFT illustrated by comparing the order N^2 of operations of multiplication and addition for the direct method of calculating DFT's with the order of operations $N\log_2 N$ for the FFT.

modifications to the butterfly formulae.† For instance if $N = 15$, the original sequence could be partitioned into three sub-sequences of five terms each or five sub-sequences of three terms each and appropriate formulae devised to relate the DFT's of the partitioned sequences to the DFT of the original sequence. However, although these more general FFT algorithms are interesting variations of the radix-2 FFT, computer programming is simpler for the $N = 2^n$ case and the possibility of artificially lengthening a sequence by adding zeros (see Chapter 11) means that the restricted choice of values of $N = 2^n$ is adequate for most spectral analysis applications.

Finally we should mention that there is still another variation of the basic FFT method in which the original time sequence is divided into sub-sequences each of which involves groups of adjacent terms (rather than alternate terms, or one term every three, etc.).‡ For example if N has a factor 2, then one sub-sequence would be the first $N/2$ terms and the other sub-sequence the second $N/2$ terms. In this case the even and odd numbered terms of the DFT of the full sequence can be expressed in terms of the DFT's of two half-sequences generated from the partitioned sub-sequences. This method is described as *decimation in frequency*, since alternate terms of the DFT (or frequency sequence) are computed from each half-sequence, while the method we have spent so much time studying is called *decimation in time*, since the DFT is calculated from half-sequences which involve alternate terms from the original time sequence. While decimation in frequency FFT algorithms are interesting in themselves, once again they do not appear to offer significant practical advantages over the basic radix-2 FFT using decimation in time.

† See problem 12.2.
‡ See, for instance, Gold *et al.* [29] Ch. 6, and problem 12.4.

Chapter 13

Pseudo random processes

All our work so far has been concerned with random processes which occur naturally. There are however two main areas in which artificial or *pseudo random* processes are used. One of these is the area of environmental testing in which naturally occurring random environments have to be simulated in the laboratory; the other is the area of system identification in which the response to an intentionally applied random input is used to predict the dynamic characteristics of a system.

The need for environmental testing arises when it is necessary to verify how a new component or device performs in a vibrating environment before putting it into service. In principle, a random signal generator is used to drive a vibration table and the spectrum and probability distribution of the signal generator are adjusted so that the required random vibration environment exists for test components mounted on the table.

The need for measuring the dynamic characteristics of a system by monitoring its response to a random input signal arises from the idea that, if these characteristics are known, the operating conditions of the system can be optimized so that it runs at peak performance. We have already seen in Chapter 7 that, if $x(t)$ is the input and $y(t)$ the output of a linear system whose complex frequency response function is $H(\omega)$, then

$$S_{xy}(\omega) = H(\omega)S_x(\omega) \qquad (7.24)$$

where $S_x(\omega)$ is the spectral density of $x(t)$ and $S_{xy}(\omega)$ is the cross-spectral density of $x(t)$ and $y(t)$. The fundamental concept of system identification by random signal testing is that, if the spectrum and cross-spectrum of a system subjected to broad band noise can be measured, then the system's frequency response over the same frequency band can be deduced from (7.24). If a random test signal is applied to a system, for instance a chemical plant, while it is in normal operation, then it is possible to monitor continuously the dynamic properties of the system and adjust its operating conditions by an "on-line" or "real time" computer control system to maintain peak performance.

Both environmental testing and system identification applications require the generation of an artificial random signal. One method of doing this is to amplify the random emission of electrons in a natural noise source such as a thyratron or a zener diode but the disadvantage of a natural source is that the statistical characteristics of the signal generated are hard to control and hold stable, particularly in the low frequency range of most mechanical equipment. An alternative method of generating artificial low-frequency noise is based on the operation of a digital waveform generator which produces a binary signal by switching randomly between two output levels. We shall now consider the properties of a random binary process and how a binary signal of this type can serve as the basis for generating continuous random signals with approximately Gaussian probability distributions.

Random binary process

Consider an artificial random process obtained as follows. A coin is tossed at regular intervals Δt. If the coin falls "head" upwards $x(t)$ is $+a$ for the next interval Δt; if it is "tails" $x(t)$ is $-a$ for the next interval. A sample function of the $x(t)$ process is then as shown in Fig. 13.1(a). It is convenient to think of many of these coin tossing experiments going on simultaneously, each one constituting a different sample function of the ensemble $\{x(t)\}$, which then appears as shown in Fig. 13.1(b). We define the "phase" of a sample ϕ as the time from $t = 0$ to the next toss of the coin, and ϕ is assumed to be randomly distributed between samples with the uniform probability distribution shown in Fig. 13.1(c). $\{x(t)\}$ is then an ergodic process and we can calculate statistical averages either by averaging across the ensemble or by averaging with respect to time along a single sample. We shall now calculate the autocorrelation function for $\{x(t)\}$ by ensemble averaging; the same result may be obtained by averaging along a single sample, and this is left as an exercise for the reader.

The autocorrelation function $R_x(\tau)$ for this process is given by

$$R_x(\tau) = E[x(t)x(t + \tau)] = \int_{-\infty}^{\infty} \int_{-\infty}^{\infty} x_1 x_2 \, p(x_1, x_2) \, dx_1 \, dx_2 \qquad (13.1)$$

where x_1 is the value of x at time t, x_2 is the value of x at time $t + \tau$ and $p(x_1, x_2)$ is the joint probability density function for x_1 and x_2 (see Chapter 2). So far we have not said what $p(x_1, x_2)$ is, but it can be worked out from the data already given. First, suppose that $\tau = 0$, so that $x_2 = x_1$. In this case x_1 and x_2 are either both $+a$ or both $-a$, and the joint probability density surface has the form of two delta functions, each of volume $\frac{1}{2}$, at $x_1 = x_2 = a$ and $x_1 = x_2 = -a$, Fig. 13.2(a). Next, consider the case when $\tau > \Delta t$. In this case there will be no correlation between x_1 and x_2 because x_2 will be the result of a different toss from x_1. The joint probability density surface is then four delta functions, each of volume $\frac{1}{4}$, at the locations (a, a), $(a, -a)$, $(-a, a)$ and $(-a, -a)$ as shown in Fig. 13.2(b). Now suppose that $\tau < \Delta t$. The values x_1 and x_2 will be correlated (in this case the same) for some samples and uncorrelated for other samples. Since the process is stationary, we may take $t = 0$ for convenience, and then x_1 and x_2 will be correlated if $\phi > \tau$ (since the next coin will not then have been tossed at time $t + \tau = \tau$) and uncorrelated if $\phi < \tau$. For the probability density $p(\phi)$

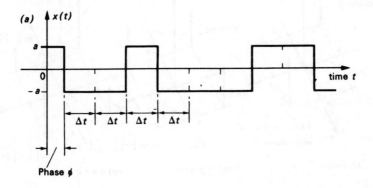

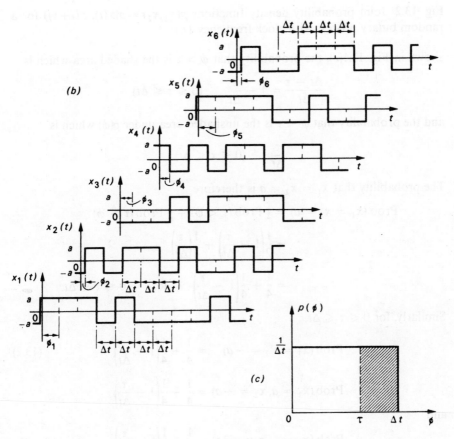

Fig. 13.1 Random binary process obtained by tossing coins at regular time intervals Δt.

170   *Pseudo random processes*

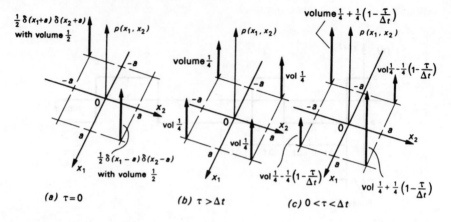

Fig. 13.2 Joint probability density functions $p(x_1,x_2) = p(x(t), x(t+\tau))$ for a random binary process with clock frequency $1/\Delta t$

shown in Fig. 13.1(c), the probability that $\phi > \tau$ is the shaded area which is

$$\frac{\Delta t - \tau}{\Delta t} = 1 - \frac{\tau}{\Delta t} \qquad (0 \leqslant \tau \leqslant \Delta t)$$

and the probability that $\phi < \tau$ is the unshaded area under $p(\phi)$ which is

$$\frac{\tau}{\Delta t} \qquad (0 \leqslant \tau \leqslant \Delta t).$$

The probability that $x_1 = x_2 = a$ is therefore

$$\text{Prob}\,(x_1 = x_2 = a) = \tfrac{1}{2} \text{Prob}\,(\tau < \phi) + \tfrac{1}{4} \text{Prob}\,(\tau > \phi)$$

$$= \frac{1}{2}\left(1 - \frac{\tau}{\Delta t}\right) + \frac{1}{4}\left(\frac{\tau}{\Delta t}\right)$$

$$= \frac{1}{4} + \frac{1}{4}\left(1 - \frac{\tau}{\Delta t}\right) \qquad \text{for} \qquad 0 \leqslant \tau \leqslant \Delta t.$$

Similarly, for $0 \leqslant \tau \leqslant \Delta t$,

$$\text{Prob}\,(x_1 = x_2 = -a) = \frac{1}{4} + \frac{1}{4}\left(1 - \frac{\tau}{\Delta t}\right) \qquad (13.2)$$

$$\text{Prob}\,(x_1 = a, x_2 = -a) = \frac{1}{4} - \frac{1}{4}\left(1 - \frac{\tau}{\Delta t}\right)$$

and

$$\text{Prob}\,(x_1 = -a, x_2 = a) = \frac{1}{4} - \frac{1}{4}\left(1 - \frac{\tau}{\Delta t}\right)$$

as shown in Fig. 13.2(c).

Substituting the delta function probability densities from Fig. 13.2 into equation (13.1) and integrating gives

$$R_x(\tau) = a^2\left\{\frac{1}{2} + \frac{1}{2}\left(1 - \frac{\tau}{\Delta t}\right)\right\} - a^2\left\{\frac{1}{2} - \frac{1}{2}\left(1 - \frac{\tau}{\Delta t}\right)\right\}$$

$$= a^2\left(1 - \frac{\tau}{\Delta t}\right) \qquad \text{for} \qquad 0 \leqslant \tau \leqslant \Delta t \qquad (13.3)$$

and

$$R_x(\tau) = 0 \qquad\qquad \text{for} \qquad \tau > \Delta t$$

and so the shape of the autocorrelation function for the $x(t)$ process is as shown in Fig. 13.3(a).

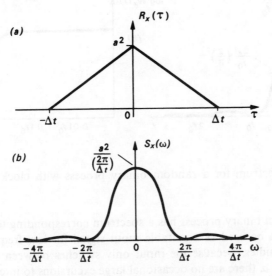

Fig. 13.3 Autocorrelation and spectral density for a random binary process with clock frequency $1/\Delta t$

The corresponding spectral density is

$$S_x(\omega) = \frac{1}{2\pi}\int_{-\infty}^{\infty} R_x(\tau)\,e^{-i\omega\tau}\,d\tau \qquad (5.1)$$

which in this case becomes

$$S_x(\omega) = \frac{1}{\pi}\int_0^{\Delta t} a^2\left(1 - \frac{\tau}{\Delta t}\right)\cos\omega\tau\,d\tau$$

$$= \frac{a^2}{\pi}\left(\frac{1}{\omega^2\Delta t}\right)(1 - \cos\omega\Delta t)$$

$$= \frac{a^2}{\pi}\frac{\Delta t}{2}\left(\frac{\sin\omega\Delta t/2}{\omega\Delta t/2}\right)^2$$

which is shown in Fig. 13.3(*b*). It is usual to express this spectrum as a single-sided function of frequency in Hz, $W_x(f)$, where if $f_0 = 1/\Delta t$ is the "clock frequency" then, using (5.25),

$$W_x(f) = \frac{2a^2}{f_0}\left(\frac{\sin \pi f/f_0}{\pi f/f_0}\right)^2.$$ (13.4)

This function is plotted to linear and log scales in Figs. 13.4(*a*) and (*b*). Up to a frequency of about $\frac{1}{2}f_0 = 1/2\Delta t$ Hz, the spectrum is approximately flat. Provided that the clock frequency (i.e. the frequency of tossing coins) is sufficiently fast,

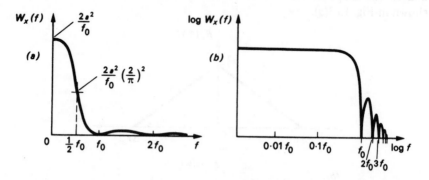

Fig. 13.4 Spectrum for a random binary process with clock frequency $f_0 = 1/\Delta t$.

then this random binary process has a spectrum corresponding to broad band noise and it can be used as a suitable input signal for low frequency system identification studies. Because the input only switches between two constant levels $+a$ and $-a$, there are no occasional large excursions to interfere with the steady state operation of a large plant, and a binary signal is therefore especially suitable for on-line process control applications.

Pseudo random binary signals

The concept of generating a random signal by tossing a coin is a helpful way of understanding the properties of a random binary process, but it is obviously not a practical means of producing such a signal. However one way of doing this would be to prepare a binary sequence in advance by tossing a coin and recording the results and then making a punched paper tape or other recording of this signal. To avoid difficulties with starting and stopping, the two ends of the recording can be joined together to make a loop which runs continuously. The signal then becomes periodic and repeats itself with period $N\Delta t$ (say). Such a signal is described as a *pseudo random binary signal* or *p.r.b.s.* for short and is in fact the type of output produced by binary noise generators by different means which we shall describe briefly later.

The spectrum of a pseudo random binary process can be calculated by first determining the autocorrelation function $R_x(\tau)$ and then transforming it as we have just done for a non-repetitive binary process. In this case we shall assume that the random process $x(t)$ consists of an ensemble of sample functions obtained by running many different binary recordings simultaneously. Each tape loop is assumed to be of the same length, with period $N\Delta t$, and the recordings are started at different randomly distributed times and then allowed to run continuously. Provided that the mean value $E[x]$ is zero, which means that the sequence must have the same number of "heads" as "tails", then the ensemble autocorrelation function will be the same as Fig. 13.3(a) except that it will repeat itself with period $N\Delta t$ as shown in Fig. 13.5(a).†

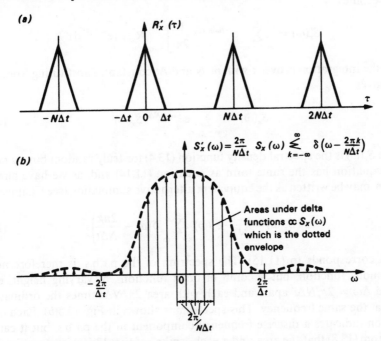

Fig. 13.5 Autocorrelation function and spectral density for a pseudo random binary process with clock frequency $1/\Delta t$ and length of sequence N

Therefore we can write

$$R'_x(\tau) = \sum_{m=-\infty}^{\infty} R_x(\tau - mN\Delta t) \qquad (13.5)$$

where $R'_x(\tau)$ is the autocorrelation· function for the p.r.b.s. and $R_x(\tau)$ is the

†Note that there will be small (random) differences between the sample autocorrelation functions and the ensemble autocorrelation function. See problem 13.3 for analysis of an ensemble consisting of the same binary tape loops.

autocorrelation function for the original truly random binary signal. Taking
Fourier transforms of both sides of (13.5) gives

$$S'_x(\omega) = \frac{1}{2\pi} \int_{-\infty}^{\infty} R'_x(\tau) e^{-i\omega\tau} d\tau$$

$$= \sum_{m=-\infty}^{\infty} \frac{1}{2\pi} \int_{-\infty}^{\infty} R_x(\tau - mN\Delta t) e^{-i\omega\tau} d\tau \qquad (13.6)$$

and changing the time variable to

$$\tau' = \tau - mN\Delta t$$

this becomes

$$S'_x(\omega) = \sum_{m=-\infty}^{\infty} e^{-i\omega mN\Delta t} \frac{1}{2\pi} \int_{-\infty}^{\infty} R_x(\tau') e^{-i\omega\tau'} d\tau'$$

since the integration is over τ with m, N and Δt constant. Substituting from (5.1)
then gives

$$S'_x(\omega) = \sum_{m=-\infty}^{\infty} e^{-i\omega mN\Delta t} S_x(\omega) \qquad (13.7)$$

where $S_x(\omega)$ is the spectral density function (13.4) for truly random binary noise.
This equation has the same form as equation (11.14) and, as we have already
shown, may be written as the equivalent alternative summation (see Chapter 11),

$$S'_x(\omega) = \frac{2\pi}{N\Delta t} S_x(\omega) \sum_{k=-\infty}^{\infty} \delta\left(\omega - \frac{2\pi k}{N\Delta t}\right) \qquad (13.8)$$

which corresponds to (11.15). The spectrum of the p.r.b.s. is therefore not a
continuous function, but a comb of delta functions of varying height, each
spaced $\Delta\omega = 2\pi/N\Delta t$ apart and each with area $2\pi/N\Delta t$ times the ordinate of
$S_x(\omega)$ at the same frequency. This spectrum is shown in Fig. 13.5(b). Each delta
function indicates a discrete frequency component in the p.r.b.s. but it can be
seen from (13.8) that the area under each "spike" of Fig. 13.5(b) is approximately
equal to the area under the continuous spectrum $S(\omega)$ of Fig. 13.3(b) over a frequency band $\Delta\omega = 2\pi/N\Delta t$ whose centre frequency corresponds to this
"spike". Provided that the sequence length N is long enough, the spacing
between adjacent spikes, $2\pi/N\Delta t$ rad/s, is much less than the clock frequency
$2\pi/\Delta t$ rad/s. Although only discrete frequencies occur, they are then so closely
spaced that the characteristics of broad band noise are approximately preserved for frequencies less than about half the clock frequency ($f_0/2$ Hz $= \pi/\Delta t$
rad/s).

The spectrum of a pseudo random binary signal is therefore an approximation for broad band noise provided that its clock frequency is fast enough
and its sequence length long enough. However its amplitude distribution is
just a two-level function which is very different from the Gaussian distribution

common in naturally occurring environments and which must be simulated in environmental testing.

Random multi-level process

In order to generate an approximately Gaussian random signal from a binary signal it is possible to take practical advantage of the central limit theorem (already briefly discussed in Chapter 7). For our present purpose, this may be stated in the following form. If a random variable y is the result of summing many statistically independent variables x_1, x_2, x_3, ..., etc., so that

$$y = \sum_{r=1}^{n} x_r \tag{13.9}$$

then the probability distribution for y tends to become Gaussian as the number n of independent variables approaches infinity. This result applies regardless of the probability distributions of the x variables so long as y has finite mean and variance.[†]

Consider, for example, what happens when we define y as the sum of two coin tossing experiments in which the outcome of each may be either $+a$ or $-a$. There are four equally likely possibilities for y: $(a + a)$, $(a - a)$, $(-a + a)$, $(-a - a)$ so that the probability of $y = 2a$ is $\frac{1}{4}$, of $y = 0$ is $\frac{1}{2}$, and of $y = -2a$ is $\frac{1}{4}$. The probability density function for y is therefore just three delta functions of areas $\frac{1}{4}$, $\frac{1}{2}$ and $\frac{1}{4}$ as shown in Fig. 13.6(b). If we go one stage further and define y as the sum of three coin tossing experiments, there are eight equally likely possible outcomes: $y = 3a$, a, a, a, $-a$, $-a$, $-a$, $-3a$ and the corresponding probability density function has the form shown in Fig. 13.6(c). In general when the results of n separate coin tossing experiments are summed, there are 2^n equally likely possible outcomes and y can hold any of the values $y = na$, $(n - 2)a$, $(n - 4)a$, ..., $(-n + 2)a$, $-na$. The probability that $y = (-n + 2r)a$ is $\{n!/r!(n - r)!\}\,(1/2^n)$ (where by 0! we understand unity) and the possible values of y are said to follow a binomial distribution. When n is large, the discrete values extend over a long length of the y axis in Fig. 13.6, but if the scale is compressed, then the case of large n looks as shown in Fig. 13.6(d). In the limit when $n \to \infty$, the variance of y is $\sigma_y^2 = na^2$ and becomes infinite, but if we reduce a as $n \to \infty$ so that σ_y^2 remains finite, then y becomes continuously variable and it can be shown that the binomial distribution becomes indistinguishable from a Gaussian distribution as required by the central limit theorem.[‡]

Noise generators can in principle take advantage of this result by summing the results of the equivalent of n independent coin tossing experiments to generate a multi-level random signal ranging from $-na$ to $+na$, in steps

[†] See, for instance, Davenport, Jr., *et al.* [22] Ch. 5.
[‡] See, for example, Feller [27].

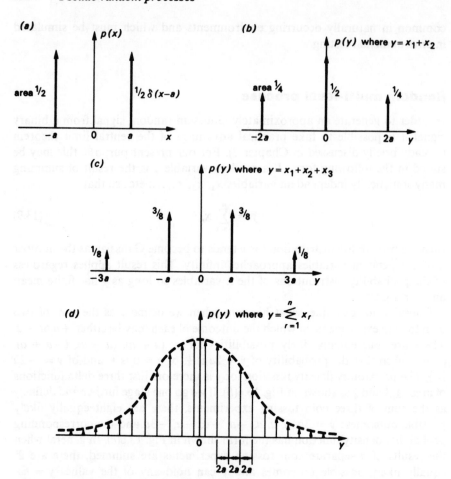

Fig. 13.6 Binomial probability distributions for multi-level random sequences. (Note change of vertical scale for (*d*))

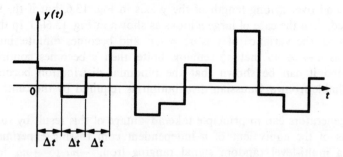

Fig. 13.7 Multi-level random signal obtained by summing the result +*a* or −*a* of each of *n* independent coin tossing experiments

of 2a, Fig. 13.7. This has an amplitude distribution which is approximately Gaussian when n is large and, after filtering to remove the sharp edges of the steps, can be used as an approximation for broad-band continuous noise.

Example

A multi-level random sequence is generated as already described by tossing n coins. Determine the probability distribution for the $(n + 1)$ different levels of the sequence when $n = 9$ and compare this distribution with the corresponding Gaussian distribution.

We can either work out values of $\{n!/r!(n - r)!\}\,(1/2^n)$ for $r = 0$ to $r = n$, or, alternatively, construct the following triangle of numbers, Table 13.1, which represents the evolving probability distribution as n increases.

TABLE 13.1 Coefficients of the binomial distribution represented by
Pascal's triangle

$$y = \sum_{r=1}^{n} x_r$$

n	2^n	$-9a$	$-8a$	$-7a$	$-6a$	$-5a$	$-4a$	$-3a$	$-2a$	$-a$	0	a	$2a$	$3a$	$4a$	$5a$	$6a$	$7a$	$8a$	$9a$
1	2									1		1								
2	4								1		2		1							
3	8							1		3		3		1						
4	16						1		4		6		4		1					
5	32					1		5		10		10		5		1				
6	64				1		6		15		20		15		6		1			
7	128			1		7		21		35		35		21		7		1		
8	256		1		8		28		56		70		56		28		8		1	
9	512	1		9		36		84		126		126		84		36		9		1

By comparing the first three rows of the triangle with Figs. 13.6(a), (b) and (c), it will be clear that the probability that y has a specific value ma (say) is the number under this column of figures divided by 2^n. For instance, if $n = 9$, the probability that

$$y = \sum_{r=1}^{9} x_r = 7a$$

is 9/512. The probability that $y = 5a$ is 36/512, and so on.

We can therefore draw up Table 13.2 which shows the probability distribution for a multi-level sequence with $n = 9$ (n is the number of independent coin tossing experiments $+a$ or $-a$ whose results are added to give y).

TABLE 13.2 Binomial probability distribution of $y = \sum_{r=1}^{9} x_r$, where $x_r = +a$ or $-a$ with equal probability, with corresponding values derived from a Gaussian distribution shown in brackets.

				Prob($y = ma$) where $m =$					
$-9a$	$-7a$	$-5a$	$-3a$	$-a$	a	$3a$	$5a$	$7a$	$9a$
0·002	0·018	0·070	0·164	0·246	0·246	0·164	0·070	0·018	0·002
(0·003)	(0·017)	(0·067)	(0·161)	(0·252)	(0·252)	(0·161)	(0·067)	(0·017)	(0·003)

In order to determine the corresponding values for a Gaussian distribution, we assume that the Gaussian probability density curve is divided into a series of vertical strips of width $2a$ as shown in Fig. 13.8.

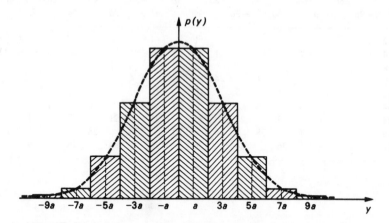

Fig. 13.8 Calculation of an approximate distribution for a multi-level sequence from a Gaussian curve

To compare with the binomial probability that $y = ma$ we shall calculate the area of the strip whose centre is at $y = ma$, which is given by

$$p(y = ma)2a \tag{13.10}$$

where $p(y)$ is the Gaussian probability density function

$$p(y) = \frac{1}{\sqrt{2\pi}\,\sigma}e^{-y^2/2\sigma^2}.$$ (13.11)

Since the x_r random variables obtained from tossing coins are un-correlated, and each has zero mean,

$$\sigma^2 = E[y^2] = nE[x^2] = na^2.$$ (13.12)

Substituting (13.11) and (13.12) into (13.10) then gives the probability of $y = ma$ as

$$\sqrt{\frac{2}{\pi n}}e^{-m^2/2n}$$

values of which are included in brackets in Table 13.2 for comparison with the coefficients of the binomial distribution.

Even for n equal to only 9 (i.e. a 10-level sequence) the probability distribution is almost Gaussian over the 0 to 3σ range, and, unless the occasional very large excursion (which cannot occur for the multi-level sequence) is important, the experimental error in assuming that the smoothed sequence is Gaussian is probably negligible.

Spectrum of a multi-level process

We have described how a multi-level signal with $(n + 1)$ separate levels may be generated by tossing n coins and adding the results $x_r = +a$ or $-a$ to define

$$y = \sum_{r=1}^{n} x_r$$ (13.9)

where it is assumed that the n coins are tossed once every interval Δt and that y remains constant during each interval, Fig. 13.7. Once again we could carry out the coin tossing experiments in advance and prepare a tape recording of the multi-level signal. If we tossed the coin afresh n times for each extra interval on the tape recording, the values of y in successive intervals would be completely uncorrelated and the autocorrelation function (for an infinitely long tape) would have the same form as Fig. 13.3(a) except that the value of $R(0)$ would be na^2 instead of a^2. However suppose that we choose not to toss the coin n more times for each new interval on the recording, but instead use the results of some of the earlier tosses to save time. The multi-level random variable y will still be the result of adding n contributions which are independent of each other, but nearby values of y will now be correlated since they are obtained by adding the results of some of the same experiments.

For instance, suppose that we make a four-level tape recording as follows. First we repeatedly toss a coin to generate a random binary sequence, from which we construct the random signal shown in Fig. 13.9(a). Then from this we generate a multi-level sequence y_r by summing the results of the three last coin tossing experiments x_r, x_{r-1} and x_{r-2} so that

$$y_r = \sum_{s=0}^{2} x_{r-s} \qquad (13.13)$$

to obtain the four-level signal shown in Fig. 13.9(b) (which corresponds to the values of x_r in the binary signal above). The signal in the lower figure will have a

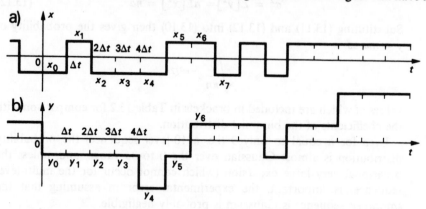

Fig. 13.9 Generation of a four-level signal by summing three adjacent values of a binary signal

binomial distribution according to the third line of Table 13.1, but its spectrum will now differ from that of the original binary signal from which it was derived. In fact, by looking at Figs. 13.9(a) and (b), it is not difficult to be convinced that lower frequency components predominate more in the multi-level sequence, and we shall now see that this is indeed always the case.

Consider the autocorrelation function $R_y(\tau)$ of the four-level sequence y_r. Clearly if $\tau \geqslant 3\Delta t$ there will be no correlation between $y(t)$ and $y(t + \tau)$ so that

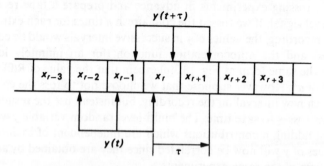

Fig. 13.10 Illustrating correlation between nearby terms of a four-level signal derived by summing three adjacent values of a binary signal

$R_y(\tau)$ will be zero. Also if $\tau = 0$, $R(0) = E[y^2] = 3a^2$. For values of τ between 0 and $3\Delta t$, $y(t)$ and $y(t + \tau)$ will be partially correlated because the components x_r, x_{r-1} and x_{r-2} of $y(t)$ and $y(t + \tau)$ will then sometimes be the same, Fig. 13.10. By making a calculation along the same lines as that used earlier to derive the autocorrelation function for a binary signal (see problem 13.2), we can show in the same way that the correlation function again falls linearly from its initial value $3a^2$ to zero at $\tau = 3\Delta t$. The correlation function for a multi-level signal of $(n + 1)$ levels, derived by progressively summing n adjacent values of a binary

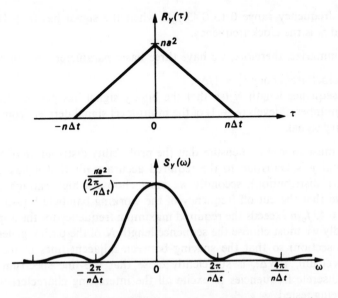

Fig. 13.11 Autocorrelation function and spectral density for an $(n+1)$-level signal derived by summing n adjacent values of a binary signal

signal, has the same form and falls linearly from na^2 to zero at $\tau = n\Delta t$ as shown in Fig. 13.11(a). The corresponding spectrum is given by

$$S_y(\omega) = \frac{na^2}{\pi}\left(\frac{n\Delta t}{2}\right)\left(\frac{\sin \omega n\Delta t/2}{\omega n\Delta t/2}\right)^2 \qquad (13.14)$$

which is illustrated in Fig. 13.11(b). This spectrum first falls to zero at frequency

$$f = \frac{\omega}{2\pi} = \frac{1}{n\Delta t} = \frac{f_0}{n} \qquad (13.15)$$

where f_0 is the clock frequency of the constituent binary signal. The range of frequencies over which the process y can be considered to have broad band characteristics is accordingly divided by the number n and very much reduced from the frequency range of the original binary signal.

For $f = 0.45 f_0/n$, the function

$$\left|\frac{\sin \omega n \Delta t/2}{\omega n \Delta t/2}\right|^2 = \left|\frac{\sin(0.45\pi)}{0.45\pi}\right|^2 = \left|\frac{0.99}{1.41}\right|^2 = 0.5 \text{ (approx.)}$$

so that the spectrum falls to half its initial level in the frequency range 0 to $0.45 f_0/N$. Alternatively, in the dB scale, there is a change of

$$10\log_{10}\left(\frac{1}{2}\right) = -3 \text{ dB}$$

over the frequency range 0 to $0.45 f_0/n$ when the signal has $(n + 1)$ discrete levels and f_0 is the clock frequency.

To summarize, therefore, we have three basic parameters to control:

(i) the clock frequency $f_0 = 1/\Delta t$;
(ii) the sequence length N (so that the binary signal has period $N\Delta t$);
(iii) the number of levels $(n + 1)$ of the multi-level signal derived from the basic binary signal.

First we must choose n to ensure that the probability distribution of the multi-level signal y is Gaussian to the required accuracy (or that it has a suitable alternative distribution); secondly we must choose a high enough clock frequency so that the cut-off frequency of the working bandwidth (-3 dB level) given by $0.45 f_0/n$ exceeds the required maximum frequency for the application; and thirdly we must choose the sequence length N of the p.r.b.s. generator (see the next section) so that the spacing between adjacent lines of the periodic noise spectrum, $1/N\Delta t$, is sufficiently close that the line spectrum includes enough discrete frequencies to excite all the interesting characteristics of the system being tested.

The calculation of y_r as the sum of adjacent values of x_r according to (13.13) is an example of a *digital filtering* operation and we have seen that

$$y_r = \sum_{s=0}^{n-1} x_{r-s} \tag{13.16}$$

defines an imperfect low pass filter. By assigning different weights α_s to the contributions from individual values of the x sequence, to give

$$y_r = \sum_{s=0}^{n-1} \alpha_s x_{r-s} \tag{13.17}$$

the frequency response of the digital filter can be improved so that it has a sharper cut-off frequency and so that it suppresses unwanted high-frequency components more effectively than the crude filter of (13.16). If a precisely shaped noise spectrum is required, such additional digital filtering may be necessary and the reader is referred to specialist books on filter design for more information on this subject.[†] However digital noise generators for environmental

† See, for instance, Gold *et al.* [29].

testing and system identification work generally already incorporate sufficiently precise spectrum shaping filters for most practical applications.

Generation of random numbers

We shall now briefly investigate the way in which a digital noise generator produces its basic p.r.b. signal.

Fundamentally there are two different methods of "tossing coins" electronically. In one method the output of a controlled semiconductor noise source is sampled just before each clock pulse and, depending on the instantaneous level of natural noise, the output signal is switched to $+a$ or $-a$ for the next interval. In this case the binary signal produced is theoretically truly random and has the continuous spectrum shown in Fig. 13.4. The spectrum and probability distribution of the output can be controlled independently of the characteristics of the source of random noise and some of the disadvantages of "natural noise" are overcome. However it is still important to maintain the noise source's stability in order to ensure that both output levels occur with equal probability. In order to avoid this difficulty, a common alternative approach is to use a discrete random number generator as the noise source. This is a digital circuit for producing a chain of numbers in a seemingly random order.

Suppose, for instance, that we want to generate four-digit binary numbers. There are 2^4 possible numbers: 0000, 0001, 0010, 0011, 0100, 0101, 0110, 0111, 1000, 1001, 1010, 1011, 1100, 1101, 1110 and 1111. It turns out that all these numbers except 0000 can be successively generated by the simple logical circuit shown in Fig. 13.12 which involves a four-stage "shift register", an adding device

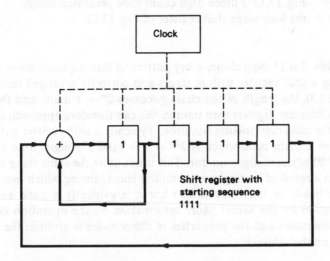

Fig. 13.12 Generation of 15 digit chain code using a four-stage shift register

and a clock. These components work in the following way. Each stage of the shift register stores a digit 0 or 1. At every pulse from the clock, all the digits are moved one place to the right. The digit in the last (right-hand) stage is abandoned while the empty place in the first stage is filled by a new digit obtained by adding the digits previously present in the first and last stages of the shift register (before the clock pulse arrived). When the digits to be added are both 1, then 0 is entered as the new number in the first stage of the register (this is called "modulo 2" addition). Suppose we begin with four 1's in the shift register corresponding to the binary number 1111. On arrival of the first clock pulse, this becomes 0111. When the second pulse arrives it becomes 1011, and so on. The full sequence is 1111, 0111, 1011, 0101, 1010, 1101, 0110, 0011, 1001, 0100, 0010, 0001, 1000, 1100, 1110, 1111, 0111, etc., with all 15 of the possible non-zero numbers appearing before the sequence begins to repeat itself. If we consider the four digits in the shift register as digits on a tape which is incremented one stage at every clock pulse, the digits on this tape will be as shown in Fig. 13.13. The overall pattern of digits repeats itself every $2^4 - 1 = 15$ digits,

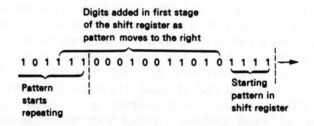

Fig. 13.13 Fifteen digit chain code generated using the four-stage shift register of Fig. 13.12

while, within this 15-digit chain, every pattern of four adjacent digits is unique.

By using a shift register with m stages and correctly arranged feedback (see problem 13.3), the length of the chain becomes $2^m - 1$ digits and the pattern of every m adjacent digits is then unique. We can therefore approach closer and closer to the ideal coin tossing situation. Typically a shift register with $m = 20$ stages gives a chain code which is $2^{20} - 1 = 1\,048\,575$ digits long with every pattern of 20 adjacent digits unique. This is not quite the same thing as a signal built up as a result of tossing over a million coins (during which more than 20 successive heads or tails may occur) but it is evidently a good engineering approximation for the same! More information on the operation of random number generators and the properties of chain codes is given in the specialist literature on the subject†.

---

†This includes Briggs *et al.* [6], Golomb [30], Hartley [33, 34], Heath *et al.* [35], Huffman [37], MacLaren *et al.* [47] and Vincent [98, 99].

Synthesis of correlated noise sources

So far we have been concerned with generating artificially a random signal which reproduces the statistical characteristics of an ergodic random process $x(t)$ (say) and so may be thought of as a member function of the ensemble $\{x(t)\}$. We have seen how a random binary sequence obtained by "tossing coins" can serve as the basic mechanism for artificial random signal generation, and how, after manipulating this to form a multi-level sequence, an approximately Gaussian signal with broad band frequency composition is produced. By filtering the broad band output signal, any desired output spectrum $S_x(\omega)$ can then be obtained.

For environmental testing work, more than one artificial random input may be needed and it may be necessary for some or all of these separate inputs to be correlated with each other. In this case the simulation problem becomes more complex because not only must the artificial noise sources have (auto) spectral densities and amplitude distributions which are representative of the environment being simulated, but they must also have appropriate cross-spectral densities. There are a number of ways of attacking this problem.

One way is to begin by generating statistically independent signals using the p.r.b.s. approach, but using a different binary chaincode for each signal, and then combining these independent signals to produce correlated signals with the required characteristics.† For instance, if $x(t)$ and $y(t)$ are two independent (uncorrelated) random signals, and a new process $z(t)$ is generated such that

$$z(t) = \alpha x(t - t_1) + \beta y(t - t_2) \tag{13.18}$$

where α, β, t_1 and t_2 are all constants, then the autocorrelation function for z is

$$\begin{aligned} R_z(\tau) &= E[z(t)z(t + \tau)] \\ &= \alpha^2 R_x(\tau) + \beta^2 R_y(\tau) \end{aligned} \tag{13.19}$$

and the cross-correlation functions of z with x and y are

$$\begin{aligned} R_{zx}(\tau) &= E[z(t)x(t + \tau)] \\ &= \alpha R_x(\tau + t_1) \end{aligned}$$

and $$\tag{13.20}$$

$$R_{zy}(\tau) = \beta R_y(\tau + t_2)$$

since $R_{xy}(\tau) = R_{yx}(\tau) = 0$ because x and y are uncorrelated. If necessary more terms can be included in (13.18) to permit more latitude in tailoring the statistical characteristics of $z(t)$, but the approach lacks flexibility and a completely different method of noise simulation based on the FFT is more suitable for practical application.

---

† See, for instance, Otnes *et al.* [52] Ch. 12; and Dodds [23]. (The latter paper shows how the two correlated vibration inputs to the two front wheels of a motor vehicle may be simulated by combining two independent noise sources.)

This alternative method involves using the FFT to generate discrete time series which, after smoothing, serve as the required artificial random signals. The basic idea is to specify discrete values of the spectra and cross-spectra and then to use the FFT to calculate their IDFT's (inverse discrete Fourier transforms) and so generate discrete time sequences $\{x_r\}$ and $\{y_r\}$ (say) which are regarded as sampled values of the continuous random signals $x(t)$ and $y(t)$ required.

We know from Chapters 10 and 11 that the DFT's of $\{x_r\}$ and $\{y_r\}$, $r = 0, 1, 2, ..., (N - 1)$, are defined as

$$X_k = \frac{1}{N} \sum_{r=0}^{N-1} x_r e^{-i(2\pi kr/N)}$$

$$\qquad\qquad\qquad k = 0, 1, 2, ..., (N - 1) \qquad (10.8)$$

$$Y_k = \frac{1}{N} \sum_{r=0}^{N-1} y_r e^{-i(2\pi kr/N)}$$

and that discrete values of the auto- and cross-spectra are given by

$$S_{xx_k} = X_k^* X_k \qquad\qquad S_{xy_k} = X_k^* Y_k$$

$$S_{yx_k} = Y_k^* X_k \qquad\qquad S_{yy_k} = Y_k^* Y_k \qquad (10.26)$$

where the S_k give estimates of the corresponding smoothed continuous spectrum† by the formula (from (11.33))

$$S_k \simeq \frac{2\pi}{N\Delta} \tilde{S}(\omega_k) \qquad (13.21)$$

at angular frequency

$$\omega_k = \frac{2\pi k}{N\Delta} \text{ rad/s.}$$

Consider the case when discrete values of the frequency series $\{S_{xx_k}\}$, $\{S_{xy_k}\} = \{S_{yx_k}^*\}$ and $\{S_{yy_k}\}$, $k = 0, 1, 2, ..., (N - 1)$, are all specified and we want to generate the original time series $\{x_r\}$ and $\{y_r\}$, $r = 0, 1, 2, ..., (N - 1)$, from which they have been derived. If we put

$$X_k = |X_k| e^{i\theta_k}$$

and

$$Y_k = |Y_k| e^{i\phi_k} \qquad (13.22)$$

then, by substituting into (10.26),

$$|X_k|^2 = S_{xx_k}$$

$$|Y_k|^2 = S_{yy_k} \qquad (13.23)$$

---

† We saw in Ch. 11 that $\tilde{S}(\omega_k)$ is the result of smoothing the time spectrum $S(\omega)$ according to the formula

$$\tilde{S}(\omega_k) = \int_{-\infty}^{\infty} W(\Omega - \omega_k) S(\Omega) \, d\Omega \qquad (11.29)$$

where $W(\Omega)$ is the basic spectral window function defined by (11.32).

and

$$|X_k||Y_k|e^{i(\phi_k - \theta_k)} = S_{xy_k} = S_{yx_k}^* . \qquad (13.24)$$

We can therefore determine the amplitudes $|X_k|$ and $|Y_k|$ and the phase difference $(\phi_k - \theta_k)$, but not the absolute values of θ_k and ϕ_k. Hence the spectral data is not sufficient to allow X_k and Y_k to be uniquely determined and there is not a unique solution to our problem. This is hardly surprising when we recall that two completely different random processes – for instance a binary process and a Gaussian process – can have identical spectral densities.

The choice of one or other (but not both) of the phase angles θ_k and ϕ_k is arbitrary, but if we want to generate time series $\{x_r\}$ and $\{y_r\}$ which represent Gaussian signals $x(t)$ and $y(t)$, then the phases must be randomly distributed between 0 and 2π. In this case we know from the central limit theorem that the distributions of the terms in $\{x_r\}$ and $\{y_r\}$ approach Gaussian distributions when the number of terms becomes very large. The reason for this is that, from the IDFT relations,

$$x_r = \sum_{k=0}^{N-1} X_k e^{i(2\pi kr/N)}$$

$$y_r = \sum_{k=0}^{N-1} Y_k e^{i(2\pi kr/N)} \qquad r = 0, 1, 2, \ldots, (N-1), \qquad (10.7)$$

each term of the $\{x_r\}$ and $\{y_r\}$ series is the result of summing many statistically independent contributions and so becomes a sample from a population with Gaussian characteristics.

In synthesizing the signals $x(t)$ and $y(t)$, the procedure is to calculate each of the $|X_k|$, $|Y_k|$ and $(\phi_k - \theta_k)$ from the specified spectral data, using (13.23) and (13.24). Then each value of ϕ_k (say) is chosen at random from the range 0 to 2π, using a random number generator as further described below, and specific values of ϕ_k and θ_k obtained. Although the difference between each pair of values $(\phi_k - \theta_k)$ must satisfy (13.24), both ϕ_k and θ_k may be thought of as samples drawn from populations in which any value between 0 and 2π is equally likely. The discrete time series $\{x_r\}$ and $\{y_r\}$ are calculated from (10.7) and, after smoothing, serve as an approximation for the correlated random signals required.

In the last section we discussed briefly the way in which an apparently random sequence of the numbers 0 and 1 can be obtained by using a logical circuit incorporating a shift register. The operation of this system involving a four-stage shift register as shown in Fig. 13.12 can be described by the equation

$$x_n = (x_{n-1} + x_{n-4}) \text{ modulo } 2 \qquad (13.25)$$

where x_n denotes the number in the first (left-hand) stage of the register at time $n\Delta t$, x_{n-1} is the number there at time $(n-1)\Delta t$, and x_{n-4} is the number in the first stage at time $(n-4)\Delta t$ which becomes the number in the fourth (right-hand) stage at time $(n-1)\Delta t$ since the numbers move one stage to the right at every clock pulse. The addition is "modulo 2" because x_n is the remainder after dividing $(x_{n-1} + x_{n-m})$ by 2. This scheme may be extended to generate random numbers distributed uniformly between 0 and 1 by modifying the equation to

$$x_n = (x_{n-1} + x_{n-m}) \text{ modulo } 1 \qquad (13.26)$$

where the equivalent shift register now has m stages and the initial m numbers in the stages are distributed between 0 and 1 (Green *et al.* [31]). For instance, if $x_{n-1} = 0.739$ and $x_{n-m} = 0.527$, then $x_n = 0.266$ which is the remainder after dividing $(x_{n-1} + x_{n-m})$ by 1. In this case, any number $x = x_n$ can be thought of as a randomly chosen sample from a uniformly distributed population in the range 0 to 1, so we can choose the random phase angle ϕ_k by putting

$$\phi_k = x(2\pi).$$

The corresponding phase angle θ_k is then calculated from the known value of $(\phi_k - \theta_k)$, equation (13.24).

When two pseudo-random signals are generated in this way from (13.23) and (13.24), they have the special property that their coherence (see Chapter 14) is unity. To generate two signals which are not fully coherent, a third time series $\{z_r\}$, generated from a spectrum with an independent set of random phase angles, must be added to either $\{x_r\}$ or $\{y_r\}$ (problem 15.1).

The digital simulation of pseudo-random processes is an interesting if specialized field in which there is a growing literature (Shinozuka *et al.* [62, 63]).

Chapter 14

Application notes

In this book we have concentrated on the fundamental ideas of random vibrations and random process analysis rather than on specific applications of the theory. The purpose of this final chapter is to add some brief notes on aspects of the subject which relate more particularly to its practical application and have not so far been covered in the text.

Response of a resonant mode to broad band excitation

The first note concerns the response of a single resonant mode to random excitation. Consider the single degree-of-freedom vibrating system shown in

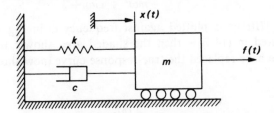

Fig. 14.1 Damped linear oscillator representing the response of a resonant mode.

Fig. 14.1. If its excitation is a force applied in the direction shown, then the equation of motion is

$$m\ddot{x} + c\dot{x} + kx = f(t). \tag{14.1}$$

According to the normal mode theory of linear systems,† this equation may also be considered to represent the response of a single resonant mode of a multi-degree-of-freedom system when the modal mass m, stiffness k, equivalent viscous damping c and modal excitation $f(t)$ are all correctly defined. Our problem is to determine the mean square response of the mode, $E[x^2]$, when it is excited by broad band excitation.

† See, for instance, Halfman [32]

For a theoretical white noise input, we have in fact already carried out this calculation (in Example 1 of Chapter 7) and found that

$$E[x^2] = \frac{\pi S_0}{kc} \tag{14.2}$$

when S_0 is the constant (double-sided) spectral density of the excitation $f(t)$. Consider how this result may be interpreted and extended for a resonant mode whose (undamped) natural frequency is

$$\omega_N = \sqrt{\frac{k}{m}} \tag{14.3}$$

and whose (small) damping ratio ζ is defined by

$$2\zeta\omega_N = \frac{c}{m}. \tag{14.4}$$

From (7.18),

$$E[x^2] = \int_{-\infty}^{\infty} |H(\omega)|^2 S_0 \, d\omega = 2S_0 \int_0^{\infty} |H(\omega)|^2 \, d\omega \tag{14.5}$$

where $H(\omega)$ is the complex frequency response function given in this case by

$$H(\omega) = \frac{1}{-m\omega^2 + ci\omega + k}. \tag{14.6}$$

The function $|H(\omega)|^2$ is plotted against frequency ω in Fig. 14.2(a) and the integral required in (14.5) is then the shaded area shown. In the adjoining Fig. 14.2(b) we have replaced the true response curve (now shown as a broken

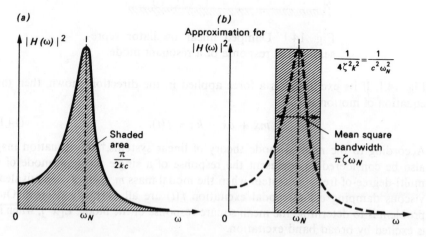

Fig. 14.2 Mean square bandwidth for approximate calculations of the mean square response of a resonant mode

line) by a rectangular approximation which has the same area. If the height of the rectangle is $1/4\zeta^2 k^2$ which is the same as the peak height of $|H(\omega)|^2$ for small damping, then its width must be $\pi\zeta\omega_N$ in order to give the correct area. This bandwidth, which is called the *mean square bandwidth* of the mode, allows a rapid approximate calculation of $E[x^2]$ to be made whenever the excitation bandwidth includes ω_N and is reasonably flat.

Instead of making a precise calculation of the mean square response, which probably means numerically integrating (7.18) when $S_f(\omega)$ is not a constant, all we need do is to put

$$E[x^2] = \sigma_x^2 \text{ (since we are assuming } E[f] = 0)$$

$$\simeq 2 \begin{pmatrix} \text{Average value of } S_f(\omega) \\ \text{in the region of } \omega_N \end{pmatrix} \begin{pmatrix} \text{Peak value} \\ \text{of } H(\omega) \end{pmatrix}^2 \begin{pmatrix} \text{Mean square} \\ \text{bandwidth} \end{pmatrix} \quad (14.7)$$

to find a result which is accurate enough for most requirements. Notice that the mean square bandwidth $\pi\zeta\omega_N$ is wider than the "half-power bandwidth" which is the bandwidth between the two points where

$$|H(\omega)|^2 = \tfrac{1}{2}|H(\omega_N)|^2 \quad (14.8)$$

and comes out to be approximately $2\zeta\omega_N$.

We saw in Chapter 8 that, for Gaussian white noise excitation, the statistical average frequency (the frequency of zero crossings with positive slope) of $x(t)$ is given by

$$v_0^+ = \frac{\omega_N}{2\pi} \quad (14.9)$$

and, for a Gaussian narrow band process, the distribution of peaks is given, from (8.24), by

$$p_p(a) = \frac{a}{\sigma_x^2} e^{-a^2/2\sigma_x^2} \quad 0 \leqslant a \leqslant \infty \quad (14.10)$$

where $p_p(a)\,da$ is the probability that any peak of $x(t)$, chosen at random, lies in the range a to $a + da$, and σ_x^2 is the variance of x. For a highly resonant mode subjected to broad band excitation, we know from the central limit theorem that the output will tend to be Gaussian even though the input is not Gaussian (see Chapter 7). Also (14.9) still applies quite accurately when the excitation is not white noise provided that it is a broad band process covering the resonant frequency of the mode. Both (14.9) and (14.10) may therefore be used with good accuracy to describe the output characteristics of any resonant mode subjected to broad band excitation. The variance σ_x^2 required in (14.10) may be calculated with sufficient accuracy from the approximate result (14.7).

Fatigue and failure due to random vibration

The commonest form of mechanical failure due to vibration is fatigue failure caused by the gradual propagation of cracks in a region of high stress. Under high alternating stresses, cracks propagate faster than under low alternating stresses, and a material's "fatigue law" is generally expressed by means of an experimentally determined *S–N* curve. Here *S* denotes the stress amplitude and *N* the number of cycles of stress (of fixed amplitude *S*) which causes failure, Fig. 14.3. Data is usually obtained by some form of reversed bending test in

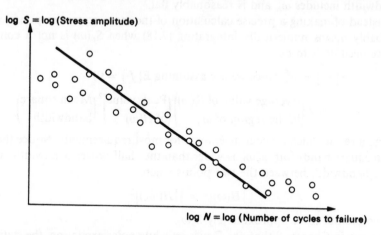

Fig. 14.3 Typical *S–N* fatigue curve plotted to log scales

which samples are alternately bent one way and then the other to generate an alternating stress of constant amplitude, *S*. The number of cycles *N* to failure for each value of *S* is recorded and used to plot the *S–N* curve.

When subjected to random excitation, failure occurs as a result of the combined effect of stress cycles of many different amplitudes, so that the appropriate *S–N* curve is not directly applicable. Unfortunately, at the present time, there is not sufficient understanding of the basic mechanism of fatigue for it to be possible to apply the results of constant stress experiments to the case of randomly varying stresses with certainty. Various hypotheses for what happens under random loading conditions have, however, been suggested and the best known of these is the Palmgren–Miner hypothesis (Palmgren [53]; Miner [50]). This applies for a narrow band process in which separate stress cycles can be identified and suggests that, if n_i cycles of stress occur at a level of stress at which N_i constant stress cycles would cause fracture, then the fractional damage done by the n_i cycles is (n_i/N_i). Failure is to be expected when the sum of all the "fractional damages" is equal to one, i.e. when

$$\sum_i \left(\frac{n_i}{N_i}\right) = 1. \tag{14.11}$$

For a narrow band process obtained as a result of subjecting a resonant system to broad band excitation, we know that in time T there will be on average $v_0^+ T$ stress cycles of which $p_p(S)\,dS$ will have peak values in the stress range S to $S + dS$. If $N(S)$ is the number of cycles of stress level S which cause failure, then one cycle at level S will cause fractional damage $1/N(S)$. Since in time T we expect $(v_0^+ T)(p_p(S)\,dS)$ cycles of this stress level, the average fractional damage done at this stress level is

$$(v_0^+ T)(p_p(S)\,dS)\frac{1}{N(S)}$$

and the average damage resulting from cycles of all stress levels occurring in time T is

$$(v_0^+ T)\int_0^\infty \frac{1}{N(S)} p_p(S)\,dS.$$

This expression may be used to predict the mean lifetime before failure occurs, which is given approximately by

$$T = \frac{1}{v_0^+ \int_0^\infty (1/N(S))p_p(S)\,dS} \tag{14.12}$$

and may be evaluated if the S–N curve is available. Values of v_0^+ and $p_p(S)$ are available from (14.9) and (14.10). The calculation is subject to statistical errors (because of the essentially random nature of the applied stress) which may be approximately calculated (Crandall *et al.* [18]) and experimental errors due to ignorance about the true mechanism of fatigue, which cannot be calculated. In practice it is found that, provided the bandwidth of the resonant peak is not too narrow (i.e. the damping ratio ζ not too small) and the number of cycles to failure (N) is large (so that $\zeta N > 10^3$, say), then the main error is the experimental one due to our ignorance of the fatigue "law". Typically the actual lifetime is then likely to be in a range of values of the order $0.3T$ to $3T$ where T is the value given by (14.12).

Other forms of failure due to random vibration can arise. For instance failure may occur if a response spends too much of its time off range, i.e. it is outside set limits for more than a minimum fraction of its lifetime, Fig. 14.4(*a*).

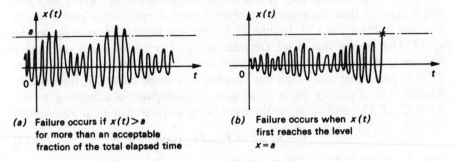

(*a*) Failure occurs if $x(t)>a$
       for more than an acceptable
       fraction of the total elapsed time

(*b*) Failure occurs when $x(t)$
       first reaches the level
       $x=a$

Fig. 14.4 Possible modes of failure under random excitation

Alternatively failure may occur when the response first crosses a set level, Fig. 14.4(*b*).

If we assume that $x(t)$ is ergodic and consider a very long (theoretically infinite) length of a sample function, then the proportion of time spent off range will be the same as the probability that, at any arbitrary time, x exceeds a. Hence, if $p(x)$ is the probability density function, then we can write

$$\binom{\text{Fraction of elapsed}}{\text{time for which } x > a} = \int_{a}^{\infty} p(x)\,dx. \qquad (14.13)$$

When $x(t)$ is a Gaussian process for which

$$p(x) = \frac{1}{\sqrt{2\pi}\,\sigma_x}\,e^{-x^2/2\sigma_x^2}$$

this becomes

$$\binom{\text{Fraction of elapsed}}{\text{time for which } x > a} = \frac{1}{2}\left(1 - \int_{x=0}^{a} \frac{2}{\sqrt{\pi}}\,e^{-x^2/2\sigma_x^2}\,d\left(\frac{x}{\sqrt{2}\,\sigma_x}\right)\right)$$

$$= \frac{1}{2}\left(1 - \operatorname{erf}\frac{a}{\sqrt{2}\,\sigma_x}\right) \qquad (14.14)$$

where $\operatorname{erf} a/\sqrt{2}\,\sigma_x$ is the error function which, when $a/\sqrt{2}\,\sigma_x$ is large, is given by

$$\operatorname{erf}\frac{a}{\sqrt{2}\,\sigma_x} \simeq 1 - \sqrt{\frac{2}{\pi}}\,\frac{\sigma_x}{a}\,e^{-a^2/2\sigma_x^2} \qquad (14.15)$$

so that in this case†

$$\binom{\text{Fraction of elapsed}}{\text{time for which } x > a} \simeq \frac{1}{\sqrt{2\pi}}\,\frac{\sigma_x}{a}\,e^{-a^2/2\sigma_x^2} \qquad \text{(for } a \gg \sigma\text{).} \qquad (14.16)$$

The first crossing problem, Fig. 14.4(*b*), is a more difficult one because the situation is fundamentally non-stationary as the random process is "switched-on" at $t = 0$. If we gloss over this point by assuming that $x(t)$ exists for $t < 0$ and is ergodic, then the average number of times it crosses the level $x = a$ in any time interval T is $v_a^+\,T$. In time dT the average number of crossings will be $v_a^+\,dT$. This means that if we consider many intervals of width dT, only a fraction $v_a^+\,dT$ of all such intervals will have a crossing (we are assuming that dT is so small that more than one crossing in a single interval is impossible). Another way of putting this is to say that the probability of a crossing in time dT is $v_a^+\,dT$. The probability of there *not* being a crossing in dT is accordingly

$$1 - v_a^+\,dT = P_0(dT) \quad \text{(say).} \qquad (14.17)$$

Now if we can make a further sweeping assumption that the crossings are randomly distributed along the time axis, so that what happens in one time

---

†Equation (14.16) gives a result which is greater than the exact value. When $a/\sigma_x = 3$, the approximation from (14.16) is about 10 per cent too high.

interval is completely independent of what happens in adjacent intervals, we can write

$$P_0(T + dT) = P_0(T)P_0(dT)$$
$$= P_0(T)(1 - v_a^+ \, dT) \qquad (14.18)$$

from (14.17) where $P_0(T)$ is the probability of there being no crossings in time T. This may be rearranged to give

$$\frac{P_0(T + dT) - P_0(T)}{dT} = \frac{dP_0(T)}{dT} = -P_0(T)v_a^+ \qquad (14.19)$$

which is a first-order differential equation for $P_0(T)$ which may be solved to give

$$P_0(T) = C\,e^{-v_a^+ T}$$

where C is an arbitrary constant. Since $P_0(0)$ is the probability that there will be no crossings in zero time interval, it must be unity, and so $C = 1$ giving

$$P_0(T) = e^{-v_a^+ T} . \qquad (14.20)$$

The probability of a failure in time T is the probability that there will be one (or more) crossings in this time which is just $(1 - P_0(T))$ and so we finally obtain the following result

$$\text{(Probability of a failure in time } T) = 1 - e^{-v_a^+ T} \qquad (14.21)$$

which is sketched in Fig. 14.5.

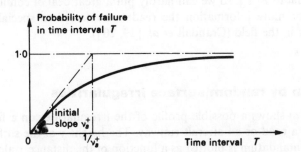

Fig. 14.5 Probability that a narrow band process $x(t)$ crosses the level $x = a$ at least once during time interval T

This result is strictly only correct for an ergodic process with crossings randomly distributed along the time axis. Because large amplitude peaks tend to occur in clumps, the intervals between clumps will be longer than the average spacing between crossings and the probability of failure in time T will therefore be less than that indicated by (14.21).

A further interesting point is that the first passage probability density $p(T)$ can be deduced from (14.21). The probability $p(T)\,dT$ of a *first* crossing between T and $T + dT$ must be the same as the *increase* in the probability of failure between T and $T + dT$. So, from (14.21), we obtain

$$p(T)\,dT = \frac{d}{dT}(1 - e^{-v_a^+ T})\,dT$$

which gives

$$p(T) = v_a^+ e^{-v_a^+ T} \tag{14.22}$$

for the first passage probability density function. The mean value of the time to failure is then given by

$$E(T) = \int_0^\infty T p(T)\,dT = \frac{1}{v_a^+} \tag{14.23}$$

and, since

$$E[T^2] = \int_0^\infty T^2 p(T)\,dT = \frac{2}{(v_a^+)^2}, \tag{14.24}$$

the standard deviation of the time to failure is

$$\sigma_T = \frac{1}{v_a^+} \tag{14.25}$$

which is equal to $E[T]$, so we can hardly put a great deal of confidence in this estimate. For more information the reader is referred to specialist research publications in the field (Crandall *et al.* [19, 20]).

Excitation by random surface irregularities

Figure 14.6(*a*) shows a possible profile of the irregularities on a fixed surface, for instance a road or an aircraft runway. The height y of the surface above a fixed horizontal datum is plotted as a function of the distance x along the road or runway. Instead of varying with time, the height y is now a function of distance; long wavelength irregularities correspond to low frequency components in the time domain and short wavelength irregularities correspond to high frequency components. The angular frequency ω (rad/s) that we are accustomed to is replaced by the so-called *wavenumber* γ (rad/m or rad/ft) which expresses the rate of change with respect to distance. Just as the period T of a time-varying component with frequency ω is given by

$$T = \frac{2\pi}{\omega} \quad \text{(units of time)} \tag{14.26}$$

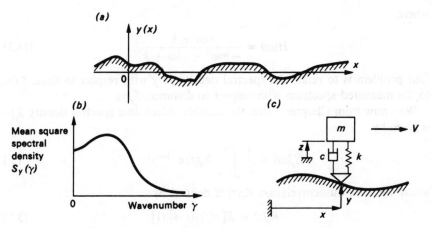

Fig. 14.6 Profile and spectrum of a rough surface with a primitive model of a moving vehicle

so the wavelength λ of a spatially-varying component with wavenumber γ is given by

$$\lambda = \frac{2\pi}{\gamma} \text{ (units of distance).} \tag{14.27}$$

The mean square spectral density of the height variable y is therefore now a function of wavenumber (instead of angular frequency) and typically has the form shown in Fig. 14.6(b). In practice $S_y(\gamma)$ may be determined by measuring y at closely spaced intervals along the x axis and then employing the calculation procedure described in Chapters 10 and 11. In Fig. 14.6(b), $S_y(\gamma)$ is drawn as a single-sided function of γ and therefore

$$E[y^2] = \int_0^\infty S_y(\gamma)\,d\gamma. \tag{14.28}$$

An interesting application of random vibration theory involves calculating how a moving vehicle responds to road surface irregularities. In Fig. 14.6(c), a one degree-of-freedom model of a "vehicle" with a simple suspension system is shown moving at constant speed V along a rough surface. With respect to an an observer on the moving vehicle, coordinates y and z are functions only of time, and the equation of motion of the vehicle is

$$m\ddot{z} + c\dot{z} + kz = c\dot{y} + ky. \tag{14.29}$$

The input to the suspension system is the time varying parameter $y(t)$, and if this has spectral density $S_y(\omega)$ with respect to time, then the spectral density of the vehicle response $S_z(\omega)$ is given by (see Chapter 7)

$$S_z(\omega) = |H(\omega)|^2 S_y(\omega) \tag{14.30}$$

where

$$H(\omega) = \frac{ci\omega + k}{-m\omega^2 + ci\omega + k}. \tag{14.31}$$

Our problem is to relate the spectral density of y with respect to time, $S_y(\omega)$, to the measured spectrum with respect to distance, $S_y(\gamma)$.

We know from Chapter 5 that the (double-sided) time spectral density $S_y(\omega)$ is given by

$$S_y(\omega) = \frac{1}{2\pi} \int_{-\infty}^{\infty} R_y(\tau) e^{-i\omega\tau} d\tau \tag{5.1}$$

where the time autocorrelation $R_y(\tau)$ is defined by

$$R_y(\tau) = E[y(t)y(t + \tau)]. \tag{3.13}$$

A corresponding spatial autocorrelation function $R_y(X)$ is defined by

$$R_y(X) = E[y(x)y(x + X)] \tag{14.32}$$

and this leads to the formal definition of the (double-sided) spatial spectral density

$$S_y(\gamma) = \frac{1}{2\pi} \int_{-\infty}^{\infty} R_y(X) e^{-i\gamma X} dX. \tag{14.33}$$

Assuming that the vehicle speed V is constant, we can relate $S_y(\omega)$ and $S_y(\gamma)$ by some simple substitutions. First there is a relation between the time lag τ and the space lag X appearing in (3.13) and (14.32), since a vehicle travelling at speed V takes time

$$\tau = \frac{X}{V} \tag{14.34}$$

to travel between two points distance X apart on the road. Secondly, since a cycle of wavelength $\lambda = 2\pi/\gamma$ is covered in period T, where

$$T = \frac{\lambda}{V},$$

we know, from (14.26) and (14.27), that the time angular frequency

$$\omega = \frac{2\pi}{T} = V\frac{2\pi}{\lambda} = V\gamma. \tag{14.35}$$

Substituting these results (14.34) and (14.35) into (5.1) gives

$$S_y(\omega = V\gamma) = \frac{1}{2\pi} \int_{-\infty}^{\infty} R_y\left(\tau = \frac{X}{V}\right) e^{-i(V\gamma)(X/V)} \cdot \frac{1}{V} dX$$

$$= \frac{1}{V} \cdot \frac{1}{2\pi} \int_{-\infty}^{\infty} R_y(X) e^{-i\gamma X} dX$$

which, from (14.33), gives

$$S_y(\omega = V\gamma) = \frac{1}{V} S_y(\gamma) \tag{14.36}$$

or, its equivalent,

$$S_y(\omega) = \frac{1}{V} S_y\left(\gamma = \frac{\omega}{V}\right) \tag{14.37}$$

and allows the time spectral density $S_y(\omega)$ to be obtained directly from the spatial spectral density $S_y(\gamma)$ of the surface profile. Both spectral densities in (14.36) and (14.37) are double-sided functions (i.e. defined for ω and γ running from $-\infty$ to ∞) or, alternatively, both are single-sided functions (ω and γ running from 0 to ∞). Also if we are dealing with single-sided spectra whose units are those of (mean square)/(Hz), $W_y(f)$, and (mean square)/(cycle per unit length), $W_y(1/\lambda)$, then, from Chapter 5,

$$W_y(f) = 4\pi S_y(\omega = 2\pi f) \tag{5.25}$$

when $S_y(\omega)$ is a double-sided function, and, similarly,

$$W_y\left(\frac{1}{\lambda}\right) = 4\pi S_y\left(\gamma = \frac{2\pi}{\lambda}\right) \tag{14.38}$$

when $S_y(\gamma)$ is double sided. In this case (14.37) gives

$$W_y(f) = \frac{1}{V} W_y\left(\frac{1}{\lambda} = \frac{f}{V}\right). \tag{14.39}$$

Example

Estimate the r.m.s. displacement response of the "vehicle" shown in Fig. 14.6(c) when its suspension has a natural frequency

$$\frac{1}{2\pi}\sqrt{\frac{k}{m}} = 1.5 \text{ Hz}$$

and a damping ratio

$$\zeta = \frac{1}{2}\frac{c}{\sqrt{mk}} = 0.1$$

and it is travelling at (a) 30 km/h and (b) 100 km/h over a road surface whose one-sided spatial spectral density is 100 mm² per cycle/m (constant).

First note that the units of spatial spectral density are mm² per (cycle/m). The corresponding one-sided time spectral density is from (14.39),

$$W_y(f) = \frac{100}{V} \text{ mm}^2 \text{ per cycle/s}$$

when V is in m/s.

Rather than make an exact calculation, we can approximate by using the results of the first section of this chapter. In this case the frequency response function is

$$\frac{ci\omega + k}{-m\omega^2 + ci\omega + k}$$

rather than

$$\frac{1}{-m\omega^2 + ci\omega + k}$$

but since the resonant peak (Fig. 14.2) is determined almost entirely by the denominator (we are assuming small damping), the approximate result (14.7) is still a first approximation for the variance, which in this example is σ_z^2, and so, from (14.7),†

$$\sigma_z^2 \simeq \begin{pmatrix} \text{Average value of } W_y(f) \text{ in} \\ \text{the region of } f = f_N = \omega_N/2\pi \end{pmatrix} \begin{pmatrix} \text{Peak value} \\ \text{of } H(f) \end{pmatrix}^2 \begin{pmatrix} \text{Mean square} \\ \text{bandwidth in Hz} \end{pmatrix}$$

so that

$$\sigma_z^2 \simeq \left(\frac{100}{V}\right)\left(\frac{k}{c\omega}\right)^2 (\pi \zeta f_N) \text{ mm}^2.$$

Substituting from (14.3) and (14.4) then gives

$$\sigma_z^2 \simeq \left(\frac{100}{V}\right)\left(\frac{1}{2\zeta}\right)^2 (\pi \zeta f_N) \text{ mm}^2$$

which, putting $f_N = 1 \cdot 5$ and $\zeta = 0 \cdot 1$, becomes

$$\sigma_z^2 \simeq \frac{375\pi}{V} \text{ mm}^2.$$

Hence, for $V = 30 \text{ km/h}$, $\sigma_z = 12 \text{ mm}$ and, for $V = 100 \text{ km/h}$, $\sigma_z = 6 \cdot 5 \text{ mm}$.

A real vehicle is of course subjected to inputs at more than one point. Normally there will be at least four separate inputs from the road but, for simplicity, consider the case when the response $z(t)$ arises from two inputs $y_1(t)$ and $y_2(t)$, rather than from a single input $y(t)$. In this case, from (7.13), the spectrum of the response $S_z(\omega)$ is given by

$$S_z(\omega) = H_1^*(\omega)H_1(\omega)S_{y_1}(\omega) + H_1^*(\omega)H_2(\omega)S_{y_1y_2}(\omega) +$$
$$+ H_2^*(\omega)H_1(\omega)S_{y_2y_1}(\omega) + H_2^*(\omega)H_2(\omega)S_{y_2}(\omega) \qquad (14.40)$$

where, if $y_1(t) = e^{i\omega t}$ with $y_2(t) = 0$,

$$z(t) = H_1(\omega)e^{i\omega t}$$

---

† Note that, since $W_y(f)$ is defined as a *one-sided* function of frequency, the factor of 2 in equation (14.7) is omitted here.

and, if $y_2(t) = e^{i\omega t}$ with $y_1(t) = 0$,

$$z(t) = H_2(\omega) e^{i\omega t}$$

and $S_{y_1}(\omega)$, $S_{y_1 y_2}(\omega)$, $S_{y_2 y_1}(\omega)$ and $S_{y_2}(\omega)$ are the time spectral and cross-spectral densities of the excitation. For a linear system, the frequency response functions $H_1(\omega)$ and $H_2(\omega)$ can in theory be calculated, or possibly they can be measured if a test vehicle is available, and the time spectra of the excitation can be calculated from measured road spectra by using (14.37). Provided that its suspension has substantially linear dynamic characteristics, which means that the amount of Coulomb friction present must be small, the vibration response of a moving vehicle to road surface irregularities can then be calculated from (14.40) and the effect of different suspension elements and different structural features assessed theoretically. There is an extensive literature† on the response of wheeled vehicles to road surface undulations and this subject has now become an important branch of the subject of vehicle dynamics.

Simulation of random environments

In Chapter 13 we referred to the need to simulate random vibration environments in order to test how a new product performs before putting it into service, or, possibly, to study faults that are known to occur in already established products which fail in service. Since the operating environment of such a product probably arises from its excitation at more than one point by separate (and probably correlated) noise sources, and may involve excitation over finite areas of its surface by randomly varying pressure fields, a complete simulation of typical service conditions will very likely be impracticable. However, it turns out‡ that resonant systems and structures can sometimes be excited satisfactorily by inputs at just one or two points and, when the necessary conditions are met, considerable simplifications (and cost savings) can then be achieved.

Consider a linear system which is excited at two points 1 and 2, Fig. 14.7, by separate ergodic random inputs $x_1(t)$ and $x_2(t)$ and which responds at two output points A and B with responses $y_A(t)$ and $y_B(t)$. If complex frequency response functions $H_{A1}(\omega)$, $H_{A2}(\omega)$, $H_{B1}(\omega)$ and $H_{B2}(\omega)$ are defined so that

when $\qquad x_1(t) = e^{i\omega t}$ and $\quad x_2(t) = 0$

then $\qquad y_A(t) = H_{A1}(\omega) e^{i\omega t}$ and $\quad y_B(t) = H_{B1}(\omega) e^{i\omega t}$ $\qquad$ (14.41a)

and, when $x_1(t) = 0$ and $\quad x_2(t) = e^{i\omega t}$

then $\qquad y_A(t) = H_{A2}(\omega) e^{i\omega t}$ and $\quad y_B(t) = H_{B2}(\omega) e^{i\omega t}$ $\qquad$ (14.41b)

and if the spectral and cross-spectral densities associated with the inputs and

---

† See, for instance, Dodds *et al.* [24] and Virchis *et al.* [66].
‡ Robson [59] Ch. VII. The treatment here follows Muster *et al.* [51] (which is itself based on Robson's analysis).

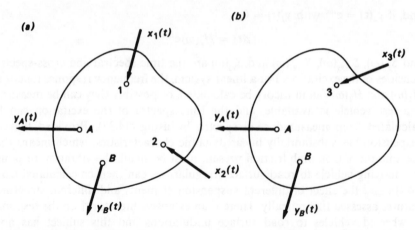

Fig. 14.7 Simulation of two-point excitation $x_1(t)$ and $x_2(t)$ by single point excitation $x_3(t)$

outputs are $S_{x_1}(\omega)$, $S_{x_2}(\omega)$, $S_{x_1x_2}(\omega) = S^*_{x_2x_1}(\omega)$, $S_{y_A}(\omega)$, $S_{y_B}(\omega)$ and $S_{y_Ay_B}(\omega) = S^*_{y_By_A}(\omega)$, then, from (7.13),

$$S_{yA} = H^*_{A1}H_{A1}S_{x_1} + H^*_{A1}H_{A2}S_{x_1x_2} + H^*_{A2}H_{A1}S_{x_2x_1} + H^*_{A2}H_{A2}S_{x_2}$$
and $\qquad$ (14.42a)
$$S_{yB} = H^*_{B1}H_{B1}S_{x_1} + H^*_{B1}H_{B2}S_{x_1x_2} + H^*_{B2}H_{B1}S_{x_2x_1} + H^*_{B2}H_{B2}S_{x_2}$$

and, from (7.30) (see problem 7.5),

$$S_{y_Ay_B} = H^*_{A1}H_{B1}S_{x_1} + H^*_{A1}H_{B2}S_{x_1x_2} + H^*_{A2}H_{B1}S_{x_2x_1} + H^*_{A2}H_{B2}S_{x_2}$$
and $\qquad$ (14.42b)
$$S_{y_By_A} = S^*_{y_Ay_B} = H^*_{B1}H_{A1}S_{x_1} + H^*_{B1}H_{A2}S_{x_1x_2} + H^*_{B2}H_{A1}S_{x_2x_1} + H^*_{B2}H_{A2}S_{x_2}.$$

Our problem is to find whether a single input $x_3(t)$ applied at an arbitrary point 3, Fig. 14.7(c), can, if its spectrum is appropriately adjusted, reproduce at points A and B the same response spectra which occurred as a result of the two simultaneous inputs $x_1(t)$ and $x_2(t)$.

In general terms we know that, if $S_{x_3}(\omega)$ is the spectral density of $x_3(t)$, then the response spectra at points A and B are

$$S_{yA} = H^*_{A3}H_{A3}S_{x_3}$$
$$S_{yB} = H^*_{B3}H_{B3}S_{x_3}$$
$$S_{y_Ay_B} = H^*_{A3}H_{B3}S_{x_3} \qquad (14.43)$$
and
$$S_{y_By_A} = H^*_{B3}H_{A3}S_{x_3}$$

where $H_{A3}(\omega)$ and $H_{B3}(\omega)$ are complex frequency response functions defined so that
when $\qquad\qquad x_3(t) = e^{i\omega t}$

then
$$y_A(t) = H_{A3}(\omega)\,e^{i\omega t} \tag{14.44}$$

and
$$y_B(t) = H_{B3}(\omega)\,e^{i\omega t}.$$

We want the response spectra calculated from (14.43) to be the same as those calculated from (14.42). But we have only one variable, the spectral density of $x_3(t)$, $S_{x_3}(\omega)$, to play with. Suppose we choose S_{x_3} so that $S_{y_A}(\omega)$ is the same for both two-point and single-point excitation. In this case, combining the first of (14.42) and the first of (14.43) gives

$$S_{x_3} = \frac{H_{A1}^* H_{A1}}{H_{A3}^* H_{A3}}S_{x_1} + \frac{H_{A1}^* H_{A2}}{H_{A3}^* H_{A3}}S_{x_1 x_2} + \frac{H_{A2}^* H_{A1}}{H_{A3}^* H_{A3}}S_{x_2 x_1} + \frac{H_{A2}^* H_{A2}}{H_{A3}^* H_{A3}}S_{x_2}. \tag{14.45}$$

Alternatively, if we choose S_{x_3} so that $S_{y_B}(\omega)$ is the same for both cases, then

$$S_{x_3} = \frac{H_{B1}^* H_{B1}}{H_{B3}^* H_{B3}}S_{x_1} + \frac{H_{B1}^* H_{B2}}{H_{B3}^* H_{B3}}S_{x_1 x_2} + \frac{H_{B2}^* H_{B1}}{H_{B3}^* H_{B3}}S_{x_2 x_1} + \frac{H_{B2}^* H_{B2}}{H_{B3}^* H_{B3}}S_{x_2} \tag{14.46}$$

or, if we make $S_{y_A y_B} = S_{y_B y_A}^*$ the same for both cases, then

$$S_{x_3} = \frac{H_{A1}^* H_{B1}}{H_{A3}^* H_{B3}}S_{x_1} + \frac{H_{A1}^* H_{B2}}{H_{A3}^* H_{B3}}S_{x_1 x_2} + \frac{H_{A2}^* H_{B1}}{H_{A3}^* H_{B3}}S_{x_2 x_1} + \frac{H_{A2}^* H_{B2}}{H_{A3}^* H_{B3}}S_{x_2}. \tag{14.47}$$

These three equations (14.45), (14.46) and (14.47) will only give the same result for $S_{x_3}(\omega)$, independently of $S_{x_1}(\omega)$, $S_{x_2}(\omega)$ and $S_{x_1 x_2}(\omega) = S_{x_2 x_1}^*(\omega)$, if the corresponding coefficients in the equations are the same. The S_{x_1} terms on the r.h.s. of the equations will be the same if

$$\frac{H_{A1}^* H_{A1}}{H_{A3}^* H_{A3}} = \frac{H_{B1}^* H_{B1}}{H_{B3}^* H_{B3}} = \frac{H_{A1}^* H_{B1}}{H_{A3}^* H_{B3}}. \tag{14.48}$$

The $S_{x_1 x_2}$ and $S_{x_2 x_1}$ terms will be the same if

$$\frac{H_{A1}^* H_{A2}}{H_{A3}^* H_{A3}} = \frac{H_{B1}^* H_{B2}}{H_{B3}^* H_{B3}} = \frac{H_{A1}^* H_{B2}}{H_{A3}^* H_{B3}} = \frac{H_{A2} H_{B1}^*}{H_{A3} H_{B3}^*} \tag{14.49}$$

and the S_{x_2} terms will be the same if

$$\frac{H_{A2}^* H_{A2}}{H_{A3}^* H_{A3}} = \frac{H_{B2}^* H_{B2}}{H_{B3}^* H_{B3}} = \frac{H_{A2}^* H_{B2}}{H_{A3}^* H_{B3}}. \tag{14.50}$$

In general, equations (14.48), (14.49) and (14.50) will not be true and so two-point excitation may not then be simulated by single-point excitation. However, if, as very often happens, the response is dominated by a single resonant mode, then it turns out that (14.48)–(14.50) are approximately true, and the response to single-point excitation closely resembles that for two-point excitation and, in the general case, that for multi-point excitation.

The fact that (14.48)–(14.50) are approximately true when there is a single resonant mode present depends on the fact that a lightly damped resonant mode has the same shape, independently of where it is excited. For example, if a pin jointed beam is excited at its lowest resonant frequency ω_N by a unit force applied distance s from one end, Fig. 14.8, then the half sine wave mode shape of the resulting motion is not affected by the dimension s. Although the amplitude of the resulting motion depends on s and on the inherent damping present, to a close approximation the mode shape and therefore the ratio of displacements $y_A(t)$ and $y_B(t)$ does not depend on where the force is applied to

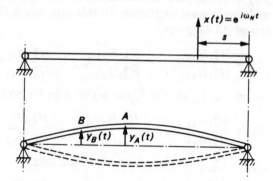

Fig. 14.8 Lowest resonant mode of a pin jointed beam

the beam. Also all points on the beam move in phase (or, for higher modes, in phase or in antiphase) with each other. Hence, if for $s = s_1$ (say)

$$\frac{y_A(t)}{y_B(t)} = \frac{H_{A1}(\omega_N)\,\mathrm{e}^{i\omega_N t}}{H_{B1}(\omega_N)\,\mathrm{e}^{i\omega_N t}} = \frac{H_{A1}(\omega_N)}{H_{B1}(\omega_N)} \tag{14.51}$$

and for $s = s_2$ (say)

$$\frac{y_A(t)}{y_B(t)} = \frac{H_{A2}(\omega_N)\,\mathrm{e}^{i\omega_N t}}{H_{B2}(\omega_N)\,\mathrm{e}^{i\omega_N t}} = \frac{H_{A2}(\omega_N)}{H_{B2}(\omega_N)}, \tag{14.52}$$

then, since these ratios of displacements will be the same (same amplitude and phase relationship), we obtain the approximate result that

$$\frac{H_{A1}(\omega_N)}{H_{B1}(\omega_N)} = \frac{H_{A2}(\omega_N)}{H_{B2}(\omega_N)}. \tag{14.53}$$

Furthermore, for broad band random excitation of a highly resonant mode, we know, from the first section of this chapter, that the response is a narrow band process whose average frequency is the natural frequency of the mode, ω_N. The response is dominated by frequency components in the region of ω_N, and so, in this case

$$\frac{H_{A1}}{H_{B1}} = \frac{H_{A1}(\omega)}{H_{B1}(\omega)} \simeq \frac{H_{A1}(\omega_N)}{H_{B1}(\omega_N)} \tag{14.54}$$

since we are concerned only with frequencies near to ω_N. Hence, finally, (14.53) may be written approximately as

$$\frac{H_{A1}}{H_{B1}} = \frac{H_{A2}}{H_{B2}} \tag{14.55}$$

where frequencies near to ω_N are understood. The same result applies also for excitation at point 3 so that

$$\frac{H_{A1}}{H_{B1}} = \frac{H_{A2}}{H_{B2}} = \frac{H_{A3}}{H_{B3}} \tag{14.56}$$

and, taking the complex conjugates of each term,

$$\frac{H_{A1}^*}{H_{B1}^*} = \frac{H_{A2}^*}{H_{B2}^*} = \frac{H_{A3}^*}{H_{B3}^*}. \tag{14.57}$$

If these results (14.56) and (14.57) are true, then it is easy to see that (14.48), (14.49) and (14.50) must also be true, in which case the spectral characteristics of the response to two-point excitation (and, in a more general case, to multi-point excitation) can be simulated by the application of a single random input $x_3(t)$ whose spectrum is defined by any one of (14.45), (14.46) or (14.47) (each of which gives the same answer to a close approximation).

We therefore conclude that, if the motion of a system is dominated by a single resonant mode, then in order to simulate the motion of this system in response to broad band multi-point random excitation, it is only necessary to excite the system at a single point. The location of this point is arbitrary, except that it may not be at a node (point of no motion) for the resonant mode. In that case $H_{A3}(\omega)$ and $H_{B3}(\omega)$ would be zero in the above analysis and so the required input spectral density $S_{x_3}(\omega)$ given by (14.45), (14.46) or (14.47) would be infinite. If this condition is met, then the response of a highly resonant system to multi-point excitation may be simulated by using a single vibrator (or input exciter) which has to be set so that the response at any one point on the system corresponds to its response under service conditions (multi-point excitation). The response at all other points will then be correctly set automatically.

The above analysis may be extended to the case where there is more than one highly resonant mode present (Robson *et al.* [60]), and it turns out that if there are n highly resonant modes, then the system's response can be accurately simulated by exciting it at a minimum of n separate points. The required spectral characteristics of the necessary excitation can be calculated by extending the analysis already described. Furthermore, if two or more modes have natural frequencies which are sufficiently far apart, it may even be possible to excite more than one mode correctly from a single point. In this case the spectrum of the excitation is adjusted so that each mode has the correct response in its own frequency band. Provided that these frequency bands do not overlap, each mode can be "tuned" to its correct response (by adjusting the excitation spectrum) independently of the other modes.

Frequency response function and coherency measurements

We have seen how random vibration is transmitted through linear systems and, from Chapter 7, if a linear system has two inputs $x_1(t)$ and $x_2(t)$ and a

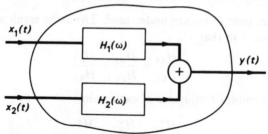

Fig. 14.9 Constant parameter linear system with two inputs and one output

single output $y(t)$, Fig. 14.9, then we know that the input and output spectra are related by the equations

$$S_y(\omega) = H_1^*(\omega)H_1(\omega)S_{x_1}(\omega) + H_1^*(\omega)H_2(\omega)S_{x_1x_2}(\omega) +$$
$$+ H_2^*(\omega)H_1(\omega)S_{x_2x_1}(\omega) + H_2^*(\omega)H_2(\omega)S_{x_2}(\omega) \tag{7.13}$$

and

$$S_{x_1y}(\omega) = H_1(\omega)S_{x_1}(\omega) + H_2(\omega)S_{x_1x_2}(\omega)$$
$$S_{x_2y}(\omega) = H_2(\omega)S_{x_2}(\omega) + H_1(\omega)S_{x_2x_1}(\omega). \tag{7.22}$$

Suppose that the frequency response functions $H_1(\omega)$ and $H_2(\omega)$ are initially unknown and we want to calculate them from measured values of the spectra and cross-spectra of the inputs and output. Basically we have three equations involving only two unknowns, $H_1(\omega)$ and $H_2(\omega)$, and we can in fact find $H_1(\omega)$ and $H_2(\omega)$ from the two equations (7.22) without ever using the first equation (7.13). However this additional equation serves as a useful check on the accuracy of our calculation because, after finding $H_1(\omega)$ and $H_2(\omega)$, we can substitute into (7.13) and check that this extra equation is correctly satisfied.

Following the methods described in Chapters 10 and 11, the input spectra and cross-spectra $S_{x_1}(\omega)$, $S_{x_2}(\omega)$ and $S_{x_1x_2}(\omega) = S_{x_2x_1}^*(\omega)$ and the cross-spectra between each input and the output $S_{x_1y}(\omega)$ and $S_{x_2y}(\omega)$ can all be measured. Assume that we know values of all the spectra at a series of closely spaced discrete frequencies, $\omega = \omega_k$ (say). Actually, of course, we can only determine estimates of the true values, but for the present assume that we have exact values of the spectra.

At frequency ω_k (7.22) become

$$S_{x_1y}(\omega_k) = H_1(\omega_k)S_{x_1}(\omega_k) + H_2(\omega_k)S_{x_1x_2}(\omega_k)$$

and

$$S_{x_2y}(\omega_k) = H_2(\omega_k)S_{x_2}(\omega_k) + H_1(\omega_k)S_{x_2x_1}(\omega_k)$$

and, solving for $H_1(\omega_k)$ and $H_2(\omega_k)$ we obtain

$$H_1(\omega_k) = \frac{S_{x_2}S_{x_1y} - S_{x_2y}S_{x_1x_2}}{S_{x_1}S_{x_2} - S_{x_1x_2}S_{x_2x_1}}$$

and

$$H_2(\omega_k) = \frac{S_{x_1}S_{x_2y} - S_{x_1y}S_{x_2x_1}}{S_{x_1}S_{x_2} - S_{x_1x_2}S_{x_2x_1}}$$

(14.58)

where the argument ω_k is omitted from each S term for brevity.

Now suppose that the spectral density of the output process, $S_y(\omega)$, is not at first measured, but that we calculate it from (7.13). First substituting from (7.22) into (7.13) gives

$$S_y(\omega) = H_1^*(\omega)S_{x_1y}(\omega) + H_2^*(\omega)S_{x_2y}(\omega).$$ (14.59)

Then, since $S_y(\omega)$ is a real quantity,† its complex conjugate is equal to itself, and so if we take the complex conjugates of both sides of (14.59), we obtain

$$
\begin{aligned}
S_y(\omega) &= H_1(\omega)S_{x_1y}^*(\omega) + H_2(\omega)S_{x_2y}^*(\omega) \\
&= H_1(\omega)S_{yx_1}(\omega) + H_2(\omega)S_{yx_2}(\omega)
\end{aligned}
$$ (14.60)

from (5.23). Finally, at frequency ω_k,

$$S_y(\omega_k) = H_1(\omega_k)S_{yx_1}(\omega_k) + H_2(\omega_k)S_{yx_2}(\omega_k) = \tilde{S}_y(\omega_k) \quad \text{(say)}$$ (14.61)

in which we can substitute for $H_1(\omega_k)$ and $H_2(\omega_k)$ from (14.58). The symbol $\sim$ on $\tilde{S}_y(\omega_k)$ is introduced to indicate a *calculated* rather than a measured value of $S_y(\omega_k)$.

Provided that we are dealing with a constant parameter linear system for which $x_1(t)$ and $x_2(t)$ are the only inputs, and assuming that the spectra and cross-spectra are accurately known, then, if we now also measure the output spectral density function $S_y(\omega_k)$, this should agree with the value $\tilde{S}_y(\omega_k)$ we have just calculated. However owing to the presence of unwanted noise in the system and due to inherent nonlinearities, the measured spectrum $S_y(\omega_k)$ will in general differ slightly from the calculated value based on linear theory. The ratio $\tilde{S}_y(\omega_k)/S_y(\omega_k)$ is defined as the *multiple coherence function* between the output process $y(t)$ and the input processes $x_1(t)$ and $x_2(t)$, and we shall represent this‡ by the symbol $\eta_{y.x}^2(\omega)$. If the output $y(t)$ is completely accounted for by a linear response to the recognized inputs then $\eta_{y.x}^2(\omega)$ will be unity (non-dimensional) at all frequencies. From (14.61) we can see that,

$$\eta_{y.x}^2(\omega_k) = \frac{\tilde{S}_y(\omega_k)}{S_y(\omega_k)} = \frac{H_1(\omega_k)S_{yx_1}(\omega_k) + H_2(\omega_k)S_{yx_2}(\omega_k)}{S_y(\omega_k)}$$ (14.62)

for two inputs, while for more than two inputs the numerator is just the sum of more terms so that, for n inputs, the multiple coherence between the output

---

† See Ch. 5, equation (5.7) and the discussion which follows.
‡ Other authors use $\gamma_{y.x}^2(\omega)$ or $\kappa_{y.x}^2(\omega)$ for the same quantity.

and the n separate inputs at frequency ω_k is given by

$$\eta^2_{y.x}(\omega_k) = \frac{\sum_{r=1}^{n} H_r(\omega_k)S_{yx_r}(\omega_k)}{S_y(\omega_k)} \qquad (14.63)$$

This is an important result because $\eta^2_{y.x}$ will only be unity if all the separate inputs have been taken account of and if the system being studied is a linear system free from extraneous noise.

We can express (14.63) entirely in terms of measured auto- and cross-spectra by substituting for $H_1(\omega_k)$ and $H_2(\omega_k)$ from (14.58). These equations can be extended to the case of n inputs by noting that, in this case, from (7.23),

$$\begin{aligned}
S_{x_1 y} &= H_1 S_{x_1 x_1} + H_2 S_{x_1 x_2} + H_3 S_{x_1 x_3} + \cdots + H_n S_{x_1 x_n} \\
S_{x_2 y} &= H_1 S_{x_2 x_1} + H_2 S_{x_2 x_2} + H_3 S_{x_2 x_3} + \cdots + H_n S_{x_2 x_n} \\
&\ \vdots \\
S_{x_n y} &= H_1 S_{x_n x_1} + H_2 S_{x_n x_2} + H_3 S_{x_n x_3} + \cdots + H_n S_{x_n x_n}
\end{aligned} \qquad (7.23)$$

where the arguments (ω_k) are omitted for brevity and we write the auto-spectra $S_{x_1} = S_{x_1 x_1}$, $S_{x_2} = S_{x_2 x_2}$, etc., to keep the notation uniform. This set of equations may be written more conveniently in the matrix form

$$\begin{bmatrix} S_{x_1 y} \\ S_{x_2 y} \\ \vdots \\ S_{x_n y} \end{bmatrix} = \begin{bmatrix} S_{x_1 x_1} & S_{x_1 x_2} & \cdots & S_{x_1 x_n} \\ S_{x_2 x_1} & S_{x_2 x_2} & \cdots & S_{x_2 x_n} \\ \vdots & \vdots & & \vdots \\ S_{x_n x_1} & S_{x_n x_2} & \cdots & S_{x_n x_n} \end{bmatrix} \begin{bmatrix} H_1 \\ H_2 \\ \vdots \\ H_n \end{bmatrix} \qquad (14.64)$$

or, more briefly, in the shorthand equivalent form

$$[S_{xy}] = [S_{xx}][H]. \qquad (14.65)$$

By premultiplying by the inverse matrix $[S_{xx}]^{-1}$, we can solve for the column matrix $[H]$ which is then given by

$$[H] = [S_{xx}]^{-1}[S_{xy}]. \qquad (14.66)$$

The numerator of (14.63) may also be written in matrix notation, since

$$\sum_{r=1}^{n} H_r S_{yx_r} = \begin{bmatrix} S_{yx_1} & S_{yx_2} & \cdots & S_{yx_n} \end{bmatrix} \begin{bmatrix} H_1 \\ H_2 \\ \vdots \\ H_n \end{bmatrix} \qquad (14.67)$$

and, if we define

$$[S_{yx}] = \begin{bmatrix} S_{yx_1} \\ S_{yx_2} \\ \vdots \\ S_{yx_n} \end{bmatrix} \qquad (14.68)$$

its transpose is the row matrix

$$[S_{yx}]^t = [S_{yx_1} \quad S_{yx_2} \quad \cdots \quad S_{yx_n}]. \tag{14.69}$$

Substituting from (14.66) and (14.69) into (14.67) now gives

$$\sum_{r=1}^{n} H_r S_{yx_r} = [S_{yx}]^t [S_{xx}]^{-1} [S_{xy}] \tag{14.70}$$

so that finally we can express the multiple coherence function in the matrix form

$$\eta_{y.x}^2 = \frac{[S_{yx}]^t [S_{xx}]^{-1} [S_{xy}]}{S_{yy}} \tag{14.71}$$

where the spectral matrices $[S_{yx}]$, $[S_{xx}]$ and $[S_{xy}]$ are defined by reference to equations (14.68), (14.64) and (14.65).

We have seen that $\eta_{y.x}^2$ should be unity for a constant parameter, linear, noise-free system. It can be shown (Jenkins *et al.* [39] p. 467) that this is its maximum value and that $\eta_{y.x}^2$ will always be less than 1 (but greater than zero) in all other cases, so that

$$0 \leqslant \eta_{y.x}^2 \leqslant 1. \tag{14.72}$$

If $\eta^2 = 0$, then the measured output at the frequency being investigated is entirely due to noise and cannot be predicted from the known inputs, whereas if $\eta^2 = 1$, then the output can be completely accounted for by passing the known inputs through a linear noise-free system.

Consider now the special case when there is only one input $x_1(t)$. The two input problem studied above reduces to this case if we put $x_2(t) = 0$, in which case

$$S_{x_1 x_2}(\omega) = S_{x_2 x_1}(\omega) = S_{x_2}(\omega) = S_{x_2 y}(\omega) = 0.$$

The first of (7.22) then gives

$$H_1(\omega_k) = \frac{S_{x_1 y}(\omega_k)}{S_{x_1}(\omega_k)} \tag{14.73}$$

and (14.63) becomes

$$\eta_{yx_1}^2(\omega_k) = \frac{H_1(\omega_k) S_{yx_1}(\omega_k)}{S_y(\omega_k)} \tag{14.74}$$

which, after substituting from (14.73), gives

$$\eta_{yx_1}^2(\omega_k) = \frac{S_{x_1 y}(\omega_k) S_{yx_1}(\omega_k)}{S_{x_1}(\omega_k) S_y(\omega_k)}. \tag{14.75}$$

In this case $\eta_{yx_1}^2$ is called the single or *ordinary coherence function* between $y(t)$ and $x_1(t)$ at frequency $\omega = \omega_k$. For a constant parameter linear system (14.75) will again always work out to unity unless either (a) there is unwanted noise present in the measurements or (b) there is an input other than $x_1(t)$ which has not been accounted for.

Estimates of frequency response and coherence functions should be supported by information about the confidence limits which can be associated with these estimates. We saw in Chapter 9 how confidence limits can be calculated for a spectral estimate by assuming that the distribution of the estimate is of a chi-square type. It is possible to extend this approach to determine approximate confidence limits for frequency response and coherence estimates and the methods for doing this are discussed in the specialist literature (Bendat *et al.* [3]; Otnes *et al.* [52]). However there is one very important point about calculating frequency response and coherence function estimates which should be mentioned here. It concerns estimates found by using the FFT approach to calculate discrete Fourier transforms of sample functions of the input and output.

Consider a system with a single input $x(t)$ and a single output $y(t)$. Assume that both $x(t)$ and $y(t)$ are recorded during the same time period T and then passed through an analogue-to-digital converter to generate the discrete time series $\{x_r\}$ and $\{y_r\}$, $r = 0, 1, 2, ..., (N - 1)$. We can now use the FFT to calculate the DFT's of these two time series, $\{X_k\}$ and $\{Y_k\}$, $k = 0, 1, 2, ..., (N - 1)$, and hence find the spectral estimates (see Chapter 11)

$$\tilde{S}_{xx}(\omega_k) \simeq \frac{T}{2\pi} X_k^* X_k \qquad \tilde{S}_{xy}(\omega_k) \simeq \frac{T}{2\pi} X_k^* Y_k$$

$$\tilde{S}_{yx}(\omega_k) \simeq \frac{T}{2\pi} Y_k^* X_k \qquad \tilde{S}_{yy}(\omega_k) \simeq \frac{T}{2\pi} Y_k^* Y_k. \tag{14.76}$$

The spectral estimates will satisfy (14.76) whether there is noise present in the output process $y(t)$ or not and whether the system is linear or not. However, on substituting (14.76) into (14.75) to determine the coherence function η_{yx}^2, this comes out to be exactly 1 since

$$\eta_{yx}^2(\omega_k) = \frac{S_{xy}(\omega_k) S_{yx}(\omega_k)}{S_{xx}(\omega_k) S_{yy}(\omega_k)}$$

$$\simeq \frac{\tilde{S}_{xy}(\omega_k) \tilde{S}_{yx}(\omega_k)}{\tilde{S}_{xx}(\omega_k) \tilde{S}_{yy}(\omega_k)} \tag{14.77}$$

$$\simeq \frac{X_k^* Y_k \, Y_k^* X_k}{X_k^* X_k \, Y_k^* Y_k} = 1.$$

The reason for this apparent paradox is explained by the fact that the coherence function estimate η_{yx}^2 calculated from (14.77) suffers from a bias error which is severe when the equivalent number of statistical degrees of freedom of the spectral estimates is small. It turns out that values of η_{yx}^2 calculated by using (14.76) always come out to be unity, whatever the true value. The difficulty may be overcome by smoothing adjacent spectral estimates (as described in Chapter 11) to improve their statistical reliability *before* substituting these estimates into (14.77) to calculate the coherence function. Then η_{yx}^2 does not come out to be unity unless the system being investigated is a linear, noise-free system.

For accurate measurements of frequency response functions, we want η^2 to be unity, but only when it is calculated from properly smoothed spectral estimates. Digital calculations, without final smoothing, indicate falsely that a system is linear and noise-free when this is not the case, so caution is needed when calculating and interpreting coherence function estimates.

Nonstationary processes

In Chapter 2 we discussed the concept of ensemble averaging, based on the notion that a random process $\{x(t)\}$ can be thought of as an infinite collection of sample functions $x(t)$ all occurring simultaneously, Fig. 2.4. If the process is stationary, its ensemble averages are independent of absolute time t. If, in addition, the process is ergodic, then any one sample function is completely representative of the process as a whole. Ensemble averages and sample averages are then the same, so that, for instance,

$$E[x(t)x(t + \tau)] = R_x(\tau) = \lim_{T \to \infty} \frac{1}{T} \int_{-T/2}^{T/2} x(t)x(t + \tau)\,dt.$$

Throughout this book we have been talking about stationary processes and, most of the time, about ergodic processes. This is a necessary assumption in order to make progress with the theory and it is an acceptable approximation to make for many practical calculations. However the question obviously arises: is a given random function a sample from a stationary process, and, if so, is this an ergodic process?

It is never possible to answer this question with certainty unless we have infinitely many sample functions available and they last for ever. Otherwise there is always the possibility that the statistical characteristics of the process will change for the next sample or for the next time interval. The best practical guide is probably just common sense. If a process has a beginning or ending, then theoretically it is not stationary. However if it lasts for a long time (long compared with the period of its lowest frequency spectral components) then it may be sensible to assume that the process is approximately stationary over most of its lifetime. Strictly this is a contradiction in terms, but we mean that sample records of the process in question may be thought of as finite lengths cut from records of sample functions of a stationary process which are infinite in length. In other cases it may only be reasonable to assume that a process is stationary over part of its lifetime. For instance the random pressure fluctuation on the casing of a rocket is an example of a case in which there are initially very violent fluctuations on launching, but these die down to a much lower level when the rocket leaves the earth's atmosphere and thrust is reduced. Obviously it would not be reasonable to assume that this process was stationary for most of its duration. However, possibly we could divide the record into sections, one section for lift-off, one section for accelerating through the earth's atmosphere, one section for coasting in space, and so on, each of which may be thought of as finite lengths of sample functions from different random processes, each approximately stationary.

If it is reasonable to think of a length of sample function as being representative of a stationary process, then it is usually easier to say whether this process is likely to be ergodic or not. For instance if we are concerned with launching rockets, and if all the rockets are the same and are launched in the same way under similar conditions, then we could expect each section of the record of random pressure fluctuations on any one rocket to be representative of the pressure fluctuations recorded at about the same time on other rockets. In this case, each section of the pressure–time history could be expected to be representative of an ergodic (and therefore stationary process). However if we are concerned with recording the instantaneous sound pressure level as aircraft take off from an airport, then the record obtained when a small private plane takes off will not represent the record when a jumbo jet takes off. In this case separate time histories are not samples from the same ergodic random process. Instead we should have to group aircraft according to their different engine characteristics, and also take account of wind and weather conditions, in order to be able to group the sample time histories in order that each might approximately represent the characteristics of other sample functions in the same group.

Suppose that we believe a given length of a sample function comes from an ensemble of sample functions which together constitute a stationary, ergodic random process. How can this hypothesis be tested? If we only have one sample to examine, as is very often the case in practice, all we can do is to cut this up into a series of short lengths, calculate sample averages for each short length, and examine how these short-sample averages compare with each other and with the corresponding average for the whole sample. Obviously we would not expect the short-sample mean-square values to be exactly the same as the overall average value, because the sample average over length T only approaches the ensemble average (constant) when $T \to \infty$. For finite T we would naturally expect some difference from the ensemble average. This raises the question: how much variation from one short sample to the next is acceptable while still being consistent with our hypothesis that the full length sample is taken from a stationary ergodic ensemble?

In fact we considered this problem in Chapter 9 when discussing the distribution of

$$z(t) = \frac{1}{T} \int_0^T x^2(t)\, dt$$

in connection with the accuracy of spectral analysis. The conclusion was that, for a Gaussian process,

$$\frac{\sigma}{m} = \frac{\sigma_z}{E[z]} = \frac{1}{\sqrt{(BT)}} \qquad \text{for} \qquad BT \gg 1 \qquad (9.22)$$

where B is the bandwidth of the process $x(t)$ in Hz. If, as before, we assume that z has a chi-square distribution, then we can calculate confidence limits, as described in Chapter 9, for each of the sample mean square measurements. We can then use this information to judge whether the original hypothesis of stationarity should be accepted or not. However this approach requires knowledge of the bandwidth of $x(t)$ and assumes that it is a Gaussian process, and this information is generally not available at the initial stage of an investigation when stationarity is being tested. Hence other more direct approaches are adopted.

Suppose that the short-time averaged mean square values are arranged in time order, so that the value representing the first "slice" of the time history occurs first and other values follow in order. Then if there is a nonstationary trend in the data, we would expect the same trend to show in the ordered sequence of average values. There are many different methods of testing for non-random trends in an ordered sequence of observations (Kendall *et al.* [41c] Ch. 45) but one of the simplest methods which is effective in identifying a gradual drift is the method of counting "reverse arrangements" or "inversions".

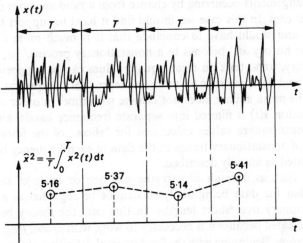

Fig. 14.10 Testing for nonstationary trend by dividing a sample record into "slices" of length T, estimating mean square values for each slice and counting the number of "reverse arrangements" of these values.

In order to explain this *trend test*, suppose that the time history being studied is cut into four "slices", Fig. 14.10, and the mean square value for each slice is calculated. In order of time, these values might be

$$5 \cdot 16, \quad 5 \cdot 37, \quad 5 \cdot 14, \quad 5 \cdot 41.$$

Since $5 \cdot 16$ and $5 \cdot 37$ both occur before the smaller value $5 \cdot 14$ there are said to be two "reverse arrangements" in the ordered sequence. If the order was

$$5 \cdot 14, \quad 5 \cdot 16, \quad 5 \cdot 37, \quad 5 \cdot 41$$

there would have been no reverse arrangements, and if it had been

$$5\cdot41,\quad 5\cdot37,\quad 5\cdot16,\quad 5\cdot14$$

there would have been the maximum possible number of six reverse arrangements ($5\cdot41$ exceeds three following values, $5\cdot37$ exceeds two following values and $5\cdot16$ exceeds one following value). It is a relatively easy matter to calculate the probability that, assuming there is no underlying trend and that the values have been drawn in a random order, there will be two reverse arrangements or less in the resulting sequence (problem 14.4) and this comes out to be $\frac{3}{8}$. We can therefore say that, for stationary data showing no trend, in which all orders are equally likely, there is a probability of $37\cdot5$ per cent that there will be two or less reverse arrangements in the sequence of measured mean square values.

This is hardly convincing one way or the other, but suppose that we have 10 short-sample mean square values and that the ordered sequence of values has 12 reverse arrangements. From Appendix 6 the probability of 12 or less reverse arrangements occurring by chance from a random series of 10 values is only $3\cdot6$ per cent. In this case we should find it hard to support the stationary hypothesis and would have to conclude that it is much more likely that the sample time history $x(t)$ belongs to a nonstationary process.

If necessary, trend tests on mean square values can be supported by similar tests on filtered data, in order to obtain evidence that the spectral composition, as well as the mean square, does not change with time. In order to do this, the sample function $x(t)$ is filtered into separate frequency bands, and short-time averaged mean-square values calculated for "slices" of the filtered data. The possibility of nonstationary trends in the data in each frequency band can then be investigated as already described.

Suppose that, as a result of carrying out one or more of these tests, we conclude that the data being analysed cannot be regarded as even approximately stationary over short lengths. In this case the theory becomes much more complicated because it is necessary to work with *double frequency spectral density functions*. Beginning with the fundamental definition of the nonstationary autocorrelation function

$$R_x(t_1, t_2) = E[x(t_1)x(t_2)] \tag{14.78}$$

the corresponding generalised spectral density function $S_x(\omega_1, \omega_2)$ is defined by the double Fourier transform relation

$$S_x(\omega_1, \omega_2) = \frac{1}{4\pi^2} \int_{-\infty}^{\infty} dt_1 \int_{-\infty}^{\infty} dt_2 \, R_x(t_1, t_2) e^{i\omega_1 t_1} e^{-i\omega_2 t_2} \tag{14.79}$$

and is now a function of two angular frequencies ω_1 and ω_2.

In order to calculate how $S_x(\omega_1, \omega_2)$ is modified by transmission through a linear system we can follow exactly the same approach as that described in Chapter 7 for stationary processes. For a single input–output system with impulse response function $h(t)$ and frequency response function $H(\omega)$ for which

$$H(\omega) = \int_{-\infty}^{\infty} h(t) e^{-i\omega t} \, dt \tag{6.21}$$

we know that the output $y(t)$ is related to the input $x(t)$ by the convolution integral,

$$y(t) = \int_{-\infty}^{\infty} h(\theta)x(t - \theta)\,d\theta. \tag{6.27}$$

The nonstationary output autocorrelation function is

$$R_y(t_1, t_2) = E[y(t_1)y(t_2)]$$

and substituting from (6.27)

$$R_y(t_1, t_2) = E\left[\int_{-\infty}^{\infty} h(\theta_1)x(t_1 - \theta_1)\,d\theta_1 \cdot \int_{-\infty}^{\infty} h(\theta_2)x(t_2 - \theta_2)\,d\theta_2\right]$$

$$= \int_{-\infty}^{\infty} d\theta_1 \int_{-\infty}^{\infty} d\theta_2\, h(\theta_1)h(\theta_2)\, E[x(t_1 - \theta_1)x(t_2 - \theta_2)]$$

$$= \int_{-\infty}^{\infty} d\theta_1 \int_{-\infty}^{\infty} d\theta_2\, h(\theta_1)h(\theta_2)\, R_x(t_1 - \theta_1, t_2 - \theta_2). \tag{14.80}$$

Now using (14.79) to calculate the double Fourier transform of (14.80) gives

$$S_y(\omega_1, \omega_2) = \frac{1}{4\pi^2} \int_{-\infty}^{\infty} dt_1 \int_{-\infty}^{\infty} dt_2\, R_y(t_1, t_2)\, e^{i\omega_1 t_1} e^{-i\omega_2 t_2}$$

$$= \frac{1}{4\pi^2} \int_{-\infty}^{\infty} dt_1 \int_{-\infty}^{\infty} dt_2 \int_{-\infty}^{\infty} d\theta_1 \int_{-\infty}^{\infty} d\theta_2\, h(\theta_1)h(\theta_2)\, e^{i\omega_1 t_1} e^{-i\omega_2 t_2} \times$$

$$\times R_x(t_1 - \theta_1, t_2 - \theta_2). \tag{14.81}$$

By changing the order of integration, this may be written in the identical form

$$S_y(\omega_1, \omega_2) = \frac{1}{4\pi^2} \int_{-\infty}^{\infty} d\theta_1\, h(\theta_1)\, e^{i\omega_1 \theta_1} \int_{-\infty}^{\infty} d\theta_2\, h(\theta_2)\, e^{-i\omega_2 \theta_2} \times$$

$$\times \int_{-\infty}^{\infty} dt_1\, e^{i\omega_1(t_1 - \theta_1)} \int_{-\infty}^{\infty} dt_2\, e^{-i\omega_2(t_2 - \theta_2)}\, R_x(t_1 - \theta_1, t_2 - \theta_2).$$

The integrations with respect to t_1 and t_2 are made with θ_1 and θ_2 constant, so that, if we put

$$\phi_1 = t_1 - \theta_1$$

$$\phi_2 = t_2 - \theta_2$$

then

$$S_y(\omega_1, \omega_2) = \int_{-\infty}^{\infty} d\theta_1\, h(\theta_1)\, e^{i\omega_1 \theta_1} \int_{-\infty}^{\infty} d\theta_2\, h(\theta_2)\, e^{-i\omega_2 \theta_2} \times$$

$$\times \frac{1}{4\pi^2} \int_{-\infty}^{\infty} d\phi_1\, e^{i\omega_1 \phi_1} \int_{-\infty}^{\infty} d\phi_2\, e^{-i\omega_2 \phi_2}\, R_x(\phi_1, \phi_2)$$

$$= \int_{-\infty}^{\infty} d\theta_1 \, h(\theta_1) e^{i\omega_1\theta_1} \int_{-\infty}^{\infty} d\theta_2 \, h(\theta_2) e^{-i\omega_2\theta_2} \, S_x(\omega_1, \omega_2). \tag{14.82}$$

Lastly, we can introduce the frequency response function $H(\omega)$, evaluated at the two frequencies ω_1 and ω_2, from (6.21), so that (14.82) becomes

$$S_y(\omega_1, \omega_2) = H^*(\omega_1)H(\omega_2)S_x(\omega_1, \omega_2) \tag{14.83}$$

which is the nonstationary equivalent of the familiar result

$$S_y(\omega) = H^*(\omega)H(\omega)S_x(\omega) \tag{7.15}$$

for stationary processes. Notice that the form of equation (14.79) with $e^{i\omega_1 t_1}$ and $e^{-i\omega_2 t_2}$ appearing under the integrals is specially chosen so that the form of (7.15) (and of the more general equation (7.14)) is preserved for nonstationary spectra. Similar input–output relations can be derived for the case of multiple inputs and for nonstationary cross-spectra and these results parallel those for the stationary case. In fact they are surprisingly simple. The real problem is finding the double frequency input spectra in the first place and interpreting what they mean.

In general it is not possible to determine $S_x(\omega_1, \omega_2)$ from a single sample function because the ensemble average $E[x(t_1)x(t_2)]$ from which it is derived cannot be obtained by averaging over time on a single record. This introduces enormous practical complications because many individual sample records must be available and they must all be sampled simultaneously, starting from the same time origin. The only possibility of simplification is when knowledge of the physical nature of the process or inspection of sample records allows the nonstationary process to be broken down into components, one of which is a stationary random process and the others of which are deterministic slowly varying functions of time. In this case it may still be possible to estimate $S_x(\omega_1, \omega_2)$ from a single sample record with sufficient accuracy.

For example, consider a nonstationary random process $x(t)$ which may be assumed to have the form

$$x(t) = a(t)z(t) \tag{14.84}$$

where $z(t)$ is a stationary random process and $a(t)$ is an (initially unknown) deterministic function which varies with time as the mean square level of $x(t)$ changes. In this case the autocorrelation function

$$R_x(t_1, t_2) = E[x(t_1)x(t_2)] = E[a(t_1)z(t_1)a(t_2)z(t_2)]$$
$$= a(t_1)a(t_2) E[z(t_1)z(t_2)]$$
$$= a(t_1)a(t_2)R_z(t_2 - t_1) \tag{14.85}$$

since $z(t)$ is stationary. If we can also assume that the function $a(t)$ changes slowly enough with time that

$$a(t_1)a(t_2) \simeq a^2(t) \tag{14.86}$$

where

$$t = \tfrac{1}{2}(t_1 + t_2), \tag{14.87}$$

and putting

$$\tau = t_2 - t_1, \tag{14.88}$$

equation (14.85) then becomes

$$R_x(t_1, t_2) = a^2(t)R_z(\tau). \tag{14.89}$$

So-called *locally stationary* processes for which (14.89) applies have been discussed in the literature (Bendat *et al.* [3] Ch. 10; Piersol [55]) and include some important physical situations such as the vibration of a rocket where the vibration level varies at different stages of its flight and the fluctuation of certain economic indices which are nonstationary owing to the effect of inflation.

The advantage of using (14.89) is that the form of $a^2(t)$ and $R_z(\tau)$ can be estimated by time-averaging over a single sample record. First $a^2(t)$ is estimated by averaging over short "slices" of the record since, for a slice of length T which runs from $t_0 - T/2$ to $t_0 + T/2$, the time average of $x^2(t)$ is $\hat{x}^2(t_0)$ (say) where

$$\hat{x}^2(t_0) = \frac{1}{T}\int_{t_0-T/2}^{t_0+T/2} x^2(t)\,dt = \frac{1}{T}\int_{t_0-T/2}^{t_0+T/2} a^2(t)z^2(t)\,dt$$

$$\simeq a^2(t_0)\frac{1}{T}\int_{t_0-T/2}^{t_0+T/2} z^2(t)\,dt \tag{14.90}$$

provided that $a(t)$ varies slowly enough with time that it may be taken as constant over the interval T. Since $z(t)$ is stationary, the integral on the r.h.s. of (14.90) is an estimate for $E[z^2]$ and so

$$\hat{x}^2(t_0) \simeq a^2(t_0)E[z^2]. \tag{14.91}$$

The relative magnitude of $a(t)$ and $z(t)$ in (14.84) is arbitrary, so we can choose any suitable scale factor for $E[z^2]$, of which the most convenient is to put

$$E[z^2] = 1. \tag{14.92}$$

Equation (14.91) then gives an estimate for $a(t_0)$ for each slice of the segmented sample record.

Knowing $a(t)$, it is not difficult to generate $z(t)$ from $x(t)$, from (14.84), and to test $z(t)$ for nonstationary trends. The autocorrelation function of the stationary process $z(t)$ can then be found by averaging along the complete sample of of length T_n (say) to obtain

$$R_z(\tau) \simeq \frac{1}{T_n - \tau}\int_0^{T_n-\tau} z(t)z(t+\tau)\,dt \tag{14.93}$$

which can be substituted into (14.89) to find the required nonstationary autocorrelation function $R_x(t_1, t_2)$.

Local spectral density calculations

Although the mathematical approaches described above are appropriate for some non-stationary problems, there are two other approximate methods of digital spectral analysis that are used for studying the frequency content of non-stationary processes. The *short-time Fourier transform* (STFT) has a short data window centered at time t. Spectral coefficients are calculated for this short length of data. The idea is that local spectral coefficients are obtained which describe the frequency composition of the record being analysed at time t. The window is then moved to a new position and the calculation repeated. Unfortunately there is a fundamental problem with this approach, which is that high resolution cannot be obtained simultaneously in both the time domain and the frequency domain. Suppose that the width of the window is T. Then we know from Chapter 11, equation (11.35), that its frequency bandwidth is of the order of $1/T$. Using a short data window (T small) in order to compute spectral density coefficients for the length of record that is local to time t means that the bandwidth of each spectral coefficient is necessarily wide ($1/T$ large). The two requirements of a short data window and a narrow bandwidth are irreconcilable.

The *Wigner–Ville method* is a second approach. Suppose that an ensemble of sample records is available for analysis. The ensemble-averaged instantaneous correlation function at time t is

$$R_{xx}(\tau, t) = E[x(t - \tau/2)x(t + \tau/2)] \qquad (14.94)$$

and its Fourier transform is

$$S_{xx}(\omega, t) = \frac{1}{2\pi} \int_{-\infty}^{\infty} R_{xx}(\tau, t)\, e^{-i\omega\tau}\, d\tau \qquad (14.95)$$

where the instantaneous spectral density function $S_{xx}(\omega, t)$ is a function of time and of the parameter ω. Then, in theory, $S_{xx}(\omega, t)$ is a measure of the frequency content of the non-stationary random process $\{x(t)\}$ at time t. In practice, it is never possible to compute the ensemble-averaged function $R_{xx}(\tau, t)$ because the necessary data are just not available. If the ensemble averaging operation is omitted and (14.94) substituted into an equation like (14.95) then we find that

$$\Phi_{xx}(\omega, t) = \frac{1}{2\pi} \int_{-\infty}^{\infty} x(t - \tau/2)x(t + \tau/2)\, e^{-i\omega\tau}\, d\tau. \qquad (14.96)$$

This applies for a single sample record. $\Phi_{xx}(\omega, t)$ is a function of ω and t for this one record $x(t)$ and is called the Wigner distribution of $x(t)$ after E. P. Wigner [118] who used it first in quantum mechanics. The application of (14.96) to harmonic analysis was made first by J. Ville [117] which led to the name Wigner–Ville method. Unfortunately there are also difficulties with this method. Although the integral in (14.96) is centered at time t, it covers an infinite range of τ and so depends on the character of x far away from the local time t. It is impossible to have a truly local spectral density because of the continuing nature of harmonic waves. There is a fundamental uncertainty principle for time-dependent spectra which, for both methods, makes it impossible to achieve high resolution in time and frequency simultaneously.

In order to overcome this limitation of harmonic analysis, there has been considerable theoretical development of an alternative method of signal analysis. Instead of using sines and cosines as the basis functions for decomposing a general signal, alternative families of orthogonal basis functions called *wavelets* are used. Whereas harmonic functions by definition go on for ever, wavelets are local functions. By joining these together and by using different scales, it is possible to assemble a family of basis functions which are particularly suitable for examining the local character of non-stationary signals. Although wavelet theory is complex, there are numerical algorithms akin to the FFT for calculating wavelet coefficients. These discrete wavelet transforms appear to have important applications in random vibration analysis (see Chapter 17 and Appendix 7).

Weibull distribution of peaks

In Chapter 8, we derived the Rayleigh distribution of peaks for a narrow-band Gaussian process. In a number of applications of the theory, it has been found that the Gaussian assumption may not be valid and that the distribution of peaks then departs significantly from a Rayleigh distribution. Two examples are the calculation of wave-induced bending moments in ships [87] and the wind loading of buildings [88].

For a Rayleigh distribution, the probability that any peak, chosen at random, is less than level a is, from Chapter 8,

$$\text{Prob}(\text{Peak} < a) = P_p(a) = 1 - e^{-a^2/2\sigma_y^2}. \qquad (14.97)$$

Let a_0 be the median peak height, for which

$$\text{Prob}(\text{Peak} < a_0) = P_p(a_0) = 1 - e^{-a_0^2/2\sigma_y^2} = 1/2. \qquad (14.98)$$

Then a_0 can be expressed in terms of the variance of the underlying displacement distribution σ_y^2 by solving (14.97) to obtain

$$a_0 = \sigma_y\sqrt{2\ln 2}. \qquad (14.99)$$

On substituting a_0 into (8.26) by using (14.98), we obtain the following alternative description of the Rayleigh distribution

$$\text{Prob}\left(\frac{\text{Peak}}{a_0} < \frac{a}{a_0}\right) = P_p\left(\frac{a}{a_0}\right) = 1 - e^{-(\ln 2)(a/a_0)^2}. \qquad (14.100)$$

In order to approximate more general distributions, the exponent 2 in (14.99) can be replaced by the general coefficient k to give

$$P_p\left(\frac{a}{a_0}\right) = 1 - e^{-(\ln 2)(a/a_0)^k}. \qquad (14.101)$$

This is called the *Weibull* probability distribution function. When $a = a_0$, (14.100) gives

$$P_p\left(\frac{a}{a_0} = 1\right) = 1 - e^{-\ln 2} = 1/2 \qquad (14.102)$$

so that a_0 remains the median peak height independently of k. In Fig. 14.11a, the Weibull probability distribution function, $P_p(a/a_0)$, is plotted against a/a_0 for several different values of k. From (1.24), the corresponding probability density function for the distribution of peaks, $p_p(a/a_0)$, is given by differentiating (14.100) to obtain

$$p_p\left(\frac{a}{a_0}\right) = k(\ln 2)\left(\frac{a}{a_0}\right)^{k-1} e^{-(\ln 2)(a/a_0)^k}. \qquad (14.103)$$

This function is plotted against a/a_0 in Fig. 14.11b for the same values of k as for Fig. 14.11a. When $k > 2$, there are fewer large and fewer small peaks than for the Rayleigh distribution. For the case when $k = 10$, it is clear from Fig. 14.11b that most of the peak heights are close to the median peak height. When $k = \infty$, all the

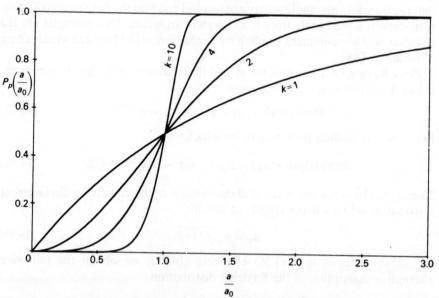

Fig. 14.11a Weibull probability distribution of peaks, $P_p(a/a_0) = \text{Prob}$ (Peak height$/a_0 < a/a_0$), plotted against a/a_0, where a_0 is the median peak height and a is an arbitrary level, for different values of the Weibull exponent, k

peaks occur at the median peak height a_0. This is the case for a random process each of whose sample functions consists of a harmonic wave of the same amplitude but random phase.

Conversely, when $k < 2$, there are more large and more small peaks than for a Rayleigh distribution. In Fig. 14.11b, the probability density function for $k = 1$ shows that the density of peaks decreases monotonically from its maximum value at zero height towards zero at large peak height. However, the probability density for the occurrence of peaks of small height and large height is greater for the case when $k = 1$ than when the exponent k is greater than 1.

Consider the highest peak that may be expected to occur in time T. The average number of cycles (and therefore of peaks) will be $v_0^+ T$, where v_0^+ is the frequency of positive slope zero crossings. Let the level a_{max} be set so that the average number of positive slope crossings of a_{max} in time T is exactly 1, giving

$$v_{a_{max}}^+ T = 1. \tag{14.104}$$

On average, there will be only one peak greater than a_{max} in every period T. We do not know the height of this one peak. We know only that all the other peaks in time T are less than a_{max}. However, a_{max} is easy to calculate and it gives a lower

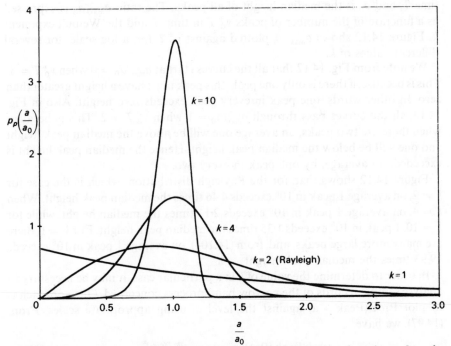

Fig. 14.11b Corresponding Weibull probability density functions for the distribution of peaks, obtained by differentiating the functions plotted in Fig. 14.11a

bound for the highest peak which is useful in practical cases. From Chapter 8, we know from (8.23) that

$$\text{Prob}\,(\text{Peak} > a_{max}) = \frac{v_{a_{max}}^+}{v_0^+} \tag{14.105}$$

and, for the Weibull distribution,

$$\text{Prob}\,(\text{Peak} > a_{max}) = e^{-(\ln 2)(a_{max}/a_0)^k}. \tag{14.106}$$

If we eliminate $v_{a_{max}}^+$ between (14.103) and (14.104) and then eliminate $\text{Prob}(\text{Peak} > a_{max})$ between the resulting equation and (14.105), we obtain

$$\frac{1}{v_0^+ T} = e^{-(\ln 2)(a_{max}/a_0)^k} \tag{14.107}$$

which may be solved to give

$$\left(\frac{a_{max}}{a_0}\right) = \left\{\frac{\ln\,(v_0^+ T)}{\ln 2}\right\}^{1/k} \tag{14.108}$$

This equation (14.107) is a general result for any narrow band process for which there is one full cycle for every positive slope crossing of the zero axis and whose distribution of peaks is described by a Weibull distribution. In (14.107), a_{max} is the level which on average is exceeded by one peak (only) in time T, and a_0 is the median height of all peaks. The ratio a_{max}/a_0 is expressed as a function of the number of peaks $v_0^+ T$ in time T and the Weibull exponent k. Figure 14.12 shows a_{max}/a_0 plotted against $v_0^+ T$ (on a log scale) for several different values of k.

We note from Fig. 14.12 that all the curves start at $a_{max}/a_0 = 0$ when $v_0^+ T = 1$. This is because, if there is only one peak, this peak must have a height greater than zero. In other words, one peak in every peak exceeds zero height. Also in Fig. 14.12, all the curves pass through $a_{max}/a_0 = 1$ when $v_0^+ T = 2$. This is because, when there are two peaks, on average one will be above the median peak height and one will be below the median peak height. Hence the median peak height is exceeded, on average, by one peak in every two.

Figure 14.12 shows that, for the Rayleigh distribution, which is the case for $k = 2$, on average 1 peak in 10^6 exceeds 4·46 times the median peak height. When $k = 4$, on average 1 peak in 10^6 exceeds 2·11 times the median height, while for $k = 10$, 1 peak in 10^6 exceeds 1·35 times the median peak height. For $k = 1$, there are many more large peaks, and, from (14.104), on average 1 peak in 10^6 exceeds 19·93 times the median peak height.

In order to determine the value of k in a particular case, it may be necessary to analyse sample records of the narrow band process concerned. One approach is to plot $\text{Prob}(\text{Peak} > a)$ against the level a, using appropriate scales. From (14.97), we have

$$\text{Prob}\,(\text{Peak} > a) = e^{-(\ln 2)(a/a_0)^k} \tag{14.109}$$

and so

$$\ln\{\text{Prob}\,(\text{Peak} > a)\} = -(\ln 2)(a/a_0)^k \tag{14.110}$$

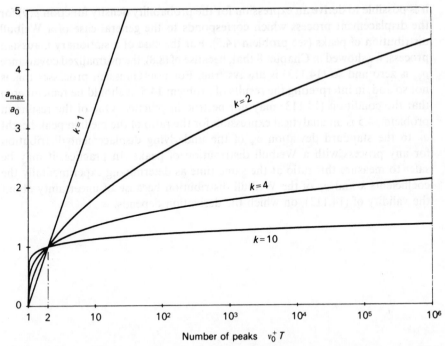

Fig. 14.12 Relative level a_{max}/a_0 exceeded on average by one peak only in every $v_0^+ T$ peaks, plotted as a function of $v_0^+ T$, for different values of the Weibull exponent, k. The median height of peaks is a_0

and

$$\ln[-\ln\{\text{Prob}(\text{Peak} > a)\}] = \ln\ln 2 + k\ln a - k\ln a_0. \qquad (14.111)$$

Hence the slope of a graph of $\ln[-\ln\{\text{Prob}(\text{Peak} > a)\}]$ against $\ln a$ is the Weibull exponent k. The median peak height can also be obtained from a zero intercept on this graph, since, if

$$\ln[-\ln\{\text{Prob}(\text{Peak} > a)\}] = 0$$

when $a = a_1$, then

$$(\ln 2)(a_1/a_0)^k = 1 \qquad (14.112)$$

and so

$$a_0 = a_1(\ln 2)^{1/k}. \qquad (14.113)$$

This method has been used by Melbourne for the analysis of the probability distributions of wind loading on buildings [88].

Subject to an assumption that the underlying distributions of the displacement process $\{y(t)\}$ and the velocity process $\{\dot{y}(t)\}$ are statistically independent so that

$$p(y, \dot{y}) = p(y)p(\dot{y}), \qquad (14.110)$$

it is possible to derive an expression for the probability density function $p(y)$ for the displacement process which corresponds to the general case of a Weibull distribution of peaks (see problem 14.5). For the case of a stationary Gaussian process, we showed in Chapter 8 that, because of (8.8), the normalized covariance ρ_{yy} is zero and so (14.113) is always true. For non-Gaussian processes this is not so and, in interpreting the results of problem 14.5, it should be remembered that the condition (14.113) may not be true in practice. One of the results in problem 14.5 is an analytical expression for the ratio of the median peak height a_0 to the standard deviation σ_y of the underlying displacement distribution for any process with a Weibull distribution of peaks. In practice, it may be safer to measure this ratio at the same time as determining experimentally the coefficients k and a_0 of the Weibull distribution because of uncertainty about the validity of (14.113), on which the derivation depends.

Chapter 15

Multi-dimensional spectral analysis

Two-dimensional Fourier series

In Chapter 4, we discussed the Fourier series expansion of an arbitrary one-dimensional periodic function $x(t)$, with period T, Fig. 4.1. The trigonometric functions

$$\cos\left(2\pi\frac{kt}{T}\right) \quad \text{and} \quad \sin\left(2\pi\frac{mt}{T}\right), \qquad k = 0, 1, \ldots; \; m = 1, 2, \ldots,$$

are orthogonal in the interval $0 \leqslant t \leqslant T$, so that the definite integral from $t = 0$ to T of the product of any two different functions from this set is zero. Therefore we can write

$$x(t) = a_0 + 2 \sum_{k=1}^{\infty} \left(a_k \cos 2\pi\frac{kt}{T} + b_k \sin 2\pi\frac{kt}{T} \right) \tag{10.1}$$

where the Fourier coefficients a_k and b_k are given by (10.2). Alternatively this result may be written in the equivalent complex exponential form

$$x(t) = \sum_{k=-\infty}^{\infty} X_k e^{i2\pi(kt/T)} \tag{15.1}$$

where the X_k are given by

$$X_k = \frac{1}{T} \int_0^T x(t)\, e^{-i2\pi(kt/T)}\, dt. \tag{15.2}$$

The complex coefficients X_k are related to the a_k and b_k in (10.1) by

$$\begin{aligned} X_k &= a_k - ib_k, & k &\geqslant 0 \\ X_k &= a_{|k|} + ib_{|k|}, & k &< 0. \end{aligned} \tag{15.3}$$

Consider now a two-dimensional periodic function $y(x_1, x_2)$ which has period L_1 in the x_1 dimension and period L_2 in the x_2 dimension, so that

$$y(x_1, x_2) = y(x_1 + pL_1, x_2 + qL_2) \tag{15.4}$$

where p and q are integers. This function can be expressed in the form of a two-dimensional Fourier series expansion involving terms which are each the product of two trigonometric functions. This is possible because the functions

$$\cos 2\pi \frac{kx_1}{L_1} \cos 2\pi \frac{mx_2}{L_2}, \qquad k = 0, 1, \ldots; \, m = 0, 1, \ldots,$$

$$\sin 2\pi \frac{kx_1}{L_1} \cos 2\pi \frac{mx_2}{L_2}, \qquad k = 1, 2, \ldots; \, m = 0, 1, \ldots,$$

$$\cos 2\pi \frac{kx_1}{L_1} \sin 2\pi \frac{mx_2}{L_2}, \qquad k = 0, 1, \ldots; \, m = 1, 2, \ldots, \qquad (15.5)$$

$$\sin 2\pi \frac{kx_1}{L_1} \sin 2\pi \frac{mx_2}{L_2}, \qquad k = 1, 2, \ldots; \, m = 1, 2, \ldots,$$

are orthogonal in the rectangle $0 \leqslant x_1 \leqslant L_1, 0 \leqslant x_2 \leqslant L_2$. We can write $y(x_1, x_2)$ as a series expansion similar in form to (10.1) with positive values of the integers k and m, but involving products of trigonometric functions. However, it is simpler to use the equivalent complex exponential form, which is

$$y(x_1, x_2) = \sum_{k=-\infty}^{\infty} \sum_{m=-\infty}^{\infty} Y_{km} \, e^{i2\pi(kx_1/L_1 + mx_2/L_2)} \qquad (15.6)$$

where

$$Y_{km} = \frac{1}{L_1 L_2} \int_0^{L_1} dx_1 \int_0^{L_2} dx_2 \, y(x_1, x_2) \, e^{-i2\pi(kx_1/L_1 + mx_2/L_2)}. \qquad (15.7)$$

These formulae (15.6) and (15.7) have a straightforward extension to the general case of n dimensions, when there will be n summations in (15.6) and n integrals in (15.7) and the exponent of the exponentials then involves the summation of n terms.

The discrete equivalent of (15.7) may be obtained by sampling the continuous function $y(x_1, x_2)$ within the rectangle $0 \leqslant x_1 \leqslant L_1, 0 \leqslant x_2 \leqslant L_2$. Suppose that length L_1 is divided into N_1 equal intervals Δ_1 and length L_2 into N_2 equal intervals Δ_2, so that

$$L_1 = N_1 \Delta_1 \quad \text{and} \quad L_2 = N_2 \Delta_2. \qquad (15.8)$$

Then if we put

$$x_1 = r\Delta_1 \quad \text{and} \quad x_2 = s\Delta_2 \qquad (15.9)$$

and write

$$y(x_1, x_2) = y_{rs}, \qquad (15.10)$$

(15.7) is replaced by

$$Y_{km} = \frac{1}{N_1 N_2} \sum_{r=0}^{N_1-1} \sum_{s=0}^{N_2-1} y_{rs} \, e^{-i2\pi(kr/N_1 + ms/N_2)}, \qquad k = 0, 1, 2, \ldots, N_1 - 1$$

$$m = 0, 1, 2, \ldots, N_2 - 1.$$

$$(15.11)$$

As in the case of a one-dimensional transform, Chapter 10, the two-dimensional transform defined by (15.11) has an exact inverse which allows the two-dimensional array of numbers y_{rs} to be recovered exactly (problem 15.2). The two-dimensional inverse discrete Fourier transform is

$$y_{rs} = \sum_{k=0}^{N_1-1} \sum_{m=0}^{N_2-1} Y_{km} e^{i2\pi(kr/N_1 + ms/N_2)}, \quad \begin{array}{l} r = 0, 1, 2, \ldots, N_1 - 1 \\ s = 0, 1, 2, \ldots, N_2 - 1. \end{array} \tag{15.12}$$

As for the one-dimensional case, Chapter 10, the upper limits of $N_1 - 1$ and $N_2 - 1$ on k and m limit the maximum frequencies of the orthogonal functions in the series expansion. We shall consider this limitation shortly. Also we note that there are no negative integers in the summation (15.12), although there are negative integers in the complex exponential series expansion (15.6). We shall see that the symmetry properties of Y_{km} calculated from (15.11) permit only positive integers to be used without losing information but at the expense of the reduced number of frequency components that can be represented in an $N_1 \times N_2$ point transform.

In order to compute the two-dimensional discrete transform, we can make repeated application of the one-dimensional FFT algorithm. This may be seen, for example, from (15.11), which can be written in the equivalent form

$$Y_{km} = \frac{1}{N_1} \sum_{r=0}^{N_1-1} e^{-i2\pi(kr/N_1)} \left(\frac{1}{N_2} \sum_{s=0}^{N_2-1} y_{rs} e^{-i2\pi(ms/N_2)} \right). \tag{15.13}$$

Consider calculating the summation in brackets on the r.h.s. of (15.13). The integer r is held constant and the summation made over s. This involves a one-dimensional transformation of row r of the $N_1 \times N_2$ array of numbers y_{rs}, and the one-dimensional FFT algorithm in Appendix 2 may be used for the calculation. The calculation has to be repeated N_1 times in order to transform each of the N_1 rows.

Then the summation outside the brackets in (15.13) has to be made, involving summing over the range of values of r. In this summation, integer s is held constant, and the calculation involves a one-dimensional transformation of column s of the $N_1 \times N_2$ array of numbers obtained from the first part of the calculation. This time the one-dimensional FFT algorithm must be used N_2 times, one for each of the N_2 columns of the partly transformed array.

This procedure can be programmed quite simply, using the FFT algorithm in Appendix 2, and a two-dimensional discrete Fourier transform based on this approach is included in Appendix 2. It is listed in both Fortran and Basic. The Basic language version can be implemented on desk-top equipment for study purposes. This program has been used for the example which follows, and it can be applied to the computer problems for this chapter. For many practical problems, the major difficulty of computing two-dimensional (or higher order) transforms is the great amount of storage space needed. For example, if $N_1 = N_2 = N = 512$, then a two-dimensional $N \times N$ array with real and imaginary parts requires $2N^2 = 524{,}288$ registers of storage. Practical problems require considerable random-access memory or, alternatively, the calculation has to be programmed to load and unload mass storage devices as it proceeds.

Example

Consider an 8×8 discrete Fourier transform of the array

$$y_{rs} = \cos 2\pi\left(\frac{r}{8}\right)\cos 2\pi\left(\frac{3s}{8}\right), \qquad r, s = 0, 1, 2, ..., 7.$$

The two-dimensional DFT is a new array Y_{km} in which all the elements are zero except for

$$Y_{1,3} = Y_{1,5} = Y_{7,3} = Y_{7,5} = 1/4.$$

This result has been obtained by using the two-dimensional DFT computer program in Appendix 2, and it can be checked by applying (15.12) which becomes in this case

$$y_{rs} = \sum_{k=1,7} \sum_{m=3,5} \frac{1}{4} e^{i2\pi(kr+ms)/8}$$

$$= \frac{1}{4} \sum_{k=1,7} e^{i(\pi/4)kr} \sum_{m=3,5} e^{i(\pi/4)ms}$$

$$= \frac{1}{4}(e^{i(\pi/4)r} + e^{i(7\pi/4)r})(e^{i(3\pi/4)s} + e^{i(5\pi/4)s})$$

$$= \frac{1}{4}(e^{i(\pi/4)r} + e^{i[2\pi-(\pi/4)]r})(e^{i(3\pi/4)s} + e^{i[2\pi-(3\pi/4)]s})$$

$$= \frac{1}{4}(e^{i(\pi/4)r} + e^{-i(\pi/4)r})(e^{i(3\pi/4)s} + e^{-i(3\pi/4)s})$$

$$= \cos\frac{\pi}{4}r\cos\frac{3\pi}{4}s$$

since we know that $e^{i2\pi} = 1$ and $\cos\theta = \frac{1}{2}(e^{i\theta} + e^{-i\theta})$.

In Fig. 15.1, the non-zero values of Y_{km} are shown graphically, with Y_{km} plotted vertically on a grid base of integer values of k and m.

For the case when

$$y_{rs} = \cos 2\pi(r/8)\sin 2\pi(3s/8), \qquad r, s = 0, 1, 2, ..., 7,$$

we find that all the Y_{km} are zero except for

$$Y_{1,3} = Y_{7,3} = -i(1/4) \quad \text{and} \quad Y_{1,5} = Y_{7,5} = i(1/4).$$

When

$$y_{rs} = \sin 2\pi(r/8)\cos 2\pi(3s/8), \qquad r, s = 0, 1, 2, ..., 7,$$

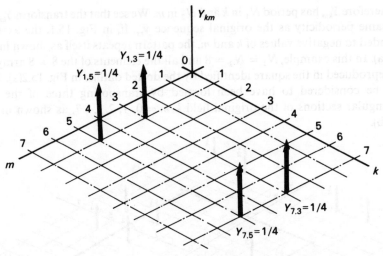

Fig. 15.1 Non-zero values of $Y_{km} = \dfrac{1}{64} \displaystyle\sum_{r=0}^{7} \sum_{s=0}^{7} y_{rs}\, e^{-i2\pi(kr/8\,+\,ms/8)}$ when

$y_{rs} = \cos(2\pi\, r/8)\cos(2\pi\, 3s/8)$, plotted on a grid base of the integer values of k and m

then all the Y_{km} are zero except

$$Y_{1,3} = Y_{1,5} = -i(1/4) \quad \text{and} \quad Y_{7,3} = Y_{7,5} = i(1/4),$$

and when

$$y_{rs} = \sin 2\pi(r/8)\sin 2\pi(3s/8), \qquad r, s = 0, 1, 2, \ldots, 7,$$

then all the Y_{km} are zero except

$$Y_{1,3} = Y_{7,5} = -1/4 \quad \text{and} \quad Y_{1,5} = Y_{7,3} = 1/4.$$

Properties of the two-dimensional DFT

Consider what happens if we use (15.11) to calculate Y_{km} for values of k greater than $N_1 - 1$ and values of m greater than $N_2 - 1$. If p and q are two new integers, then

$$
\begin{aligned}
Y_{pN_1+k,\,qN_2+m} &= \frac{1}{N_1 N_2} \sum_{r=0}^{N_1-1} \sum_{s=0}^{N_2-1} y_{rs}\, e^{-i2\pi\{(pN_1+k)r/N_1 \,+\, (qN_2+m)s/N_2\}} \\
&= \frac{1}{N_1 N_2} \sum_{r=0}^{N_1-1} \sum_{s=0}^{N_2-1} y_{rs}\, e^{-i2\pi(pr+qs)}\, e^{-i2\pi(kr/N_1 \,+\, ms/N_2)} \\
&= Y_{km},
\end{aligned}
\tag{15.14}
$$

since $e^{-i2\pi(pr+qs)} = 1$ when $(pr+qs)$ is an integer.

Therefore Y_{km} has period N_1 in k and N_2 in m. We see that the transform Y_{km} has the same periodicity as the original sequence y_{rs}. If, in Fig. 15.1, the axes are extended to negative values of k and m, the pattern repeats itself as shown in Fig. 15.2(a). In this example, $N_1 = N_2 = 8$ and all the elements of the 8×8 array Y_{km} are reproduced in the square identified by the dotted outline in Fig. 15.2(a). They may be considered to have been formed by transposing three of the four rectangular sections of the original field $k, m = 0, 1, 2, \ldots, 7$, as shown in Fig. 15.2(b).

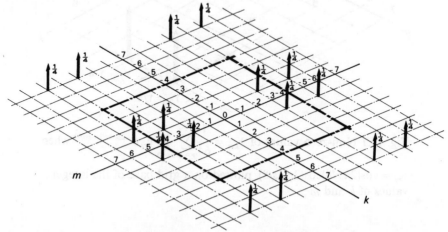

Fig. 15.2(a) Periodic repetition of values calculated by the two-dimensional DFT, Y_{km}, when the values of k and m are taken negative. In this example, Y_{km} is the DFT of $y_{rs} = \cos(2\pi r/8)\cos(2\pi 3s/8)$, where $r, s = 0, 1, 2, \ldots, 7$

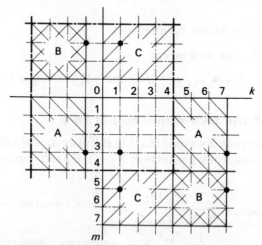

Fig. 15.2(b) Showing how values within the square drawn in heavy broken line in Fig. 15.2(a) may be formed by transposing parts of the field $k, m = 0, 1, 2, \ldots, 7$

If the values of y_{rs} are real, then most of the Y_{km} occur in complex conjugate pairs. Consider the case when N_1 and N_2 are both even. First we note that four members of the sequence $\{Y_{km}\}$ will always be real. These are Y_{00}, $Y_{N_1/2,0}$, $Y_{0,N_2/2}$ and $Y_{N_1/2,N_2/2}$. This result may be verified from the definition of Y_{km}, (15.11). For example, put $k = N_1/2$, $m = 0$ in (15.11) to find

$$Y_{N_1/2,0} = \frac{1}{N_1 N_2} \sum_{r=0}^{N_1-1} \sum_{s=0}^{N_2-1} y_{rs} e^{-i\pi r}. \tag{15.15}$$

Since

$$e^{-i\pi r} = \cos \pi r - i \sin \pi r = \cos \pi r,$$

the imaginary part of (15.15) is always zero and so, for y_{rs} real, $Y_{N_1/2,0}$ must be real.

The remaining $(N_1 N_2 - 4)$ elements of $\{Y_{km}\}$ occur in complex conjugate pairs. For y_{rs} real, we have

$$Y_{N_1-k,0} = Y_{k,0}^* \tag{15.16}$$

$$Y_{0,N_2-m} = Y_{0,m}^* \tag{15.17}$$

and

$$Y_{N_1-k,N_2-m} = Y_{k,m}^*. \tag{15.18}$$

The proof of each of (15.16), (15.17) and (15.18) follows directly from (15.11). For example, putting $k = N_1 - k$, $m = N_2 - m$ in (15.11) we have

$$Y_{N_1-k,N_2-m} = \frac{1}{N_1 N_2} \sum_{r=0}^{N_1-1} \sum_{s=0}^{N_2-1} y_{rs} e^{-i2\pi(r+s)} e^{i2\pi(kr/N_1 + ms/N_2)}$$

$$= Y_{k,m}^*$$

since $e^{-i2\pi(r+s)} = 1$ for all integer values of r and s.

Equation (15.16) defines $N_1/2 - 1$ elements as complex conjugates of other corresponding elements with $m = 0$ (note that Y_{00} and $Y_{N/2,0}$ are real, so that there are only $N_1 - 2$ elements with $m = 0$ which may be complex). Equation (15.17) defines $N_2/2 - 1$ elements in the same way as complex conjugates of other elements with $k = 0$. Equation (15.18) relates elements within the range $k = 1, 2, ..., N_1 - 1, m = 1, 2, ..., N_2 - 1$. There are $(N_1 - 1)(N_2 - 1)$ such elements, but one of these, $Y_{N_1/2,N_2/2}$, is always real. Hence $(N_1 N_2 - N_1 - N_2)/2$ elements are defined by (15.18) as complex conjugates of other corresponding elements.

We conclude that, for y_{rs} real and N_1 and N_2 both even, we shall always find that the four elements $Y_{0,0}$, $Y_{N_1/2,0}$, $Y_{0,N_2/2}$ and $Y_{N_1/2,N_2/2}$ are real and that the remaining elements occurs in complex conjugate pairs of which $N_1/2 - 1$ are defined by (15.16), $N_2/2 - 1$ by (15.17) and $(N_1 N_2 - N_1 - N_2)/2$ by (15.18). As a check, we see that

$$4 + 2\{(N_1/2 - 1) + (N_2/2 - 1) + (N_1 N_2 - N_1 - N_2)/2\} = N_1 N_2$$

so that all the elements of Y_{km}, $k = 0, 1, 2, ..., N_1 - 1$, $m = 0, 1, 2, ..., N_2 - 1$, have been included.

When considering the one-dimensional DFT in Chapter 10, we saw that the maximum frequency that can be detected from a real continuous series which is sampled at interval Δ is the *Nyquist* or *folding frequency*,

$$f_0 = \frac{1}{2\Delta} \text{ cycles/sampling interval.} \tag{15.19}$$

The same limitation applies to the multi-dimensional DFT of sequences obtained by sampling real continuous multi-dimensional functions. For two dimensions, the maximum frequency of the trigonometric functions (15.5) in the series expansion represented by the DFT is

$$f_1 = \frac{1}{2\Delta_1} \quad \text{and} \quad f_2 = \frac{1}{2\Delta_2} \text{ cycles/interval.} \tag{15.20}$$

Consider components in the x_1 direction, to which the subscript k relates. Although the maximum value of k is N_1, by virtue of (15.16) and (15.18), for $N_1/2 < k \leqslant N_1 - 1$ the frequency coefficients Y_{km} are related to values in the range $0 < k < N_1/2$. This may be seen in the above example for an 8×8 array in which the input sequence y_{rs} consists of values obtained by sampling the product of two harmonic functions. The value $k = N_1/2$ corresponds to a frequency

$$k/L_1 = (N_1/2)/(N_1\Delta_1) = \frac{1}{2\Delta_1}. \tag{15.21}$$

Similarly $m = N_2/2$ corresponds to a maximum frequency of $f_2 = 1/2\Delta_2$ in the x_2 direction. When the input array y_{rs} includes components with frequencies above the Nyquist frequencies (15.20), then these components introduce errors by distorting falsely the values of Y_{km} at integer values which correspond to frequencies below the frequencies (15.20). This is illustrated in the following example.

Example

Consider an 8×8 array derived by sampling the continuous surface

$$y(x_1, x_2) = \cos 2\pi \frac{x_1}{L} \cos 2\pi \frac{5x_2}{L}$$

at a sampling interval $\Delta = L/8$. For this interval the Nyquist frequency is $4/L$ (cycles/distance), which is less than the frequency of $5/L$ in the second of the two cosine functions. Therefore there will be distortion of the values of Y_{km} due to aliasing.

By using the two-dimensional DFT program in Appendix 2, the two-dimensional discrete transform of

$$y_{rs} = \cos 2\pi\left(\frac{r}{8}\right)\cos 2\pi\left(\frac{5s}{8}\right), \qquad r, s = 0, 1, 2, \ldots, 7,$$

is Y_{km} where all the values are zero except

$$Y_{1,3} = Y_{1,5} = Y_{7,3} = Y_{7,5} = 1/4.$$

This is the same as the transform of

$$y_{rs} = \cos 2\pi\left(\frac{r}{8}\right)\cos 2\pi\left(\frac{3s}{8}\right), \qquad r, s = 0, 1, 2, ..., 7$$

calculated for the previous example.

One of the important results from Fourier analysis is expressed by Parseval's theorem, which says that the mean square value of a periodic function is equal to the sum of the squares of all its Fourier coefficients. For a one-dimensional series, this was the subject of problems 4.3 and 10.6. When the series expansion for $y(t) = y(t + T)$ is written in the complex form (15.1), then

$$\frac{1}{T}\int_0^T y^2(t)\,dt = \sum_{k=-\infty}^{\infty} Y_k^* Y_k. \tag{15.22}$$

For the two-dimensional case when $y(x_1,x_2) = y(x_1 + L_1, x_2 + L_2)$, the corresponding result is

$$\frac{1}{L_1 L_2}\int_0^{L_1} dx_1 \int_0^{L_2} dx_2\, y^2(x_1,x_2) = \sum_{k=-\infty}^{\infty}\sum_{m=-\infty}^{\infty} Y_{km}^* Y_{km}. \tag{15.23}$$

Similar results apply for the discrete Fourier transform. Problem 10.6(ii) was concerned with proving that

$$\frac{1}{N}\sum_{r=0}^{N-1} y_r^2 = \sum_{k=0}^{N-1} Y_k^* Y_k, \tag{15.24}$$

and the corresponding two-dimensional result is

$$\frac{1}{N_1 N_2}\sum_{r=0}^{N_1-1}\sum_{s=0}^{N_2-1} y_{rs}^2 = \sum_{k=0}^{N_1-1}\sum_{m=0}^{N_2-1} Y_{km}^* Y_{km}. \tag{15.25}$$

We leave the proof of (15.23) and (15.25) to problem 15.2.

Lastly we may note that the two-dimensional discrete Fourier transform relationships (15.11) and (15.12) can be extended to the n-dimensional case without difficulty: there will be n summations in each of the equations and the exponent of the exponential function which appears in both equations will itself consist of the summation of n terms of the form kr/N.

Spectral density of a multi-dimensional random process

So far in this book we have been concerned with random processes whose sample functions are each real continuous functions of time, or, in the section on surface irregularities in Chapter 14, continuous functions of distance in one direction. We turn now to the more general case, when a random process depends on more than one independent variable. For example, suppose that y measures the height of a continuous surface above a horizontal datum plane. If OX_1 and OX_2 are axes in the horizontal base plane, then y is a function of the two coordinates X_1 and X_2, $y(X_1, X_2)$. Suppose also that the height of the surface at a given location X_1, X_2 may change with time. Perhaps y measures the changing height of the sea at different locations. Then y is a function of t also, $y(X_1, X_2, t)$.

The correlation function for y is defined, as before, as the ensemble average of the product of the height of the surface at coordinates X_1, X_2, t and the height of the surface at a different set of coordinates $X_1 + x_1, X_2 + x_2, t + \tau$. The concept of an ensemble average has now to be stretched to include a theoretically infinite collection of sample functions in space as well as an infinite collection in time. For each moment of time, we assume that we have an infinite collection of sample functions each of which is a different surface; and we assume that we have this information for an infinite collection of different times. Then we can calculate

$$E[y(X_1, X_2, t)y(X_1 + x_1, X_2 + x_2, t + \tau)]$$

and, if the process is stationary in space and time, this average depends only on the separation between the coordinates, x_1, x_2 and τ, so that

$$E[y(X_1, X_2, t)y(X_1 + x_1, X_2 + x_2, t + \tau)] = R_{yy}(x_1, x_2, \tau). \tag{15.26}$$

In Chapter 5, for a one-dimensional stationary random process which is a function of time and has autocorrelation function $R(\tau)$, we defined its mean-square spectral density by the Fourier integral

$$S(\omega) = \frac{1}{2\pi} \int_{-\infty}^{\infty} R(\tau) e^{-i\omega\tau} \, d\tau. \tag{15.27}$$

In Chapter 14, when the random process is a function of distance, with autocorrelation function $R(x)$, the spectral density is

$$S(\gamma) = \frac{1}{2\pi} \int_{-\infty}^{\infty} R(x) e^{-i\gamma x} \, dx. \tag{15.28}$$

In (15.27) the time frequency is measured in radians/time; in (15.28) the spatial frequency (or wavenumber) is measured in radians/distance. When the correlation function depends on more than one variable, the spectral density is defined in a similar way. Just as the one-dimensional Fourier integral and its inverse were derived from the one-dimensional Fourier series equations (see Chapter 4), the multi-dimensional Fourier integral and its inverse can be derived from the corresponding equations for multi-dimensional Fourier analysis, which for two dimensions are (15.6) and (15.7). For a three-dimensional stationary random process whose correlation function is $R(x_1, x_2, \tau)$, the corresponding three-dimensional spectral density is defined as

$$S(\gamma_1, \gamma_2, \omega) = \frac{1}{(2\pi)^3} \int_{-\infty}^{\infty} dx_1 \int_{-\infty}^{\infty} dx_2 \int_{-\infty}^{\infty} d\tau \, R(x_1, x_2, \tau) e^{-i(\gamma_1 x_1 + \gamma_2 x_2 + \omega t)}.$$

$$\tag{15.29}$$

The inverse three-dimensional Fourier transform corresponding to (15.29) is

$$R(x_1, x_2, \tau) = \int_{-\infty}^{\infty} d\gamma_1 \int_{-\infty}^{\infty} d\gamma_2 \int_{-\infty}^{\infty} d\omega\, S(\gamma_1, \gamma_2, \omega)\, e^{i(\gamma_1 x_1 + \gamma_2 x_2 + \omega\tau)}.$$

$$(15.30)$$

In the classical development of Fourier integral theory, there are conditions for the existence of the integrals in (15.29) and (15.30) although the introduction of delta functions in the generalized theory allows these conditions to be largely relaxed. We shall assume in what follows that correlation functions possess Fourier transforms and vice versa. Since, for practical calculations, we can use only discrete Fourier transforms which always exist because they operate on a limited length of data only, there is no practical difficulty in making this assumption.

For a stationary process, we must have

$$R(x_1, x_2, \tau) = R(-x_1, -x_2, -\tau). \qquad (15.31)$$

Therefore $R(x_1, x_2, \tau)$ is a real, even function of its variables, and it follows from (15.29) that $S(\gamma_1, \gamma_2, \omega)$ is then also real and even. From (15.30), by putting $x_1 = x_2 = \tau = 0$, we have

$$E[y^2] = R_{yy}(0, 0, 0) = \int_{-\infty}^{\infty} d\gamma_1 \int_{-\infty}^{\infty} d\gamma_2 \int_{-\infty}^{\infty} d\omega\, S(\gamma_1, \gamma_2, \omega) \qquad (15.32)$$

and we can argue heuristically that the elemental contribution to the mean square $S(\gamma_1, \gamma_2, \omega)\, d\gamma_1\, d\gamma_2\, d\omega$ must be positive and so $S(\gamma_1, \gamma_2, \omega)$ must be a real, even and non-negative quantity. Problem 15.4 considers a formal proof of the last part of this result.

Suppose that the three-dimensional spectral density $S(\gamma_1, \gamma_2, \omega)$ is known for a stationary random process $\{y(X_1, X_2, t)\}$. Imagine that we can take a "snapshot" of y at a particular instant of time. The ensemble of sample functions $\{y(X_1, X_2)\}$ at this moment of time constitutes a stationary two-dimensional random process. It does not matter when the snapshot is taken because, if $\{y(X_1, X_2, t)\}$ is stationary, the statistical averages of y are independent of t; although each sample function $y(X_1, X_2)$ will be different at different t, the average values for the ensemble $\{y(X_1, X_2)\}$ will be the same. We can calculate the two-dimensional spectral density $S(\gamma_1, \gamma_2)$ for the stationary process $\{y(X_1, X_2)\}$ by simply integrating out the dependence of the three-dimensional spectral density $S(\gamma_1, \gamma_2, \omega)$ on the unwanted time frequency ω. The proof of this statement is as follows.

If $S(\gamma_1, \gamma_2)$ is the spectral density of the two-dimensional process $\{y(X_1, X_2)\}$ which has correlation function $R(x_1, x_2)$ then, from (15.30),

$$R(x_1, x_2) = \int_{-\infty}^{\infty} d\gamma_1 \int_{-\infty}^{\infty} d\gamma_2\, S(\gamma_1, \gamma_2)\, e^{i(\gamma_1 x_1 + \gamma_2 x_2)}. \qquad (15.33)$$

But, from (15.26), the two-dimensional correlation function $R(x_1, x_2)$ is the same as the three-dimensional function $R(x_1, x_2, \tau)$ evaluated at zero time separation $\tau = 0$. In other words,

$$R(x_1, x_2) = R(x_1, x_2, 0). \tag{15.34}$$

By putting $\tau = 0$ in (15.30), we obtain

$$R(x_1, x_2, 0) = \int_{-\infty}^{\infty} d\gamma_1 \int_{-\infty}^{\infty} d\gamma_2 \int_{-\infty}^{\infty} d\omega \, S(\gamma_1, \gamma_2, \omega) e^{i(\gamma_1 x_1 + \gamma_2 x_2)} \tag{15.35}$$

and, on comparing the r.h.s. of (15.33) with the r.h.s. of (15.35), both of which are expressions for the same function, we conclude that

$$S(\gamma_1, \gamma_2) = \int_{-\infty}^{\infty} d\omega \, S(\gamma_1, \gamma_2, \omega) \tag{15.36}$$

thus confirming the result just stated.

One-dimensional spectral densities can be obtained from two-dimensional spectral density functions by the same method.

Suppose that we want to find the spatial cross-spectral density between the one-dimensional stationary random processes $\{y(X_1, X_2 = 0\}$ and $\{y(X_1, X_2 = x_0)\}$. This is the cross-spectral density between the height $\{y_1(X_1)\}$ measured at $X_2 = 0$ and the height $\{y_2(X_1)\}$ measured at $X_2 = x_0$. The cross-correlation function between these two one-dimensional processes is, from (15.26),

$$R_{yy}(x_1, x_2 = x_0, \tau = 0) = E[y(X_1, X_2, t)y(X_1 + x_1, X_2 + x_0, t)] \tag{15.37}$$

which, from (15.30), is given by

$$R_{yy}(x_1, x_2 = x_0, \tau = 0) = \int_{-\infty}^{\infty} d\gamma_1 \int_{-\infty}^{\infty} d\gamma_2 \int_{-\infty}^{\infty} d\omega \, S(\gamma_1, \gamma_2, \omega) e^{i(\gamma_1 x_1 + \gamma_2 x_0)}. \tag{15.38}$$

But, from Chapter 3, we know also that

$$R_{y_1 y_2}(x_1) = E[y_1(X_1)y_2(X_1 + x_1)] = E[y(X_1, X_2 = 0, t)y(X_1 + x_1, X_2 = x_0, t)]$$
$$= R_{yy}(x_1, x_2 = x_0, \tau = 0) \tag{15.39}$$

and that, from (5.21),

$$R_{y_1 y_2}(x_1) = \int_{-\infty}^{\infty} d\gamma_1 \, S_{y_1 y_2}(\gamma_1) e^{i\gamma_1 x_1}. \tag{15.40}$$

By comparing the r.h.s. of (15.38) with the r.h.s. of (15.40), both of which, from (15.39), are expressions for the same correlation function, we see that

$$S_{y_1 y_2}(\gamma_1) = \int_{-\infty}^{\infty} d\gamma_2 \int_{-\infty}^{\infty} d\omega \, S(\gamma_1, \gamma_2, \omega) e^{i\gamma_2 x_0} \tag{15.41}$$

which, on account of (15.36), becomes

$$S_{y_1 y_2}(\gamma_1) = \int_{-\infty}^{\infty} d\gamma_2\, S(\gamma_1, \gamma_2)\, e^{i\gamma_2 x_0}. \tag{15.42}$$

This is an expression for the cross-spectral density $S_{y_1 y_2}(\gamma_1)$ between the height profile $y(X_1, X_2 = 0) = y_1(X_1)$ at $X_2 = 0$ and $y(X_1, X_2 = x_0) = y_2(X_1)$ at $X_2 = x_0$. In (15.41), the cross-spectral density is expressed in terms of the three-dimensional spectral density $S(\gamma_1, \gamma_2, \omega)$; in (15.42), the dependence on time frequency has been integrated out by using (15.36) and the cross-spectral density is expressed in terms of the two-dimensional spectral density $S(\gamma_1, \gamma_2)$.

There are two final points to be made about multi-dimensional spectral density functions. The first is a mathematical one. The multi-dimensional equations (15.29) and (15.30) can be condensed by writing

$$\mathbf{x} = x_1 \mathbf{e}_1 + x_2 \mathbf{e}_2$$
$$\boldsymbol{\gamma} = \gamma_1 \mathbf{e}_1 + \gamma_2 \mathbf{e}_2 \tag{15.43}$$

where $\mathbf{e}_1$ and $\mathbf{e}_2$ are unit vectors in the x_1 and x_2 directions, and by defining

$$d\mathbf{x} = dx_1\, dx_2$$
$$d\boldsymbol{\gamma} = d\gamma_1\, d\gamma_2 \tag{15.44}$$

so that the Fourier transform (15.29) becomes

$$S(\boldsymbol{\gamma}, \omega) = \frac{1}{(2\pi)^3} \int_\infty d\mathbf{x} \int_{-\infty}^{\infty} d\tau\, R(\mathbf{x}, \tau)\, e^{-i(\boldsymbol{\gamma}\cdot\mathbf{x} + \omega\tau)} \tag{15.45}$$

where the integration $\int_\infty$ ranges over the infinite two-dimensional space, and the inverse Fourier transform becomes

$$R(\mathbf{x}, \tau) = \int_\infty d\boldsymbol{\gamma} \int_{-\infty}^{\infty} d\tau\, S(\boldsymbol{\gamma}, \omega)\, e^{i(\boldsymbol{\gamma}\cdot\mathbf{x} + \omega\tau)}. \tag{15.46}$$

The second point is that, for a process which is stationary (in space and time), the properties of the correlation and spectral density functions allow the same equations to be written in a simpler, although less compact version. Because $R(x_1, x_2, \tau)$ and $S(\gamma_1, \gamma_2, \omega)$ are both real, even functions, (15.29) and (15.30) can be written without using complex exponential functions, when they may have the following form:

$$S(\gamma_1, \gamma_2, \omega) = \frac{1}{4\pi^3} \int_{-\infty}^{\infty} dx_1 \int_{-\infty}^{\infty} dx_2 \int_0^\infty d\tau\, R(x_1, x_2, \tau)\cos(\gamma_1 x_1 + \gamma_2 x_2 + \omega\tau) \tag{15.47}$$

and

$$R(x_1, x_2, \tau) = 2\int_{-\infty}^{\infty} d\gamma_1 \int_{-\infty}^{\infty} d\gamma_2 \int_0^\infty d\omega\, S(\gamma_1, \gamma_2, \omega)\cos(\gamma_1 x_1 + \gamma_2 x_2 + \omega\tau). \tag{15.48}$$

Example

The two-dimensional spectral density for a homogeneous (stationary in space), isotropic surface is given by

$$S_{yy}(\gamma_1, \gamma_2) = \frac{S_0}{\sqrt{2\pi}\,\gamma_0}\, e^{-(\gamma_1^2 + \gamma_2^2)/2\gamma_0^2}. \tag{15.49}$$

The corresponding two-dimensional correlation function is given by (15.30), and the integrals may be evaluated by making use of the standard results (1.20) from Chapter 1, to give

$$R_{yy}(x_1, x_2) = \sigma_y^2\, e^{-(x_1^2 + x_2^2)\gamma_0^2/2} \tag{15.50}$$

where

$$\sigma_y^2 = \sqrt{2\pi}\, S_0\gamma_0. \tag{15.51}$$

The one-dimensional autocorrelation function for the height of the profile of the intersection of the surface with any vertical plane is

$$R_{yy}(x) = R_{yy}(x_1 = x, x_2 = 0) = \sigma_y^2\, e^{-x^2\gamma_0^2/2}. \tag{15.52}$$

The corresponding one-dimensional spectral density may be obtained either by integrating out from (15.49) the dependence of $S_{yy}(\gamma_1, \gamma_2)$ on γ_2 or by taking the Fourier transform of $R_{yy}(x)$ using (15.28), and will be found to be

$$S_{yy}(\gamma) = S_0\, e^{-\gamma^2/2\gamma_0^2}. \tag{15.53}$$

The cross-spectral density between the profiles of two parallel tracks distance x_0 apart can be obtained either by taking the Fourier transform of $R_{yy}(x_1, x_2 = x_0)$ or by using (15.42). If we do the first of these, we have

$$S_{y_1 y_2}(\gamma) = \frac{1}{2\pi} \int_{-\infty}^{\infty} R_{yy}(x_1, x_2 = x_0)\, e^{-i\gamma x_1}\, dx_1$$

$$= \frac{\sigma_y^2}{2\pi} e^{-x_0^2\gamma_0^2/2} \int_{-\infty}^{\infty} e^{-(\gamma_0^2 x_1^2/2 + i\gamma x_1)}\, dx_1.$$

This integral can be evaluated after first completing the square of the exponent of e to obtain

$$S_{y_1 y_2}(\gamma) = \frac{\sigma_y^2}{2\pi} e^{-x_0^2\gamma_0^2/2} \int_{-\infty}^{\infty} e^{-(\gamma_0 x_1/\sqrt{2} + i[\gamma/\sqrt{2}\gamma_0])^2}\, e^{-\gamma^2/2\gamma_0^2}\, dx_1 \tag{15.54}$$

and then using the standard integral which is the first of (1.20) in Chapter 1 where

$$y = \left(\frac{\gamma_0}{\sqrt{2}} x_1 + \frac{i\gamma}{\sqrt{2}\gamma_0}\right) \quad \text{and} \quad dy = \frac{\gamma_0}{\sqrt{2}} dx_1.$$

The result after substituting for σ_y^2 from (15.51) is

$$S_{y_1 y_2}(\gamma) = S_0\, e^{-\gamma^2/2\gamma_0^2}\, e^{-\gamma_0^2 x_0^2/2}. \tag{15.55}$$

Lastly, in this example, we calculate the coherence, $\eta(\gamma)$, between two parallel tracks which, from (14.75), is defined by

$$\eta^2(\gamma) = \frac{S_{y_1 y_2}(\gamma) S_{y_2 y_1}(\gamma)}{S_{y_1 y_1}(\gamma) S_{y_2 y_2}(\gamma)}. \tag{15.56}$$

In this case, we have

$$S_{y_1 y_2}(\gamma) = S_{y_2 y_1}(\gamma) = S_0 \, e^{-\gamma^2/2\gamma_0^2} e^{-\gamma_0^2 x_0^2/2}$$

and

$$S_{y_1 y_1}(\gamma) = S_{y_2 y_2}(\gamma) = S_0 \, e^{-\gamma^2/2\gamma_0^2}$$

so that

$$\eta(\gamma) = e^{-\gamma_0^2 x_0^2/2} \tag{15.57}$$

which, in this special case, comes out to be independent of γ.

The cross-spectral density function $S_{y_1 y_2}(\gamma)$ relates to two tracks $y_1(X_1) = y(X_1, X_2 = 0)$ and $y_2(X_1) = y(X_1, X_2 = x_0)$ which are parallel to the X_1 axis. From (15.42), we note that $S_{y_1 y_2}(\gamma)$ will be real whenever $S(\gamma_1, \gamma_2)$ is even in γ_2 so that $S(\gamma_1, \gamma_2) = S(\gamma_1, -\gamma_2)$. If we had begun by choosing two tracks parallel to the X_2 axis, their cross-spectral density would have been real if $S(\gamma_1, \gamma_2)$ had been even in γ_1. If, as in the above example, $S(\gamma_1, \gamma_2)$ has rotational symmetry about its origin, the cross-spectral density between parallel tracks in any direction will be real. A random surface whose spectrum is defined by such a rotationally symmetrical two-dimensional spectral density function is said to be *isotropic* because then its spectral characteristics are independent of direction.

Discrete spectral density and circular correlation functions for a two-dimensional random process

We shall consider now the problem of calculating discrete values of a multi-dimensional spectral density by using the multi-dimensional discrete Fourier transform. For simplicity in the derivations that follow, it is convenient to use the two-dimensional case to develop the theory, but the extension to more than two dimensions follows the same approach and is not difficult.

We begin by writing down the two-dimensional version of (15.29):

$$S(\gamma, \omega) = \frac{1}{(2\pi)^2} \int_{-\infty}^{\infty} dx \int_{-\infty}^{\infty} d\tau \, R(x, \tau) e^{-i(\gamma x + \omega \tau)}. \tag{15.59}$$

This gives the two-dimensional spectral density $S(\gamma, \omega)$ as a double integral of the two-dimensional correlation function $R(x, \tau)$.

We now assume that $S(\gamma, \omega)$ and $R(x, \tau)$ are periodic with one period of $R(x, \tau)$ occupying the rectangle $0 \leqslant x \leqslant L, 0 \leqslant \tau \leqslant T$. If we divide the periodic length L into N_1 equal intervals Δ_1 and the periodic time T into N_2 equal intervals Δ_2, then (15.11) applies, and the discrete version of (15.59) becomes

$$S_{km} = \frac{1}{N_1 N_2} \sum_{r=0}^{N_1-1} \sum_{s=0}^{N_2-1} R_{rs}\, e^{-i2\pi(kr/N_1 + ms/N_2)}, \qquad k = 0, 1, 2, \ldots, N_1 - 1$$

$$m = 0, 1, 2, \ldots, N_2 - 1$$

$$(15.60)$$

where

$$R_{rs} = R(x = r\Delta_1, \tau = s\Delta_2). \qquad (15.61)$$

From the definition of the correlation function (15.26), we can see that we should have

$$R(x = r\Delta_1, \tau = s\Delta_2) = E[y(X,t)y(X + r\Delta_1, t + s\Delta_2)] \qquad (15.62)$$

but the ensemble average is not possible in practice and, instead, the r.h.s. of (15.62) is obtained by summing products of the discrete values of $R(x, \tau)$ in order to obtain

$$R_{rs} = \frac{1}{N_1 N_2} \sum_{u=0}^{N_1-1} \sum_{v=0}^{N_2-1} y_{u,v} y_{u+r,v+s}, \qquad r = 0, 1, 2, \ldots, N_1 - 1$$

$$s = 0, 1, 2, \ldots, N_2 - 1 \qquad (15.63)$$

where

$$y_{uv} = y(X = u\Delta_1, t = v\Delta_2)$$

$$y_{u+r,v+s} = y(X = u\Delta_1 + r\Delta_1, t = v\Delta_2 + s\Delta_2) \qquad (15.64)$$

and where, because of the assumed periodicity of $R(x, \tau)$, we must assume also that the function $y(X, t)$ is similarly periodic so that

$$y_{u,v} = y_{u-N_1,v} = y_{u,v-N_2} = y_{u-N_1,v-N_2}. \qquad (15.65)$$

Combining (15.63) with (15.60) we can now put

$$S_{km} = \frac{1}{N_1 N_2} \sum_{r=0}^{N_1-1} \sum_{s=0}^{N_2-1} \left(\frac{1}{N_1 N_2} \sum_{u=0}^{N_1-1} \sum_{v=0}^{N_2-1} y_{uv} y_{u+r,v+s} \right) e^{-i2\pi(kr/N_1 + ms/N_2)}$$

$$(15.66)$$

which may be rearranged in the form

$$S_{km} = \left\{ \frac{1}{N_1 N_2} \sum_{u=0}^{N_1-1} \sum_{v=0}^{N_2-1} y_{uv}\, e^{i2\pi(ku/N_1 + mv/N_2)} \right\} \times$$

$$\times \left\{ \frac{1}{N_1 N_2} \sum_{r=0}^{N_1-1} \sum_{s=0}^{N_2-1} y_{u+r,v+s}\, e^{-i2\pi\{k(u+r)/N_1 + m(v+s)/N_2\}} \right\}. \qquad (15.67)$$

From (15.11), the terms in the first bracket on the r.h.s. of (15.67) are equal to Y_{km}^*. By using the substitution

$$u + r = a \quad \text{and} \quad v + s = b$$

the terms in the second bracket of the r.h.s. of (15.67) may be written as

$$\frac{1}{N_1 N_2} \sum_{a=u}^{(N_1-1+u)} \sum_{b=v}^{(N_2-1+v)} y_{a,b} e^{-i2\pi(ka/N_1 + mb/N_2)}$$

which, because of (15.65), is equal to Y_{km}. We have now arrived at the important result that the coefficients of the discrete two-dimensional spectral density are given by the formula

$$S_{km} = Y_{km}^* Y_{km} \tag{15.68}$$

where the Y_{km} are the coefficients of the discrete Fourier transform of the two-dimensional series y_{rs} calculated according to (15.11). This result is the two-dimensional equivalent of the result in Chapter 10 that

$$S_k = Y_k^* Y_k. \tag{10.26}$$

When data is analysed by using discrete Fourier transforms, there is automatically the assumption that this data is periodic, so that, for the two-dimensional case, (15.65) applies. When calculating the two-dimensional correlation function by (15.63), we make this assumption; otherwise there would be no values for $y_{u+r,v+s}$ when $u + r$ or $v + s$ are greater than $N_1 - 1$ and $N_2 - 1$ respectively. The correlation coefficients† R_{rs} calculated from (15.63) therefore differ from the true correlation function, which is not calculated from periodic data. In Chapter 11 we introduced the term *circular correlation function* to describe correlation functions calculated from periodic data, and which are therefore themselves periodic. The coefficients R_{rs} calculated from (15.63) are circular correlation coefficients, and the two-dimensional sequence repeats itself with period N_1 in one dimension and N_2 in the other dimension.

We have shown for the one-dimensional case that the circular correlation coefficients are biased estimates of the values of the true (linear) correlation function. If $\hat{R}_r$ is an unbiased estimate for $R(\tau = r\Delta)$ such that

$$E[\hat{R}_r] = R(\tau = r\Delta) \tag{15.69}$$

then for the one-dimensional case we showed in Chapter 11 that

$$R_r = \left(\frac{N-r}{N}\right)\hat{R}_r + \left(\frac{N-(N-r)}{N}\right)\hat{R}_{-(N-r)}, \qquad r = 0, 1, 2, \ldots, N-1. \tag{11.10}$$

This result may be extended to two dimensions as follows. Corresponding to (15.69), we shall use the notation $\hat{R}_{rs}$ to denote an unbiased estimate for the two-dimensional, linear correlation function $R(x = r\Delta_1, \tau = s\Delta_2)$. From (15.63), the correlation coefficient R_{rs} may be written as

$$R_{rs} = \frac{1}{N_1} \sum_{u=0}^{N_1-1} \left(\frac{1}{N_2} \sum_{v=0}^{N_2-1} y_{u,v} y_{u+r,v+s}\right). \tag{15.70}$$

---

† The term *correlation coefficient* is used here to mean the result of a discrete calculation which gives an estimate for a correlation function. It should not be confused with the normalized covariance (3.8).

Consider the terms in brackets and compare these with the definition of the one-dimensional circular correlation function

$$R_r = \frac{1}{N} \sum_{s=0}^{N-1} x_s y_{s+r} \qquad (10.20)$$

given in Chapter 10. The terms in brackets in (15.70) give the circular correlation between the y values on track u (i.e. along the line $x_1 = u\Delta_1$) and the y values on track $u + r$. If we denote this correlation function by

$$R_s^{(u,u+r)}$$

then, from equation (11.10) we find that

$$R_s^{(u,u+r)} = \left(\frac{N_2 - s}{N_2}\right)\hat{R}_s^{(u,u+r)} + \left(\frac{s}{N_2}\right)\hat{R}_{-(N_2-s)}^{(u,u+r)}, \qquad s = 0, 1, 2, \ldots, N_2 - 1. \qquad (15.71)$$

Therefore (15.70) may be written as

$$R_{rs} = \frac{1}{N_1} \sum_{u=0}^{N_1-1} \left\{\left(\frac{N_2 - s}{N_2}\right)\hat{R}_s^{(u,u+r)} + \left(\frac{s}{N_2}\right)\hat{R}_{-(N_2-s)}^{(u,u+r)}\right\}, s = 0, 1, 2, \ldots, N_2 - 1, \qquad (15.72)$$

and we can now consider the second summation, with the integer u varying from 0 to $N_1 - 1$. For u running from 0 to $N_1 - r - 1$, values of the first correlation coefficient on the r.h.s. of (15.72) are being obtained without going outside the range of the known data. On summing these values, dividing by N_1, and multiplying by the factor $(N_2 - s)/N_2$, we obtain

$$\left(\frac{N_1 - r}{N_1}\right)\left(\frac{N_2 - s}{N_2}\right)\hat{R}_{rs}.$$

For the range of values of u from $N_1 - r$ to $N_1 - 1$, values of the correlation coefficient have gone outside the range of the known data, and are obtained by assuming that the data repeats itself periodically. As explained in Chapter 11, this leads to the wrong correlation coefficient being calculated (see Fig. 11.1 and the accompanying description). On summing the latter values, dividing by N_1, and multiplying by $(N_2 - s)/N_2$, we obtain

$$\left(\frac{r}{N_1}\right)\left(\frac{N_2 - s}{N_2}\right)\hat{R}_{-(N_1-r),s}.$$

Following the same reasoning for the second correlation coefficient in (15.72), we come to the conclusion that there is the following relation between the two-dimensional circular correlation coefficient R_{rs} and unbiased estimates of the true (linear) two-dimensional correlation function calculated at four different locations (which are identified by the different integer subscripts on the r.h.s. of the equation)

$$R_{rs} = \left(\frac{N_1 - r}{N_1}\right)\left(\frac{N_2 - s}{N_2}\right)\hat{R}_{rs} + \left(\frac{r}{N_1}\right)\left(\frac{N_2 - s}{N_2}\right)\hat{R}_{-(N_1-r),s} +$$

$$+ \left(\frac{N_1 - r}{N_1}\right)\left(\frac{s}{N_2}\right)R_{r,-(N_2-s)} + \left(\frac{r}{N_1}\right)\left(\frac{s}{N_2}\right)\hat{R}_{-(N_1-r),-(N_2-s)},$$

$$r = 0, 1, 2, \ldots, N_1 - 1$$
$$s = 0, 1, 2, \ldots, N_2 - 1. \quad (15.73)$$

Example

Calculate the circular correlation coefficients which correspond to the (linear) two-dimensional correlation function defined by

$$R_{yy}(x_1, x_2) = e^{-(x_1^2 + x_2^2)} \quad (15.74)$$

when the sampling interval $\Delta = 1$.

In applying equation (15.73), we can replace the unbiased estimates of the linear correlation function, $\hat{R}_{rs}$, by the exact values of the function, and put

$$\hat{R}_{rs} = e^{-(r^2 + s^2)}. \quad (15.75)$$

Then, from (15.73), we find that the circular correlation coefficients are given by

$$R_{rs} = \left(\frac{N-r}{N}\right)\left(\frac{N-s}{N}\right)e^{-(r^2+s^2)} + \left(\frac{r}{N}\right)\left(\frac{N-s}{N}\right)e^{-\{(N-r)^2+s^2\}} +$$

$$+ \left(\frac{N-r}{N}\right)\left(\frac{s}{N}\right)e^{-\{r^2+(N-s)^2\}} + \left(\frac{r}{N}\right)\left(\frac{s}{N}\right)e^{-\{(N-r)^2+(N-s)^2\}}$$

$$r, s = 0, 1, 2, \ldots, N - 1. \quad (15.76)$$

We can now use the computer program for the two-dimensional discrete Fourier transform in Appendix 2 to calculate the two-dimensional Fourier transform of (15.76) for the case of an array size 8×8 and compare the resulting spectral coefficients with exact values obtained from the exact two-dimensional spectral density which is given in the previous example in this chapter. By comparing (15.50) with (15.74) and using (15.51), we have

$$\sigma_y^2 = \sqrt{2\pi}\, S_0 \gamma_0 = 1 \quad \text{and} \quad \gamma_0^2 = 2$$

so that (15.49) gives

$$S_{yy}(\gamma_1, \gamma_2) = \frac{1}{4\pi} e^{-(\gamma_1^2 + \gamma_2^2)/4}. \quad (15.77)$$

The input array for the two-dimensional transform subprogram D12fft in Appendix 2 is $Cr(J, K) + iCi(J, K)$, where, in this case, from (15.76)

$Cr(J, K) =$

$$\frac{(9 - J)(9 - K)}{64} e^{-\{(J-1)^2 + (K-1)^2\}} + \frac{(J - 1)(9 - K)}{64} e^{-\{(9-J)^2 + (K-1)^2\}} +$$

$$+ \frac{(9 - J)(K - 1)}{64} e^{-\{(J-1)^2 + (9-K)^2\}} + \frac{(J - 1)(K - 1)}{64} e^{-\{(9-J)^2 + (9-K)^2\}} \quad (15.78)$$

where J and K range from 1 to 8. The correlation coefficients R_{rs} are real so that $Ci(J, K) = 0$ for all J and K.

The transform of this two-dimensional 8×8 sequence is an array of 64 real numbers. There is symmetry about row number 5 and column number 5. The numbers in the first quadrant of the array, $J = 1$ to 5 and $K = 1$ to 5, are shown in Table 15.1.

Table 15.1 Spectral coefficients calculated by taking the two-dimensional DFT of the circular correlation function in (15.76)

		Column, K			
		1	2	3	4	5
	1	0·0436 5	0·0380 0	0·0254 0	0·0142 3	0·0100 2
	2	0·0380 0	0·0330 8	0·0221 1	0·0123 9	0·0087 2
Row	3	0·0254 0	0·0221 1	0·0147 8	0·0082 8	0·0058 3
J	4	0·0142 3	0·0123 9	0·0082 8	0·0046 4	0·0032 7
	5	0·0100 2	0·0087 2	0·0058 3	0·0032 7	0·0023 0

In order to calculate the theoretical values for comparison, we have to know what values of wavenumber γ_1 and γ_2 to substitute into (15.77).

The longest discrete wavelength (except for infinity) is L and so the smallest non-zero wavenumber is $2\pi/L$. From (15.8), we know that $L = N\Delta$ and we have chosen $\Delta = 1$, so that the wavenumber interval is $2\pi/N$ which, for $n = 8$, is $\pi/4$. Hence the wavenumbers are

along the x_1 axis: $(J - 1)\pi/4$
along the x_2 axis: $(K - 1)\pi/4$.

In the one-dimensional case, we saw that the continuous spectral density is related approximately to the one-dimensional spectral coefficient by

$$S(\omega_k) = \frac{T}{2\pi} S_k \quad (11.33)$$

and the two-dimensional equivalent in the terms we are now using is

$$S(\gamma_j, \gamma_k) = \left(\frac{L}{2\pi}\right)^2 S_{jk}. \quad (15.79)$$

Putting $L = 8\Delta = 8$ into (15.79) and substituting into (15.77), we obtain

$$S_{jk} = \frac{\pi}{64} e^{-(\pi/8)^2 \{(J-1)^2 + (K-1)^2\}} \qquad (15.80)$$

which has been used to calculate the results in Table 15.2. When these figures are compared with those above, the approximation involved in using the DFT becomes apparent.

Table 15.2 Spectral coefficients for the same frequencies as in Table 15.1, calculated exactly from the true (linear) correlation function of the same problem

		Column, K				
		1	2	3	4	5
	1	0·04909	0·04207	0·02649	0·01225	0·00416
	2	0·04207	0·03606	0·02270	0·01050	0·00357
Row	3	0·02649	0·00270	0·01429	0·00661	0·00225
J	4	0·01225	0·01050	0·00661	0·00306	0·00104
	5	0·00416	0·00357	0·00225	0·00104	0·00035

Because the spectral coefficients calculated by using the DFT involve averaging the spectral density in the vicinity of each coefficient, they differ from the exact values. Also because they are derived from circular rather than true (linear) correlation functions, there is the possibility of error due to aliasing. In this example, there are errors due to both causes, which we shall investigate in detail later.

In the next section, we consider the two-dimensional spectral window function which defines how the true spectral density is averaged by the two-dimensional DFT. We shall also discuss how the spectral window may be modified to allow for aliasing.

Two-dimensional windows

In Chapters 10 and 11, we showed that, if S_k is a spectral coefficient calculated by the (one-dimensional) DFT according to

$$S_k = Y_k^* Y_k, \qquad (10.26)$$

then the ensemble average is

$$E[S_k] = \frac{2\pi}{T} \bar{S}(\omega_k) \qquad (11.21)$$

where $\bar{S}(\omega_k)$ is a weighted version of the true spectral density $S(\omega)$ calculated by

$$S(\omega_k) = \int_{-\infty}^{\infty} W(\omega_k - \Omega)S(\Omega)\,d\Omega. \qquad (11.28)$$

The spectral weighting function (or spectral window) $W(\omega)$ is the Fourier transform of the data weighting function (or data window)

$$w(\tau) = \begin{cases} 1 - \dfrac{|\tau|}{T} & \text{for } 0 \leqslant \tau \leqslant T \\ 0 & \text{elsewhere} \end{cases} \qquad (11.31)$$

and we showed that

$$W(\omega) = \frac{1}{2\pi} \int_{-\infty}^{\infty} \left(1 - \frac{|\tau|}{T}\right) e^{-i\omega\tau}\,d\tau$$

$$= \frac{T}{2\pi}\left(\frac{\sin\!\left(\dfrac{\omega T}{2}\right)}{\dfrac{\omega T}{2}}\right)^{2}. \qquad (11.32)$$

Equation (11.28) is true when there is no error by aliasing. The coefficients Y_k are calculated from the discrete series y_r which is derived by sampling the continuous sample function $y(t)$. For there to be no error by aliasing, $y(t)$ must have no components whose frequencies are greater than the Nyquist frequency $1/2\Delta$. If this condition is not met, we showed in problems 11.6 and 11.7 that (11.28) remains true if the spectral window function $W(\omega)$ is replaced by the aliased spectral window function defined as

$$W_a(\omega) = \sum_{j=-\infty}^{\infty} W\!\left(\omega - j\frac{2\pi}{\Delta}\right) \qquad (11.57)$$

$$\text{(of problem 11.7)}$$

where the function $W(\omega)$ is still given by (11.32) above and where Δ is the sampling interval. Then, by combining (11.21) and (11.28), we have

$$E[S_k] = \frac{2\pi}{T} \int_{-\infty}^{\infty} W_a(\omega_k - \Omega)S(\Omega)\,d\Omega \qquad (15.81)$$

which is an important result that relates the average values of the spectral coefficients, calculated by the DFT, to the exact spectral density $S(\omega)$. T is the record length being analysed, $W_a(\omega)$ is the aliased spectral window given above, and $\omega_k = 2\pi k/T$ is the angular frequency of the kth spectral coefficient. The convolution integral in (15.81) using the aliased window function (11.57) allows for the spurious increase in the coefficients of the DFT below the Nyquist frequency due to the higher frequency components present in the signal being analysed, which appear, under an assumed lower frequency, in the DFT.

It was also proved in problem 11.7 that, by substituting for $W(\omega)$ in (11.57) by the function given in (11.32) and making the summation, we obtain the following expression for the one-dimensional aliased spectral window

$$W_a(\omega) = \frac{T}{2\pi N^2} \left(\frac{\sin \dfrac{\omega T}{2}}{\sin \dfrac{\omega T}{2N}} \right)^2 \qquad (11.58)$$

(of problem 11.7)

where the range of the DFT is from $k = 0$ to $N - 1$.

Now we turn to the two-dimensional case and we shall show that, for two dimensions, equation (15.81) is replaced by

$$E[S_{km}] = \frac{4\pi^2}{LT} \int_{-\infty}^{\infty} d\Gamma \int_{-\infty}^{\infty} d\Omega \, W_{1a}(\gamma_k - \Gamma) W_{2a}(\omega_m - \Omega) S(\Gamma, \Omega) \quad (15.82)$$

where

$$W_{1a}(\gamma) = \frac{L}{2\pi N_1^2} \left(\frac{\sin \dfrac{\gamma L}{2}}{\sin \dfrac{\gamma L}{2N_1}} \right)^2 \qquad (15.83)$$

$$W_{2a}(\omega) = \frac{T}{2\pi N_2^2} \left(\frac{\sin \dfrac{\omega T}{2}}{\sin \dfrac{\omega T}{2N_2}} \right)^2. \qquad (15.84)$$

The meaning of the various terms is as follows. We are considering a random function $y(X, t)$ which changes continuously over two independent variables X and t. The function $y(X, t)$ is sampled in a rectangle $0 \leqslant X \leqslant L, 0 \leqslant t \leqslant T$ with the sampling intervals being Δ_1 and Δ_2 so that

$$N_1 \Delta_1 = L \qquad \text{and} \qquad N_2 \Delta_2 = T. \qquad (15.85)$$

The two-dimensional DFT is used to generate the frequency series Y_{km} according to (15.11) and the spectral coefficients are found by using

$$S_{km} = Y_{km}^* Y_{km}. \qquad (15.68)$$

The true, ensemble-averaged, two-dimensional spectral density of the random process $y(X, t)$ is $S(\gamma, \omega)$ and is a function of the spatial frequency γ and the time frequency ω. The frequencies of the spectral coefficient S_{km} are

$$\gamma = \gamma_k = \frac{2\pi k}{L} \qquad \text{and} \qquad \omega = \omega_m = \frac{2\pi m}{T} \qquad (15.86)$$

and Γ and Ω are dummy variables for γ and ω in the double convolution integral (15.82).

Equation (15.82) is an important exact result and it tells us how the ensemble-averaged spectral coefficients relate to the true spectrum. If the true spectrum

$S(\gamma, \omega)$ is known, then (15.82) allows us to work out exactly what the coefficients S_{km} will be. In practice, of course, we want to do the opposite. The FFT algorithm allows us to compute S_{km} and we want to estimate the function $S(\gamma, \omega)$ from these computed S_{km}. For one-dimension, we have considered this problem already in Chapter 11, and we shall examine the two-dimensional case shortly. But first we prove equation (15.82).

From the previous section of this chapter, we know that

$$S_{km} = \frac{1}{N_1 N_2} \sum_{r=0}^{N_1-1} \sum_{s=0}^{N_2-1} R_{rs} e^{-i2\pi(kr/N_1 + ms/N_2)} \tag{15.60}$$

$$k = 0, 1, 2, \ldots, N_1 - 1$$
$$m = 0, 1, 2, \ldots, N_2 - 1.$$

We begin by taking the average of both sides of this equation to obtain

$$E[S_{km}] = \frac{1}{N_1 N_2} \sum_{r=0}^{N_1-1} \sum_{s=0}^{N_2-1} E[R_{rs}] e^{-i2\pi(kr/N_1 + ms/N_2)}. \tag{15.87}$$

Also, from (15.73), we know how the coefficients R_{rs} depend on estimates, denoted by $\hat{R}_{rs}$, of the (linear) correlation function; corresponding to (15.69) we have

$$E[\hat{R}_{rs}] = R(x = r\Delta_1, \tau = s\Delta_2) \tag{15.88}$$

so that (15.73) may be written

$$E[R_{rs}] =$$

$$\left(1 - \frac{r}{N_1}\right)\left(1 - \frac{s}{N_2}\right)R(r\Delta_1, s\Delta_2) + \left(\frac{r}{N_1}\right)\left(1 - \frac{s}{N_2}\right)R(-(N_1-r)\Delta_1, s\Delta_2) +$$

$$\left(1 - \frac{r}{N_1}\right)\left(\frac{s}{N_2}\right)R(r\Delta_1, -(N_2-s)\Delta_2) + \left(\frac{r}{N_1}\right)\left(\frac{s}{N_2}\right)R(-(N_1-r)\Delta_1, -(N_2-s)\Delta_2).$$

$$\tag{15.89}$$

When we substitute this result (15.89) into (15.87), we obtain

$$E[S_{km}] =$$

$$\frac{1}{N_1 N_2} \sum_{r=0}^{N_1-1} \sum_{s=0}^{N_2-1} \left(1 - \frac{r}{N_1}\right)\left(1 - \frac{s}{N_2}\right)R(r\Delta_1, s\Delta_2) e^{-i2\pi(kr/N_1 + ms/N_2)} +$$

$$+\frac{1}{N_1 N_2} \sum_{r=0}^{N_1-1} \sum_{s=0}^{N_2-1} \left(\frac{r}{N_1}\right)\left(1 - \frac{s}{N_2}\right)R(-(N_1-r)\Delta_1, s\Delta_2) e^{-i2\pi(kr/N_1 + ms/N_2)} +$$

$$+\frac{1}{N_1 N_2} \sum_{r=0}^{N_1-1} \sum_{s=0}^{N_2-1} \left(1 - \frac{r}{N_1}\right)\left(\frac{s}{N_2}\right)R(r\Delta_1, -(N_2-s)\Delta_2) e^{-i2\pi(kr/N_1 + ms/N_2)} +$$

$$\frac{1}{N_1 N_2} \sum_{r=0}^{N_1-1} \sum_{s=0}^{N_2-1} \left(\frac{r}{N_1}\right)\left(\frac{s}{N_2}\right)R(-(N_1-r)\Delta_1, -(N_2-s)\Delta_2) e^{-i2\pi(kr/N_1 + ms/N_2)}.$$

$$\tag{15.90}$$

Consider the second of the four terms which are summed on the r.h.s. of (15.90). If we put $t = -(N_1 - r)$, then this becomes

$$\frac{1}{N_1 N_2} \sum_{t=-N_1}^{-1} \sum_{s=0}^{N_2-1} \left(1 - \frac{|t|}{N_1}\right)\left(1 - \frac{s}{N_2}\right) R(t\Delta_1, s\Delta_2) e^{-i2\pi(kt/N_1 + ms/N_2)}.$$

Similarly, if we put $u = -(N_2 - s)$ in the third term on the r.h.s. of (15.90), this becomes

$$\frac{1}{N_1 N_2} \sum_{r=0}^{N_1-1} \sum_{u=-N_2}^{-1} \left(1 - \frac{r}{N_1}\right)\left(1 - \frac{|u|}{N_2}\right) R(r\Delta_1, u\Delta_2) e^{-i2\pi(kr/N_1 + mu/N_2)}$$

and, if we put $t = -(N_1 - r)$ and $u = -(N_2 - s)$ into the fourth term on the r.h.s. of (15.90), this becomes

$$\frac{1}{N_1 N_2} \sum_{t=-N_1}^{-1} \sum_{u=-N_2}^{-1} \left(1 - \frac{|t|}{N_1}\right)\left(1 - \frac{|u|}{N_2}\right) R(t\Delta_1, u\Delta_2) e^{-i2\pi(kt/N_1 + mu/N_2)}.$$

If we make all these substitutions in (15.90), then we obtain

$$E[S_{km}] = \frac{1}{N_1 N_2} \sum_{r=-N_1}^{N_1} \sum_{s=-N_2}^{N_2} \left(1 - \frac{|r|}{N_1}\right)\left(1 - \frac{|s|}{N_2}\right) \times$$
$$\times R(r\Delta_1, s\Delta_2) e^{-i2\pi(kr/N_1 + ms/N_2)} \tag{15.91}$$

which alternatively may be written as

$$E[S_{km}] = \frac{1}{N_1 N_2} \sum_{r=-N_1}^{N_1} \sum_{s=-N_2}^{N_2} \left(1 - \frac{|x|}{L}\right)\left(1 - \frac{|\tau|}{T}\right) R(x, \tau) e^{-i2\pi(kx/L + m\tau/T)} \tag{15.92}$$

where
$$x = r\Delta_1, \; L = N_1\Delta_1$$
$$\tau = s\Delta_2, \; T = N_2\Delta_2.$$

We now define two data window functions

$$w_1(x) = \begin{cases} 1 - \dfrac{|x|}{L} & \text{for } 0 \leqslant |x| \leqslant L \\ 0 & \text{elsewhere} \end{cases} \tag{15.93}$$

$$w_2(\tau) = \begin{cases} 1 - \dfrac{|\tau|}{T} & \text{for } 0 \leqslant |\tau| \leqslant T \\ 0 & \text{elsewhere} \end{cases} \tag{15.94}$$

with which we can write (15.92) as

$$E[S_{km}] = \frac{1}{N_1 N_2} \sum_{r=-N_1}^{N_1} \sum_{s=-N_2}^{N_2} w_1(x)w_2(\tau)R(x, \tau) e^{-i2\pi(kx/L + m\tau/T)}. \tag{15.95}$$

Since by definition $w_1(x)$ is zero for $|x| \geqslant L$ and $w_2(\tau)$ is zero for $|\tau| \geqslant T$, we have $w_1(x) = 0$ for $|r| \geqslant N_1$ and $w_2(\tau) = 0$ for $|s| \geqslant N_2$. Therefore the range of the two summations in (15.95) can be extended to infinity in each direction to give

$$E[S_{km}] = \frac{1}{N_1 N_2} \sum_{r=-\infty}^{\infty} \sum_{s=-\infty}^{\infty} w_1(x) w_2(\tau) R(x,\tau) e^{-i2\pi(kx/L + m\tau/T)}$$

where
$$x = r\Delta_1 \quad \text{and} \quad \tau = s\Delta_2. \tag{15.96}$$

By introducing the delta functions

$$\delta(x - r\Delta_1) \quad \text{and} \quad \delta(\tau - s\Delta_2)$$

this may be written in the alternative form

$$E[S_{km}] = \frac{1}{N_1 N_2} \sum_{r=-\infty}^{\infty} \sum_{s=-\infty}^{\infty} \int_{-\infty}^{\infty} dx \int_{-\infty}^{\infty} d\tau \, \delta(x - r\Delta_1) \delta(\tau - s\Delta_2) w_1(x) w_2(\tau) \times$$
$$\times R(x,\tau) e^{-i2\pi(kx/L + m\tau/T)}. \tag{15.97}$$

By exchanging the order of integration and summation in (15.97) and then using the following two identities (which are similar to (11.16))

$$\sum_{r=-\infty}^{\infty} \delta(x - r\Delta_1) = \frac{1}{\Delta_1} \sum_{p=-\infty}^{\infty} e^{-ixp(2\pi/\Delta_1)}$$

$$\sum_{s=-\infty}^{\infty} \delta(\tau - s\Delta_2) = \frac{1}{\Delta_2} \sum_{q=-\infty}^{\infty} e^{-i\tau q(2\pi/\Delta_2)} \tag{15.98}$$

we find that (15.97) becomes

$$E[S_{km}] = \frac{1}{N_1 N_2} \sum_{p=-\infty}^{\infty} \sum_{q=-\infty}^{\infty} \int_{-\infty}^{\infty} dx \int_{-\infty}^{\infty} d\tau \, w_1(x) w_2(\tau) R(x,\tau) \times$$
$$\times e^{-i2\pi\{(k/L + p/\Delta_1)x + (m/T + q/\Delta_2)\tau\}}. \tag{15.99}$$

Then if the spectral windows are

$$W_1(\gamma) = \frac{1}{2\pi} \int_{-\infty}^{\infty} dx \, w_1(x) e^{-i\gamma x}$$

$$W_2(\omega) = \frac{1}{2\pi} \int_{-\infty}^{\infty} d\tau \, w_2(\tau) e^{-i\omega\tau} \tag{15.100}$$

we can write, by the inverse Fourier transforms,

$$w_1(x) = \int_{-\infty}^{\infty} d\gamma \, W_1\left(\gamma + \frac{2\pi p}{\Delta_1}\right) e^{i(\gamma + 2\pi p/\Delta_1)x}$$

$$w_2(\tau) = \int_{-\infty}^{\infty} d\omega \, W_2\left(\omega + \frac{2\pi q}{\Delta_2}\right) e^{i(\omega + 2\pi q/\Delta_2)\tau} \tag{15.101}$$

since the factors $2\pi p/\Delta_1$ and $2\pi q/\Delta_2$ are constants during the integrations. On substituting (15.101) into (15.99), we get

$$E[S_{km}] = \frac{1}{LT} \sum_{p=-\infty}^{\infty} \sum_{q=-\infty}^{\infty} \int_{-\infty}^{\infty} d\gamma \int_{-\infty}^{\infty} d\omega\, W_1\left(\gamma + 2\pi\frac{p}{\Delta_1}\right) W_2\left(\omega + 2\pi\frac{q}{\Delta_2}\right) \times$$

$$\times \left\{ \int_{-\infty}^{\infty} dx \int_{-\infty}^{\infty} d\tau\, R(x,\tau)\, e^{-i((\gamma_k-\gamma)x + (\omega_m-\omega)\tau)} \right\} \qquad (15.102)$$

where $\gamma_k = 2\pi k/L$ and $\omega_m = 2\pi m/T$.

The term in $\{\ \}$ brackets on the r.h.s. of (15.102) is, from (15.29),

$$(2\pi)^2 S(\gamma_k - \gamma, \omega_m - \omega)$$

so that (15.102) becomes

$$E[S_{km}] = \frac{(2\pi)^2}{LT} \int_{-\infty}^{\infty} d\gamma \int_{-\infty}^{\infty} d\omega \left(\sum_{p=-\infty}^{\infty} W_1\left(\gamma + \frac{2\pi p}{\Delta_1}\right) \right) \times$$

$$\times \left(\sum_{q=-\infty}^{\infty} W_2\left(\omega + \frac{2\pi q}{\Delta_2}\right) \right) S(\gamma_k - \gamma, \omega_m - \omega). \qquad (15.103)$$

The aliased spectral windows are defined as

$$W_{1a}(\gamma) = \sum_{p=-\infty}^{\infty} W_1\left(\gamma + \frac{2\pi p}{\Delta_1}\right)$$

$$W_{2a}(\omega) = \sum_{q=-\infty}^{\infty} W_2\left(\omega + \frac{2\pi q}{\Delta_2}\right) \qquad (15.104)$$

in which terms (15.103) becomes

$$E[S_{km}] = \frac{(2\pi)^2}{LT} \int_{-\infty}^{\infty} d\gamma \int_{-\infty}^{\infty} d\omega\, W_{1a}(\gamma) W_{2a}(\omega) S(\gamma_k - \gamma, \omega_m - \omega) \qquad (15.105)$$

or, if we define new variables

$$\Gamma = \gamma_k - \gamma$$

$$\Omega = \omega_m - \omega \qquad (15.106)$$

then

$$E[S_{km}] = \frac{(2\pi)^2}{LT} \int_{-\infty}^{\infty} d\Gamma \int_{-\infty}^{\infty} d\Omega\, W_{1a}(\gamma_k - \Gamma) W_{2a}(\omega_m - \Omega) S(\Gamma, \Omega) \qquad (15.82)$$

which confirms the result (15.82) stated previously.

By following the approach outlined in problem 11.7, it can be demonstrated that the aliased spectral windows $W_{1a}(\gamma)$ and $W_{2a}(\omega)$ are given by (15.83) and (15.84).

In Fig. 15.3, $W_{1a}(\gamma)$ is shown graphically for the case when $N_1 = 8$. The graph shows the non-dimensional weighting value $W_{1a}(\gamma)(2\pi/L)$ plotted against wavenumber γ. The principal peak at $\gamma = 0$ repeats itself at intervals at which

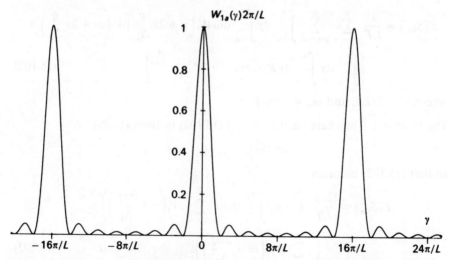

Fig. 15.3 Part of the one-dimensional aliased spectral window function given by (15.83) for the case when $N_1 = 8$

$(\gamma L/2\pi)$ is a multiple of $N_1 = 8$, and it is these additional peaks which introduce spurious contributions to the calculated spectral coefficients at the frequency of the principal peak of the window function. This is illustrated in the following example.

Example

    Consider again the two-dimensional calculation in the example at the end of the last section.

    The two-dimensional DFT yielded the spectral coefficients listed in Table 15.1. These coefficients must be given also by equation (15.82), where W_{1a} and W_{2a} are given by (15.83,84) and $S(\Gamma, \Omega)$ becomes $S(\Gamma_1, \Gamma_2)$ which, from (15.77), is

$$S_{yy}(\Gamma_1, \Gamma_2) = \frac{1}{4\pi} e^{-(\Gamma_1^2 + \Gamma_2^2)/4}. \qquad (15.107)$$

Hence (15.82) gives

$$E[S_{jk}] = \frac{(2\pi)^2}{L_1 L_2} \int_{-\infty}^{\infty} d\Gamma_1 \int_{-\infty}^{\infty} d\Gamma_2 \, W_a(\gamma_j - \Gamma_1) W_a(\gamma_k - \Gamma_2) \frac{1}{4\pi} e^{-(\Gamma_1^2 + \Gamma_2^2)/4}. \qquad (15.108)$$

In the previous example we took $N_1 = N_2 = N = 8$ and a sampling interval $\Delta = 1$, so that the two-dimensional DFT gives spectral coefficients for wavenumbers in the series

$$\gamma_j, \gamma_k = 0, \frac{\pi}{4}, \frac{\pi}{2}, \frac{3\pi}{4}.$$

Consider evaluating the integrals on the r.h.s. of (15.108) for the case when $\gamma_j = \gamma_k = 3(2\pi/8) = 3\pi/4$. In this example the spectral density can be written as the product of two identical functions, one a function of Γ_1 and the other a function of Γ_2, so that, in this case, (15.108) may be written as

$$\underset{\gamma_j=\gamma_k=3\pi/4}{E[S_{jk}]} = \left\{ \frac{\sqrt{\pi}}{8} \int_{-\infty}^{\infty} d\Gamma\, W_a\!\left(\frac{3\pi}{4} - \Gamma\right) e^{-(\Gamma/2)^2} \right\}^2. \qquad (15.109)$$

In Fig. 15.4, the spectral window $W_a(3\pi/4 - \Gamma)$ and the spectral density function

$$S(\Gamma) = e^{-(\Gamma/2)^2}$$

are plotted against Γ. We can see from Fig. 15.4 how the spectral coefficients

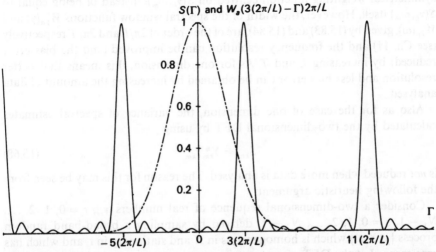

Fig. 15.4 Illustrating the interpretation of spectral coefficients calculated by the DFT as weighted averages of the underlying continuous spectral density. In this example, the continuous spectral density is shown by the broken line and the spectral window function (aliased for an 8-point transform) by the continuous line. The window function is placed for the calculation of the spectral coefficient at wavenumber $3(2\pi/L)$

calculated by the DFT are biased values of the underlying spectral density. The integral in the { } brackets on the r.h.s. of (15.109) has a main contribution from the irregularly shaped weighting function in the region of $\Gamma = 3(2\pi/L)$, with minor contributions from lesser peaks to the left and right of the principal peak, and a significant contribution from the second main peak in the weighting function at $\Gamma = -5(2\pi/L)$.

In this example, the rate of change of the spectral density in the vicinity of the frequency of $\Gamma = 3(2\pi/L)$ is not constant. Therefore there is a bias error due to the width of the spectral window. Also, with only an 8-point transform, because of aliasing there is a significant contribution to the spectral coefficient at $\Gamma = 3(2\pi/L)$ from the spectral density in the region of $\Gamma = -5(2\pi/L)$. Both these errors can be reduced by taking more points in the discrete transform.

Two-dimensional smoothing

We know from equation (15.82) that $S_{km}(LT/4\pi^2)$ is a biased estimate for the two-dimensional spectral density $S(\gamma, \omega)$ at $\gamma = \gamma_k = 2\pi k/L$ and $\omega = \omega_m = 2\pi m/T$. The example in the last section illustrated this point. The average value of $S_{km}(LT/4\pi^2)$, $E[S_{km}](LT/4\pi^2)$, is equal to a weighted average of $S(\gamma, \omega)$, with a symmetrical weighting function centred on $S(\gamma_k, \omega_m)$, instead of being equal to $S(\gamma_k, \omega_m)$ itself. However, the width of the spectral window functions $W_{1a}(\gamma)$ and $W_{2a}(\omega)$, given by (15.83) and (15.84), are of the order of $2\pi/L$ and $2\pi/T$ respectively (see Ch. 11) and the frequency resolution can be improved (and the bias error reduced) by increasing L and T. As for one dimension, this means that better resolution and less bias error can be obtained by increasing the amount of data analysed.

Also as for the case of one dimension, the variance of spectral estimates calculated by the two-dimensional DFT by using

$$S_{km} = Y_{km}^* Y_{km} \tag{15.68}$$

is *not* reduced when more data is analysed. The reason for this may be seen from the following heuristic argument.

Consider a two-dimensional sequence of real numbers y_{rs}, $r = 0, 1, 2, \ldots, N_1 - 1$, $s = 0, 1, 2, \ldots, N_2 - 1$, derived by sampling a broad-band random process $\{y(X, t)\}$ which is homogeneous in X and stationary in t, and which has zero mean. Let the DFT of y_{rs} be

$$Y_{km} = A_{km} - iB_{km} \tag{15.110}$$

where A_{km} and B_{km} are the real and imaginary parts of Y_{km}. Then, from (15.11), we have

$$A_{km} = \frac{1}{N_1 N_2} \sum_{r=0}^{N_1-1} \sum_{s=0}^{N_2-1} y_{rs} \cos 2\pi \left(\frac{kr}{N_1} + \frac{ms}{N_2} \right)$$

$$B_{km} = \frac{1}{N_1 N_2} \sum_{r=0}^{N_1-1} \sum_{s=0}^{N_2-1} y_{rs} \sin 2\pi \left(\frac{kr}{N_1} + \frac{ms}{N_2} \right) \tag{15.111}$$

$$k = 0, 1, 2, \ldots, N_1 - 1$$
$$m = 0, 1, 2, \ldots, N_2 - 1,$$

and we can think of A_{km} and B_{km} as being two different linear sums of approximately independent random variables (because of the broad-band assumption). By the central limit theorem, the A_{km} and B_{km} will have probability distributions which tend towards Gaussian distributions when N_1 and N_2 are large. The random variables A_{km} and B_{km} have zero means because $y(X, t)$ has zero mean, and it can be shown that they have variances of similar magnitude (problem 15.5). Since, from (15.68),

$$S_{km} = Y_{km}^* Y_{km} = A_{km}^2 + B_{km}^2, \tag{15.112}$$

the spectral coefficient S_{km} is a random variable which is the sum of the squares of two random variables which are approximately independent and Gaussian and which have zero means and approximately the same variance. From Chapter 9, the probability distribution for successive values of S_{km} (calculated in each case from new data) will be approximately that of a chi-square random variable with two degrees-of-freedom. Therefore, from (9.37), the ratio of the standard deviation σ to the mean m of the S_{km} random variable is $\sigma/m = 1$ approximately. As for the one-dimensional case, increasing the size of the two-dimensional sequence y_{rs} does not make any difference to the accuracy of a spectral coefficient calculated by the two-dimensional DFT.

There is an exception to the above result for the cases when $k = 0, N_1/2, m = 0, N_2/2$. For these values, we have seen already that Y_{km} is real when the y_{rs} are real, and this conclusion may be confirmed by checking that the B_{km} given by (15.111) is zero for these values of k and m. Then S_{km} is the square of only one Gaussian random variable with zero mean, A_{km}. The probability distribution for S_{km} becomes that of a chi-square random variable with only one degree-of-freedom, for which $\sigma/m = \sqrt{2}$.

To reduce the variance of spectral estimates, adjacent spectral coefficients must be averaged. Consider an unweighted arithmetic average of $(2n_1 + 1)(2n_2 + 1)$ spectral coefficients to calculate

$$S_{km} = \left(\frac{1}{2n_1 + 1}\right)\left(\frac{1}{2n_2 + 1}\right) \sum_{r=-n_1}^{n_1} \sum_{s=-n_2}^{n_2} S_{k+r, m+s}. \tag{15.113}$$

In this case, S_{km} can be thought of as being the sum of squares of $2(2n_1 + 1)(2n_2 + 1)$ independent Gaussian random variables, and then S_{km} has a chi-square probability distribution with $2(2n_1 + 1)(2n_2 + 1)$ degrees-of-freedom and the ratio of its standard deviation to its mean is, from (9.37),

$$\frac{\sigma}{m} = \left(\frac{1}{(2n_1 + 1)(2n_2 + 1)}\right)^{1/2}. \tag{15.114}$$

There is a minor change in (15.114) if the average includes one of the four spectral coefficients $S_{00}, S_{N_1/2,0}, S_{0,N_2/2}$, or $S_{N_1/2,N_2/2}$, because these coefficients have only one degree-of-freedom, not two.

We can see that there is a close parallel between calculations in the two-dimensional (or higher) case and the corresponding calculations in the one-dimensional case. The bandwidths of a single spectral coefficient are approximately $1/L$ cycles/length and $1/T$ cycles/time (compare equation (11.35) for one dimension). The number of statistical degrees-of-freedom for each spectral coefficient is 2, which is the same as for the one-dimensional case. The average values of the spectral coefficients S_{km} are related to the underlying two-dimensional spectral density by the double convolution integral (15.82) (compare the single convolution integral (15.81) in the one-dimensional case).

Because of the large amount of computer memory required for two-dimensional (and higher order) DFT calculations, it may be desirable to divide the two-dimensional data sequence of $N_1 N_2$ values into $K_1 K_2$ smaller sequences, each of $M_1 M_2$ values, where

$$K_1 M_1 = N_1 \quad \text{and} \quad K_2 M_2 = N_2. \tag{15.115}$$

Then each $M_1 M_2$ point transform will yield spectral coefficients covering bandwidths of K_1/L and K_2/T where

$$L = N_1 \Delta_1 \quad \text{and} \quad T = N_2 \Delta_2. \tag{15.116}$$

When the spectral coefficients are averaged across the $K_1 K_2$ sets of coefficients, the number of degrees-of-freedom of the averaged coefficients are $2K_1 K_2$ and so

$$\frac{\sigma}{m} = \sqrt{1/(K_2 K_2)} \tag{15.117}$$

except for $S_{0,0}, S_{N_1/2,0}, S_{0,N_2/2,} s_{N_1/2,N_2/2}$, for which

$$\sigma/m = \sqrt{2/K_1 K_2}.$$

Artificial generation of a two-dimensional random process

In Chapter 13, we discussed the artificial generation of two one-dimensional random processes. Consider a one-dimensional sequence $y_r, r = 0, 1, 2, \ldots, N - 1$, whose DFT yields $Y_k, k = 0, 1, 2, \ldots, N - 1$, and whose spectral coefficients are $S_k = Y_k^* Y_k$. If the S_k are specified, then

$$|Y_k| = \sqrt{S_k}. \tag{15.118}$$

As explained in Chapter 13, for a Gaussian sequence y_r, the phase angles ϕ_k of Y_k must be chosen randomly, with a uniform distribution between 0 and 2π, and we have then

$$Y_k = \sqrt{S_k}\, e^{i\phi_k}, \qquad\qquad k = 0, 1, 2, \ldots, N - 1 \tag{15.119}$$

and

$$y_r = \sum_{k=0}^{N-1} \sqrt{S_k}\, e^{i\{\phi_k + 2\pi(kr/N)\}}, \qquad r = 0, 1, 2, \ldots, N - 1. \tag{15.120}$$

The y_r sequence generated by (15.120) has the prescribed spectral coefficients and, in the limit when $N \to \infty$, it has a Gaussian probability distribution.

Now consider the two-dimensional case. Let the spectral coefficients S_{km} be specified. From (15.68), we have then

$$|Y_{km}| = \sqrt{S_{km}} \qquad (15.121)$$

and if ϕ_{km} is a random phase angle chosen from a population with a uniform distribution between 0 and 2π, then as before

$$Y_{km} = \sqrt{S_{km}}\, e^{i\phi_{km}} \qquad (15.122)$$

and, from (15.12),

$$y_{rs} = \sum_{k=0}^{N_1-1} \sum_{m=0}^{N_2-1} \sqrt{S_{km}}\, e^{i\{\phi_{km} + 2\pi(kr/N_1) + 2\pi(ms/N_2)\}}. \qquad (15.123)$$

The two-dimensional sequence generated by (15.123) has the prescribed two-dimensional spectral coefficients and, in the limit when $N_1 \to \infty$ and $N_2 \to \infty$, it has a second-order Gaussian probability distribution.

In specifying the spectral density coefficients S_{km} and choosing the random phase angles ϕ_{km} in (15.122), we must remember the conditions which must be satisfied if y_{rs} is to be real. We consider the case when N_1 and N_2 are both even. First of all, we know from (15.11) that $Y_{0,0}, Y_{N_1/2,0}, Y_{0,N_2/2}$ and $Y_{N_1/2,N_2/2}$ are all real, so that we must put

$$\phi_{0,0} = \phi_{N_1/2,0} = \phi_{0,N_2/2} = \phi_{N_1/2,N_2/2} = 0. \qquad (15.124)$$

Then, from (15.16), we must have

$$S_{N_1-k,0} = S_{k,0} \qquad k = 1, 2, \ldots, N_1/2 - 1 \qquad (15.125)$$
$$\phi_{N_1-k,0} = -\phi_{k,0},$$

from (15.17), we must have

$$S_{0,N_2-m} = S_{0,m} \qquad m = 1, 2, \ldots, N_2/2 - 1 \qquad (15.126)$$
$$\phi_{0,N_2-m} = -\phi_{0,m},$$

and, from (15.18), we must have

$$S_{N_1-k,N_2-m} = S_{k,m} \qquad \begin{matrix} k = 1, 2, \ldots, N_1 - 1 \\ m = 1, 2, \ldots, N_2/2 - 1 \end{matrix}$$
$$\phi_{N_1-k,N_2-m} = -\phi_{k,m} \qquad \begin{matrix} \text{and} \\ k = 1, 2, \ldots, N_1/2 - 1, m = N_2/2. \end{matrix} \qquad (15.127)$$

The total number of the S_{km} and ϕ_{km} specified by (15.125), (15.126) and (15.127) is

$$(N_1/2 - 1) + (N_2/2 - 1) + (N_1 - 1)(N_2/2 - 1) + (N_1/2 - 1) = N_1 N_2/2 - 2$$

so that, when applying (15.123), we have to specify $(N_1 N_2/2 + 2)$ spectral density coefficients and choose $(N_1 N_2/2 - 2)$ random phase angles; the remainder are then determined by using (15.124), (15.125), (15.126) and (15.127).

From (15.82), we note that

$$E[S_{km}] = \frac{4\pi^2}{LT} \bar{S}(\gamma_k, \omega_m) \qquad (15.128)$$

where $\tilde{S}(\gamma_k, \omega_m)$ is the windowed two-dimensional spectral density specified by the double integral on the r.h.s. of (15.82), which becomes indistinguishable from the true spectral density when L and T become large enough.

Generation of an isotropic surface

An interesting case occurs when it is required to generate an artificial two-dimensional isotropic random process from one-dimensional spectral density data. For example, suppose that an artificial two-dimensional random process is to describe the height $y(x_1, x_2)$ of a surface above a datum plane. The one-dimensional spectral density for the randomly varying height along any straight track across the surface is specified.

The procedure is as follows. First, generate the one-dimensional auto-correlation function from the spectrum of the one-dimensional process. Second, rotate this one-dimensional correlation function about a vertical axis through its origin to obtain the rotationally-symmetrical two-dimensional correlation function. Third, use the two-dimensional Fourier transform to generate the isotropic two-dimensional spectral density function. Fourth, apply (15.123) to generate the two-dimensional profile.

If the two-dimensional transform is an $N \times N$ transform with distance sampling interval Δ, we need values of the one-dimensional linear correlation function for separation distances up to $\sqrt{2}N\Delta$ because the distance across the diagonal of the $N \times N$ field is $\sqrt{2}N\Delta$. Therefore, for the same sampling interval, we need a one-dimensional spectral density sequence of at least $\sqrt{2}N$ points.

The key step in this procedure is step 2. Because the two-dimensional correlation function for an isotropic surface is a function which has rotational symmetry about its origin, this function may be generated by rotating the corresponding one-dimensional correlation function. But we must remember that it is the linear correlation function (rather than the circular correlation function derived by the DFT) which must be rotated and an estimate for the linear correlation function must be obtained. How this is done depends on how the initial one-dimensional data are presented. There are two different cases to be considered: case (i), when height data for a typical straight track are presented as a sequence $\{y_r\}$ from which one-dimensional spectral coefficients can be calculated, and case (ii), when a continuous one-dimensional spectral density function $S(\gamma)$ is specified.

In case (i), we can generate the one-dimensional spectral sequence $S_k = Y_k^* Y_k$ and then take the IDFT of this to obtain the sequence of circular correlation coefficients $\{R_r\}, r = 0, 1, 2, \ldots, N-1$. These circular correlation coefficients $\{R_r\}$ are related to the corresponding linear correlation coefficients $\{\hat{R}_r\}$ defined by

$$E[\hat{R}_r] = R(x = r\Delta) \tag{15.129}$$

by equation (11.10) or by (11.43) if zeros have been added to the original sequence

$\{y_r\}$ before calculating its spectral coefficients. We want to separate the two overlapping parts of the circular correlation sequence so that the linear correlation coefficients can be distinguished. This will occur if at least an equal number of zeros is added to the one-dimensional sequence $\{y_r\}$ before this sequence is transformed to yield $\{Y_k\}$ and thence the spectral coefficients by using $S_k = Y_k^* Y_k$. From (11.43), we know that, for $L \geqslant N$,

$$R_r = \left(\frac{N - |r|}{N + L}\right)\hat{R}_r, \qquad -N \leqslant r \leqslant N \qquad (15.130)$$

where L is the number of zeros added, and the estimates $\hat{R}_r$ for the linear correlation function can be calculated from this equation.

In case (ii), when a continuous one-dimensional spectral density function $S(\gamma)$ is prescribed, the approach may be as follows.

First we must generate a sequence of spectral coefficients $\{S_k\}$, $k = 0, 1, 2, \ldots,$ $N - 1$. For this to correspond to a sequence of real-valued correlation coefficients, $\{S_k\}$ must be a symmetrical sequence such that $S_k = S_{N-k}$. One way of forming such a symmetrical sequence is to sample $S(\gamma)$ over the frequency range for which it is significant, which we will assume is 0 to γ_0, to generate

$$S_k = S(\gamma_k)(\gamma_0/N), \qquad k = 0, 1, 2, \ldots, N - 1 \qquad (15.131)$$

where $\gamma_k = k\gamma_0/N$, and then to form the symmetrical sequence $\{S_{c_k}\}$, where

$$S_{c_k} = S_k(1 - k/N) + S_{N-k}(k/N), \qquad k = 0, 1, 2, \ldots, N - 1. \qquad (15.132)$$

This can be thought of as a circular spectral sequence, similar to the circular correlation sequence (11.10) which is a feature of one-dimensional spectral analysis by the direct method described in Chapters 10 and 11.

When this symmetrical sequence (15.132) is transformed by the IDFT, a sequence of real-valued correlation coefficients is generated which we show in problem 15.6 is $\{R_r\}$, $r = 0, 1, 2, \ldots, N - 1$, where

$$E[R_r] = \int_{-\infty}^{\infty} R(x)w_a(r\Delta - x)\,dx \qquad (15.133)$$

with $\Delta = 2\pi/\gamma_0$, in which the aliased data window

$$w_a(x) = \sum_{j=-\infty}^{\infty} w(x + jN\Delta) = \sum_{j=-\infty}^{\infty} w\left(x + j\frac{N2\pi}{\gamma_0}\right), \qquad (15.134)$$

where

$$w(x) = \frac{\gamma_0}{2\pi}\left(\frac{\sin\frac{\gamma_0 x}{2}}{\frac{\gamma_0 x}{2}}\right)^2. \qquad (15.135)$$

Provided that γ_0 is high enough to include all the significant spectral components, then the data window width

$$\Delta = 2\pi/\gamma_0 \tag{15.136}$$

will be small enough to recognise all the detailed features of $R(x)$. The sequence length N has to be large enough to prevent an aliasing error in the computed correlation coefficients. This is achieved by making N large enough that $R(x)$ is negligible when

$$x > N\Delta/2 = N\pi/\gamma_0. \tag{15.137}$$

In that case,

$$R_r \simeq R(x = r\Delta) = E[\hat{R}_r], \qquad r = 0, 1, 2, \ldots, N/2, \tag{15.138}$$

and the coefficients calculated by the IDFT are close approximations for the linear correlation coefficients $\hat{R}_r$.

Once the one-dimensional linear correlation coefficients $\hat{R}_r$ have been found, the two-dimensional linear correlation coefficients can be found from the condition of rotational symmetry which requires that, for the same sampling intervals,

$$\hat{R}_{st} = \hat{R}_{r=\sqrt{s^2+t^2}} \tag{15.139}$$

when $\sqrt{s^2 + t^2}$ is an integer. When $\sqrt{s^2 + t^2}$ is not an integer, $\hat{R}_{st}$ must be found by interpolation between the nearby values of $\hat{R}_r$ where r is an integer close to $\sqrt{s^2 + t^2}$. The two-dimensional circular correlation function can then be found by applying (15.73) and the rest of the calculation is straightforward, as described in the following example.

Example

Use the two-dimensional DFT to generate a typical surface profile for an isotropic road surface for which the one-dimensional spectral density is specified by

$$S(\gamma) = \frac{2\cdot3113}{1 + (2\cdot2281\gamma)^2} \text{ cm}^2/(\text{rad/m}) \tag{15.140}$$

where γ is the wavenumber in rad/m.

The corresponding one-dimensional correlation function can be found either by the method described above, or, in this example, it can be calculated exactly by taking the inverse Fourier transform of (15.140), which gives

$$R(x) = 3\cdot2589\,e^{-(x/2\cdot2281)} \text{ cm}^2, \tag{15.141}$$

x being expressed in metres. Because the surface is isotropic, its two-dimensional correlation function must be rotationally symmetrical about its origin, and so the two-dimensional correlation function is

$$R(x_1, x_2) = 3\cdot2589\,e^{-(\sqrt{x_1^2 + x_2^2}/2\cdot2281)}\ cm^2. \tag{15.142}$$

Next, we have to generate the two-dimensional circular correlation sequence from (15.142) by using (15.73). To illustrate the procedure, consider the case of a 64×64 point transform. Then $N_1 = N_2 = 64$, and we have

$$R_{rs} = 3\cdot2589\left\{\frac{(64-r)(64-s)}{64^2}\,e^{-\sqrt{r^2+s^2}(\Delta/2\cdot2281)} + \right.$$

$$+ \frac{r(64-s)}{64^2}\,e^{-\sqrt{(64-r)^2+s^2}(\Delta/2\cdot2281)} +$$

$$+ \frac{(64-r)s}{64^2}\,e^{-\sqrt{r^2+(64-s)^2}(\Delta/2\cdot2281)} +$$

$$\left. + \frac{rs}{64^2}\,e^{-\sqrt{(64-r)^2+(64-s)^2}(\Delta/2\cdot2281)}\right\} (cm^2)$$

$$r = 0, 1, 2, \ldots, 63$$
$$s = 0, 1, 2, \ldots, 63. \tag{15.143}$$

This function is plotted in Fig. 15.5(a) for the case when the sampling interval Δ is chosen to be $\Delta = 1\cdot6$ m. The maximum distance over which correlation

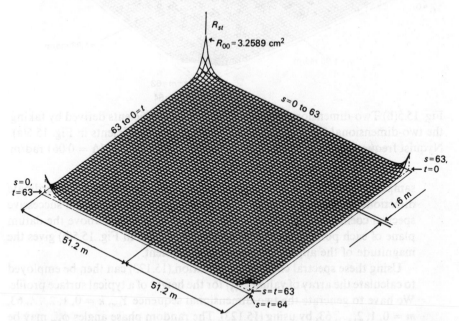

Fig. 15.5(a) Two-dimensional circular correlation sequence defined by (15.143). The height above the datum plane of the point of intersection of any pair of lines in the network gives the value of the appropriate correlation coefficient. Sampling interval $\Delta_1 = \Delta_2 = \Delta = 1\cdot6$ m. $N_1 = N_2 = N = 64$. $L = N\Delta = 64(1\cdot6) = 102\cdot4$ m

values are available is then $N = (64)(1.6) = 102.4$ m. The network of lines in Fig. 15.5(a) joins together the tops of vertical ordinates each of which is a correlation coefficient. The height of the point of intersection of any two lines above the datum plane gives the magnitude of the appropriate correlation coefficient.

By using the two-dimensional discrete Fourier transform computer program in Appendix 2, the two-dimensional sequence of spectral coefficients which corresponds to the array of correlation coefficients (15.143) has been calculated, and this is plotted in Fig. 15.5(b). For the

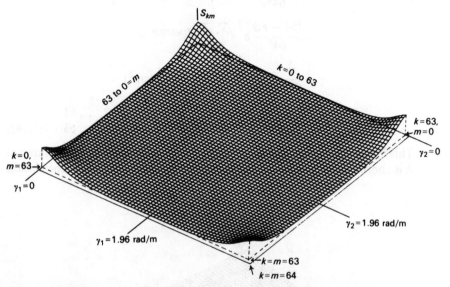

Fig. 15.5(b) Two-dimensional sequence of spectral coefficients derived by taking the two-dimensional DFT of the circular correlation coefficients in Fig. 15.5(a). Nyquist frequency $\pi/\Delta = 1.96$ rad/m. Frequency interval $2\pi/N\Delta = 0.061$ rad/m

sampling interval $\Delta = 1.6$ m, the Nyquist frequency (in both coordinate directions) is $\pi/\Delta = 1.96$ rad/m. The frequency interval between successive spectral coefficients is $2\pi/N\Delta = 0.061$ rad/m. The height above the datum plane of each point of intersection of any two lines in Fig. 15.5(b) gives the magnitude of the appropriate spectral coefficient.

Using these spectral coefficients, equation (15.123) can then be employed to calculate the array of values $\{y_{rs}\}$ for the height of a typical surface profile. We have to generate the two-dimensional sequence Y_{km}, $k = 0, 1, 2, \ldots, 63$, $m = 0, 1, 2, \ldots, 63$, by using (15.123). The random phase angles ϕ_{km} may be generated as described in problem 13.4. Alternatively they may be generated by any suitable computer function such as the Basic language function RND. For precise work, it is necessary to check that the array of random angles ϕ_{km} are uniformly distributed between 0 and 2π, which is necessary if the generated profile heights y_{rs} are to have Gaussian characteristics. We must remember, of course, that for y_{rs} to be real-valued, the conditions of symmetry (15.124)–(15.127) must be satisfied, so that we have to generate only 2046 random phase angles for the 4096 point array.

Once the array Y_{km} is available, it remains only to calculate its two-dimensional inverse Fourier transform, which may be done by using the computer programs in Appendix 2 after making the changes explained in notes 5 and 10 of the appendix. A typical result of this calculation is shown in Fig. 15.5(c). This is a sample two-dimensional isotropic surface with the prescribed spectral character. The datum plane in Fig. 15.5(c) is drawn

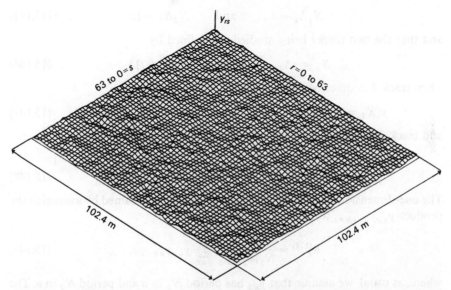

Fig. 15.5(c) Typical sample of an isotropic random surface generated by equation (15.123) for the spectral coefficients in Fig. 15.5(b). Mean height zero; mean square height $3·2589\,\mathrm{cm}^2$

slightly larger than the extent of the surface. The surface ordinates (which have not been drawn) along the two nearer edges of the base square would repeat those along the two farther edges because the surface is one period of a two-dimensional periodic function with period $102·4\,\mathrm{m}$ in both directions. The mean height of the surface is zero and its mean square height is $3·2589\,\mathrm{cm}^2$.

Cross-spectral density between parallel tracks across a random surface

Now imagine that a wheeled vehicle following a straight path runs across a homogeneous random surface and that we are interested in the surface profiles over which opposite wheels of the vehicle pass. The profile of each track is a one-dimensional random process. Consider the cross-spectral density between two such tracks.

For convenience, we will assume that the two tracks are parallel to the X_1 axis. If this is not the case, it would be easy to redefine the directions of the orthogonal axes X_1 and X_2 so that it is. Suppose that the surface profile is described by the two-dimensional sequence

$$y_{uv} = y(X_1 = u\Delta_1, X_2 = v\Delta_2) \qquad (15.144)$$

where a rectangle of plan area $L_1 L_2$ is covered at sampling intervals Δ_1 and Δ_2 so that

$$N_1\Delta_1 = L_1 \qquad \text{and} \qquad N_2\Delta_2 = L_2 \qquad (15.145)$$

and that the two tracks being studied are defined by

$$X_2 = v\Delta_2 \qquad \text{and} \qquad X_2 = (v + t)\Delta_2. \qquad (15.146)$$

Then track 1 is described by the one-dimensional sequence

$$y(X_1 = u\Delta_1, X_2 = v\Delta_2) = y_{uv}, \qquad u = 0, 1, 2, \ldots, N_1 - 1, \qquad (15.147)$$

and track 2 is described by the one-dimensional sequence

$$y(X_1 = u\Delta_1, X_2 = (v + t)\Delta_2) = y_{u,v+t} \qquad u = 0, 1, 2, \ldots, N_1 - 1. \qquad (15.148)$$

The one-dimensional (circular) correlation function is obtained by averaging the product $y_{u,v}y_{u+r,v+t}$ to obtain

$$R_r^{(1,2)} = \frac{1}{N_1 N_2} \sum_{u=0}^{N_1-1} \sum_{v=0}^{N_2-1} y_{u,v}y_{u+r,v+t} \qquad (15.149)$$

where, as usual, we assume that $y_{u,v}$ has period N_1 in u and period N_2 in v. The one-dimensional cross-spectral density between tracks 1 and 2 is then given by the one-dimensional DFT of $R_r^{(1,2)}$ and is

$$S_j^{(1,2)} = \frac{1}{N_1} \sum_{r=0}^{N_1-1} R_r^{(1,2)} e^{-i2\pi(jr/N_1)}. \qquad (15.150)$$

We shall now prove that, when the one-dimensional cross-spectral density coefficients $S_j^{(1,2)}$ for the profiles of two tracks parallel to X_1 and distance $X_2 = t\Delta_2$ apart are defined by (15.150), then they are given by the general result

$$E[S_j^{(1,2)}] = \sum_{m=0}^{N_2-1} E[S_{jm}] e^{i2\pi(mt/N_2)} \qquad (15.151)$$

where the S_{jm} are the two-dimensional spectral density coefficients for the surface over which the tracks pass. Equation (15.151) expresses the ensemble-averaged one-dimensional cross-spectral density coefficients in terms of a summation of the ensemble-averaged two-dimensional spectral density coefficients.

We begin with the definition of S_{jm} from (15.66), which becomes

$$S_{jm} = \frac{1}{N_1^2 N_2^2} \sum_{r=0}^{N_1-1} \sum_{s=0}^{N_2-1} \sum_{u=0}^{N_1-1} \sum_{v=0}^{N_2-1} y_{uv} y_{u+r,v+s} \, e^{-i2\pi(jr/N_1 + ms/N_2)} \quad (15.152)$$

$$j = 0, 1, 2, \ldots, N_1 - 1$$
$$m = 0, 1, 2, \ldots, N_2 - 1.$$

Multiplying both sides of (15.152) by $e^{i2\pi(mt/N_2)}$ and summing over m gives

$$\sum_{m=0}^{N_2-1} S_{jm} \, e^{i2\pi(mt/N_2)} = \frac{1}{N_1^2 N_2^2} \sum_{r=0}^{N_1-1} \sum_{s=0}^{N_2-1} \sum_{u=0}^{N_1-1} \sum_{v=0}^{N_2-1} y_{uv} y_{u+r,v+s} \times$$

$$\times e^{-i2\pi(jr/N_1)} \sum_{m=0}^{N_2-1} e^{-i2\pi\{m(s-t)/N_2\}}. \quad (15.153)$$

In making the summation over m on the r.h.s. of (15.153) we notice that

$$\sum_{m=0}^{N_2-1} e^{-i2\pi\{m(s-t)/N_2\}} = 1 \qquad \text{when } s = t$$

$$= 0 \qquad \text{elsewhere} \quad (15.154)$$

so that completing this summation gives

$$\sum_{m=0}^{N_2-1} S_{jm} \, e^{i2\pi(mt/N_2)} = \frac{1}{N_1^2 N_2} \sum_{r=0}^{N_1-1} \sum_{u=0}^{N_1-1} \sum_{v=0}^{N_2-1} y_{uv} y_{u+r,v+t} \, e^{-i2\pi(jr/N_1)}. \quad (15.155)$$

Taking the ensemble average of equation (15.155) gives

$$\sum_{m=0}^{N_2-1} E[S_{jm}] \, e^{i2\pi(mt/N_2)} = \frac{1}{N_1^2 N_2} \sum_{r=0}^{N_1-1} \sum_{u=0}^{N_1-1} \sum_{v=0}^{N_2-1} R_{yy}(r\Delta_1, t\Delta_2) \, e^{-i2\pi(jr/N_1)} \quad (15.156)$$

with the dependence on u and v of the terms within the summations on the r.h.s. disappearing on the assumption that $\{y(X_1, X_2)\}$ is stationary. The summations over u and v can now be completed to obtain

$$\sum_{m=0}^{N_2-1} E[S_{jm}] \, e^{i2\pi(mt/N_2)} = \frac{1}{N_1} \sum_{r=0}^{N_1-1} R_{yy}(r\Delta_1, t\Delta_2) \, e^{-i2\pi(jr/N_1)}. \quad (15.157)$$

By taking the ensemble average of (15.149), we see that

$$E[R_r^{(1,2)}] = \frac{1}{N_1 N_2} \sum_{u=0}^{N_1-1} \sum_{v=0}^{N_2-1} E[y_{uv} y_{u+r,v+t}] = R_{yy}(r\Delta_1, t\Delta_2) \quad (15.158)$$

and, by doing the same for (15.150) and substituting for $E[R_r^{(1,2)}]$ from (15.158), we obtain

$$E[S_j^{(1,2)}] = \frac{1}{N_1} \sum_{r=0}^{N_1-1} R_{yy}(r\Delta_1, t\Delta_2) \, e^{-i2\pi(jr/N_1)}. \quad (15.159)$$

The r.h.s. of (15.159) is the same as the r.h.s. of (15.157) and so we conclude that

$$E[S_j^{(1,2)}] = \sum_{m=0}^{N_2-1} E[S_{jm}] \, e^{i2\pi(mt/N_2)}. \tag{15.151}$$

This result (15.151) is the discrete equivalent of the result

$$S_{y_1 y_2}(\gamma_1) = \int_{-\infty}^{\infty} d\gamma_2 \, S(\gamma_1, \gamma_2) \, e^{i\gamma_2 x_0} \tag{15.42}$$

derived previously. When $t = 0$ in (15.151), the two tracks are the same, so that the average one-dimensional auto-spectral density coefficients are given by

$$E[S_j^{(1,1)}] = E[S_j] = \sum_{m=0}^{N_2-1} E[S_{jm}] \tag{15.160}$$

which corresponds to the result of putting $x_0 = 0$ in (15.42) to obtain

$$S_{yy}(\gamma_1) = \int_{-\infty}^{\infty} d\gamma_2 \, S(\gamma_1, \gamma_2). \tag{15.161}$$

Equations (15.151) and (15.160) are true for any homogeneous surface, which is one that is stationary in X_1 and X_2. For a surface which has been generated artificially by using (15.123), the sample spectral density coefficients are automatically the same as the ensemble-averaged coefficients, so that the ensemble average symbol on the r.h.s. of (15.151) and (15.160) can then be omitted. However, the sample surfaces must be members of an ensemble which is stationary. For a process each of whose sample functions has the same spectral density coefficients, this will be so only if the phase angles ϕ_{km} in (15.123) are chosen randomly with a uniform distribution between 0 and 2π (subject of course to the restrictions (15.124), (15.125), (15.126) and (15.127) which are needed to ensure that the artificial y_{rs} are real). In problem 15.7 we show that, for a surface generated by (15.123), the result (15.151) is true when the ϕ_{km} have such a uniform distribution.

When using (15.151), we must remember that the cross-spectral density coefficients $S_j^{(1,2)}$ between tracks 1 and 2 are given by the one-dimensional DFT of the correlation sequence $R_r^{(1,2)}$ which is defined by (15.149). Because the cross-spectral coefficients are calculated from a two-dimensional sequence $y_{u,v}$ which is assumed to be periodic (in both dimensions), these coefficients are related to the exact cross-spectral density by appropriate convolution integrals like (15.81). The details of this relationship are worked through in problem 15.8 and the result is as follows:

$$E[S_k^{(1,2)}] = \left\{ (1 - t/N_2) \int_{-\infty}^{\infty} S^{(1,2)}(\gamma) W_a(\gamma - \gamma_k) \, d\gamma + \right.$$

$$\left. + t/N_2 \int_{-\infty}^{\infty} S^{(1,3)}(\gamma) W_a(\gamma - \gamma_k) \, d\gamma \right\} \frac{2\pi}{L_1}. \tag{15.162}$$

In this result, $S^{(1,2)}(\gamma)$ is the exact cross-spectral density between the two tracks 1 and 2, which are distance $t\Delta_2$ apart. $S^{(1,3)}(\gamma)$ is the exact cross-spectral density between tracks 1 and 3, which are distance $(t - N_2)\Delta_2$ apart. This is evidently nothing to do with the cross-spectral density we are trying to calculate. It arises only because of the circular property of the correlation sequence $R_r^{(1,2)}$ defined by (15.149). The spectral window function $W_a(\gamma - \gamma_k)$ which appears in (15.162) is the aliased spectral window defined by (15.83).

Provided that $t \ll N_2$, then we can see from (15.162) that the cross-spectral coefficients $S_k^{(1,2)}$ are approximate measures of the underlying exact cross-spectral density. This means that, for (15.151) to give cross-spectral coefficients which reflect accurately the exact cross-spectral density, we must make sure that the dimension N_2 of the two-dimensional field $y_{u,v}$ is large enough that $t \ll N_2$ where the two tracks are distance $t\Delta_2$ apart.

Chapter 16

Response of continuous linear systems to stationary random excitation

Response to excitation applied at a point

Consider the case when a continuous linear system, which we shall assume to be stable and time-invariant, is subjected to random excitation. For example, the system may be a mechanical structure which is being caused to vibrate by a randomly varying surface pressure. Such a structure is represented diagrammatically in Fig. 16.1. Suppose that, at point S on the structure, defined by position vector $\mathbf{s}$ from a fixed origin O, excitation $f(\mathbf{s}, t)$ is applied along a specified direction represented by the constant unit vector $\mathbf{e}_s$ and that, at point R, position vector $\mathbf{r}$, the response in the direction of the constant unit vector $\mathbf{e}_r$ is $y(\mathbf{r}, t)$.

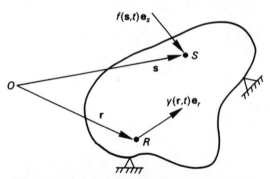

Fig. 16.1 Dynamical system whose response is measured at position $\mathbf{r}$ and to which excitation is applied at position $\mathbf{s}$

The dynamical behaviour of the system may be described either by its impulse response function $h(\mathbf{r}, \mathbf{s}, t)$ or by its frequency response function $H(\mathbf{r}, \mathbf{s}, \omega)$. The impulse response function $h(\mathbf{r}, \mathbf{s}, t)$ gives the response of the structure at position $\mathbf{r}$ in the direction $\mathbf{e}_r$ due to a unit impulse applied at time $t = 0$ at position $\mathbf{s}$ in the direction $\mathbf{e}_s$. The structure is assumed to be at rest before the impulse is applied. The frequency response function $H(\mathbf{r}, \mathbf{s}, \omega)$ gives the steady-state harmonic response at **r** in direction e_r as a result of unit amplitude harmonic excitation at frequency ω applied at **s** in direction e_s.

From Chapter 6 we know that $h(\mathbf{r}, \mathbf{s}, t)$ and $H(\mathbf{r}, \mathbf{s}, \omega)$ are related by the Fourier transform equations which are, corresponding to (6.21) and (6.22),

$$H(\mathbf{r}, \mathbf{s}, \omega) = \int_{-\infty}^{\infty} h(\mathbf{r}, \mathbf{s}, t) e^{-i\omega t} \, dt \tag{16.1}$$

and

$$h(\mathbf{r}, \mathbf{s}, t) = \frac{1}{2\pi} \int_{-\infty}^{\infty} H(\mathbf{r}, \mathbf{s}, \omega) e^{i\omega t} \, d\omega. \tag{16.2}$$

When $f(\mathbf{s}, t)$ is the only source of excitation, and this excitation is a stationary random process, then the spectral density of the response at **r** is given, from (7.15), by

$$S_{yy}(\mathbf{r}, \omega) = H^*(\mathbf{r}, \mathbf{s}, \omega) H(\mathbf{r}, \mathbf{s}, \omega) S_{ff}(\mathbf{s}, \omega). \tag{16.3}$$

The mean-square spectral density of the response at **r**, denoted by $S_{yy}(\mathbf{r}, \omega)$, is related to the mean-square spectral density of the excitation at **s**, denoted by $S_{ff}(\mathbf{s}, \omega)$, by the usual input-output relationship for the spectral densities of stationary random vibration.

Example

Consider the random vibration of a taut string when this is excited in transverse vibration by a randomly varying force applied at a fixed point.

We shall assume that the length of the string is L and that it is excited by a stationary random force $f(a, t)$ at distance a from its left-hand end, Fig. 16.2(a). Then if T is the tension in the string, which has mass per unit length m and is subjected to a viscous damping force c per unit length and velocity, the equation of motion is

$$-T \frac{\partial^2 y}{\partial x^2} + c \frac{\partial y}{\partial t} + m \frac{\partial^2 y}{\partial t^2} = f(a, t) \delta(x - a) \tag{16.4}$$

$$0 < x < L$$

where $y(x, t)$ is the transverse displacement of the string at distance x from its left-hand end.

For excitation with the broad-band spectral density shown in Fig. 16.2(b), we shall calculate the mean-square transverse displacement and velocity of the string at different points along its length.

First, we have to calculate the frequency response function $H(x, a, \omega)$. As in Chapter 6, this is done by substituting into the equation of motion the harmonic excitation

$$f(a, t) = e^{i\omega t} \tag{16.5}$$

and the corresponding steady-state harmonic response

$$y(x, t) = H(x, a, \omega) e^{i\omega t} \tag{16.6}$$

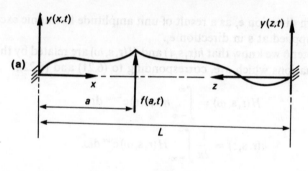

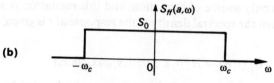

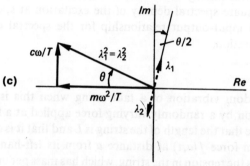

Fig. 16.2 Taut string subjected to random
excitation at one point. The spectral density
of the excitation is shown in (b) and the
calculation of the roots λ_1 and λ_2 of

$$H(x, a, \omega) = A e^{\lambda_1 x} + B e^{\lambda_2 x}$$

in (c)

and solving to find $H(x, a, \omega)$. When we do this, we find that

$$\frac{d^2 H}{dx^2} + H \left(\frac{m\omega^2}{T} - i \frac{c\omega}{T} \right) = -\frac{1}{T} \delta(x - a), \qquad 0 < x < L. \quad (16.7)$$

The solution to (16.7) is

$$H(x, a, \omega) = A e^{\lambda_1(\omega)x} + B e^{\lambda_2(\omega)x} \qquad \text{for} \qquad 0 < x < a$$

and

$$H(x, a, \omega) = C e^{\lambda_1(\omega)x} + D e^{\lambda_2(\omega)x} \qquad \text{for} \qquad a < x < L \quad (16.8)$$

where $\lambda_1(\omega)$, $\lambda_2(\omega)$ are the two roots of

$$\lambda^2 + \left(\frac{m\omega^2}{T} - i\frac{c\omega}{T}\right) = 0. \tag{16.9}$$

The solution of (16.9) is shown in the Argand diagram in Fig. 16.2(c), and the two roots are

$$\lambda_1(\omega) = -\lambda_2(\omega) = \left\{\left(\frac{m\omega^2}{T}\right)^2 + \left(\frac{c\omega}{T}\right)^2\right\}^{1/4}\left(\sin\frac{\theta}{2} + i\cos\frac{\theta}{2}\right)$$

$$\text{where} \quad \tan\theta = \frac{c}{m\omega}. \tag{16.10}$$

Hence, for $0 < x < a$, we have

$$H(x, a, \omega) = A\,e^{\lambda_1 x} + B\,e^{-\lambda_1 x}$$

and, since $H(x, a, \omega)$ at $x = 0$ must be zero for no response at the support, $(A + B) = 0$, and so

$$H(x, a, \omega) = A(e^{\lambda_1 x} - e^{-\lambda_1 x}) = 2A\sinh(\lambda_1 x), \qquad 0 < x < a, \tag{16.11}$$

in which we must remember that λ_1 is a complex quantity and so $\sinh(\lambda_1 x)$ will have real and imaginary parts. For $a < x < L$, it is convenient to express $y(x, t)$ in terms of the distance $z = L - x$ measured from the right-hand end of the string. Then we have, corresponding to (16.11),

$$H(z, a, \omega) = 2E\sinh(\lambda_1 z), \qquad 0 < z < L - a. \tag{16.12}$$

Since the string is continuous at $x = a$, we have

$$2A\sinh(\lambda_1 a) = 2E\sinh(\lambda_1(L - a)) \tag{16.13}$$

and, for equilibrium at $x = a$, the change in direction of the string at $x = a$ must be such that the transverse components of the tension in the string are together equal to the applied force $f(a, t)$. Hence, at $x = a$,

$$T\frac{\partial y}{\partial x} + T\frac{\partial y}{\partial z} = e^{i\omega t} \tag{16.14}$$

which gives

$$A\cosh(\lambda_1 a) + E\cosh(\lambda_1(L - a)) = \frac{1}{2T\lambda_1}. \tag{16.15}$$

On solving for the constants A and E from (16.13) and (16.15), we obtain

$$A = \frac{\sinh\lambda_1(L - a)}{2T\lambda_1\sinh\lambda_1 L} \tag{16.16}$$

and

$$E = \frac{\sinh \lambda_1 a}{2T\lambda_1 \sinh \lambda_1 L} \qquad (16.17)$$

and so obtain the results that

$$H(x, a, \omega) = \frac{\sinh \lambda_1(L - a)}{T\lambda_1 \sinh \lambda_1 L} \sinh \lambda_1 x \qquad \text{for} \qquad 0 < x < a$$

and

$$H(x, a, \omega) = \frac{\sinh \lambda_1 a}{T\lambda_1 \sinh \lambda_1 L} \sinh \lambda_1(L - x) \qquad \text{for} \qquad a < x < L.$$

$$(16.18)$$

Now that we have found the frequency response function $H(x, a, \omega)$, its value may be calculated numerically from (16.18) after first finding the complex root $\lambda_1 = \lambda_1(\omega)$ from (16.10). To illustrate the result of doing this, we take a specific case when

$$T = 1, \qquad L = 1, \qquad a = 1/3, \qquad m = 1/400, \qquad c = 2\pi/400$$

in consistent units.

The result of the calculation is shown in Fig. 16.3(a). Values of $|H(x, a, \omega)|$

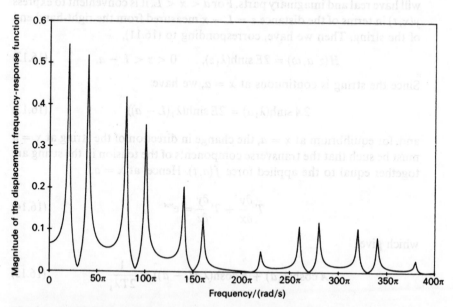

Fig. 16.3(a) Magnitude of the frequency response function for the displacement of a taut string at $x = 0 \cdot 1L$ when the string is excited at $x = L/3$ (where x is the distance from one end of the string). The equation of motion is (16.4) where $T = 1$, $m = 1/400$, $c = 2\pi/400$ and $L = 1$ in consistent units

have been calculated at 401 discrete frequencies evenly spaced between 0 and 400π rad/s. The string has 20 natural frequencies in this range, with the undamped 20th natural frequency being 400π rad/s. Not all the modes show as resonant peaks in Fig. 16.3(a). This is because when $a = 1/3L$, as in this case, the point of excitation lies at a nodal point for modes 3, 6, 9, 12, 15 and 18, and so these modes do not show in the response curve of Fig. 16.3(a); also when the point at which the response is being measured is $x = 0 \cdot 1L$, as in this case, the response is being found at a point on the string which is a node for modes 10 and 20, so that these two modes do not appear as resonant peaks in Fig. 16.3(a).

Figure 16.3(a) shows the magnitude of the complex frequency response function for the displacement response at $x = 0 \cdot 1L$ when there is a steady-state harmonic input of unit amplitude and frequency ω applied at $x = a = L/3$. From it, we can calculate also the magnitude of the complex frequency response function for the velocity response at $x = 0 \cdot 1L$ for the same excitation. By differentiating (16.6) with respect to time, we obtain

$$\dot{y}(x, t) = i\omega H(x, a, \omega) e^{i\omega t} \tag{16.19}$$

and so the amplitude of the frequency response function for the velocity is just $\omega|H(x, a, \omega)|$ and this is plotted in Fig. 16.3(b), for the same frequency range as before. By comparing Figs. 16.3(a) and 16.3(b), we can see that the higher frequency modes are more pronounced in the velocity response because of the multiplication by frequency involved in going from

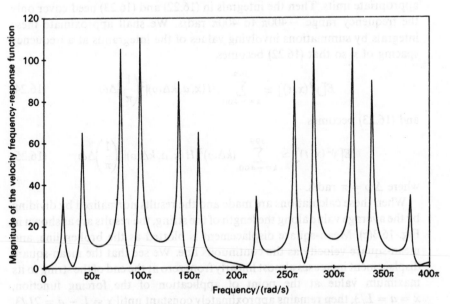

Fig. 16.3(b) The same as Fig. 16.3(a), except that the magnitude of the velocity response is plotted

displacement in Fig. 16.3(a) to velocity in Fig. 16.3(b). This leads to a significant difference between the mean-square displacement response and the mean-square velocity response, as we shall now see.

The mean-square displacement at position x, which we shall denote by $E[y^2(x, t)]$ where the symbol E means the result of an ensemble average over time, is given by

$$E[y^2(x, t)] = \int_{-\infty}^{\infty} S_{yy}(x, \omega) \, d\omega \tag{16.20}$$

and the mean-square velocity at position x is given by

$$E[\dot{y}^2(x, t)] = \int_{-\infty}^{\infty} S_{\dot{y}\dot{y}}(x, \omega) \, d\omega. \tag{16.21}$$

By using (16.3), we then obtain the results that

$$E[y^2(x, t)] = \int_{-\infty}^{\infty} |H(x, a, \omega)|^2 S_{ff}(\omega) \, d\omega \tag{16.22}$$

and

$$E[\dot{y}^2(x, t)] = \int_{-\infty}^{\infty} \omega^2 |H(x, a, \omega)|^2 S_{ff}(\omega) \, d\omega. \tag{16.23}$$

We shall now assume that the broad-band excitation shown in Fig. 16.2(b) has a cut-off frequency $\omega_c = 400\pi$ rad/s and that the level $S_0 = 1/\pi$ in the appropriate units. Then the integrals in (16.22) and (16.23) need cover only the frequency range -400π to 400π rad/s. We shall approximate these integrals by summations involving values of the integrands at a frequency spacing of π so that (16.22) becomes

$$E[y^2(x, t)] \simeq \sum_{k=-400}^{399} |H(x, a, k\Delta\omega)|^2 \left(\frac{1}{\pi}\right) \Delta\omega \tag{16.24}$$

and (16.23) becomes

$$E[\dot{y}^2(x, t)] \simeq \sum_{k=-400}^{399} (k\Delta\omega)^2 |H(x, a, k\Delta\omega)|^2 \left(\frac{1}{\pi}\right) \Delta\omega \tag{16.25}$$

where $\Delta\omega = \pi$ rad/s.

When these calculations are made and the results normalized by dividing by the average value along the length of the string, the results are as shown in Fig. 16.3(c). Mean-square displacement is plotted as the broken line, and mean-square velocity as the continuous line. We see that the mean-square displacement increases almost linearly from zero at the end of the string to its maximum value at the point of application of the forcing function, $x = a = L/3$, then remains approximately constant until $x = L - a = 2L/3$, and then falls almost linearly back to zero at $x = L$. In contrast, the mean-square velocity, which is a measure of the kinetic energy of the string per unit

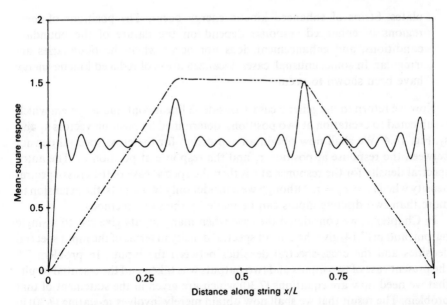

Fig. 16.3(c) Response of a taut string of length L to broad-band random excitation applied at distance $L/3$ from the left-hand end. The broken line gives the mean-square displacement response; the continuous line gives the mean-square velocity response. Both responses have been normalized by dividing by their average value along the length of the string. The cut-off frequency ω_c (see Fig. 16.2(b)) is 400π rad/s which is equal to the 20th natural frequency of the string. The parameter values are $T = 1$, $m = 1/400$, $c = 2\pi/400$ and $L = 1$ in consistent units

length, is approximately constant over most of the length of the string away from its ends, with the exception of the two regions near $x = a = L/3$ and $x = L - a = 2L/3$. Here the average kinetic energy is significantly greater than its value at other points on the string.

In problem 16.1, the above calculation is taken further to consider cases of different damping and to investigate the distribution of mean-square acceleration along the string. For lighter damping than that chosen for Fig. 16.3(c), the two principal peaks in the mean-square velocity response are found to have the same height to a close approximation. The mean-square acceleration response is broadly similar to the mean-square velocity response except for more pronounced ripples, particularly near the two ends of the string.

This behaviour has been discussed in detail by Crandall *et al.* [75–78] for a number of different lightly-damped systems, including strings. The occurrence of localized variations in their otherwise substantially uniform mean-square velocity response has been found to occur also for two-dimensional structures with geometrical symmetry, such as rectangular plates. Strips of enhanced kinetic energy occur. The positions of these regions of enhanced response depend on the nature of the boundary conditions, and enhancement does not occur when the boundaries are irregular. In some, unusual, cases, localized areas of reduced kinetic energy have been shown to occur.

Now we return to the general case. Consider a linear continuous system which is subjected to excitation at two positions, determined by position vectors s_1 and s_2, Fig. 16.4. We shall now obtain an expression for the cross-spectral density between the response at position r_1 and the response at position r_2. The auto-spectral density for the response at r is then the special case of the cross-spectral density when $r_1 = r_2 = r$. Although we consider only two inputs, the extension to more than two discrete inputs can be made by the same methods.

In Chapter 7, we considered the case when many inputs give rise to a single output, and in (7.14) give the output spectral density in terms of the input spectral densities and the cross-spectral densities between the inputs. In problem 7.5 the present case of two inputs and two outputs was taken and the essential results that we need now are equations (7.30) which are given in the statement of that problem. The result that we shall now obtain merely involves re-stating (7.30) in terms of a notation which is appropriate for the analysis of a continuous system.

We define the following parameters:

$S_{yy}(r_1, r_2, \omega)$ = cross-spectral density between the responses $y(r_1, t)$ at r_1 and $y(r_2, t)$ at r_2

$S_{ff}(s_1, s_2, \omega)$ = cross-spectral density between the excitation $f(s_1, t)$ at s_1 and $f(s_2, t)$ at s_2

$S_{ff}(s_1, s_1, \omega)$ = auto-spectral density of the excitation at position s_1

$S_{ff}(s_2, s_2, \omega)$ = auto-spectral density of the excitation at position s_2

$H(r_1, s_1, \omega)$ = frequency response function for the response at r_1 to a unit harmonic input at s_1

and similarly for the other three frequency response functions, $H(r_2, s_1, \omega)$, $H(r_1, s_2, \omega)$ and $H(r_2, s_2, \omega)$.

In terms of these parameters, the first of (7.30) may be written as

$$S_{yy}(r_1, r_2, \omega) = H^*(r_1, s_1, \omega)H(r_2, s_1, \omega)S_{ff}(s_1, s_1, \omega) +$$
$$+ H^*(r_1, s_1, \omega)H(r_2, s_2, \omega)S_{ff}(s_1, s_2, \omega) +$$
$$+ H^*(r_1, s_2, \omega)H(r_2, s_1, \omega)S_{ff}(s_2, s_1, \omega) +$$
$$+ H^*(r_1, s_2, \omega)H(r_2, s_2, \omega)S_{ff}(s_2, s_2, \omega). \quad (16.26)$$

The other response cross-spectral density $S_{yy}(r_2, r_1, \omega)$ comes from the second of (7.30), or it can obtained either (a) by exchanging r_1 and r_2 in (16.26), or (b) by using the result from (5.23) that

$$S_{yy}(r_2, r_1, \omega) = S_{yy}^*(r_1, r_2, \omega). \quad (16.27)$$

If we take the complex conjugates of all the terms on the r.h.s. of (16.26), remembering that

$$S_{ff}(\mathbf{s}_2, \mathbf{s}_1, \omega) = S_{ff}^*(\mathbf{s}_1, \mathbf{s}_2, \omega), \tag{16.28}$$

then it is easy to see that both (a) and (b) yield the same result for $S_{yy}(\mathbf{r}_2, \mathbf{r}_1, \omega)$.

We shall now consider extending (16.26) to the case of infinitely many infinitesimal inputs which is the case when a continuous structure is subjected to a randomly varying surface pressure or other distributed loading.

Response to distributed excitation

For the case when there are N separate inputs, equation (16.26) becomes

$$S_{yy}(\mathbf{r}_1, \mathbf{r}_2, \omega) = \sum_{j=1}^{N} \sum_{k=1}^{N} H^*(\mathbf{r}_1, \mathbf{s}_j, \omega) H(\mathbf{r}_2, \mathbf{s}_k, \omega) S_{ff}(\mathbf{s}_j, \mathbf{s}_k, \omega). \tag{16.29}$$

This result may be proved by following the method described in problem 7.5 for two inputs, or it may be deduced by studying the form of (16.26).

Now we shall consider the limiting case when the number of separate inputs becomes infinite, as would be the case for a distributed random loading such as a fluctuating pressure field, for example a turbulent boundary-layer on the skin of an aircraft. We shall assume that the distributed random excitation, $p(\mathbf{s}, t)$, is stationary in space and time (or, in other words, that it is homogeneous in space and stationary in time) and that its mean-square spectral density is $S_{pp}(\boldsymbol{\gamma}, \omega)$ where $\boldsymbol{\gamma}$ is the wave vector (see Ch. 15) and ω is the time frequency.

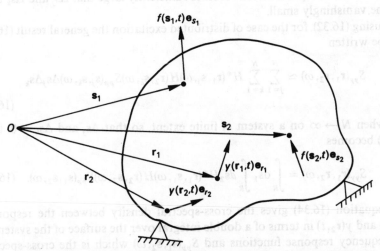

Fig. 16.4 Dynamical system whose response is measured at two positions $\mathbf{r}_1$ and $\mathbf{r}_2$ and to which excitation is applied at two positions $\mathbf{s}_1$ and $\mathbf{s}_2$

First we have to relate the applied force acting at an elemental surface area to the pressure which is applied to that elemental area. This is just a statement that the magnitude of the applied force is pressure (or stress) multiplied by area. It is expressed by the following equation

$$f(\mathbf{s}, t) = p(\mathbf{s}, t)\, d\mathbf{s} \qquad (16.30)$$

where $f(\mathbf{s}, t)$ is the force exerted by pressure $p(\mathbf{s}, t)$ on an elemental area $d\mathbf{s}$. In this equation, we recall that $f(\mathbf{s}, t)$ is by definition a scalar quantity (see Fig. 16.4), $p(\mathbf{s}, t)$ is pressure which is a scalar, and $d\mathbf{s}$ is also a scalar quantity (which is $d\mathbf{s} = d\mathbf{s}_1\, d\mathbf{s}_2$ for two dimensions). The different directions of the excitation at different positions are taken care of by the frequency response functions, each of which gives the response at a specific location and direction to excitation applied at another specific location and direction.

In (16.29), $S_{ff}(\mathbf{s}_j, \mathbf{s}_k, \omega)$ is the cross-spectral density between the excitation $f(\mathbf{s}_j, t)$ at location $\mathbf{s}_j$ and the excitation $f(\mathbf{s}_k, t)$ at $\mathbf{s}_k$. When this excitation is derived from a surface pressure, then we can write, from (16.30),

$$f(\mathbf{s}_j, t) \simeq p(\mathbf{s}_j, t)\Delta\mathbf{s}_j$$

$$f(\mathbf{s}_k, t) \simeq p(\mathbf{s}_k, t)\Delta\mathbf{s}_k \qquad (16.31)$$

where the surface areas occupied by the two forces are assumed to be $\Delta\mathbf{s}_j$ and $\Delta\mathbf{s}_k$. Hence the cross-spectral density between the forces at $\mathbf{s}_j$ and $\mathbf{s}_k$ is related to the cross-spectral density between the pressures at the same points by the result that

$$S_{ff}(\mathbf{s}_j, \mathbf{s}_k, \omega) \simeq S_{pp}(\mathbf{s}_j, \mathbf{s}_k, \omega)\Delta\mathbf{s}_j\Delta\mathbf{s}_k. \qquad (16.32)$$

The approximately equal sign becomes an equality in the limit when the number of discrete point forces becomes sufficiently large that $\Delta\mathbf{s}_j$ and $\Delta\mathbf{s}_k$ both become vanishingly small.

By using (16.32), for the case of distributed excitation the general result (16.29) may be written

$$S_{yy}(\mathbf{r}_1, \mathbf{r}_2, \omega) \simeq \sum_{j=1}^{N} \sum_{k=1}^{N} H^*(\mathbf{r}_1, \mathbf{s}_j, \omega) H(\mathbf{r}_2, \mathbf{s}_k, \omega) S_{pp}(\mathbf{s}_j, \mathbf{s}_k, \omega)\Delta\mathbf{s}_j\Delta\mathbf{s}_k \qquad (16.33)$$

and, when $N \to \infty$ on a system of finite extent, so that $\Delta\mathbf{s}_j$ and $\Delta\mathbf{s}_k \to 0$, then (16.33) becomes

$$S_{yy}(\mathbf{r}_1, \mathbf{r}_2, \omega) = \int_R d\mathbf{s}_1 \int_R d\mathbf{s}_2 H^*(\mathbf{r}_1, \mathbf{s}_1, \omega) H(\mathbf{r}_2, \mathbf{s}_2, \omega) S_{pp}(\mathbf{s}_1, \mathbf{s}_2, \omega). \qquad (16.34)$$

This equation (16.34) gives the cross-spectral density between the responses $y(\mathbf{r}_1, t)$ and $y(\mathbf{r}_2, t)$ in terms of a double integral over the surface of the system of the frequency response functions and $S_{pp}(\mathbf{s}_1, \mathbf{s}_2, \omega)$ which is the cross-spectral density between the pressure $p(\mathbf{s}_1, t)$ at $\mathbf{s}_1$ and $p(\mathbf{s}_2, t)$ at $\mathbf{s}_2$.

An alternative derivation, using impulse response functions, is the subject of problem 16.2.

Eqn (16.34) is true for any time-stationary excitation. Now consider the case when the distributed excitation $p(\mathbf{s}, t)$ is a sample function from a process which is also homogeneous in space with multi-dimensional spectral density $S_{pp}(\mathbf{\gamma}, \omega)$. Then the time cross-spectral density $S_{pp}(\mathbf{s}_1, \mathbf{s}_2, \omega)$ can be calculated from $S_{pp}(\mathbf{\gamma}, \omega)$ by using the methods described in Chapter 15.

For example, suppose that we want to calculate the time cross-spectral density between values of the random variable $y(X_1, X_2, t)$ at two points which are distance x_{10} apart in the X_1 direction and distance x_{20} apart in the X_2 direction. From (15.30), we can write the cross-correlation function as

$$R_{yy}(x_{10}, x_{20}, \tau) = \int_{-\infty}^{\infty} d\omega \int_{-\infty}^{\infty} d\gamma_1 \int_{-\infty}^{\infty} d\gamma_2 \, S_{yy}(\gamma_1, \gamma_2, \omega) \, e^{i(\gamma_1 x_{10} + \gamma_2 x_{20} + \omega\tau)}, \tag{16.35}$$

while, from (5.21), the equivalent one-dimensional result may be written

$$R_{y_1 y_2}(\tau) = \int_{-\infty}^{\infty} d\omega \, S_{y_1 y_2}(\omega) \, e^{i\omega\tau} \tag{16.36}$$

where $y_1(t) = y(X_1, X_2, t)$ and $y_2(t + \tau) = y(X_1 + x_{10}, X_2 + x_{20}, t + \tau)$. Since the correlation functions on the l.h.s. of these two equations are alternative expressions for the same result, by comparing terms on the r.h.s. of each equation, we see that we must have

$$S_{y_1 y_2}(\omega) = \int_{-\infty}^{\infty} d\gamma_1 \int_{-\infty}^{\infty} d\gamma_2 \, S_{yy}(\gamma_1, \gamma_2, \omega) \, e^{i(\gamma_1 x_{10} + \gamma_2 x_{20})} \tag{16.37}$$

or, in more general terms,

$$S_{y_1 y_2}(\omega) = \int_{\infty} d\mathbf{\gamma} \, S_{yy}(\mathbf{\gamma}, \omega) \, e^{i(\mathbf{\gamma} \cdot \mathbf{x}_0)} \tag{16.38}$$

where $\mathbf{\gamma}$ is the wave vector and $\mathbf{x}_0$ is the position vector from point 1 to point 2. This is an important result and its meaning is as follows. For a homogeneous and stationary random process $y(\mathbf{x}, t)$ whose multi-dimensional spectral density is $S_{yy}(\mathbf{\gamma}, \omega)$, the cross-spectral density between any two points 1 and 2 separated by the position vector $\mathbf{x}_0$ from 1 to 2 can be expressed in terms of a weighted integral of $S_{yy}(\mathbf{\gamma}, \omega)$ where the weighting function is $e^{i(\mathbf{\gamma} \cdot \mathbf{x}_0)}$ and the integration is taken over all $\mathbf{\gamma}$.

In the nomenclature of our problem in this section, we can use (16.38) to write

$$S_{pp}(\mathbf{s}_j, \mathbf{s}_k, \omega) = \int_{\infty} d\mathbf{\gamma} \, S_{pp}(\mathbf{\gamma}, \omega) e^{i(\mathbf{\gamma} \cdot (\mathbf{s}_k - \mathbf{s}_j))} \tag{16.39}$$

because the position vector between the two points at which the excitation cross-spectral density $S_{ff}(\mathbf{s}_j, \mathbf{s}_k)$ is being calculated is $(\mathbf{s}_k - \mathbf{s}_j)$. Now, substituting (16.39) into (16.34) gives

$$S_{yy}(\mathbf{r}_1, \mathbf{r}_2, \omega) = \int_{\infty} d\mathbf{\gamma} \, S_{pp}(\mathbf{\gamma}, \omega) G^*(\mathbf{r}_1, \mathbf{\gamma}, \omega) G(\mathbf{r}_2, \mathbf{\gamma}, \omega) \tag{16.40}$$

where

$$G(\mathbf{r}, \boldsymbol{\gamma}, \omega) = \int_R H(\mathbf{r}, \mathbf{s}, \omega)\, e^{i(\boldsymbol{\gamma} \cdot \mathbf{s})}\, d\mathbf{s}. \tag{16.41}$$

This is the final result. It gives the ensemble-average cross-spectral density between the stationary response $y(\mathbf{r}_1, t)$ at position vector $\mathbf{r}_1$ and the stationary response $y(\mathbf{r}_2, t)$ at position vector $\mathbf{r}_2$, Fig. 16.4, for a stable, linear, time-invariant continuous structure. The structure is assumed to be subjected to distributed excitation $p(\mathbf{s}, t)$ which is homogeneous in space and stationary in time and for which the multi-dimensional spectral density is $S_{pp}(\boldsymbol{\gamma}, \omega)$. The function $G(\mathbf{r}, \boldsymbol{\gamma}, \omega)$ has been called by Lin [84] the *sensitivity function* for the structure. It gives the sensitivity of the structure at $\mathbf{r}$ to a distributed excitation which is harmonic in space and time at wave vector $\boldsymbol{\gamma}$ and time frequency ω. $G(\mathbf{r}, \boldsymbol{\gamma}, \omega)$ is a weighted integral of the frequency response function $H(\mathbf{r}, \mathbf{s}, \omega)$ which, by comparison with (15.45), is similar in form to a Fourier transform equation (except for the sign of the exponent).

Normal mode analysis

In the previous section, we derived two equations which, in theory, enable the response of a continuous system to distributed random excitation to be calculated. The first result, equation (16.34), gives the output cross-spectral density $S_{yy}(\mathbf{r}_1, \mathbf{r}_2, \omega)$ as a double integral involving the frequency response function $H(\mathbf{r}, \mathbf{s}, \omega)$ and the cross-spectral density of the excitation $S_{pp}(\mathbf{s}_1, \mathbf{s}_2, \omega)$. The second result, equation (16.40), gives the output cross-spectral density in terms of a treble integral of the frequency response function and the multi-dimensional spectral density $S_{pp}(\boldsymbol{\gamma}, \omega)$. Which of these two equations is used for a calculation depends on the information available. When the input cross-spectral density $S_{pp}(\mathbf{s}_1, \mathbf{s}_2, \omega)$ is available, then (16.34) is applicable directly, and obviously there is no point is labouring with (16.40). In any case, exact integrations will be possible only in the simplest cases, and in most practical problems some form of approximate numerical integration is unavoidable.

In many examples, it is easier to make approximations for the input spectral density than it is to find expressions for the frequency response function $H(\mathbf{r}, \mathbf{s}, \omega)$. In this section, we shall be concerned with an approximate method of calculating $H(\mathbf{r}, \mathbf{s}, \omega)$. The method depends on the experimental finding that, in many practical vibration problems, vibration is confined mainly to a limited number of natural modes. With this knowledge, we shall now seek to express $H(\mathbf{r}, \mathbf{s}, \omega)$ as a series expansion, each of whose terms represents the contribution to $H(\mathbf{r}, \mathbf{s}, \omega)$ of a separate mode. It will then be possible to approximate $H(\mathbf{r}, \mathbf{s}, \omega)$ by including the modes that are of interest and excluding those that are not.

We shall begin by taking a specific example of a simple beam, and then, later, move on to the general case.

Consider a simply-supported elastic beam which is subjected to a distributed transverse loading represented by $p(x, t)$ per unit length of the beam. We shall assume that its equation of motion is given by

$$m\frac{\partial^2 y}{\partial t^2} + c\frac{\partial y}{\partial t} + EI\frac{\partial^4 y}{\partial x^4} = p(x, t), \qquad 0 < x < L \tag{16.42}$$

where $y(x, t)$ is the deflection of the beam at time t and distance x from its end, m is its mass per unit length, c is the viscous damping coefficient per unit length, and EI is the bending stiffness of the beam (E is the modulus of elasticity and I is the second moment of area of the cross-section of the beam about its neutral axis). We shall seek a solution in terms of the undamped natural modes of the beam, of the form

$$y(x, t) = \sum_{j=1}^{\infty} \Psi_j(x) y_j(t) \tag{16.43}$$

where $\Psi_j(x)$ are the natural modes of the simply-supported beam defined by

$$\Psi_j(x) = \sqrt{2}\sin\left(j\frac{\pi x}{L}\right) \tag{16.44}$$

and satisfying the condition of normality that

$$\int_0^L \Psi_j(x)\Psi_k(x)\,\mathrm{d}x = L\delta_{jk} \tag{16.45}$$

where δ_{jk} is zero except when $j = k$, when it is unity.

If we substitute the assumed solution (16.43) into the equation of motion (16.42), we obtain

$$\sum_{j=1}^{\infty}\left\{m\Psi_j(x)\frac{\mathrm{d}^2 y_j}{\mathrm{d}t^2} + c\Psi_j(x)\frac{\mathrm{d}y_j}{\mathrm{d}t} + EI\frac{\mathrm{d}^4}{\mathrm{d}x^4}\Psi_j(x)y_j\right\} = p(x, t). \tag{16.46}$$

On multiplying through by $\Psi_j(x)$ (keeping it outside the summation) and integrating over x, we find, after using (16.45), that

$$\frac{\mathrm{d}^2 y_j}{\mathrm{d}t^2} + \frac{c}{m}\frac{\mathrm{d}y_j}{\mathrm{d}t} + \frac{EI}{m}\left(j\frac{\pi}{L}\right)^4 y_j = \frac{1}{mL}\int_0^L \Psi_j(x)p(x, t)\,\mathrm{d}x. \tag{16.47}$$

The undamped natural frequencies of the beam are given by

$$\omega_j = \left(j\frac{\pi}{L}\right)^2 \sqrt{\frac{EI}{m}} \tag{16.48}$$

and the modal bandwidth is

$$\beta_j = \frac{c}{m}. \tag{16.49}$$

When we define the modal exciting force as

$$p_j(t) = \int_0^L \Psi_j(x)p(x, t)\,\mathrm{d}x \tag{16.50}$$

and substitute (16.48)–(16.50) in (16.47), then we find that

$$\frac{d^2y_j}{dt^2} + \beta_j \frac{dy_j}{dt} + \omega_j^2 y_j = \frac{1}{mL} p_j(t). \tag{16.51}$$

We see that the response of each mode of the beam is similar to the response of a single degree-of-freedom oscillator. Now consider the frequency response function. In this example, when the response $y(x, t)$ is measured at x for unit harmonic excitation $e^{i\omega t}$ applied at $x = s$, it is $H(x, s, \omega)$. For such a point excitation, the distributed load per unit length is

$$p(x, t) = e^{i\omega t} \delta(x - s) \tag{16.52}$$

and, from (16.44) and (16.50),

$$p_j(t) = \int_0^L \sqrt{2} \sin\left(j\frac{\pi x}{L}\right) e^{i\omega t} \delta(x - s)\, dx = \sqrt{2} \sin\left(j\frac{\pi s}{L}\right) e^{i\omega t} \tag{16.53}$$

so that (16.51) becomes

$$\frac{d^2y_j}{dt^2} + \beta_j \frac{dy_j}{dt} + \omega_j^2 y_j = \frac{\sqrt{2}}{mL} \sin\left(j\frac{\pi s}{L}\right) e^{i\omega t} \tag{16.54}$$

and the response $y(x, t)$ is, from (16.43),

$$y(x, t) = H(x, s, \omega) e^{i\omega t} = \sum_{j=1}^{\infty} \Psi_j(x) y_j(t) = \sum_{j=1}^{\infty} \sqrt{2} \sin\left(j\frac{\pi x}{L}\right) y_j(t). \tag{16.55}$$

The steady-state solution $y_j(t)$ of (16.54) is

$$y_j(t) = \left(\frac{1}{-\omega^2 + \beta_j i\omega + \omega_j^2}\right) \frac{\sqrt{2}}{mL} \sin\left(j\frac{\pi s}{L}\right) e^{i\omega t} \tag{16.56}$$

and so we conclude from (16.55) and (16.56) that $H(x, s, \omega)$ may be written as the summation

$$H(x, s, \omega) = \sum_{j=1}^{\infty} \frac{2\sin\left(j\frac{\pi x}{L}\right)\sin\left(j\frac{\pi s}{L}\right)}{mL(-\omega^2 + \beta_j i\omega + \omega_j^2)}. \tag{16.57}$$

From (16.41), the sensitivity function, which in this example is $G(x, \gamma, \omega)$, may then be written as

$$G(x, \gamma, \omega) = \sum_{j=1}^{\infty} \int_0^L ds \frac{2\sin\left(j\frac{\pi x}{L}\right)\sin\left(j\frac{\pi s}{L}\right)}{mL(-\omega^2 + \beta_j i\omega + \omega_j^2)} e^{i\gamma s} \tag{16.58}$$

for a uniform simply-supported beam whose equation of motion is given by (16.42).

Now we move from the particular to the general, and repeat the above analysis in more general terms.

Following Lin [84], we assume a general equation of motion

$$m(\mathbf{r})\frac{\partial^2 y(\mathbf{r},t)}{\partial t^2} + c(\mathbf{r})\frac{\partial y(\mathbf{r},t)}{\partial t} + L(y(\mathbf{r},t)) = p(\mathbf{r},t) \tag{16.59}$$

where now m and c are in general functions of $\mathbf{r}$ and L is a linear differential operator in the spatial variables. For example, for a taut string

$$L = -T\frac{\partial^2}{\partial x^2} \tag{16.60}$$

and for a uniform beam

$$L = EI\frac{\partial^4}{\partial x^4}. \tag{16.61}$$

For a thin plate,

$$L = D\nabla^4 = D\left(\frac{\partial^4}{\partial x_1^4} + 2\frac{\partial^4}{\partial x_1^2 \partial x_2^2} + \frac{\partial^4}{\partial x_2^4}\right) \tag{16.62}$$

where D is the plate stiffness defined by

$$D = \frac{Et^3}{12(1-\mu^2)} \tag{16.63}$$

where E is the modulus of elasticity, t is the plate thickness and μ is Poisson's ratio.

We shall assume now that we can find normal modes such that the jth normal mode of the system for undamped free vibration is $\Psi_j(\mathbf{r})$ where the modes are normalized so that

$$\int_R m(\mathbf{r})\Psi_j(\mathbf{r})\Psi_k(\mathbf{r})\,d\mathbf{r} = m_j\delta_{jk}. \tag{16.64}$$

Also we shall assume that the damping coefficient $c(\mathbf{r})$ is such that

$$\int_R c(\mathbf{r})\Psi_j(\mathbf{r})\Psi_k(\mathbf{r})\,d\mathbf{r} = c_j\delta_{jk}. \tag{16.65}$$

This condition restricts the generality of the systems considered. It will be true, for example, if the damping coefficient $c(\mathbf{r})$ in (16.59) is always proportional to the mass coefficient $m(\mathbf{r})$, but for arbitrarily chosen $c(\mathbf{r})$, (16.65) will not be true. In practical vibration problems, damping is usually small and its exact distribution is not then important. In any case, we do not usually know how the damping is distributed. Therefore it is customary to assume that the damping is distributed in such a way that (16.65) is satisfied and the consequences of this assumption do not appear to detract seriously from the usefulness of the results so obtained.

On account of (16.64) and (16.65), we can substitute an assumed solution

$$y(\mathbf{r},t) = \sum_{j=1}^{\infty} \Psi_j(\mathbf{r})y_j(t) \tag{16.66}$$

into (16.59), multiply both sides of the resulting equation by $\Psi_j(\mathbf{r})$, and integrate over the surface of the system, to obtain

$$\ddot{y}_j + \beta_j \dot{y}_j + \omega_j^2 y_j = \frac{1}{m_j} \int_R \Psi_j(\mathbf{r}) p(\mathbf{r},t)\, d\mathbf{r} \qquad (16.67)$$

where ω_j is the jth undamped natural frequency of the system and β_j is the modal bandwidth defined by

$$\beta_j = \frac{c_j}{m_j} \qquad (16.68)$$

with c_j and m_j being obtained from (16.65) and (16.64). Equation (16.67) is one of a series of uncoupled linear differential equations, one for each mode of the system, which describes motion in the jth mode of the system.

As before, the frequency response function $H(\mathbf{r}, \mathbf{s}, \omega)$ is obtained by putting

$$p(\mathbf{r},t) = e^{i\omega t}\, \delta(\mathbf{r} - \mathbf{s}) \qquad (16.69)$$

when we find that, from (16.67),

$$\ddot{y}_j + \beta_j \dot{y}_j + \omega_j^2 y_j = \frac{1}{m_j} \Psi_j(\mathbf{s})\, e^{i\omega t}. \qquad (16.70)$$

The solution of (16.70) may be written

$$y_j(t) = H_j(\omega)\Psi_j(\mathbf{s})\, e^{i\omega t} \qquad (16.71)$$

which, on substitution into (16.70), gives

$$H_j(\omega) = \frac{1}{m_j(-\omega^2 + \beta_j i\omega + \omega_j^2)}. \qquad (16.72)$$

Hence, from (16.66),

$$H(\mathbf{r}, \mathbf{s}, \omega) = \sum_{j=1}^{\infty} \Psi_j(\mathbf{r}) H_j(\omega) \Psi_j(\mathbf{s}) \qquad (16.73)$$

where $\mathbf{r}$ and $\mathbf{s}$ are the position vectors of points on the surface of the system. Lastly, from (16.41), the sensitivity function $G(\mathbf{r}, \boldsymbol{\gamma}, \omega)$ is, after substituting for $H(\mathbf{r}, \mathbf{s}, \omega)$ from (16.73),

$$G(\mathbf{r}, \boldsymbol{\gamma}, \omega) = \sum_{j=1}^{\infty} \int_R \Psi_j(\mathbf{r}) H_j(\omega) \Psi_j(\mathbf{s})\, e^{i(\boldsymbol{\gamma}\cdot\mathbf{s})}\, d\mathbf{s}. \qquad (16.74)$$

Equation (16.74) gives the sensitivity of the system at $\mathbf{r}$ to harmonic excitation at time frequency ω and wave-number vector $\boldsymbol{\gamma}$ expressed as a summation over all the modes of the sensitivity in each mode.

The modal shape functions $\Psi_j(\mathbf{r})$ describe the modes of damped as well as undamped free vibration of the system and it is assumed that the damping satisfies (16.65) so that such uncoupled modes exist. The frequency response function $H_j(\omega)$ is a measure of the response of the jth mode to harmonic excitation at frequency ω and is defined by (16.72).

Example 1

Find the cross-spectral density of the response of a simply-supported beam when this is subjected to a distributed random loading which is homogeneous and stationary and for which the two-dimensional spectral density is S_0 (constant). Hence find the mean-square displacement and mean-square velocity response along the length of the beam.

We shall assume that the beam is of length L and that distances x, r and s are all measured from one end of the beam. The displacement of the beam is $y(x, t)$ as a result of the excitation which is the force per unit length $p(x, t)$.

We shall apply equation (16.34) which requires the cross-spectral density function $S_{pp}(s_1, s_2, \omega)$ to be found. This is the cross-spectral density between the excitation $p(s_1, t)$ at $x = s_1$ and $p(s_2, t)$ at $x = s_2$. From (16.39), we can calculate this from the two-dimensional spectral density $S_{pp}(\gamma, \omega) = S_0$ by the integral

$$S_{pp}(s_1, s_2, \omega) = \int_{-\infty}^{\infty} d\gamma \, S_0 \, e^{i\gamma(s_2 - s_1)}. \tag{16.75}$$

To evaluate this integral, we recall the result in Chapter 5 that the autocorrelation function for stationary random white noise of spectral density S_0 is

$$R_x(\tau) = 2\pi S_0 \delta(\tau). \tag{5.10}$$

From (5.2), we know that

$$R_x(\tau) = \int_{-\infty}^{\infty} S_x(\omega) e^{i\omega\tau} \, d\omega = \int_{-\infty}^{\infty} S_0 \, e^{i\omega\tau} \, d\omega \tag{16.76}$$

so that we must have

$$2\pi\delta(\tau) = \int_{-\infty}^{\infty} e^{i\omega\tau} \, d\omega. \tag{16.77}$$

Using (16.77), equation (16.75) gives

$$S_{pp}(s_1, s_2, \omega) = 2\pi S_0 \delta(s_2 - s_1). \tag{16.78}$$

For excitation which is white in space, we see that its cross-spectral density is zero everywhere except at $s_2 = s_1$. Here the cross-spectral density becomes the auto-spectral density, which is infinite.

Now we can use (16.34) to calculate the cross-spectral density of the response $y(r_1, t)$ at $x = r_1$ and $y(r_2, t)$ at $x = r_2$ by substituting from (16.78) to obtain

$$S_{yy}(r_1, r_2, \omega) = \int_0^L ds_1 \int_0^L ds_2 \, H^*(r_1, s_1, \omega) H(r_2, s_2, \omega) 2\pi S_0 \delta(s_2 - s_1). \tag{16.79}$$

One of the two integrals can be evaluated at once to obtain

$$S_{yy}(r_1, r_2, \omega) = \int_0^L ds\, H^*(r_1, s, \omega) H(r_2, s, \omega) 2\pi S_0. \tag{16.80}$$

We have already worked out $H(x, s, \omega)$ for this case and the result is given in (16.57). On substituting from (16.57) into (16.80), we get

$$S_{yy}(r_1, r_2, \omega) = \sum_{j=1}^{\infty} \sum_{k=1}^{\infty} \frac{8\pi S_0}{(mL)^2} \frac{\sin\left(j\dfrac{\pi r_1}{L}\right)}{(-\omega^2 - \beta_j i\omega + \omega_j^2)} \times$$

$$\times \frac{\sin\left(k\dfrac{\pi r_2}{L}\right)}{(-\omega^2 + \beta_k i\omega + \omega_k^2)} \int_0^L ds\, \sin\left(j\frac{\pi s}{L}\right)\sin\left(k\frac{\pi s}{L}\right). \tag{16.81}$$

The integral over s in (16.81) is zero except when $j = k$, when it is $L/2$, and so we arrive at the following result for the response cross-spectral density

$$S_{yy}(r_1, r_2, \omega) = \sum_{j=1}^{\infty} \frac{4\pi S_0}{m^2 L} \sin\left(j\frac{\pi r_1}{L}\right)\sin\left(j\frac{\pi r_2}{L}\right) \left|\frac{1}{-\omega^2 + \beta_j i\omega + \omega_j^2}\right|^2. \tag{16.82}$$

In order to find the mean-square responses, we use the results (16.20) and (16.21) which become, in the nomenclature of this problem

$$E[y^2(x, t)] = \int_{-\infty}^{\infty} S_{yy}(x, x, \omega)\, d\omega \tag{16.83}$$

and

$$E[\dot{y}^2(x, t)] = \int_{-\infty}^{\infty} \omega^2 S_{yy}(x, x, \omega)\, d\omega. \tag{16.84}$$

where E denotes the ensemble average over time.

Substituting (16.82) into (16.83) gives

$$E[y^2(x, t)] = \sum_{j=1}^{\infty} \frac{4\pi S_0}{m^2 L} \sin^2\left(j\frac{\pi x}{L}\right) \int_{-\infty}^{\infty} \left|\frac{1}{-\omega^2 + \beta_j i\omega + \omega_j^2}\right|^2 d\omega \tag{16.85}$$

and the integral may be evaluated with the help of Appendix 1, case 2, to obtain

$$E[y^2(x, t)] = \sum_{j=1}^{\infty} \frac{4\pi^2 S_0}{m^2 L \beta_j} \left(\frac{\sin\left(j\dfrac{\pi x}{L}\right)}{\omega_j}\right)^2. \tag{16.86}$$

On substituting for ω_j from (16.48) and for β_j from (16.49) into (16.86), we obtain the result that

$$E[y^2(x,t)] = \frac{4S_0L^3}{\pi^2EIc} \sum_{j=1}^{\infty} \frac{\sin^2\left(j\frac{\pi x}{L}\right)}{j^4}. \tag{16.87}$$

To find the mean-square velocity response from (16.84), the calculation is exactly the same except that a different integral has to be found from Appendix 1, case 2, and the result corresponding to (16.87) is

$$E[\dot{y}^2(x,t)] = \frac{4\pi^2S_0}{mLc} \sum_{j=1}^{\infty} \sin^2\left(j\frac{\pi x}{L}\right). \tag{16.88}$$

We can see that the series in (16.87) will converge because of the j^4 factor in the denominator.† In (16.88), this factor is absent, and the series will not converge. As more modes are included in the summation, the mean-square velocity response increases without bound. This is because the theoretical excitation covers all wavenumbers and all frequencies.

Example 2

Assuming that simple beam theory applies for the beam analysed in Example 1, find the distribution along its length of the mean-square bending moment.

If $y(x,t)$ is the displacement of the centre-line of the beam, then we know from simple beam theory that, if $M(x,t)$ is the bending moment at section x, then

$$M(x,t) = EI\frac{\partial^2}{\partial x^2}y(x,t). \tag{16.89}$$

As before, we assume that the damping is distributed in such a way that (16.65) is true. Then, for steady-state unit harmonic excitation at $x = s$, we have, from (16.73),

$$y(x,t) = \sum_{j=1}^{\infty} \Psi_j(x)H_j(\omega)\psi_j(s)e^{i\omega t}. \tag{16.90}$$

Combining (16.89) and (16.90) gives

$$M(x,t) = EI\sum_{j=1}^{\infty} \Psi_j''(x)H_j(\omega)\Psi_j(s)e^{i\omega t} \tag{16.91}$$

so that, for unit harmonic excitation, we can say that

$$M(x,t) = H_M(x,s,\omega)e^{i\omega t} \tag{16.92}$$

†In problem 16.5, the result of the summation is shown to be

$$E[y^2(x,t)] = \frac{2\pi^2S_0L^3}{3EIc}\left(\frac{x}{L}\right)^2\left(1-\frac{x}{L}\right)^2, \quad 0 \leqslant \frac{x}{L} \leqslant 1.$$

where

$$H_M(x, s, \omega) = \sum_{j=1}^{\infty} EI\Psi_j''(x)H_j(\omega)\Psi_j(s).$$ (16.93)

This result for $H_M(x, s, \omega)$ may now be substituted into (16.80) and the same steps followed as in Example 1 to obtain the following expression for the mean-square bending moment in the beam:

$$E[M^2(x, t)] = \frac{4\pi^2 S_0 EI}{Lc} \sum_{j=1}^{\infty} \sin^2\left(j\frac{\pi x}{L}\right).$$ (16.94)

The summation in (16.94) is the same as in (16.88) which gives the mean-square velocity of the beam. As noted already, this summation does not converge to a limit, and as more modes are added, the mean-square bending moment and the mean-square velocity both increase without limit.

Figure 16.5 shows the distribution of mean-square displacement and of

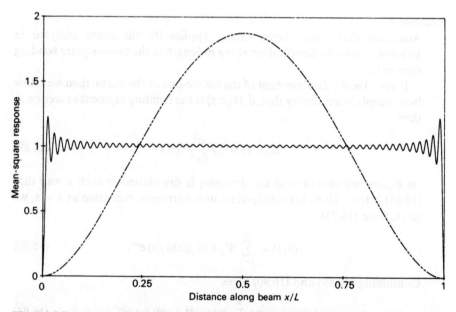

Fig. 16.5 Response of a simply-supported uniform beam to distributed random excitation which is uncorrelated in space and in time. The broken line gives the mean-square displacement response. The continuous line gives the contribution to the mean-square velocity and mean-square bending moment (which are given by the same summation) of the first 60 modes. The responses have been normalized by dividing by their average value along the length of the beam. Classical normal mode theory is assumed to apply. The absolute levels are then inversely proportional to the level of viscous damping present but the normalized levels are independent of damping

mean-square bending moment and velocity for the case when the summations extend to include the 60th mode. The results are normalized by dividing by their average value along the length of the beam. For the mean-square displacement, calculated from (16.87) with the summation extending from $j = 1$ to $j = 60$, and plotted as the broken line in Fig. 16.5, the result has converged to very close to its mathematical limit. For the mean-square bending moment and mean-square velocity, calculated from the summation in (16.88) and (16.94) with j extending from 1 to 60, the corresponding result is shown as the continuous curve. Although the summation has not converged, the contributions of the first 60 modes produce a level which is approximately constant along the length of the beam. The ripples on this constant level have a wavelength of approximately $L/60$ where L is the length of the beam. When more than 60 modes are included in the summation, this wavelength decreases in proportion to the number of modes, and the amplitudes of the ripples in the centre of the beam diminish. However, the amplitudes of the ripples at the two ends of the beam remain approximately constant.

The average value along the length of the beam of the summation

$$\sum_{j=1}^{N} \sin^2\left(j\frac{\pi x}{L}\right)$$

is

$$\sum_{j=1}^{N} \frac{1}{L} \int_0^L \sin^2\left(j\frac{\pi x}{L}\right) dx = \frac{N}{2} \qquad (16.95)$$

so that the levels of the mean-square bending moment and mean-square velocity increase in proportion to the number of modes included in the summations (16.88) and (16.94). Therefore, although the amplitudes of the end ripples in Fig. 16.5 are approximately independent of the number of modes, N, included in the summation, the amplitudes of the same ripples on the absolute level of mean-square bending moment and velocity increase in proportion to N.

Crandall [76] has pointed out that this behaviour is analagous to Gibbs' Phenomenon in the theory of Fourier series.† In the immediate vicinity of a jump discontinuity, the sum of Fourier terms contains oscillations whose frequency increases as the number of terms in the summation increases but whose amplitude remains finite.

The practical conclusion from Fig. 16.5 is that, while the mean-square displacement increases monotonically to a maximum at the centre of the beam, the mean-square velocity and mean-square bending moment are substantially constant along the length of the beam away from the immediate vicinity of its ends. For excitation which is substantially uncorrelated in space and time, the mean-square velocity (and therefore the

†See, for example, Courant and Hilbert [74], p. 105.

average kinetic energy per unit length) and the mean-square bending moment (and therefore the mean-square stress in the beam) are approximately the same at all points on the beam.

Kinetic energy of a flat plate subjected to uncorrelated random excitation

We can now make a general calculation of the average kinetic energy per unit area of a flat plate subjected to random excitation which is uncorrelated in space and time. The assumption that the distributed excitation is uncorrelated is useful because it is an approximation for excitation whose spectral density covers broad bands of spatial and time frequencies, and it is possible to produce some simple answers that may be used as a guide to the response of systems subjected to less idealized excitation.

The vibration of a thin, flat plate subjected to random excitation will be assumed to be described by the partial differential equation

$$m\frac{\partial^2 y}{\partial t^2} + c\frac{\partial y}{\partial t} + D\left(\frac{\partial^4 y}{\partial x_1^4} + 2\frac{\partial^4 y}{\partial x_1^2 \partial x_2^2} + \frac{\partial^4 y}{\partial x_2^4}\right) = p(x_1, x_2, t). \qquad (16.96)$$

We shall assume that the plate has constant mass per unit area m and that it is subjected to a viscous damping force c per unit of area of the plate and per unit velocity. D is the flexural rigidity of the plate defined by (16.63). Its transverse displacement is $y(x_1, x_2, t)$ as a result of the distributed excitation $p(x_1, x_2, t)$. The boundary conditions will be assumed to be arbitrary. The excitation $p(x_1, x_2, t)$ is homogeneous and stationary and uncorrelated in space and time. The excitation's three-dimensional spectral density is then

$$S_{pp}(\boldsymbol{\gamma}, \omega) = S_{pp}(\gamma_1, \gamma_2, \omega) = S_0 \text{ (constant)}. \qquad (16.97)$$

The cross-spectral density between $p(\mathbf{s}_1, t)$, which is the excitation at position vector $\mathbf{s}_1$, and $p(\mathbf{s}_2, t)$, which is the excitation at $\mathbf{s}_2$, is then, from (16.39)

$$S_{pp}(\mathbf{s}_1, \mathbf{s}_2, \omega) = \int_\infty d\boldsymbol{\gamma}\, S_0\, e^{i(\boldsymbol{\gamma} \cdot (\mathbf{s}_1 - \mathbf{s}_2))}. \qquad (16.98)$$

If we define unit vectors $\mathbf{e}_1$ and $\mathbf{e}_2$ to identify two rectangular coordinate axes on the surface of the plate, then we can write

$$\mathbf{s}_1 = s_{11}\mathbf{e}_1 + s_{12}\mathbf{e}_2$$
$$\mathbf{s}_2 = s_{21}\mathbf{e}_1 + s_{22}\mathbf{e}_2 \qquad (16.99)$$

and

$$\boldsymbol{\gamma} = \gamma_1\mathbf{e}_1 + \gamma_2\mathbf{e}_2 \qquad (16.100)$$

and then (16.98) gives

$$S_{pp}(\mathbf{s}_1, \mathbf{s}_2, \omega) = \int_{-\infty}^{\infty} d\gamma_1 \int_{-\infty}^{\infty} d\gamma_2 \, S_0 \, e^{i\{\gamma_1(s_{11}-s_{21})+\gamma_2(s_{12}-s_{22})\}} \tag{16.101}$$

which, using (16.77) gives

$$S_{pp}(\mathbf{s}_1, \mathbf{s}_2, \omega) = 4\pi^2 S_0 \delta(s_{11} - s_{21}) \delta(s_{12} - s_{22}) \tag{16.102}$$

or, in shorter form,

$$S_{pp}(\mathbf{s}_1, \mathbf{s}_2, \omega) = 4\pi^2 S_0 \delta(\mathbf{s}_1 - \mathbf{s}_2). \tag{16.103}$$

Now we can apply (16.34) to obtain

$$S_{yy}(\mathbf{r}, \mathbf{r}, \omega) = \int_R d\mathbf{s}_1 \int_R d\mathbf{s}_2 \, H^*(\mathbf{r}, \mathbf{s}_1, \omega) H(\mathbf{r}, \mathbf{s}_2, \omega) S_{pp}(\mathbf{s}_1, \mathbf{s}_2, \omega). \tag{16.104}$$

On substituting for the frequency response functions from (16.73) and using (16.84) to calculate the mean-square velocity, we arrive at

$$E[\dot{y}^2(\mathbf{r}, t)] = \int_{-\infty}^{\infty} d\omega \, \omega^2 \int_R d\mathbf{s}_1 \int_R d\mathbf{s}_2 \left(\sum_{j=1}^{\infty} \Psi_j(\mathbf{r}) H_j^*(\omega) \Psi_j(\mathbf{s}_1) \right) \times$$

$$\times \left(\sum_{k=1}^{\infty} \Psi_k(\mathbf{r}) H_k(\omega) \Psi_k(\mathbf{s}_2) \right) S_{pp}(\mathbf{s}_1, \mathbf{s}_2, \omega). \tag{16.105}$$

The assumption of uncorrelated excitation allows an immediate simplification because, on substituting for the multi-dimensional spectral density $S_{pp}(\mathbf{s}_1, \mathbf{s}_2, \omega)$ from (16.103) and carrying out the integration over $\mathbf{s}_2$, this equation reduces to

$$E[\dot{y}^2(\mathbf{r}, t)] = \sum_{j=1}^{\infty} \sum_{k=1}^{\infty} \int_{-\infty}^{\infty} d\omega \, \omega^2 H_j^*(\omega) H_k(\omega) \Psi_j(\mathbf{r}) \Psi_k(\mathbf{r}) \times$$

$$\times \int_R d\mathbf{s} \, \Psi_j(\mathbf{s}) \Psi_k(\mathbf{s}) 4\pi^2 S_0. \tag{16.106}$$

The variable $\mathbf{s}_1$ has been written without its subscript now that $\mathbf{s}_2$ has been integrated out.

We shall now introduce the symbol $u(\mathbf{r})$ to denote the average kinetic energy per unit area of the plate at position $\mathbf{r}$. If m is the mass per unit area of the plate, then

$$u(\mathbf{r}) = \tfrac{1}{2} m E[\dot{y}^2(\mathbf{r}, t)]. \tag{16.107}$$

From (16.106), we find that

$$u(\mathbf{r}) = \sum_{j=1}^{\infty} \sum_{k=1}^{\infty} \int_{-\infty}^{\infty} d\omega \, \omega^2 H_j^*(\omega) H_k(\omega) \Psi_j(\mathbf{r}) \Psi_k(\mathbf{r}) \times$$

$$\times \int_R d\mathbf{s} \, m \Psi_j(\mathbf{s}) \Psi_k(\mathbf{s}) 2\pi^2 S_0 \tag{16.108}$$

and, from (16.64), we know that, because of the assumed orthogonality of the normal modes, the integral over s is zero unless $j = k$ when it is equal to m_j. Hence we can write

$$u(\mathbf{r}) = 2\pi^2 S_0 \sum_{j=1}^{\infty} m_j \Psi_j^2(\mathbf{r}) \int_{-\infty}^{\infty} |i\omega H_j(\omega)|^2 \, d\omega. \tag{16.109}$$

The integral over ω can be evaluated by using Appendix 1, case 2, with the result that (16.109) becomes

$$u(\mathbf{r}) = 2\pi^3 S_0 \sum_{j=1}^{\infty} \frac{1}{\beta_j m_j} \Psi_j^2(\mathbf{r}). \tag{16.110}$$

Now, if the average value of $u(\mathbf{r})$ over the area of the plate A is denoted by u_{av}, then we have

$$u_{av} A = \int_R u(\mathbf{r}) \, d\mathbf{r} = 2\pi^3 S_0 \sum_{j=1}^{\infty} \frac{1}{\beta_j m_j} \int_R \Psi_j^2(\mathbf{r}) \, d\mathbf{r}. \tag{16.111}$$

Since the mass and damping coefficients m and c are constants, we can use (16.64) to replace the integral in (16.111) by m_j/m and also use (16.68) with (16.65) to write c/m for β_j. Then (16.111) becomes

$$u_{av} A = 2\pi^3 S_0 \sum_{j=1}^{\infty} \frac{1}{c}. \tag{16.112}$$

This summation obviously does not converge to a limit. We saw in Example 1 in the previous section that the mean-square velocity for a beam subjected to uncorrelated excitation behaved similarly. For a beam, when the summation includes a large number of terms, the distribution given by (16.88) is approximately flat except for ripples near the ends of the beam (see Fig. 16.5). The same conclusion applies in the present case. We can think of each term in (16.112) giving the contribution of one mode to the total kinetic energy. We conclude that, for each mode, the average kinetic energy per unit area of the plate is

$$u_{av} = (\text{Average kinetic energy per unit area per mode}) = \frac{2\pi^3 S_0}{Ac}. \tag{16.113}$$

This result, that energy is distributed equally between modes, is the basis of the approximate method of statistical energy analysis for calculating the distribution of vibrational energy in structures subjected to random excitation (Lyon [86]).

Single degree-of-freedom analogy

We conclude by specializing the problem to the case of a simply-supported rectangular plate, for which we can show that (16.113) may be obtained by drawing an analogy with the response of a single degree-of-freedom system.

Consider a system with deterministic excitation whose equation of motion is

$$m\ddot{y} + c\dot{y} + ky = p(t) = p_0 e^{i\omega t}. \qquad (16.114)$$

The time-average of the excitation squared, calculated over a complete cycle, is

$$\langle p^2(t) \rangle = \tfrac{1}{2} p_0^2. \qquad (16.115)$$

At resonance, when $\omega = \sqrt{k/m}$, the steady-state velocity response is

$$\dot{y}(t) = \frac{p_0}{c} e^{i\omega t} \qquad (16.116)$$

and the time-average velocity squared is

$$\langle \dot{y}^2(t) \rangle = \frac{1}{2}\left(\frac{p_0}{c}\right)^2. \qquad (16.117)$$

The average kinetic energy, u, of a single degree-of-freedom system subjected to steady-state harmonic excitation at resonance is therefore given in terms of its average excitation squared by

$$u = \tfrac{1}{2} m \langle \dot{y}^2(t) \rangle = \frac{m}{2c^2} \langle p^2(t) \rangle. \qquad (16.118)$$

Now, for the same single degree-of-freedom system, suppose that there is broad-band random excitation. From the first section of Chapter 14, we can use (14.7) to calculate the approximate average kinetic energy. In this case, we have

$$u = \tfrac{1}{2} m E[\dot{y}^2(t)] = \tfrac{1}{2} m \,(2) \begin{pmatrix} \text{Average value of} \\ S_p(\omega) \text{ in the} \\ \text{region of } \omega = \sqrt{\dfrac{k}{m}} \end{pmatrix} \begin{pmatrix} \text{Peak value} \\ \text{of } H(\omega) \text{ for} \\ \text{the velocity} \\ \text{response} \end{pmatrix}^2 \begin{pmatrix} \text{Mean square} \\ \text{bandwidth} \end{pmatrix}$$

$$(16.119)$$

But, if we let $E[p^2(t)]_1$ denote that part of the total mean-square excitation which may be considered to excite the resonance,

$$E[p^2(t)]_1 = (2) \begin{pmatrix} \text{Average value of} \\ S_p(\omega) \text{ in the} \\ \text{region of } \omega = \sqrt{\dfrac{k}{m}} \end{pmatrix} \begin{pmatrix} \text{Mean square} \\ \text{bandwidth} \end{pmatrix}$$

so that

$$u = \tfrac{1}{2} m \begin{pmatrix} \text{Peak value} \\ \text{of } H(\omega) \text{ for} \\ \text{the velocity} \\ \text{response} \end{pmatrix}^2 E[p^2(t)]_1 \qquad (16.121)$$

which gives

$$u = \frac{m}{2c^2} E[p^2(t)]_1 \qquad (16.122)$$

corresponding to (16.118) for deterministic excitation.

Now we return to the case of a rectangular plate. We assume, as before, that the plate is subjected to excitation which has a spectral density S_0 that is uniform over all frequencies and all wavenumbers. We want to estimate that part of the mean-square value of the excitation $p(x_1, x_2, t)$ which excites any individual mode of the plate. The mean-square bandwidth in time frequency is still given by the formula in the first section of Chapter 14

$$\text{(Mean square bandwidth)} = \pi \zeta \omega_n = \frac{\pi}{2} \left(\frac{c}{m} \right). \qquad (16.123)$$

Now we have to calculate the bandwidth of wavenumbers associated with a single mode. The normal modes of a simply-supported rectangular plate of area $L_1 \times L_2$ are given by

$$\Psi_{jm}(x_1, x_2) = 2 \sin \left(j \frac{\pi x_1}{L_1} \right) \sin \left(m \frac{\pi x_2}{L_2} \right) \qquad (16.124)$$

where we now use double indices to represent each normal mode. As integer j increases by 1, wavenumber γ_1 increases by π/L_1; as integer m increases by 1, wavenumber γ_2 increases by π/L_2. Hence the wavenumber increments are π/L_1 and π/L_2. Remembering that S_0 is a three-dimensional spectral density function, we have to integrate S_0 over time frequency ω and both wavenumbers γ_1 and γ_2 in order to calculate mean-square level. Using (15.48) to do this, we find that the three-dimensional spectral density S_0 contributes a mean-square excitation per mode of

$$E[p^2(t)]_1 = 8S_0 \left(\frac{\pi}{L_1} \right) \left(\frac{\pi}{L_2} \right) \left(\frac{\pi c}{2m} \right). \qquad (16.125)$$

In this equation (16.125), $E[p^2(t)]_1$ is the mean-square excitation per unit area of the plate, and so we now consider the average kinetic energy per mode for a unit area of plate, which is what we have calculated in (16.113). Rewriting (16.113) and then using (16.125) gives

$$u_{av} = \frac{2\pi^3 S_0}{L_1 L_2 c} = \frac{1}{2} \frac{m}{c^2} \left(4S_0 \frac{\pi^3 c}{L_1 L_2 m} \right) = \frac{m}{2c^2} E[p^2(t)]_1 \qquad (16.126)$$

which is the same result as for the single degree-of-freedom case. Its validity depends on the assumption that the plate's distributed excitation is uncorrelated in space and time.

Chapter 17

Discrete wavelet analysis

A disadvantage of Fourier analysis is that frequency information can only be extracted for the complete duration of a signal $f(t)$. Since the integral in the Fourier transform equation (4.10) extends over all time, from $-\infty$ to $+\infty$, the information it provides arises from an average over the whole length of the signal. If, at some point in the lifetime of $f(t)$, there is a local oscillation representing a particular feature, this will contribute to the calculated Fourier transform $F(\omega)$, but its location on the time axis will be lost. There is no way of knowing whether the value of $F(\omega)$ at a particular ω derives from frequencies present throughout the life of $f(t)$ or during just one or a few selected periods. This disadvantage is overcome in *wavelet analysis* which provides an alternative way of breaking a signal down into its constituent parts. The original impetus for wavelets came from the analysis of earthquake records (Goupillaud *et al.* [107]), but wavelet analysis has found important applications in speech and image processing and is now a significant new tool in signal analysis generally.

Basic ideas

To illustrate how wavelet decomposition works, Fig. 17.1 shows the decomposition of a signal into its wavelet components. The signal is a square wave, two cycles of which are shown at the top of the figure. Below it are eight separate sub-signals which have been obtained by decomposing the square wave into its wavelet components. For reasons that will become apparent later in this chapter, each component is called a *level* and the levels are numbered from -1 upwards. In this example, levels -1 and 0 are both zero but in general that is not the case.

When the separate wavelet levels are added together, the original signal is regained. This is shown in Fig. 17.2. Starting at the top left-hand diagram, which shows level -1 alone, and moving down the left-hand side of the figure, successive levels are added until finally, at the bottom right-hand diagram, the original signal has been regained.

We are concerned with discrete wavelet analysis and the signal to be analysed is assumed to have been sampled at equally spaced intervals as described in

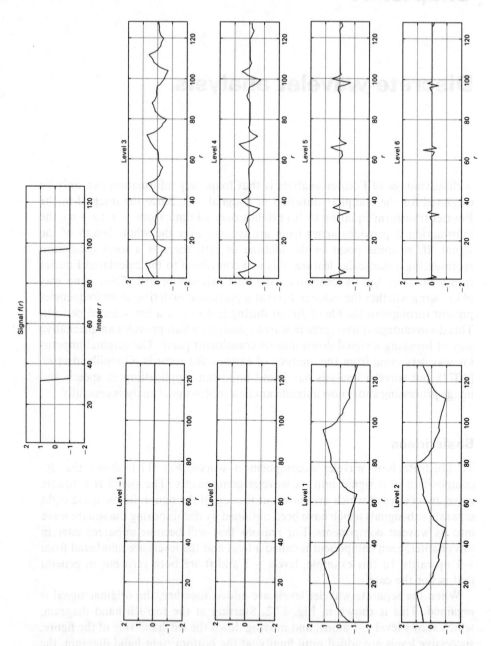

Fig. 17.1 Analysis of a square wave $f(r)$, $r = 1$ to 128, into its D4 wavelet components

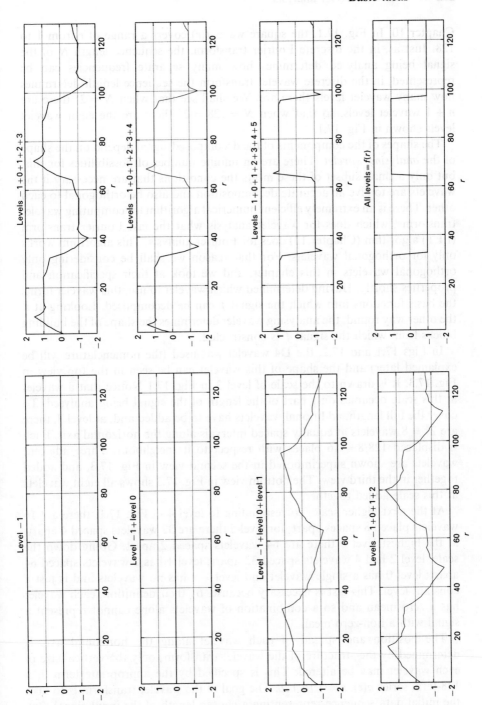

Fig. 17.2 Reconstruction of the square wave from its D4 wavelet components

Chapter 10. In Fig. 17.1, the square wave $f(r)$ covers a range of r from 1 to 128. Just as, in the discrete Fourier transform, the sequence length N of the signal being analysed determines how many separate frequencies can be represented, in the discrete wavelet transform the sequence length determines how many wavelet levels there are. We shall see that when $N = 2^n$ there are $n + 1$ wavelet levels, so that when $N = 128 = 2^7$ there are the eight wavelet levels shown in Fig. 17.1.

The shapes of the components of the decomposed signal depend on the shape of the *analysing wavelet*. There are an infinite number of possibilities for this, but only a small subset of these meets the conditions that are necessary if the wavelets are to give an accurate decomposition and also be orthogonal to each other. There is an extremely efficient numerical algorithm for computing wavelet transforms, which does for wavelet analysis what the fast Fourier transform (FFT) algorithm (Chapter 12) does for Fourier analysis. This algorithm works only for orthogonal wavelets. For that reason we shall be considering only orthogonal wavelets in this chapter, and we look at their specification and properties shortly. Having determined which wavelet to use, this wavelet forms the basis functions into which the signal f can be decomposed. Looking at it the other way round, the analysing wavelet determines the shape of the building blocks from which the signal f is constructed.

In Figs 17.1 and 17.2, the D4 wavelet was used (the nomenclature will be explained later) and the shape of this wavelet can be seen in the top view in Fig. 17.3. It is drawn to the scale of level 3 in Fig. 17.1 Notice that a wavelet at this scale occupies only part of the length of the signal being analysed. To cover the full length, additional wavelets have to be added and, at level 3, there are $2^3 = 8$ wavelets at equally spaced intervals along the horizontal axis. Each is displaced $128/8 = 16$ places with respect to its neighbour. Three adjacent wavelets are shown superimposed in the second view in Fig. 17.3, and added together in the third view. The bottom view in Fig. 17.3 shows all eight wavelets at this scale added together.

At the next higher scale, corresponding to level 4 in Fig. 17.1, there are 16 wavelets placed 8 spaces apart; for level 5 there are 32 wavelets spaced 4 apart; at the highest level 6 there are 64 wavelets spaced 2 apart. Going down the scale, level 2 had 4 wavelets spaced 32 apart; level 1 has 2 wavelets spaced 64 apart; level 0 has a single wavelet and level -1 has no wavelets and is just a constant level. This last is necessary because, by their definition, each wavelet has a zero mean and so a combination of wavelets alone cannot represent a signal with a non-zero mean.

The position and spread of each wavelet along the horizontal axis is determined by the structure of the wavelet transform; only the vertical size of each wavelet may be altered. This is specified by the appropriate term in a 128-term series $a(r)$, $r = 1$ to 128. The goal of the wavelet transform is to take the initial data sequence representing a chosen length of the input signal $f(r)$, $r = 1$ to 128, and convert this into a new sequence of (real) numbers $a(r)$, $r = 1$ to 128, which define the vertical size of the wavelets at each of the set horizontal

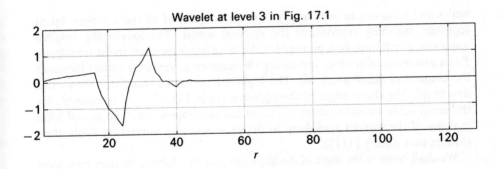

Wavelet at level 3 in Fig. 17.1

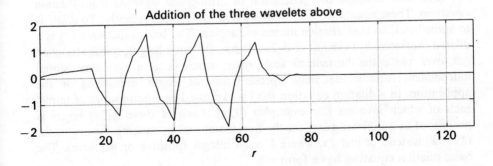

Three adjacent wavelets at level 3 superimposed

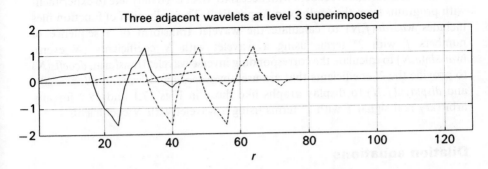

Addition of the three wavelets above

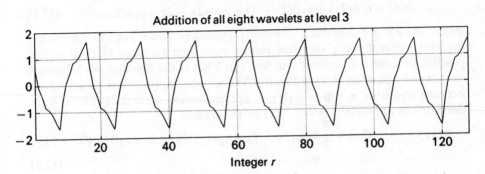

Addition of all eight wavelets at level 3

Integer r

Fig. 17.3 D4 wavelets at the scale of level 3 in Fig. 17.1

scales and positions in such a way that the addition of all the wavelets, taken together, faithfully reproduces the original signal. The longer the original sequence, which must be a power of 2, the more levels there are in the transform. For a given time duration, increasing the number of sampling points increases the amount of detail available. The highest level of the transform shows this fine detail. The sharp edges of the square wave in Fig. 17.1 are responsible for the small-scale wavelets which occur in the decomposition. The use of local maxima of the wavelet transform to detect edges is an interesting application (Mallat and Zhong [111]).

We shall come to the logic of the algorithm for the discrete wavelet transform and its inverse later, but readers with access to MATLAB may like to experiment with programs from the toolbox in Appendix 7. This collection of function files includes *wavedn(f,N)* to calculate the wavelet transform of a sequence of numbers f with 2^n terms using a wavelet with N coefficients (N even), *iwavedn(a,N)* to calculate the corresponding inverse wavelet transform, *dcoeffs(N)* to provide the N coefficients that are used by *wavedn* and *iwavedn* ($N < =20$); and *displayn(f,N)* to display graphs like those in Figs 17.1 and 17.2 for any arbitrary real signal f with 2^n terms using a wavelet with N coefficients.

Dilation equations

We shall now examine the generation of orthogonal wavelets from *dilation equations*. These equations have been studied widely only recently. To *dilate* is to spread out, so that *dilation* means expansion. The basic function $\phi(x)$ is a dilated (horizontally) version of $\phi(2x)$. It has the same height, but is stretched out over twice the horizontal scale of x, where x is a non-dimensional independent variable that may represent time or length, depending on the application. In a dilation equation $\phi(x)$ is expressed as a finite series of terms, each of which involves (for example) $\phi(2x)$. Each of these $\phi(2x)$ terms is positioned at a different place on the horizontal axis by making the argument $(2x - k)$ instead of just $2x$, where k is an integer (positive or negative). The basic dilation equation has a form

$$\phi(x) = c_0\phi(2x) + c_1\phi(2x - 1) + c_2\phi(2x - 2) + c_3\phi(2x - 3) \quad (17.1)$$

where the c's are numerical constants (generally some positive and some negative). Except for a very few simple cases (see problem 17.1), it is not possible to solve (17.1) directly to find out what is the function $\phi(x)$. Instead we have to construct $\phi(x)$ indirectly. The simplest approach is to set up an iterative algorithm in which each new approximation $\phi_j(x)$ is calculated from the previous approximation $\phi_{j-1}(x)$ by the scheme

$$\phi_j(x) = c_0\phi_{j-1}(2x) + c_1\phi_{j-1}(2x - 1) + c_2\phi_{j-1}(2x - 2) + c_3\phi_{j-1}(2x - 3).$$

$$(17.2)$$

The iteration is continued until $\phi_j(x)$ becomes indistinguishable from $\phi_{j-1}(x)$.

Consider starting from a box function $\phi_0(x) = 1$, $0 \leqslant x < 1$, $\phi_0(x) = 0$ elsewhere. After one iteration the box function over interval $x = 0$ to 1 has developed into a staircase function over the interval $x = 0$ to 2 as shown in the upper view in Fig. 17.4. Each contribution to $\phi_1(x)$ is shown separately and then the four contributions are shown added together. A particular set of the coefficients c_0, c_1, c_2, c_3 has been used, defined as follows:

$$c_0 = (1 + \sqrt{3})/4 \qquad c_1 = (3 + \sqrt{3})/4$$

$$c_2 = (3 - \sqrt{3})/4 \qquad c_3 = -(\sqrt{3} - 1)/4. \qquad (17.3)$$

These we shall see later generate an orthogonal D4 wavelet. The D stands for Daubechies who first discovered their properties (Daubechies [103–5]).

When the iterative process is continued, the function $\phi(x)$ approaches a limiting shape as shown in the lower view in Fig. 17.4. An unusual feature is the discontinuous nature of this shape. The same function is shown to larger scale (see problem 17.2) in Fig. 17.5. The graph has a fractal nature so that when drawn to larger scale its irregular outline remains. In order to obtain a smoother function, it is necessary to include more terms in the dilation equation (17.1).

The function $\phi(x)$ generated from a unit box (unit height and length) is called the *scaling function* and a corresponding wavelet function will be constructed from it in the next section. However, first we consider in more detail how the iterative scheme used in Fig. 17.4 works. Representing the initial box function by the ordinate 1 at $x = 0$, the first iteration (upper view of Fig. 17.4) produces four new ordinates c_0, c_1, c_2, c_3 at $x = 0$, 0.5, 1, 1.5. At the second iteration, ordinate c_0 at $x = 0$ contributes to four new ordinates c_0^2, c_0c_1, c_0c_2, c_0c_3 at $x = 0$, 0.25, 0.5, 0.75; ordinate c_1 at $x = 0.5$ contributes to four new ordinates c_1c_0, c_1^2, c_1c_2, c_1c_3 at $x = 0.5$, 0.75, 1, 1.25; and so on. After the second iteration is complete, the resulting ordinates (third row of the lower view in Fig. 17.4) are c_0^2, c_0c_1, $c_0c_2 + c_1c_0$, $c_0c_3 + c_1^2$, $c_1c_2 + c_2c_0$, $c_1c_3 + c_2c_1$, $c_2^2 + c_3c_0$, $c_2c_3 + c_3c_1$, c_3c_2, c_3^2 at $x = 0$, 0.25, 0.5, 0.75, ..., 2.25. This calculation follows the matrix scheme

$$[\phi_2] = \begin{bmatrix} c_0 & & & \\ c_1 & & & \\ c_2 & c_0 & & \\ c_3 & c_1 & & \\ & c_2 & c_0 & \\ & c_3 & c_1 & \\ & & c_2 & c_0 \\ & & c_3 & c_1 \\ & & & c_2 \\ & & & c_3 \end{bmatrix} \begin{bmatrix} c_0 \\ c_1 \\ c_2 \\ c_3 \end{bmatrix} [1] = M_2 M_1 [1] \qquad (17.4)$$

Construction of $\phi_1(x) = c_0\phi_0(2x) + c_1\phi_0(2x-1) + c_2\phi_0(2x-2) + c_3\phi_0(2x-3)$

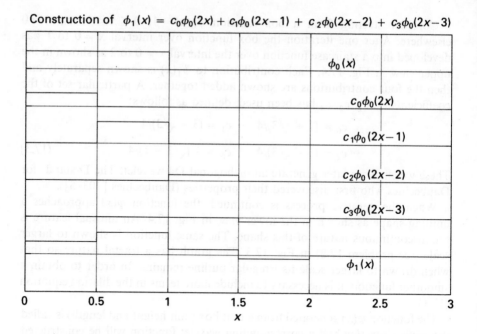

First eight iterations of the development of the D4 scaling function

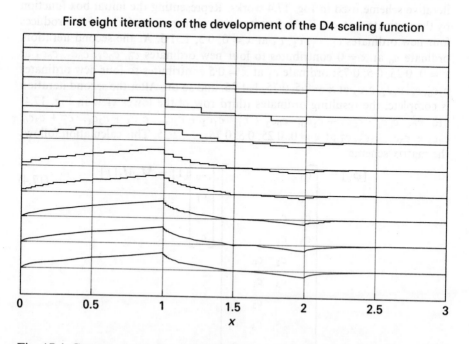

Fig. 17.4 Construction of the D4 scaling function by iteration from a box function over the interval $x = 0$ to 1

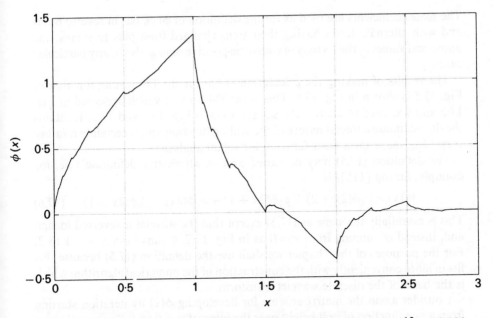

Fig. 17.5 Scaling function for the D4 wavelet calculated for $3 \times 2^{12} = 12\,288$ points using the recursive method described in problem 17.2

where M_r denotes a matrix of order $(2^{r+1} + 2^r - 2) \times (2^r + 2^{r-1} - 2)$ in which each column has a submatrix of the coefficients c_0, c_1, c_2, c_3 positioned two places below the submatrix to its left. The number of points on the graph increases in the sequence 1, 4, 10, 22, 46, ..., $2^{r+1} + 2^r - 2$ so that after eight iterations it reaches $2^9 + 2^8 - 2 = 766$ with each point spaced $1/2^8 = 1/256$ units apart along the horizontal axis. The graph begins at $x = 0$ and almost (but not quite) reaches $x = 3$. The implementation of this scheme of continued matrix multiplication lies at the heart of the discrete wavelet transform. Iteration is not the most efficient method of generating the scaling function ϕ (the recursive method described in problem 17.2 is better in that respect), but it is simple and easily programmed and, for orthogonal wavelets (particular choices of the coefficients c_0, c_1, ...), the matrix M_r has special properties which are essential for the discrete wavelet transform to work.

Dilation wavelets

So far we have not defined a *wavelet*. This is described by its *wavelet function* $W(x)$ which is derived from the corresponding scaling function by taking differences. For the four-coefficient scaling function defined by (17.1), the dilation wavelet function is

$$W(x) = -c_3\phi(2x) + c_2\phi(2x - 1) - c_1\phi(2x - 2) + c_0\phi(2x - 3). \quad (17.5)$$

The *same* coefficients are used as for the definition of $\phi(x)$, but in reverse order and with alternate terms having their signs changed from plus to minus (the numerical values of the c's may of course be positive or negative in any particular case).

The results of making the calculation (17.5) for the D4 scaling function in Fig. 17.5 is shown in Fig. 17.6. This is the D4 wavelet which appeared in Fig. 17.3 and was used to analyse the square wave in Figs 17.1 and 17.2. It retains the discontinuous, fractal nature of the scaling function and is certainly a rather surprising shape for a basis function for signal analysis.

The definition (17.5) may be varied and an alternative definition (see, for example, Strang [115]) is

$$W(x) = c_3\phi(2x + 2) - c_2\phi(2x + 1) + c_1\phi(2x) - c_0\phi(2x - 1). \quad (17.6)$$

This is essentially the same as (17.5) except that the wavelet is reversed in sign and, instead of running from $x = 0$ as in Fig. 17.7, it runs from $x = -1$ to 2. For the purposes of this chapter we shall use the definition (17.5) because this fits in more conveniently with the construction of the numerical algorithm which is the basis of the discrete wavelet transform.

Consider again the matrix scheme for developing $\phi(x)$ by iteration starting from a box function of unit height over the interval $x = 0$ to 1. Suppose that we have taken the iteration to a stage where $\phi(x)$ is fully developed and that we want to generate $W(x)$ from this $\phi(x)$ by using (17.5). For simplicity, imagine that only one iteration leads to the final $\phi(x)$ so that this is represented by

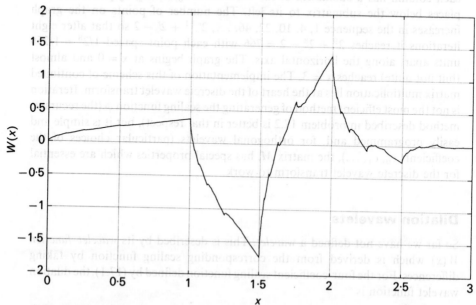

Fig. 17.6 D4 wavelet according to (17.5) from the scaling function in Fig. 17.5

just four ordinates c_0, c_1, c_2, c_3 at $x = 0, 0.5, 1, 1.5$ as in the upper view of Fig. 17.4. According to (17.5), these 4 ordinates generate 10 new ordinates spaced 0.25 apart. The term $-c_3\phi(2x)$ in (18.5) gives $-c_3c_0, -c_3c_1, -c_3c_2, -c_3^2$; the term $c_2\phi(2x - 1)$ gives $c_2c_0, c_2c_1, c_2^2, c_2c_3$ moved two places to the right; and so on for the other terms, so that the new ordinates (now for the wavelet) are

$$-c_3c_0, \ -c_3c_1, \ -c_3c_2 + c_2c_0, \ -c_3^2 + c_2c_1, \ c_2^2 - c_1c_0, \ c_2c_3 - c_1^2, \ -c_1c_2 + c_0^2,$$
$$-c_1c_3 + c_0c_1, \ c_0c_2, \ c_0c_3.$$

These are generated by the matrix scheme

$$[W_2] = \begin{bmatrix} -c_3 & & & \\ 0 & -c_3 & & \\ c_2 & 0 & -c_3 & \\ 0 & c_2 & 0 & -c_3 \\ -c_1 & 0 & c_2 & 0 \\ 0 & -c_1 & 0 & c_2 \\ c_0 & 0 & -c_1 & 0 \\ & c_0 & 0 & -c_1 \\ & & c_0 & 0 \\ & & & c_0 \end{bmatrix} \begin{bmatrix} c_0 \\ c_1 \\ c_2 \\ c_3 \end{bmatrix} [1] \qquad (17.7)$$

First eight iterations of the development of the D4 wavelet

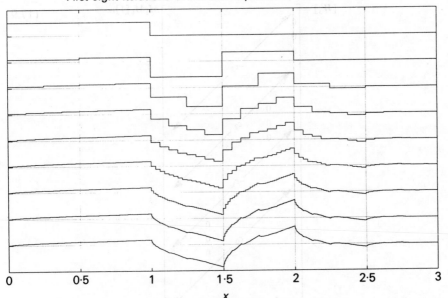

0   0.5   1   1.5   2   2.5   3

x

Fig. 17.7 D4 wavelet calculated by iteration

or, alternatively, by

$$[W_2] = \begin{bmatrix} c_0 & & & & & & & -c_3 \\ c_1 & & & & & & & c_2 \\ c_2 & c_0 & & & & & & -c_1 \\ c_3 & c_1 & & & & & & c_0 \\ & c_2 & c_0 & & & & & \\ & c_3 & c_1 & & & & & \\ & & c_2 & c_0 & & & & \\ & & c_3 & c_1 & & & & \\ & & & c_2 & & & & \\ & & & c_3 & & & & \end{bmatrix}[1]. \qquad (17.8)$$

Notice the zeros between the diagonal lines of coefficients in (17.7). If the wavelet were calculated after first completing two iterations of the scaling function, there would be three zeros between each diagonal line. This is because the ordinates after two iterations of the scaling function are spaced $x = 0.25$ apart. As a result of the $2x$ in the terms on the right-hand side of (17.5), the ordinates of the wavelet function will be $0.25/2 = 0.125$ apart. However, the translations required by (17.5) are $x = 0.5$ ($\phi(2x - 1)$ is translated 0.5 with respect to $\phi(2x)$, and so on). The upshot is that the diagonal lines of coefficients in (17.9) are each $0.5/0.125 = 4$ places apart:

$$[W_3] = \qquad\qquad\qquad\qquad\qquad M_2 M_1 [1]. \qquad (17.9)$$

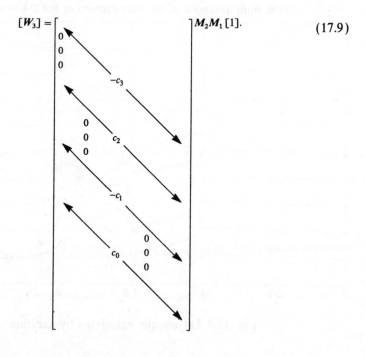

The reader can verify (problem 17.3) that just as (17.7) and (17.8) are equivalent forms, so (17.9) is the same as

$$W_3 = M_3 M_2 \begin{bmatrix} -c_3 \\ c_2 \\ -c_1 \\ c_0 \end{bmatrix} [1] \tag{17.10}$$

where M_3 is a matrix of order 22×10 with 10 submatrices $[c_0\ c_1\ c_2\ c_3]^t$ each arranged two places below its left-hand neighbour.

Equation (17.10) is important because it is the scheme by which wavelets are generated in the inverse discrete wavelet transform (IDWT). The calculation is illustrated in Fig. 17.7. This shows the result of a calculation identical to that which led to Fig. 17.4 except that the first step is $[-c_3\ c_2\ -c_1\ c_0]^t\ [1]$ instead of $[c_0\ c_1\ c_2\ c_3]^t\ [1]$. All later steps of the iteration use the matrices M_r consisting of submatrices $[c_0\ c_1\ c_2\ c_3]^t$ staggered vertically two places each. After eight steps leading to 766 ordinates as before, the calculated wavelet is very close to that in Fig. 17.6.

Properties of the wavelet coefficients

In order to generate good wavelets, their coefficients have to be chosen carefully. This section, which is in six parts, sets out the conditions that a good set of wavelet coefficients have to satisfy and the consequences of these conditions. The reader who is in a hurry may want to miss out the first five parts, which are necessarily detailed, and go straight to part 6 where the results are collected together in the form of a practical recipe to choose the coefficients. Even there the conditions are complicated and need thinking through carefully. But the outcome is that there are as many independent equations to choose coefficients as there are coefficients to be chosen, so that if we decide to use only four coefficients, there are four equations to be solved and so the values are known; if we want six coefficients, there are six equations and so again there is only one set of values. These have been calculated already and published in the literature by Daubechies [104] and so the numerical values to use in any practical case are already available. For wavelets with 2, 4, 6, 8, 10, 12, 14, 16, 18 and 20 coefficients the values of the coefficients are stored in the function file $c = dcoeffs(N)$ where N is the number of coefficients (always even) and c is an array of order N carrying the coefficients in the order $c_0, c_1, c_2, \ldots, c_{N-1}$. With this information, the reader who wants to can jump to the next main section *Circular wavelet transforms* and start considering how the discrete wavelet transform is implemented in practice. But for those who are interested in knowing how the coefficients have been calculated, we now examine the conditions that they have to satisfy.

There are three different categories of conditions. The first category has just one condition which is that the sum of the coefficients must always equal 2. This is called the *conversation of area* condition; it has the consequence that, as the scaling function is developed by iteration, its area remains constant. The second category of conditions, which contains $N/2 - 1$ independent conditions when there are N coefficients, arises from the need to ensure that the expansion of a length of signal in a finite number of wavelets represents the underlying signal as closely as possible. In other words we want the expansion to be accurate. This produces a hierarchy of conditions called the *accuracy conditions*. Thirdly there is the category of conditions required to ensure that the scaling function and wavelets generated from it are orthogonal. When they are orthogonal, the orthogonality equations listed in part 5 of this section are satisfied. Without these orthogonality conditions the discrete wavelet transform will not work, so for present purposes the *orthogonality conditions* are essential. However, it should be noted that expansions in terms of non-orthogonal wavelets are used in special cases, but without the advantages of the discrete wavelet transform described in this chapter. For N wavelet coefficients, there are $N/2$ orthogonality conditions.

We now look at all these conditions in detail.

1. Conservation of area

As we have said, a wavelet may have more or less than four coefficients. So-called *compactly supported* wavelets have a finite number but, subject to that restriction, there may be any number of coefficients. To allow for this we write (17.1) in the general form

$$\phi(x) = \sum_k c_k \phi(2x - k) \tag{17.11}$$

where k is any integer, positive, zero or negative. For reasons that will become apparent, we shall only consider cases with an even number N of non-zero coefficients. In (17.1) $N = 4$ so that all the c's were zero except c_0, c_1, c_2, c_3. Corresponding to this general scaling function (17.11), the wavelet may be written as

$$W(x) = \sum_k (-1)^k c_k \phi(2x + k - N + 1). \tag{17.12}$$

Putting $N = 4$ and $k = 0$ to 3 we see that (17.12) reduces to (17.5), as it should.

Consider the result of integrating both sides of (17.11) over all x to get

$$\int_{-\infty}^{\infty} \phi(x)\,dx = \sum_k c_k \int_{-\infty}^{\infty} \phi(2x - k)\,dx. \tag{17.13}$$

Changing the variable of integration on the r.h.s. to $y = 2x - k$, this gives

$$\int_{-\infty}^{\infty} \phi(x)\,dx = \sum_k c_k \frac{1}{2} \int_{-\infty}^{\infty} \phi(y)\,dy \tag{17.14}$$

from which we find that

$$\sum_k c_k = 2 \tag{17.15}$$

so that the sum of all the wavelet coefficients must always equal 2. The reader can quickly check that the four coefficients defined by (17.3) satisfy this condition.

Applying (17.15) to the iterative scheme (17.2), we see that $\int_{-\infty}^{\infty} \phi_j(x)\,dx$ remains the same as $\int_{-\infty}^{\infty} \phi_{j-1}(x)\,dx$, so that area is conserved as the iteration proceeds. Since a unit area box is used to start the iteration, it follows that the area under the scaling function remains equal to unity and so

$$\int_{-\infty}^{\infty} \phi(x)\,dx = 1 \tag{17.16}$$

is a consequence of (17.15) and the construction of $\phi(x)$ by iteration from a unit box.

2. Accuracy

In addition to (17.15), the wavelet coefficients have to satisfy two other generic conditions which Strang, in his excellent introduction [115], has called *condition A* (for accuracy) and *condition O* (for orthogonality). We consider first condition A.

In Fig. 17.1, a square wave is shown broken down into its wavelet components, using a wavelet with the four coefficients in (17.3). The square wave was sampled at 128 points to yield $f(r)$, $r = 1$ to 128, and the working of the discrete wavelet transform (DWT) and its inverse (IDWT) is such that these values are exactly regenerated by the IDWT. But although $f(r)$ is faithfully regenerated at its sampling positions, in between these positions there may not very good correspondence between the original signal and its representation by a summation of wavelets. This is illustrated in Fig. 17.8. Two cycles of a sine wave are sampled at 32 positions per cycle to give a 64-term series. Using a two-term wavelet (in which the only two non-zero wavelet coefficients are $c_0 = c_1 = 1$), the DWT generates a new 64-term series from which the original signal can be reconstructed. Successive levels of reconstruction are shown in Fig. 17.8, including the most exact reconstruction (at the bottom right-hand side of the figure). It is obvious from 17.8 that, in between sampling points, reconstruction using a two-term wavelet gives an inexact representation of the underlying continuous signal.

Strang [115, 116] has studied the conditions that the wavelet coefficients must satisfy in order to achieve a more faithful representation of the signal being analysed. He has shown that this depends on the form of the Fourier transform of the scaling function. In order for a wavelet expansion to represent accurately a function $f(x)$ that is described by a polynomial expansion with terms like 1, x, x^2, x^3, ..., x^n, Strang has shown that the Fourier transform of its scaling function must be periodically zero in a particular way. This condition can be

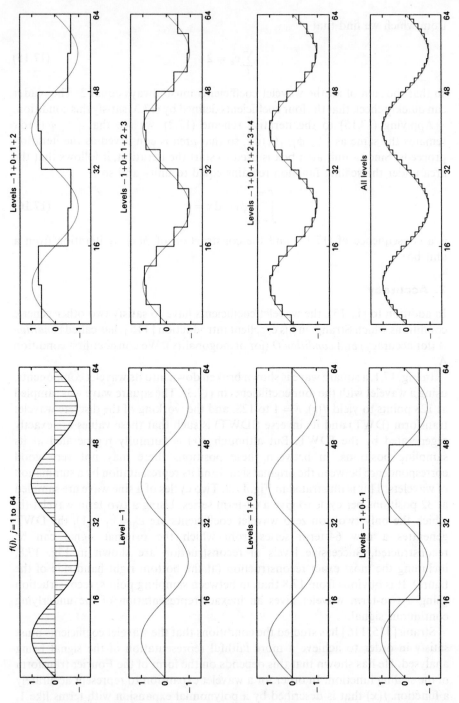

Fig. 17.8 Reconstruction of a sine wave from its D2 wavelet components obtained after sampling at 32 points per cycle

assured if the wavelet coefficients (which are also the scaling function's coefficients) are chosen appropriately.

The Fourier transform of the scaling function is

$$P(\xi) = \frac{1}{2\pi} \int_{-\infty}^{\infty} \phi(x) e^{-i\xi x} dx. \tag{17.17}$$

On account of the definition (17.11)

$$P(\xi) = \frac{1}{2\pi} \sum_k c_k \int_{-\infty}^{\infty} \phi(2x - k) e^{-i\xi x} dx \tag{17.18}$$

and, after changing the variable of integration to $y = (2x - k)$, this becomes

$$P(\xi) = \frac{1}{2} \sum_k c_k e^{-i(\xi/2)k} \frac{1}{2\pi} \int_{-\infty}^{\infty} \phi(y) e^{-i(\xi/2)y} dy$$

$$= \frac{1}{2} \sum_k c_k e^{-i(\xi/2)k} P(\xi/2) \tag{17.19}$$

or, putting

$$p(\xi/2) = \frac{1}{2} \sum_k c_k e^{-i(\xi/2)k}, \tag{17.20}$$

then

$$P(\xi) = p(\xi/2)P(\xi/2). \tag{17.21}$$

Substituting $\xi/2$ for ξ in this expression gives

$$P(\xi/2) = p(\xi/4)P(\xi/4) \tag{17.22}$$

and so we see that

$$P(\xi) = p(\xi/2)p(\xi/4)P(\xi/4) \tag{17.23}$$

and so on until the factor on the right of the r.h.s. approaches $P(0)$ which, from (17.17), is

$$P(0) = \frac{1}{2\pi} \int_{-\infty}^{\infty} \phi(x) dx = \frac{1}{2\pi} \tag{17.24}$$

since the area under the scaling function is unity according to (17.16). We conclude that the Fourier transform $P(\xi)$ of the scaling function $\phi(x)$ is given by the infinite product

$$P(\xi) = \frac{1}{2\pi} \prod_{j=1}^{\infty} p(\xi/2^j) \tag{17.25}$$

where, from (17.20),

$$p(\xi/2^j) = \frac{1}{2} \sum_k c_k e^{-i(\xi/2^j)k}. \tag{17.26}$$

Strang's accuracy condition (Strang [115] and Strang and Fix [116]) is that this Fourier transform $P(\xi)$ must have zeros of the highest possible order when $\xi = 2\pi$, 4π, 6π and so on (see problem 17.4).

Consider $P(\xi = 2\pi)$. From (17.25) we have

$$P(2\pi) = \frac{1}{2\pi} p(\pi) p(\pi/2) p(\pi/4) p(\pi/8) \ldots \tag{17.27}$$

The first of these factors will be zero to order n if

$$\frac{d^m p(\xi)}{d\xi^m} = 0 \quad \text{when } \xi = \pi \text{ for } m = 0, 1, 2, \ldots, n-1 \tag{17.28}$$

which, from (17.26), will be the case if

$$\sum_k c_k (-ik)^m e^{-i\pi k} = 0 \quad \text{for } m = 0, 1, 2, \ldots, n-1. \tag{17.29}$$

Since $e^{-i\pi k} = (-1)^k$, it follows that $p(\pi)$ will be zero to order n if

$$\sum_k (-1)^k k^m c_k = 0 \quad \text{for } m = 0, 1, 2, \ldots, n-1. \tag{17.30}$$

If this condition is satisfied, one of the factors of $P(2\pi)$ is zero to order n, and therefore $P(2\pi)$ must be zero to the same order (or higher if other factors are also zero).

Now consider $P(\xi = 4\pi)$. From (17.25) this factorizes into

$$P(4\pi) = \frac{1}{2\pi} p(2\pi) p(\pi) p(\pi/2) p(\pi/4) \ldots \tag{17.31}$$

Since $p(\pi)$ is a factor, it follows that $P(4\pi)$ is zero to at least order n. For $P(6\pi)$, the factors are

$$P(6\pi) = \frac{1}{2\pi} p(3\pi) p(3\pi/2) p(3\pi/4) \ldots \tag{17.32}$$

and, because of periodicity $p(3\pi) = p(2\pi + \pi) = p(\pi)$, so that $p(\pi)$ is again a factor. The same conclusion holds for $P(8\pi)$, $P(10\pi)$, etc. and the upshot is that (17.30) is a sufficient condition for the Fourier transform $P(\xi)$ to be zero to order n at $\xi = 2\pi$ and at all multiples of 2π. Therefore (17.30) is a recipe for choosing the wavelet coefficients in such a way that the best possible accuracy is achieved.

An alternative method of arriving at (17.30) is given by the author in a paper [113] which considers the properties of the discrete wavelet transform. This alternative analysis does not make use of the Fourier transform of the scaling function. Instead it depends directly on the conditions which must be satisfied if the DWT is to have certain accuracy properties. However we cannot follow that route until we come to the DWT later in this chapter. Interested readers are referred to the paper [113].

Writing out the recipe (17.30) in detail, the wavelet coefficients should satisfy as many as possible of the following equations:

$$m = 0: \qquad c_0 - c_1 + c_2 - c_3 + c_4 - \cdots = 0 \qquad (17.33)$$

$$m = 1: \qquad -c_1 + 2c_2 - 3c_3 + 4c_4 - \cdots = 0 \qquad (17.34)$$

$$m = 2: \qquad -c_1 + 4c_2 - 9c_3 + 16c_4 - \cdots = 0 \qquad (17.35)$$

$$m = 3: \quad -c_1 + 8c_2 - 27c_3 + 64c_4 - \cdots = 0 \qquad (17.36)$$

and so on for larger values of m. The number of equations that can be satisfied depends on how many coefficients there are and on the other conditions that have to be satisfied: the conservation of area condition (17.15) and the orthogonality conditions to which we come shortly. The combination of accuracy with stability in the numerical algorithms appears to be best satisfied by choosing wavelet coefficients which satisfy (17.30) for as many values of m as possible (Strang [115]).

3. Conditions for M_r to be an orthogonal matrix

We have seen that the development of the scaling function $\phi(x)$ by iteration from a unit box depends on the properties of the matrix M_r whose columns carry submatrices of the wavelet coefficients $[c_0 \; c_1 \; c_2 \; c_3 \ldots]^t$, each submatrix being two places below that to its left. Consider running this iterative process backwards. Suppose that there are only two steps so that (17.4) applies and we have

$$\phi_2 = M_2 M_1 \,[1]. \qquad (17.37)$$

Premultiplying by αM_2^t and αM_1^t where M_2^t, M_1^t are the transposes of M_2, M_1 and α is at present an undetermined scalar factor,

$$\alpha M_1^t \alpha M_2^t \phi_2 = \alpha M_1^t \alpha M_2^t M_2 M_1 \,[1] \qquad (17.38)$$

and if the M's are orthogonal so that

$$\alpha M_2^t M_2 = I; \qquad \alpha M_1^t M_1 = I \qquad (17.39)$$

then

$$\alpha^2 M_1^t M_2^t \phi_2 = [1] \qquad (17.40)$$

and the iteration can be reversed.

What are the conditions for (17.39) to apply? In the case of M_1 we have

$$\alpha M_1^t M_1 = \alpha [c_0 \; c_1 \; c_2 \; c_3 \ldots][c_0 \; c_1 \; c_2 \; c_3 \ldots]^t = 1 \qquad (17.41)$$

which leads to the condition that

$$\sum_k c_k^2 = 1/\alpha. \qquad (17.42)$$

For M_2, taking as an example the case of four wavelet coefficients,

$$\alpha M_2^t M_2 = \alpha \begin{bmatrix} c_0 & c_1 & c_2 & c_3 & & & & \\ & c_0 & c_1 & c_2 & c_3 & & & \\ & & c_0 & c_1 & c_2 & c_3 & & \\ & & & c_0 & c_1 & c_2 & c_3 \end{bmatrix} \begin{bmatrix} c_0 & \\ c_1 & \\ c_2 & c_0 \\ c_3 & c_1 \\ & c_2 & c_0 \\ & c_3 & c_1 \\ & & c_2 & c_0 \\ & & c_3 & c_1 \\ & & & c_2 \\ & & & c_3 \end{bmatrix}.$$

$$(17.43)$$

Multiplying out the r.h.s. gives

$$\alpha \begin{bmatrix} c_0^2 + c_1^2 + c_2^2 + c_3^2 & c_2 c_0 + c_3 c_1 & 0 & 0 \\ c_0 c_2 + c_1 c_3 & c_0^2 + c_1^2 + c_2^2 + c_3^2 & c_2 c_0 + c_3 c_1 & 0 \\ 0 & c_0 c_2 + c_1 c_3 & c_0^2 + c_1^2 + c_2^3 + c_3^2 & c_2 c_0 + c_3 c_1 \\ 0 & 0 & c_0 c_2 + c_1 c_3 & c_0^2 + c_1^2 + c_2^2 + c_3^2 \end{bmatrix}$$

$$(17.44)$$

which leads to the extra condition that

$$c_0 c_2 + c_1 c_3 = 0. \qquad (17.45)$$

In the general case, each row of M_r^t has to premultiply every column of M_r and produce a zero result except for the diagonal terms. In other words

$$[[2i \text{ zeros}][c_0\ c_1\ c_2\ \ldots][2j \text{ zeros}]] \begin{bmatrix} \begin{bmatrix} 2(i+m) \\ \text{zeros} \end{bmatrix} \\ \\ \begin{bmatrix} c_0 \\ c_1 \\ c_2 \\ \vdots \end{bmatrix} \\ \\ \begin{bmatrix} 2(j-m) \\ \text{zeros} \end{bmatrix} \end{bmatrix} = 0 \text{ unless } m = 0$$

$$(17.46)$$

and this gives the condition that

$$\sum_k c_k c_{k+2m} = 0 \quad \text{for all } m \text{ except } m = 0. \tag{17.47}$$

Suppose that there are six coefficients, c_0 to c_5; then the orthogonality conditions are

$$c_0 c_2 + c_1 c_3 + c_2 c_4 + c_3 c_5 = 0 \quad \text{for } m = 1 \tag{17.48}$$

$$c_0 c_4 + c_1 c_5 = 0 \quad \text{for } m = 2. \tag{17.49}$$

For $m = 3$, there are no non-zero products since by definition c_0 to c_5 are the only non-zero coefficients.

4. Consequences of these conditions

If (17.47) is true, then the scaling function $\phi(x)$ is orthogonal to itself in any translated position $\phi(x - m)$ so that

$$\int_{-\infty}^{\infty} \phi(x)\phi(x - m)\,dx = 0 \quad \text{for } m \neq 0. \tag{17.50}$$

The proof of this result is rather subtle. Think of how $\phi(x)$ can be constructed by iteration from a box function. The box is orthogonal to its own translates because each box only occupies a unit interval so a translated box never overlaps the original box (remember that translation means translation by one or more unit intervals). Call the box function $\phi_0(x)$. After the first iteration it becomes $\phi_1(x)$ where, according to (17.2),

$$\phi_1(x) = \sum_k c_k \phi_0(2x - k). \tag{17.51}$$

Then

$$\int_{-\infty}^{\infty} \phi_1(x)\phi_1(x - m)\,dx = \int_{-\infty}^{\infty} \sum_k c_k \phi_0(2x - k) \sum_j c_j \phi_0(2x - 2m - j)\,dx$$

$$= \sum_k \sum_j c_k c_j \int_{-\infty}^{\infty} \phi_0(2x - k)\phi_0(2x - 2m - j)\,dx. \tag{17.52}$$

Because $\phi_0(x)$ is orthogonal with respect to unit translations, the integral is zero unless $k = 2m + j$ which gives $j = k - 2m$ and so

$$\int_{-\infty}^{\infty} \phi_1(x)\phi_1(x - m)\,dx = \sum_k c_k c_{k-2m} \int_{-\infty}^{\infty} \phi_0^2(2x)\,dx \tag{17.53}$$

which is zero for all m except $m = 0$ on account of (17.47). Therefore the function produced as a result of one iteration of an orthogonal function is itself orthogonal. This argument can be repeated for the second and later iterations and leads to the conclusion that the scaling function $\phi(x)$ must itself be orthogonal, thereby proving (17.50).

A consequence of (17.50) is that the scaling coefficient α in (17.42) must be $1/\sqrt{2}$. This can be seen by substituting for $\phi(x)$ from (17.11) to get

$$\int_{-\infty}^{\infty} \phi^2(x)\,dx = \sum_k \sum_j c_k c_j \int_{-\infty}^{\infty} \phi(2x - k)\phi(2x - j)\,dx \qquad (17.54)$$

which is zero on account of (17.50) unless $j = k$, giving

$$\int_{-\infty}^{\infty} \phi^2(x)\,dx = \sum_k c_k^2 \int_{-\infty}^{\infty} \phi^2(2x - k)\,dx$$

$$= \frac{1}{2} \sum_k c_k^2 \int_{-\infty}^{\infty} \phi^2(y)\,dy \qquad (17.55)$$

from which it follows that

$$\sum_k c_k^2 = 2. \qquad (17.56)$$

We shall now show that, whatever its coefficients, the wavelet function $W(x)$ is automatically orthogonal to its scaling function $\phi(x)$ and translates of the scaling function $\phi(x - m)$ so that

$$\int_{-\infty}^{\infty} W(x)\phi(x - m)\,dx = 0 \quad \text{for all } m. \qquad (17.57)$$

Substituting for $W(x)$ from (17.12) and $\phi(x - m)$ from (17.11), we have

$$\int_{-\infty}^{\infty} W(x)\phi(x - m)\,dx$$

$$= \sum_k (-1)^k c_k \sum_j c_j \int_{-\infty}^{\infty} \phi(2x + k - N + 1)\phi(2x - 2m - j)\,dx \qquad (17.58)$$

and using (17.50) this integral is zero unless $2m + j = N - 1 - k$, so that $j = N - 1 - k - 2m$ and

$$\int_{-\infty}^{\infty} W(x)\phi(x - m)\,dx = \sum_k (-1)^k c_k c_{N-1-k-2m} \int_{-\infty}^{\infty} \phi^2(2x)\,dx. \qquad (17.59)$$

Suppose that $N = 4$ and, as for (17.5), there are four coefficients c_0, c_1, c_2, c_3. Then

$$\sum_k (-1)^k c_k c_{3-k-2m} = c_0 c_{3-2m} - c_1 c_{2-2m} + c_2 c_{1-2m} - c_3 c_{-2m}. \qquad (17.60)$$

When $m = 0$, we have

$$\sum_k (-1)^k c_k c_{3-k} = c_0 c_3 - c_1 c_2 + c_2 c_1 - c_3 c_0 = 0; \qquad (17.61)$$

when $m = 1$,

$$\sum_k (-1)^k c_k c_{1-k} = c_0 c_1 - c_1 c_0 = 0; \qquad (17.62)$$

and, for $m > 1$, the coefficients c_{3-k-2m} are all zero, so that $\sum\limits_k (-1)^k c_k c_{3-k-2m}$
is identically zero for all m. This same symmetry holds whatever the value of N because the product

$$(-1)^{k_1} c_{k_1} c_{N-1-k_1-2m} = -(-1)^{k_2} c_{k_2} c_{N-1-k_2-2m} \qquad (17.63)$$

when $k_2 = N - 1 - k_1 - 2m$, provided that N is an even number (so there is an even number of coefficients). Then if k_1 is even, k_2 will be odd, and vice versa.

The conclusion is that (17.57) is always true whatever the wavelet coefficients, provided only that there is an even number of them; then N is even and matching pairs of terms always cancel. We can also see that it follows from (17.57) that $\phi(x - m)$ is orthogonal to $W(2x)$ for all m. This is because, from (17.11), $\phi(x - m)$ is a summation of terms $\phi(2x - 2m - k)$ and, because of (17.57), these are each orthogonal to $W(2x)$ so that it follows that their summation is orthogonal to $W(2x)$. Therefore

$$\int_{-\infty}^{\infty} W(2x)\phi(x - m)\,dx = 0 \quad \text{for all } m. \qquad (17.64)$$

By the same approach, we can show that $W(2x - m)$ is orthogonal to $W(x)$ for all m. This is because, according to (17.12), $W(x)$ is a summation of scaling functions $\phi(2x + k - N + 1)$ and because of (17.57) these are all orthogonal to $W(2x - m)$. Therefore $W(x)$ must be orthogonal to $W(2x - m)$ so that

$$\int_{-\infty}^{\infty} W(x)W(2x - m)\,dx = 0 \quad \text{for all } m \text{ except } m = 0. \qquad (17.65)$$

Lastly, consider the conditions for $W(x)$ to be orthogonal to its own translates $W(x - m)$ in order that

$$\int_{-\infty}^{\infty} W(x)W(x - m)\,dx = 0 \quad \text{for all } m \text{ except } m = 0. \qquad (17.66)$$

From (17.12), we have

$$\int_{-\infty}^{\infty} W(x)W(x - m)\,dx =$$

$$\sum_k \sum_j (-1)^k (-1)^j c_k c_j \int_{-\infty}^{\infty} \phi(2x + k - N + 1)\phi(2x + j - N + 1 - 2m)\,dx.$$

$$(17.67)$$

The integral on the r.h.s. is zero by (17.50) unless $k = j - 2m$, so that

$$\int_{-\infty}^{\infty} W(x)W(x - m)\,dx = \sum_k (-1)^{2(k+m)} c_k c_{k+2m} \int_{-\infty}^{\infty} \phi^2(2x)\,dx \qquad (17.68)$$

and this will always be zero provided that

$$\sum_k c_k c_{k+2m} = 0 \qquad (17.69)$$

which is the same condition as (17.47) and is true for all m except $m = 0$. Therefore the one set of conditions (17.47) ensures that wavelets with these coefficients are orthogonal (i) to their own translates (with or without dilation), (ii) to their underlying scaling function and its translates.

5. Summary of orthogonal properties

Provided that there is an even number of wavelet coefficients and that these satisfy the conditions required by

$$\sum_k c_k c_{k+2m} = 0 \qquad m \neq 0 \qquad (17.47)$$

then

$$\int_{-\infty}^{\infty} \phi(x)\phi(x-m)\,\mathrm{d}x = 0 \qquad m \neq 0 \qquad (17.50)$$

$$\int_{-\infty}^{\infty} W(x)\phi(x-m)\,\mathrm{d}x = 0 \qquad (17.57)$$

$$\int_{-\infty}^{\infty} W(2x)\phi(x-m)\,\mathrm{d}x = 0 \qquad (17.64)$$

$$\int_{-\infty}^{\infty} W(x)W(2x-m)\,\mathrm{d}x = 0 \qquad (17.65)$$

$$\int_{-\infty}^{\infty} W(x)W(x-m)\,\mathrm{d}x = 0 \qquad m \neq 0. \qquad (17.66)$$

Equations (17.50) and (17.66) depend on (17.47); (17.57), (17.64) and (17.65) are identically zero provided only that the number of wavelet coefficients N is even. Where there are terms $W(2x)$ in (17.64) and $W(2x - m)$ in (17.65), these may be replaced by $W(2^n x)$ and $W(2^n x - m)$, $n \geqslant 0$, to give

$$\int_{-\infty}^{\infty} W(2^n x)\phi(x-m)\,\mathrm{d}x = 0 \qquad n \geqslant 0 \qquad (17.70)$$

and

$$\int_{-\infty}^{\infty} W(x)W(2^n x - m)\,\mathrm{d}x = 0 \qquad n \geqslant 0, \text{ subject to } m \neq 0 \text{ when } n = 0. \qquad (17.71)$$

Their proofs are the same as those for (17.64) and (17.65) except that $\phi(x - m)$ in (17.70) and $W(x)$ in (17.71) have to be expanded n times by repeated use of (17.11) and (17.12) before applying (17.57).

It is a remarkable result that one simple generic condition (17.47) on the wavelet coefficients allows the orthogonality of the wavelet system to be demonstrated without ever having an explicit expression for the wavelet or its scaling function.

Lastly, in this section, there is a final detail in respect to equations (17.50) and (17.66), which is their values when $m = 0$. We saw earlier that because $\sum_k c_k = 2$, area is conserved during iteration and therefore the area under the scaling function remains that of the unit box, namely 1 as specified by (17.16). Because of the additional condition (17.56) that $\sum_k c_k^2 = 2$, a similar result applies for the scaling function squared. From (17.53) with $m = 0$, we see that

$$\int_{-\infty}^{\infty} \phi_1^2(x)\,dx = \int_{-\infty}^{\infty} \phi_0^2(x)\,dx$$

which for a unit box as the starting function is 1. The same reasoning applies for each continuing iteration and leads to the general result that

$$\int_{-\infty}^{\infty} \phi^2(x)\,dx = 1. \tag{17.72}$$

Also, since from (17.68) with $m = 0$,

$$\int_{-\infty}^{\infty} W^2(x)\,dx = \int_{-\infty}^{\infty} \phi^2(x)\,dx,$$

it follows immediately that

$$\int_{-\infty}^{\infty} W^2(x)\,dx = 1. \tag{17.73}$$

So a particular feature of the scaling function and its corresponding wavelet function (assuming that these are orthogonal) is that they remain normalized so that (17.72) and (17.73) are always true.

The area under the wavelet function is always zero on account of the accuracy condition on the wavelet coefficients (17.30) which, for $m = 0$, gives

$$\sum_k (-1)^k c_k = 0. \tag{17.74}$$

Integrating both sides of (17.12) we get

$$\int_{-\infty}^{\infty} W(x)\,dx = \sum_k (-1)^k c_k \frac{1}{2} \int_{-\infty}^{\infty} \phi(y)\,dy \tag{17.75}$$

which is zero on account of (17.74) and so we always must have

$$\int_{-\infty}^{\infty} W(x)\,dx = 0. \tag{17.76}$$

6. Collected wavelet conditions

The even number N of wavelet coefficients c_k, $k = 0$ to $N - 1$, which define the scaling function $\phi(x)$ by

$$\phi(x) = \sum_{k=0}^{N-1} c_k \phi(2x - k) \tag{17.11}$$

and the corresponding wavelet by

$$W(x) = \sum_{k=0}^{N-1} (-1)^k c_k \phi(2x + k - N + 1) \tag{17.12}$$

must satisfy the following conditions:

(i)
$$\sum_{k=0}^{N-1} c_k = 2 \tag{17.15}$$

so that the scaling function is unique and retains unit area during iteration;

(ii)
$$\sum_{k=0}^{N-1} (-1)^k k^m c_k = 0 \tag{17.30}$$

for integer $m = 0, 1, 2, \ldots, N/2 - 1$ (as high as the available number of coefficients will allow), in order to achieve accuracy;

(iii)
$$\sum_{k=0}^{N-1} c_k c_{k+2m} = 0 \qquad m \neq 0 \tag{17.47}$$

for $m = 1, 2, \ldots, N/2 - 1$, in order to generate an orthogonal wavelet system, with the additional condition that

$$\sum_{k=0}^{N-1} c_k^2 = 2 \tag{17.56}$$

which arises as a consequence of the scaling function being orthogonal, (17.50), and gives constant mean square during iteration.

For N coefficients, the number of equations to be satisfied are as follows:

from	(i)	1
	(ii)	$N/2$
	(iii)	$N/2$
Total		$N + 1$

so that there is one more equation than there are coefficients.

The reason that there are $N + 1$ equations for N coefficients is because the first of the accuracy conditions ($m = 0$ in (17.30)) is redundant. It reproduces a condition derivable from the constant area and orthogonality conditions (17.15), (17.47) and (17.56). This may be seen as follows. Putting $m = 0$ in

(17.30) gives (17.33) which can be written as

$$\sum_{k\,\text{even}} c_k - \sum_{k\,\text{odd}} c_k = 0. \tag{17.77}$$

Combining this with (17.15) gives

$$\sum_{k\,\text{even}} c_k = \sum_{k\,\text{odd}} c_k = 1 \tag{17.78}$$

so that

$$\left(\sum_{k\,\text{even}} c_k\right)^2 + \left(\sum_{k\,\text{odd}} c_k\right)^2 = 2. \tag{17.79}$$

After multiplying out

$$\sum_{k=0}^{N-1} c_k^2 + 2 \sum_{m=1}^{N/2-1} \sum_{k=0}^{N-1} c_k c_{k+2m} = 2 \tag{17.80}$$

which, on account of (17.56), means that

$$\sum_m \sum_k c_k c_{k+2m} = 0, \qquad m \neq 0. \tag{17.81}$$

The same result is obtained by writing out each of the orthogonality conditions (17.47) and adding them together. Hence the result of putting $m = 0$ in (17.30) does not produce a new equation but only reproduces a condition derivable by combining (17.15), (17.47) and (17.56).

The *Haar wavelet* (named after A. Haar [108]) has only two wavelet coefficients, $N = 2$, and we have the following:

from (i) $\qquad\qquad\qquad c_0 + c_1 = 2$

from (ii) $\qquad\qquad\qquad c_0 - c_1 = 0$ $\qquad\qquad\qquad$ (17.82)

from (iii) $\qquad\qquad\qquad c_0^2 + c_1^2 = 2$

whose solution is just

$$c_0 = c_1 = 1. \tag{17.83}$$

For four wavelet coefficients, $N = 4$ and we have the following:

from (i) $\qquad\qquad c_0 + c_1 + c_2 + c_3 = 2$ $\qquad\qquad$ (17.84)

from (ii) $\qquad\qquad c_0 - c_1 + c_2 - c_3 = 0$

$$-c_1 + 2c_2 - 3c_3 = 0 \tag{17.85}$$

from (iii) $\qquad\qquad c_0 c_2 + c_1 c_3 = 0$

$$c_0^2 + c_1^2 + c_2^2 + c_3^2 = 2 \tag{17.86}$$

whose solution is

$$c_0 = (1 + \sqrt{3})/4 \qquad c_1 = (3 + \sqrt{3})/4$$
$$c_2 = (3 - \sqrt{3})/4 \qquad c_3 = (1 - \sqrt{3})/4. \tag{17.87}$$

Wavelets with these four coefficients are often called *D4 wavelets* after **Ingrid Daubechies** [103, 104, 105] who first discussed their properties.

For six wavelet coefficients, $N = 6$ and

from (i)
$$c_0 + c_1 + c_2 + c_3 + c_4 + c_5 = 2 \qquad (17.88)$$

from (ii)
$$c_0 - c_1 + c_2 - c_3 + c_4 - c_5 = 0$$
$$-c_1 + 2c_2 - 3c_3 + 4c_4 - 5c_5 = 0 \qquad (17.89)$$
$$-c_1 + 4c_2 - 9c_3 + 16c_4 - 25c_5 = 0$$

from (iii)
$$c_0c_2 + c_1c_3 + c_2c_4 + c_3c_5 = 0$$
$$c_0c_4 + c_1c_5 = 0 \qquad (17.90)$$
$$c_0^2 + c_1^2 + c_2^2 + c_3^2 + c_4^2 + c_5^2 = 2.$$

Their solution is

$$c_0 = (1 + \sqrt{10} + \sqrt{5 + 2\sqrt{10}})/16$$
$$c_1 = (5 + \sqrt{10} + 3\sqrt{5 + 2\sqrt{10}})/16$$
$$c_2 = (5 - \sqrt{10} + \sqrt{5 + 2\sqrt{10}})/8 \qquad (17.91)$$
$$c_3 = (5 - \sqrt{10} - \sqrt{5 + 2\sqrt{10}})/8$$
$$c_4 = (5 + \sqrt{10} - 3\sqrt{5 + 2\sqrt{10}})/16$$
$$c_5 = (1 + \sqrt{10} - \sqrt{5 + 2\sqrt{10}})/16.$$

These values define the *D6* wavelet.

For more than six wavelet coefficients, the solutions have to be computed numerically. Results of such calculations are given in the literature. In Appendix 7, the function file *dcoeffs(N)* returns a matrix $[c(1)\, c(2) \ldots c(N)]$ of the wavelet coefficients for $N = 2, 4, 6, \ldots, 20$ using the computed results given by Daubechies [104] who first discovered the properties of these sets of coefficients.

Circular wavelet transforms

The goal of the wavelet transform is to decompose any arbitrary signal $f(x)$ into an infinite summation of wavelets at different scales according to the expansion

$$f(x) = \sum_{j=-\infty}^{\infty} \sum_{k=-\infty}^{\infty} c_{j,k}W(2^jx - k). \qquad (17.92)$$

The interpretation of this expansion can be seen by reference to Fig. 17.9. Here the Haar wavelet with $N = 2$ coefficients is used for simplicity and all the wavelet amplitudes $c_{j,k}$ are taken to be unity. The diagram shows levels $-3, -2, -1, 0, 1, 2$ corresponding to $j = -3, -2, -1, 0, 1, 2$. At level 0 there is $2^0 = 1$ Haar wavelet in each unit interval; at level 1 there are two wavelets per unit interval, at level 2 there are four wavelets per unit interval and so on upwards. At level -1 there is one wavelet every two intervals, at level -2 there is one wavelet every four intervals and so on downwards. From Fig. 17.9 we can see that, for $j \leqslant -1$, the contribution of each wavelet level is constant over every

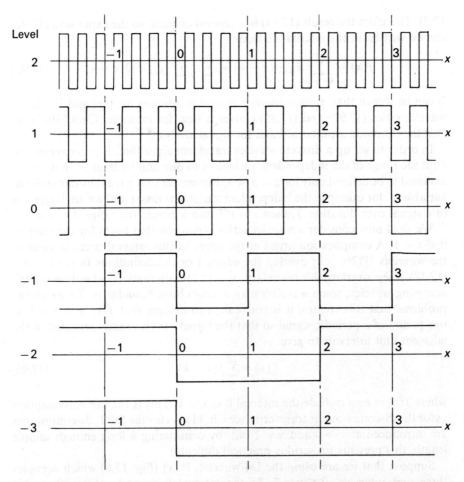

Fig. 17.9 Haar wavelets with unit amplitude plotted for levels -3, -2, -1, 0, 1, 2; for these levels there are $\frac{1}{8}$, $\frac{1}{4}$, $\frac{1}{2}$, 1, 2, 4 wavelets per unit interval of x

unit interval. It follows that the sum of the contributions from all these levels is also constant. In problem 17.1 we saw that the scaling function $\phi(x)$ for $N = 2$ coefficients, which is the Haar wavelet case, is itself a constant so that

$$\phi(x) = 1 \qquad 0 \leqslant x < 1. \tag{17.93}$$

Hence, in this example

$$\sum_{j=-\infty}^{-1} \sum_{k=-\infty}^{\infty} c_{j,k} W(2^j x - k) = \sum_{k=-\infty}^{\infty} c_{\phi,k} \phi(x - k) \tag{17.94}$$

where the $c_{\phi,k}$, $k = -\infty$ to ∞ is a new set of coefficients. Because of the way in which wavelets are defined by (17.11) and (17.12), when j is negative, $W(2^j x - k)$ can always be expressed as a sum of terms like $\phi(x - k)$ (see problem

17.5). Therefore the result (17.94) is a general one and so the expansion (17.92) can be written in the alternative form:

$$f(x) = \sum_{k=-\infty}^{\infty} c_{\phi,k}\phi(x-k) + \sum_{j=0}^{\infty} \sum_{k=-\infty}^{\infty} c_{j,k}W(2^{j}x-k). \qquad (17.95)$$

It can be shown that, subject to very general conditions on $f(x)$ and $W(x)$, the wavelet series (17.92) and (17.95) converge (see, for example, Chui [101]), so that they provide a practical basis for signal analysis.

In order to set up a discrete wavelet transform algorithm, it is convenient to limit the range of the independent variable x to one unit interval so that $f(x)$ is assumed to be defined only for $0 \leqslant x < 1$. Remember that x is a non-dimensional variable; if, for example, the independent variable is time t and we are interested in a signal over duration T, then $x = t/T$ and x covers the range $0 \leqslant x < 1$.

We shall now consider a wavelet series expansion that holds for the interval $0 \leqslant x < 1$. A complication arises at the edges of this interval because some of the wavelets $W(2^{j}x - k)$ overlap the edges. For the definitions in (17.11) and (17.12), they overlap the boundary $x = 1$; for alternative definitions of the analysing wavelet, some wavelets may overlap both boundaries. To avoid the problems that this creates, it is convenient to assume that $f(x)$, $0 \leqslant x < 1$, is one period of a periodic signal so that the signal $f(x)$ is exactly repeated in the adjacent unit intervals to give

$$F(x) = \sum_{k} f(x-k) \qquad (17.96)$$

where $f(x)$ is zero outside the interval $0 \leqslant x < 1$. This is the same assumption as for the discrete Fourier transform (see Ch. 11). As for the DFT, discontinuities are introduced at $x = 0$ and $x = 1$ but, by considering a long enough sample length, this presents no serious practical difficulty.

Suppose that we are using the D4 wavelet, $W(x)$ (Fig. 17.6) which occupies three unit intervals, $0 \leqslant x < 3$. In the interval $0 \leqslant x < 1$, $f(x)$ will receive contributions from the first third of $W(x)$, the middle third of $W(x + 1)$, and the last third of $W(x + 2)$. This is the same as if $W(x)$ is "wrapped around" the unit interval as described in Chapter 11. Therefore, when any wavelet which starts in the interval $0 \leqslant x < 1$ runs off the end at $x = 1$, it may be assumed to be wrapped around the interval (if necessary several times if there are many coefficients so that the wavelet extends over many intervals). With that assumption, from (17.95) the wavelet expansion of $f(x)$ in $0 \leqslant x < 1$ can be written as

$$f(x) = a_0\phi(x) + a_1 W(x) + \begin{bmatrix} a_2 & a_3 \end{bmatrix} \begin{bmatrix} W(2x) \\ W(2x-1) \end{bmatrix} + \begin{bmatrix} a_4 & a_5 & a_6 & a_7 \end{bmatrix} \begin{bmatrix} W(4x) \\ W(4x-1) \\ W(4x-2) \\ W(4x-3) \end{bmatrix} +$$

$$+ \ldots + a_{2^{j}+k}W(2^{j}x-k) + \ldots . \qquad (17.97)$$

The coefficients $a_1, a_2, a_3, a_4, \ldots$ give the amplitudes of each of the contributing wavelets (after wrapping) to one cycle of the periodic function (17.96) in the interval $0 \leqslant x < 1$. We shall see later that, because of wrapping, the scaling function $\phi(x)$ always becomes a constant. The second term $a_1 W(x)$ is a wavelet of scale zero; the third and fourth terms $a_2 W(2x)$ and $a_3 W(2x - 1)$ are wavelets of scale one, the second being translated $\Delta x = \frac{1}{2}$ with respect to the first; the next four terms are wavelets of scale 2, and so on for wavelets of increasingly higher scale. The higher the scale, the finer the detail and the more coefficients there are, so that at scale j there are 2^j wavelets each spaced $\Delta x = 2^{-j}$ apart along the x-axis.

Because of the orthogonality conditions (17.50) to (17.71), the general coefficient $a_{2^j + k}$ can be found by multiplying (17.97) by $W(2^j x - k)$ and integrating to get

$$\int f(x) W(2^j x - k)\, dx = a_{2^j + k} \int W^2(2^j x - k)\, dx \qquad (17.98)$$

and hence

$$a_{2^j + k} = \frac{\int f(x) W(2^j x - k)\, dx}{\frac{1}{2^j} \int W^2(x)\, dx} = 2^j \int f(x) W(2^j x - k)\, dx \qquad (17.99)$$

on account of (17.73).

By similar reasoning it follows that

$$a_0 = \int f(x) \phi(x)\, dx \qquad (17.100)$$

after using (17.72).

The limits of integration in the orthogonality conditions (17.50) to (17.71) are given as $-\infty$ to $+\infty$, but the integrand in each case is only non-zero for the finite length of the shortest wavelet or scaling function involved. If the wavelets and scaling function are not wrapped, the limits of integration in (17.99) and (17.100) may extend over several intervals. But because we are assuming that $f(x)$ is one cycle of a periodic function, which repeats itself in adjacent intervals, these contributions to the integrals from outside the interval $0 \leqslant x < 1$ are included automatically by integrating from $x = 0$ to $x = 1$ for the *wrapped* functions. Therefore the formulae (17.99) and (17.100) can be written as

$$a_{2^j + k} = 2^j \int_0^1 f(x) W(2^j x - k)\, dx \qquad (17.101)$$

and

$$a_0 = \int_0^1 f(x) \phi(x)\, dx \qquad (17.102)$$

when it is understood that in (17.97) and in (17.101) and (17.102) the scaling function $\phi(x)$ and the wavelet functions $W(2^j x - k)$ are all wrapped around the interval $0 \leqslant x < 1$ as many times as are necessary to ensure that their whole length is included in this unit interval. The orthogonality conditions (17.70) and (17.71) still apply for the wrapped functions when the limits of integration are changed to $x = 0$ to 1 (see problem 17.7). This is because they are constructed by iteration from box functions that are orthogonal to their own translates. Although the iterative steps for the wrapped case differ from those without wrapping, the same basis results are arrived at except that the integration only has to cover the interval $0 \leqslant x < 1$. It is of course still necessary for the wavelet coefficients to satisfy the orthogonality conditions required by (17.47).

Discrete wavelet transforms

The discrete wavelet transform (DWT) is an algorithm for computing (17.101) and (17.102) when $f(x)$ is sampled at equally spaced intervals over $0 \leqslant x < 1$. In its embodiment here, it is assumed that $f(x)$ is one period of a periodic signal and that the scaling and wavelet functions wrap around the interval as described above. Obviously the integrals in (17.101) and (17.102) can be computed to whatever accuracy is required after first generating $\phi(x)$ and $W(2^j x - k)$. However, a remarkable feature of the DWT algorithm is that this is not necessary. Instead, the convolutions (17.101) and (17.102) can be computed without ever finding $\phi(x)$ and $W(2^j x - k)$ explicitly. The DWT algorithm was discovered by Mallat [110] and is called *Mallat's pyramid algorithm* or sometimes *Mallat's tree algorithm*.

We shall approach the algorithm by considering first its inverse. Suppose that the DWT has been computed to generate the sequence

$$a = [a_0 \ a_1 \ a_2 \ a_3 \ a_4 \ a_5 \ a_6 \ a_7 \ \ldots \ a_{2^j + k} \ \ldots] \qquad (17.103)$$

which are the coefficients of the wavelet expansion (17.97). Suppose, for example, that we consider an expansion with the primary scaling function $\phi(x)$ and wavelets of scale 0, 1 and 2. Then a will have $(1 + 1 + 2 + 4) = 2^3$ terms. In order to include all the wavelets at any particular scale, the total number of terms in the transform must always be a power of 2. Consider the case when there are only eight terms so that

$$a = [a_0 \ a_1 \ a_2 \ a_3 \ a_4 \ a_5 \ a_6 \ a_7]. \qquad (17.104)$$

The first element a_0 is the amplitude of the scaling function term $\phi(x)$. Since $\phi(x)$ can be generated by iteration from a unit box over the integral $0 \leqslant x < 1$ (see Fig. 17.4), $a_0 \phi(x)$ can be generated by iteration starting from a box of height a_0. Suppose that we have chosen a wavelet with four coefficients. Then the first step in the iteration is, from (17.4),

$$\phi_1 = \begin{bmatrix} c_0 \\ c_1 \\ c_2 \\ c_3 \end{bmatrix} [a_0]. \qquad (17.105)$$

The initial box function occupied the interval $0 \leqslant x < 1$, but we see in Fig. 17.4 that the first iteration extends over $0 \leqslant x < 2$. If the part that lies outside the interval is wrapped round to fall back into the unit interval, we get

$$\phi_1 = \begin{bmatrix} c_0 + c_2 \\ c_1 + c_2 \end{bmatrix} [a_0]. \qquad (17.106)$$

On taking the second interative step, without including wrap-around, we have, from (17.4),

$$\phi_2 = \begin{bmatrix} c_0 & & & \\ c_1 & & & \\ c_2 & c_0 & & \\ c_3 & c_1 & & \\ & c_2 & c_0 & \\ & c_3 & c_1 & \\ & & c_2 & c_0 \\ & & c_3 & c_1 \\ & & & c_2 \\ & & & c_3 \end{bmatrix} \begin{bmatrix} c_0 \\ c_1 \\ c_2 \\ c_3 \end{bmatrix} [a_0]. \qquad (17.107)$$

From the third row of the lower diagram in Fig. 17.4, this gives a function with 10 ordinates spread over the interval $0 \leqslant x < 2.5$. These 10 terms are

$$\phi_2 = [\quad c_0^2 \qquad c_0 c_1 \qquad c_0 c_2 + c_1 c_0 \qquad c_0 c_3 + c_1^2 \ldots \qquad (17.108)$$
$$c_1 c_2 + c_2 c_0 \quad c_1 c_3 + c_2 c_1 \quad c_2^2 + c_3 c_0 \quad c_2 c_3 + c_3 c_1 \ldots$$
$$c_3 c_2 \qquad c_3^2]^t \, [a_0].$$

Allowing for wrap-around and adding terms at the same position in $0 \leqslant x < 1$, we then get only four terms which are

$$\phi_2 = \begin{bmatrix} c_0^2 + c_1 c_2 + c_2 c_0 + c_3 c_2 \\ c_0 c_1 + c_1 c_3 + c_2 c_1 + c_3^2 \\ c_0 c_2 + c_1 c_0 + c_2^2 + c_3 c_0 \\ c_0 c_3 + c_1^2 + c_2 c_3 + c_3 c_1 \end{bmatrix} [a_0].$$

The reader can check by multiplication that this is the same as

$$\phi_2 = \begin{bmatrix} c_0 & c_2 \\ c_1 & c_3 \\ c_2 & c_0 \\ c_3 & c_1 \end{bmatrix} \begin{bmatrix} c_0 + c_2 \\ c_1 + c_3 \end{bmatrix} [a_0]. \tag{17.109}$$

Notice how the left-hand matrix is formed, by taking a submatrix of order 4×2 from the left-hand matrix in (17.107) and then transposing the c_2 and c_3 elements that fell below this matrix as follows:

$$\begin{bmatrix} c_0 & \\ c_1 & \\ c_2 & c_0 \\ c_3 & c_1 \end{bmatrix} . \tag{17.110}$$

This recipe also applies for the matrix in (17.106) which is formed by transposing the same two elements

$$\begin{bmatrix} c_0 \\ c_1 \end{bmatrix} . \tag{17.111}$$

For the third iterative step, the calculation is

$$\phi_3 = \begin{bmatrix} c_0 & & & c_2 \\ c_1 & & & c_3 \\ c_2 & c_0 & & \\ c_3 & c_1 & & \\ & c_2 & c_0 & \\ & c_3 & c_1 & \\ & & c_2 & c_0 \\ & & c_3 & c_1 \end{bmatrix} \begin{bmatrix} c_0 & c_2 \\ c_1 & c_3 \\ c_2 & c_0 \\ c_3 & c_1 \end{bmatrix} \begin{bmatrix} c_0 + c_2 \\ c_1 + c_3 \end{bmatrix} [a_0] \tag{17.112}$$

and this generates the eight ordinates in the interval $0 \leqslant x < 1$ (see the fourth row of the lower diagram in Fig. 17.4) for the wrapped-around scaling function.

If we revise the definitions of the M matrices of the wavelet coefficients so that

$$M_1 = \begin{bmatrix} c_0 + c_2 \\ c_1 + c_3 \end{bmatrix} \text{ of order } 2 \times 1 \tag{17.113}$$

$$M_2 = \begin{bmatrix} c_0 & c_2 \\ c_1 & c_3 \\ c_2 & c_0 \\ c_3 & c_1 \end{bmatrix} \text{ of order } 2^2 \times 2 \tag{17.114}$$

$$M_3 = \begin{bmatrix} c_0 & & & c_2 \\ c_1 & & & c_3 \\ c_2 & c_0 & & \\ c_3 & c_1 & & \\ & c_2 & c_0 & \\ & c_3 & c_1 & \\ & & c_2 & c_0 \\ & & c_3 & c_1 \end{bmatrix} \text{ of order } 2^3 \times 2^2 \tag{17.115}$$

then the algorithm for generating the contribution of $a_0\phi(x)$ to $f(x)$ is

$$f^\phi(x) = M_3 M_2 M_1 a_0 \tag{17.116}$$

or, in diagrammatic form,

$$a_0 = f^\phi(1) \xrightarrow{M_1} f^\phi(1:2) \xrightarrow{M_2} f^\phi(1:4) \xrightarrow{M_3} f(1:8) \tag{17.117}$$

where f^ϕ (1:8) means an array of eight elements that represents the contribution of $a_0\phi(x)$ to $f(x)$ at $x = 0, \frac{1}{8}, \frac{1}{4}, \ldots, \frac{7}{8}$.

Returning to the sequence (17.104), consider the second term, a_1. This is the amplitude of the wavelet function $W(x)$ which is generated from a unit box by iteration as shown in Fig. 17.7. The matrix operations for doing this are the same as for generating the scaling function $\phi(x)$ except that the first step involves replacing

$$\begin{bmatrix} c_0 \\ c_1 \\ c_2 \\ c_3 \end{bmatrix} \text{ by } \begin{bmatrix} -c_3 \\ c_2 \\ -c_1 \\ c_0 \end{bmatrix} \tag{17.118}$$

according to (17.10). The procedure for allowing for wrap-around is exactly the same as for the scaling function except that, analogously with (17.111),

$$\begin{bmatrix} -c_3 \\ c_2 \\ -c_1 \\ c_0 \end{bmatrix} \quad \text{becomes} \quad \begin{bmatrix} -c_3 \\ c_2 \\ -c_1 \\ c_0 \end{bmatrix} = \begin{bmatrix} -c_3 - c_1 \\ c_2 + c_0 \end{bmatrix}. \tag{17.119}$$

Defining

$$G_1 = \begin{bmatrix} -c_3 - c_1 \\ c_2 + c_0 \end{bmatrix}$$

the algorithm for generating the contribution of $a_1 W(x)$ to $f(x)$ is

$$f^{(0)}(1:8) = M_3 M_2 G_1 a_1 \tag{17.120}$$

or, in diagrammatic form,

$$a_1 = f^{(0)}(1) \xrightarrow{\; G_1 \;} f^{(0)}(1:2) \xrightarrow{\; M_2 \;} f^{(0)}(1:4) \xrightarrow{\; M_3 \;} f^{(0)}(1:8). \tag{17.121}$$

The third term in (17.104), a_2, is the amplitude of $W(2x)$. Instead of extending over $0 \leqslant x < 3$ as for $W(x)$ in Fig. 17.7, this extends only over $0 \leqslant x < 3/2$ as shown in Fig. 17.10. Instead of the second iteration being reached by

$$\begin{bmatrix} c_0 & c_2 \\ c_1 & c_3 \\ c_2 & c_0 \\ c_3 & c_1 \end{bmatrix} \begin{bmatrix} -c_3 - c_1 \\ c_2 + c_0 \end{bmatrix} [a_2], \tag{17.122}$$

the first operation is omitted and we go straight to

$$\begin{bmatrix} -c_3 \\ c_2 \\ -c_1 \\ c_0 \end{bmatrix} [a_2] \tag{17.123}$$

as in Fig. 17.10. Thereafter the iteration proceeds as before to get

$$f^{(1,1)}(1:8) = M_3 \begin{bmatrix} -c_3 \\ c_2 \\ -c_1 \\ c_0 \end{bmatrix} [a_2]. \tag{17.124}$$

First eight iterations of the development of the D4 wavelet $W(2x)$

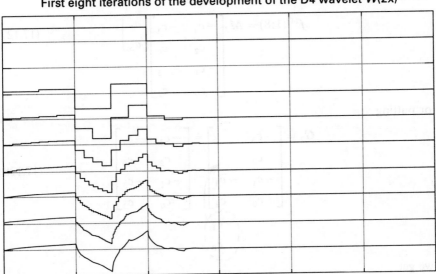

Fig. 17.10 Generation of the (unwrapped) wavelet $W(2x)$ from a unit box by omitting the first iterative step (compare with Fig. 17.7)

The fourth term in (17.104), a_3, is the amplitude of the translated wavelet $W(2x - 1)$. This is translated $x = \frac{1}{2}$ to the right in Fig. 17.10, and so it occupies the interval $0.5 \leqslant x < 2$. Allowing for wrap-around, the procedure for computing the contribution that this makes to $f(x)$ is exactly the same as for a_2 except that the elements in the first matrix are arranged in the order

$$
\begin{bmatrix}
-c_1 \\
c_0 \\
-c_3 \\
c_2
\end{bmatrix}
\tag{17.125}
$$

so that (17.124) becomes

$$
f^{(1,2)}(1:8) = M_3 \begin{bmatrix} -c_1 \\ c_0 \\ -c_3 \\ c_2 \end{bmatrix} [a_3].
\tag{17.126}
$$

Combining (17.124) and (17.126) then gives

$$f^{(1)}(1:8) = M_3 \begin{bmatrix} -c_3 & -c_1 \\ & c_2 & c_0 \\ & -c_1 & -c_3 \\ & c_0 & c_2 \end{bmatrix} \begin{bmatrix} a_2 \\ a_3 \end{bmatrix} \tag{17.127}$$

or putting

$$G_2 = \begin{bmatrix} -c_3 & \\ c_2 & \\ -c_1 & -c_3 \\ c_0 & c_2 \end{bmatrix} = \begin{bmatrix} -c_3 & -c_1 \\ c_2 & c_0 \\ -c_1 & -c_3 \\ c_0 & c_2 \end{bmatrix} \tag{17.128}$$

we get

$$f^{(1)}(1:8) = M_3 G_2 \begin{bmatrix} a_2 \\ a_3 \end{bmatrix} \tag{17.129}$$

and, in a diagram,

$$\begin{bmatrix} a_2 \\ a_3 \end{bmatrix} \xrightarrow{\ G_2\ } f^{(1)}(1:4) \xrightarrow{\ M_3\ } f^{(1)}(1:8). \tag{17.130}$$

The remaining four elements of (17.104), a_4, a_5, a_6, a_7 are the amplitudes of wavelets $W(4x)$, $W(4x - 1)$, $W(4x - 2)$, $W(4x - 3)$ which are translated $\Delta x = \frac{1}{4}$ with respect to each other. In this case the first two steps of the iteration in Fig. 17.7 are omitted and, for eight terms in the sequence, there is only one step to the calculation. There are eight points in the interval $0 \leqslant x < 1$, so that each wavelet is translated two positions with respect to its neighbour. Each wavelet has the elements $[-c_3 \ \ c_2 \ \ -c_1 \ \ c_0]^t$ and so the single stage of calculation is

$$f^{(2)}(1:8) = \begin{bmatrix} -c_3 & & & -c_1 \\ c_2 & & & c_0 \\ -c_1 & -c_3 & & \\ c_0 & c_2 & & \\ & -c_1 & -c_3 & \\ & c_0 & c_2 & \\ & & -c_1 & -c_3 \\ & & c_0 & c_2 \end{bmatrix} \begin{bmatrix} a_4 \\ a_5 \\ a_6 \\ a_7 \end{bmatrix} \tag{17.131}$$

or, diagrammatically,

$$\begin{bmatrix} a_4 \\ a_5 \\ a_6 \\ a_7 \end{bmatrix} \xrightarrow{G_3} f^{(2)}(1{:}8) \qquad (17.132)$$

Lastly, combining the four diagrams (17.117), (17.121), (17.130) and (17.132) to express the result that

$$f(1{:}8) = f^{\phi}(1{:}8) + f^{(0)}(1{:}8) + f^{(1)}(1{:}8) + f^{(2)}(1{:}8) \qquad (17.133)$$

we have the final diagram below:

$$f'(1) \xrightarrow{M_1} f'(1{:}2) \xrightarrow{M_2} f'(1{:}4) \xrightarrow{M_3} f(1{:}8) = f \qquad (17.134)$$

$$\uparrow \qquad G_1\uparrow \qquad G_2\uparrow \qquad G_3\uparrow$$

$$a = [a(1) \qquad a(2) \qquad a(3{:}4) \qquad a(5{:}8)]$$

where $a(1) = a_0$, $a(2) = a_1$, $a(3{:}4) = [a_2\ a_3]^t$ and $a(5{:}8) = [a_4\ a_5\ a_6\ a_7]^t$.

This is the inverse of Mallat's tree algorithm [110]. As drawn above, the tree has been felled because the trunk is horizontal! The roots of the tree are at the r.h.s. and the branches are vertical with many branches near the roots and a decreasing number going up the tree until the branch at the top of the tree (before it was felled) reaching out to the first coefficient of the wavelet transform $a(1)$.

This algorithm is implemented in the function file *iwavedn(a,N)* in Appendix 7. The program takes any array a of numbers of length 2^n where n is an arbitrary integer and computes the function $f(r)$, $r = 1$ to 2^n, for which a is the wavelet transform. The calculation is made for a wavelet with N coefficients (N even) using coefficients imported from the separate file *dcoeffs(N)* ($N \leqslant 20$). The reader may like to experiment with this program (see problems 17.5, 17.6 and 17.8).

Now consider how to break down an arbitrary function $f(1{:}2^n)$ into its wavelet transform $a(1{:}2^n)$. Because of the conditions imposed on the wavelet coefficients to ensure that the wavelets and their scaling function are all mutually orthogonal, we have seen that the M matrices are orthogonal matrices. The proof of this was shown by (17.46) and (17.47) and that conclusion is not altered by the reorganization of the matrices to account for wrap-around. Because of the orthogonality conditions (17.47) and (17.56), we have

$$\tfrac{1}{2}M_r^t M_r = I \qquad M_r^t G_r = 0$$

$$G_r^t M_r = 0 \qquad \tfrac{1}{2}G_r^t G_r = I \qquad (17.135)$$

as can be proved easily by substituting for a typical row of M_r^t

$$[[2i\ \text{zeros}][c_0\ c_1\ c_2\ \ldots][2j\ \text{zeros}]] \qquad (17.136)$$

and corresponding columns of M_r

$$[[2(i + m) \text{ zeros}][c_0 \ c_1 \ c_2 \ \ldots][2(j - m) \text{ zeros}]]' \qquad (17.137)$$

and the same for G_r, for example the column

$$[[2(i + m) \text{ zeros}][-c_N \quad c_{N-1} \quad -c_{N-2} \quad \ldots][2(j - m) \text{ zeros}]]' \qquad (17.138)$$

and multiplying out using (17.47) and (17.56).

If we define new symbols H and L for the transposes of G and M, so that

$$H = G^t \quad \text{and} \quad L = M^t \qquad (17.139)$$

then on account of (17.135) we have

$$\tfrac{1}{2}LM = I \qquad LG = 0$$

$$HM = 0 \qquad \tfrac{1}{2}HG = I \qquad (17.140)$$

where the dimensions of the matrices are

$$G_r = G(2^r \times 2^{r-1}) \qquad H_r = H(2^{r-1} \times 2^r)$$

$$L_r = L(2^{r-1} \times 2^r) \qquad M_r = M(2^r \times 2^{r-1}). \qquad (17.141)$$

Because of (17.140), each of the steps that make up the tree algorithm (17.134) can be simply reversed. For example (17.116) becomes

$$a_0 = \tfrac{1}{2}L_1\tfrac{1}{2}L_2\tfrac{1}{2}L_3 f^\phi. \qquad (17.142)$$

Consider replacing f^ϕ by

$$f = f^\phi + f^{(0)} + f^{(1)} + f^{(2)} \qquad (17.143)$$

and calculating

$$\tfrac{1}{2}L_1\tfrac{1}{2}L_2\tfrac{1}{2}L_3 f. \qquad (17.144)$$

Substituting for $f^{(0)}$ from (17.120), $f^{(1)}$ from (17.129) and $f^{(2)}$ from (17.132) we have

$$f = f^\phi + M_3M_2G_1[a_1] + M_3G_2\begin{bmatrix} a_2 \\ a_3 \end{bmatrix} + G_3\begin{bmatrix} a_4 \\ a_5 \\ a_6 \\ a_7 \end{bmatrix}. \qquad (17.145)$$

Then

$$\tfrac{1}{2}L_1\tfrac{1}{2}L_2\tfrac{1}{2}L_3 f = \tfrac{1}{2}L_1\tfrac{1}{2}L_2\tfrac{1}{2}L_3 f^\phi + 0 + 0 + 0 \qquad (17.146)$$

on account of (17.140). We conclude that f^ϕ in (17.142) may be replaced by f to give

$$a_0 = \tfrac{1}{2}L_1\tfrac{1}{2}L_2\tfrac{1}{2}L_3 f. \qquad (17.147)$$

Similar conclusions hold for the other steps so that, instead of (17.120) we have

$$a_1 = \tfrac{1}{2}H_1\tfrac{1}{2}L_2\tfrac{1}{2}L_3f, \tag{17.148}$$

instead of (17.129),

$$\begin{bmatrix} a_2 \\ a_3 \end{bmatrix} = \tfrac{1}{2}H_2\tfrac{1}{2}L_3f \tag{17.149}$$

and instead of (17.132),

$$\begin{bmatrix} a_4 \\ a_5 \\ a_6 \\ a_7 \end{bmatrix} = \tfrac{1}{2}H_3f. \tag{17.150}$$

Therefore the whole of the tree algorithm can be reversed so that (17.134) becomes

$$f'(1) \xleftarrow{\tfrac{1}{4}L_1} f'(1:2) \xleftarrow{\tfrac{1}{4}L_2} f'(1:4) \xleftarrow{\tfrac{1}{4}L_3} f(1:8) = f.$$
$$\downarrow \qquad \tfrac{1}{2}H_1 \downarrow \qquad \tfrac{1}{2}H_2 \downarrow \qquad \tfrac{1}{2}H_3 \downarrow \tag{17.151}$$
$$a = [a(1) \qquad a(2) \qquad a(3:4) \qquad a(5:8)]$$

This is Mallat's algorithm for the discrete wavelet transform [110] and it allows the signal $f(1:2^n)$ to be transformed to its wavelet coefficients $a(1:2^n)$ in a very efficient way. The notation L and H is used because L can be interpreted as a low-pass filter and H as a high-pass filter [113, 115]. H filters out the fine structure at each level to generate the wavelet coefficients; L gives a progressively coarser representation of the detail as the fine structure disappears. This process is illustrated by the following example.

Example
   Find by hand calculation the wavelet transform of the sequence $[-9 \ -7 \ 4 \ 4 \ 4 \ 12 \ 2 \ 6]$ using the Haar wavelet for which $c_0 = c_1 = 1$. For this, the simplest wavelet, the scaling function remains a unit box and the wavelet it gives is two half-boxes of opposite sign. If the wavelet is generated by iteration following the scheme in Fig. 17.7, its shape remains unchanged by successive steps. The transformation matrices are

$$L_1 = [1 \ \ 1]$$

$$L_2 = \begin{bmatrix} 1 & 1 & & \\ & & 1 & 1 \end{bmatrix}$$

$$L_3 = \begin{bmatrix} 1 & 1 & & & & & & \\ & & 1 & 1 & & & & \\ & & & & 1 & 1 & & \\ & & & & & & 1 & 1 \end{bmatrix} \tag{17.152}$$

$$H_1 = [-1 \quad 1]$$

$$H_2 = \begin{bmatrix} -1 & 1 & & \\ & & -1 & 1 \end{bmatrix}$$

$$H_3 = \begin{bmatrix} -1 & 1 & & & & \\ & & -1 & 1 & & \\ & & & & -1 & 1 & \\ & & & & & & -1 & 1 \end{bmatrix}. \qquad (17.153)$$

Because there are only two coefficients, there are no wrapped-around elements in the matrices. The matrix operations in the tree algorithm (17.151) then give

$$\tfrac{1}{2}H_3 f = \tfrac{1}{2}\begin{bmatrix} -1 & 1 & & & & & & \\ & & -1 & 1 & & & & \\ & & & & -1 & 1 & & \\ & & & & & & -1 & 1 \end{bmatrix}\begin{bmatrix} -9 \\ -7 \\ 4 \\ 4 \\ 4 \\ 12 \\ 2 \\ 6 \end{bmatrix} = \begin{bmatrix} 1 \\ 0 \\ 4 \\ 2 \end{bmatrix} = \begin{bmatrix} a_4 \\ a_5 \\ a_6 \\ a_7 \end{bmatrix}$$

$$\tfrac{1}{2}L_3 f = \tfrac{1}{2}\begin{bmatrix} 1 & 1 & & & & & & \\ & & 1 & 1 & & & & \\ & & & & 1 & 1 & & \\ & & & & & & 1 & 1 \end{bmatrix}\begin{bmatrix} -9 \\ -7 \\ 4 \\ 4 \\ 4 \\ 12 \\ 2 \\ 6 \end{bmatrix} = \begin{bmatrix} -8 \\ 4 \\ 8 \\ 4 \end{bmatrix}$$

$$\tfrac{1}{2}H_2 \tfrac{1}{2}L_3 f = \tfrac{1}{2}\begin{bmatrix} -1 & 1 & & \\ & & -1 & 1 \end{bmatrix}\begin{bmatrix} -8 \\ 4 \\ 8 \\ 4 \end{bmatrix} = \begin{bmatrix} 6 \\ -2 \end{bmatrix} = \begin{bmatrix} a_2 \\ a_3 \end{bmatrix}$$

$$\tfrac{1}{2}L_2 \tfrac{1}{2}L_3 f = \tfrac{1}{2}\begin{bmatrix} 1 & 1 & & \\ & & 1 & 1 \end{bmatrix}\begin{bmatrix} -8 \\ 4 \\ 8 \\ 4 \end{bmatrix} = \begin{bmatrix} -2 \\ 6 \end{bmatrix}$$

$$\tfrac{1}{2}H_1\tfrac{1}{2}L_2\tfrac{1}{2}L_3 f = \tfrac{1}{2}[-1 \quad 1]\begin{bmatrix} -2 \\ 6 \end{bmatrix} = [4] = [a_1]$$

$$\tfrac{1}{2}L_1\tfrac{1}{2}L_2\tfrac{1}{2}L_3 f = \tfrac{1}{2}[1 \quad 1]\begin{bmatrix} -2 \\ 6 \end{bmatrix} = [2] = [a_0] \qquad (17.154)$$

so the transform is

$$[2 \quad 4 \quad 6 \quad -2 \quad 1 \quad 0 \quad 4 \quad 2].$$

Figure 17.11 illustrates these results. In the upper graph the expansion (17.143) is shown where

$$f^{(\phi)} = 2\begin{bmatrix} 1 \\ 1 \\ 1 \\ 1 \\ 1 \\ 1 \\ 1 \\ 1 \end{bmatrix} \quad f^{(0)} = 4\begin{bmatrix} -1 \\ -1 \\ -1 \\ -1 \\ 1 \\ 1 \\ 1 \\ 1 \end{bmatrix} \quad f^{(1)} = 6\begin{bmatrix} -1 \\ -1 \\ 1 \\ 1 \\ 0 \\ 0 \\ 0 \\ 0 \end{bmatrix} - 2\begin{bmatrix} 0 \\ 0 \\ 0 \\ 0 \\ -1 \\ -1 \\ 1 \\ 1 \end{bmatrix}$$

and

$$f^{(2)} = 1\begin{bmatrix} -1 \\ 1 \\ 0 \\ 0 \\ 0 \\ 0 \\ 0 \\ 0 \end{bmatrix} + 0\begin{bmatrix} 0 \\ 0 \\ -1 \\ 1 \\ 0 \\ 0 \\ 0 \\ 0 \end{bmatrix} + 4\begin{bmatrix} 0 \\ 0 \\ 0 \\ 0 \\ -1 \\ 1 \\ 0 \\ 0 \end{bmatrix} + 2\begin{bmatrix} 0 \\ 0 \\ 0 \\ 0 \\ 0 \\ 0 \\ -1 \\ 1 \end{bmatrix}. \qquad (17.155)$$

The result of the low-pass filtering of f that occurs going up the trunk of the tree is shown in the lower graph in Fig. 17.11. By stretching out the filtered sequences and plotting these on a common base the relationship between the high- and low-pass filtering operations can be seen. By subtracting $f^{(2)}$ in the upper graph from $f(1:8)$ in the lower graph, $f(1:4)$ is obtained; by subtracting $f^{(1)}$ from $f(1:4)$, $f(1:2)$ is obtained; by subtracting $f^{(0)}$ from $f(1:2)$, $f(1)$ in the lower graph is obtained. The wavelet coefficients are therefore generated by filtering off the high-frequency components at each scale, allowing the residual signal to pass up the tree for filtering at the next coarser scale at each stage of the transform.

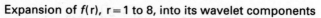

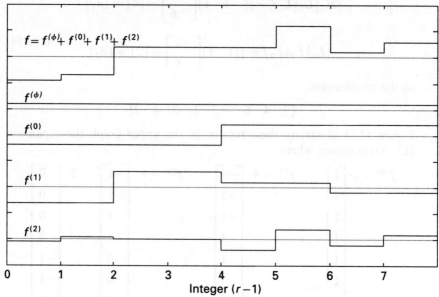

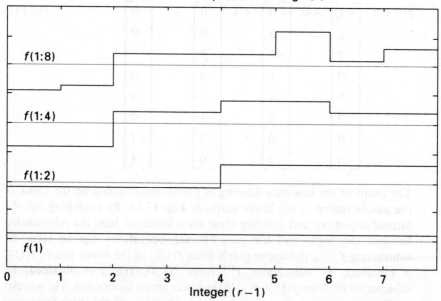

Fig. 17.11 Example using the Haar wavelet to demonstrate the working of Mallat's tree algorithm

Appendix 7 has the function file *wavedn(f,N)* which takes the array $f(1:2^n)$ and generates its wavelet transform $a(1:2^n)$, where n is any integer $n \geqslant 1$ and N is the required number of wavelet coefficients (N even). The numbers in the above example can be easily checked by this program.

The DWT based on the tree algorithm is extremely efficient computationally. For a starting array of length $2^n = M$, the first low-pass filtering operation involves $\frac{1}{2}M \times N$ multiplications when N is the number of wavelet coefficients. There will be the same number for the first high-pass filtering, and so there are $M \times N$ multiplications at the first stage of the transform, $\frac{1}{2}M \times N$ multiplications at the second stage and so on. Neglecting the reduction that comes as a result of the wavelet coefficients overlapping each other, as in (17.119), there will be a total of $(2M - 1)N$ multiplications, so that the operations count is approximately proportional to M. This compares with $M \log_2 M$ for the FFT (see Ch. 10). The number of operations increases in proportion to the number of wavelet coefficients, but this leads to greater accuracy from a limited number of terms in the wavelet expansion (17.97) and to wavelets with properties much closer to the harmonic functions we are used to in vibration analysis.

Properties of the DWT

As the number of its coefficients increases a wavelet becomes smoother and closer to a smoothly windowed harmonic function. Figure 17.12 shows wavelets with 2, 8, 14 and 20 coefficients plotted alongside each other for comparison. Even for 20 coefficients the wavelet still has some residual irregularities, as can be seen on the r.h.s. of the larger view in Fig. 17.13.

The wavelets in Fig. 17.12 are for a sequence length of $2^9 = 512$; they have been obtained by computing the IDWT of $a(1:512)$ when all the a's are zero except $a(33) = 1$. By comparison with the expansion (17.97), $a(1)$ is the coefficient a_0, $a(2)$ is a_1, and so on, making $a(33)$ the coefficient a_{32}. This is the first of the set of 16 elements which define the contributions of wavelets $W(16x - k)$. There are 32 of these in the full interval, each spaced $512/32 = 16$ points apart, corresponding to $k = 0$ to 31. The width of each wavelet depends on the number of its coefficients; if there are N coefficients, the wavelet is $N - 1$ intervals wide. Hence the D2 wavelet occupies 1 interval, or 16 points in Fig. 17.12; the D8 wavelet occupies 7 intervals or $7 \times 16 = 112$ points; the D14 wavelet occupies 13 intervals or $13 \times 16 = 208$ points and the D20 wavelet occupies 19 intervals or $19 \times 16 = 304$ points. In Fig. 17.13 the same calculation has been made for the D20 wavelet, but for a sequence length of $2^{13} = 8192$ points. In that case the interval length for $W(16x - k)$ wavelets is $8192/32 = 256$ points so that the full length of the D20 wavelet $W(16x)$ in Fig. 17.13 occupies $19 \times 256 = 4864$ points and extends from 1 to 4864. The magnitude of $W(16x)$ outside the range shown in Fig. 17.13 is very small as may be seen from the smaller view of the same wavelet in Fig. 17.12.

In Fig. 17.12 the magnitude of the discrete Fourier transforms of the four wavelets are also plotted. Because the DFT repeats itself, only values for integers

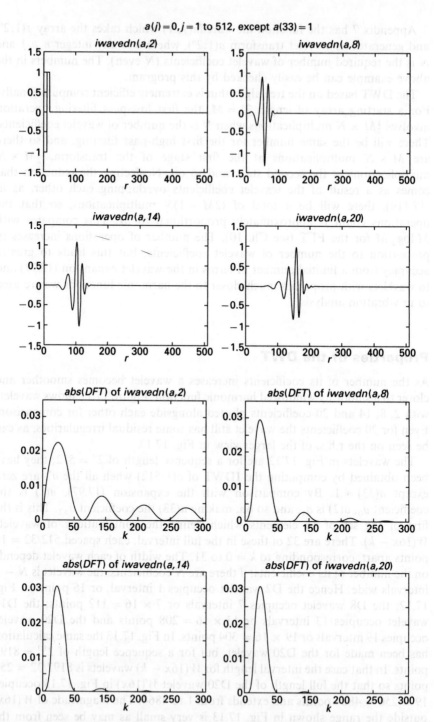

Fig. 17.12 Wavelets $W(16x)$ with 2, 8, 14 and 20 coefficients for a sequence length of 512 ($x = 1/512$ for integer 1 extending to $x = 1$ for integer 512); their corresponding discrete Fourier transforms are plotted below

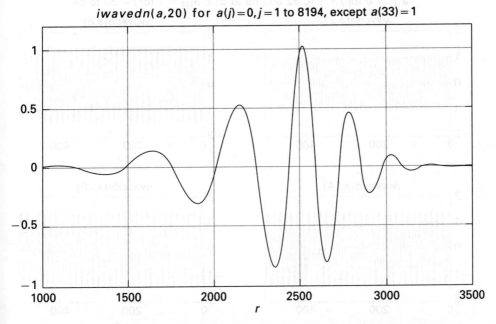

iwavedn(a,20) for *a(j)*=0,*j*=1 to 8194, except *a*(33)=1

Fig. 17.13 Part of the D20 wavelet $W(16x)$ for a sequence length of 8192; the full wavelet occupies 19 intervals from integer 1 to $19(8192/32) = 4864$

1 to $512/2 = 256$ are shown. Even for the D20 wavelet, the main frequency peak is not sharp and there are discernible humps at higher frequencies. For comparison, a Gaussian-windowed harmonic function has only a single Gaussian peak in its transform (see problem 17.8). Because dilation wavelets are compact in the sense that they have a definite beginning and ending in the x-domain, they cannot also be compact in the frequency domain so that their Fourier transforms extend over an infinite frequency range. Harmonic wavelets, which are discussed later in this chapter, are designed to be compact in the frequency domain, but they extend over an infinite range in the x-domain.

Figure 17.14 shows graphs similar to Fig. 17.12 except that all the wavelets at level 6 are included so that instead of $a(33) = 1$, with all the other a's zero, this time we have $a(33:64) = 1$. There is then a succession of similar overlapping wavelets 16 points apart. The magnitude of the Fourier transforms of these graphs is plotted below, this time using a log scale vertically. The predominant frequency is that of one cycle every 16 points which can be determined from the horizontal scale. In any specific application this will be known, as the following example explains.

Figure 17.15 shows, in the upper view, four simultaneous vibration records. They show vibration in a building close to an underground railway. The building is isolated above its ground floor so there is reduced vibration in higher floors.

$a(j) = 0$ for $j = 1$ to 32 and 65 to 512; $a(j) = 1$ for $j = 33$ to 64

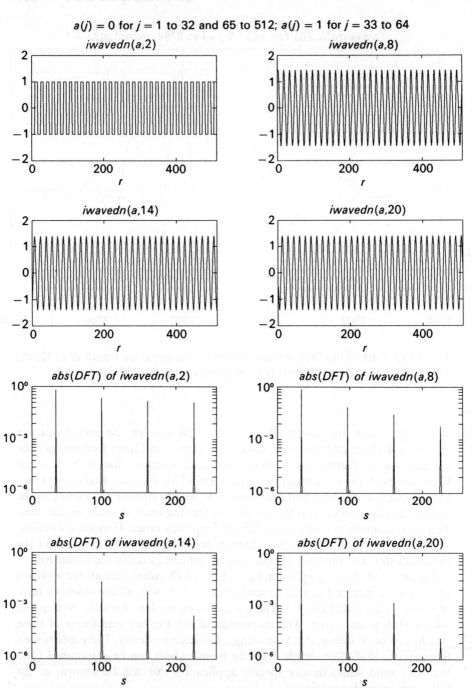

Fig. 17.14 Wavelets $W(16x - k)$ with 2, 8, 14, and 20 coefficients for $k = 0$ to 32 (for a sequence length of 512); their corresponding discrete Fourier transforms are plotted below

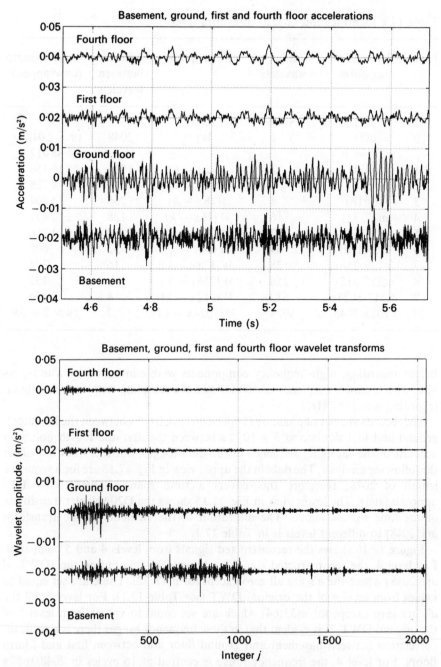

Fig. 17.15 Four simultaneous vibration records with their D20 wavelet transforms plotted below; the sequence length is 2048 corresponding to a sampling interval of 6×10^{-4} s. For clarity separate records are displaced vertically by arbitrary amounts; all records have zero mean

Table 17.1 Identification of D20 transform levels

Level	Elements of transform	Number of wavelets	Description	Spacing between wavelets	Wavelet length (unwrapped)
−1	$a(1)$	None	$\phi(x)$	—	—
0	$a(2)$	1	$W(x)$	2048	$19 \times 2048 = 38\,912$
1	$a(3:4)$	2	$W(2x - k)$	1024	19 456
2	$a(5:8)$	4	$W(4x - k)$	512	9 728
3	$a(9:16)$	8	$W(8x - k)$	256	4 864
4	$a(17:32)$	16	$W(16x - k)$	128	2 432
5	$a(33:64)$	32	$W(32x - k)$	64	1 216
6	$a(65:128)$	64	$W(64x - k)$	32	608
7	$a(129:256)$	128	$W(128x - k)$	16	304
8	$a(257:512)$	256	$W(256x - k)$	8	152
9	$a(513:1024)$	512	$W(512x - k)$	4	76
10	$a(1025:2048)$	1024	$W(1024x - k)$	2	$19 \times 2 = 38$

Before recording, high-frequency components were eliminated by a low-pass filter with a cut-off frequency of 500 Hz and the sampling frequency for digital recording was 1667 Hz.

The records are out of phase by $(1/6)(1/1667) = 10^{-4}$ s between the basement, ground and first floors and 3×10^{-4} s between the first and fourth floors on account of the sampling sequence of the data logger, but this does not affect the following analysis. The data in the upper view in Fig. 17.15 are for a sequence length of 2048; therefore this covers a time span of $2048/1667 = 1\cdot2$ s approximately. The lower view in Fig. 17.15 shows the D20 wavelet transforms of the four signals above. The allocation of the elements of the transform $a(1:2048)$ to different levels is in Table 17.1.

Figure 17.16 shows the reconstructed signals from levels 4 and 5 compared. For level 4, the reconstructed signal is obtained by computing the IDWT of $a(1:2048)$ where the a's are all zero except for $a(17:32)$. These are set equal to values from level 4 of the original DWT (see Table 17.1). For level 5, all the a's are zero except for $a(33:64)$ which are set equal to values from level 5 of the original DWT. It is evident that in these frequency ranges there is significant correlation between basement and ground floor and between first and fourth floors. For level 4, the frequency range is centred at 16 cycles in $2048/1667$ s which is a frequency of $16(1667/2048) = 13$ Hz approximately; for level 5 it is twice this, namely 26 Hz approximately. Figure 17.17 shows similar results but for levels 8 and 9. There is now very little apparent correlation between these reconstructed signals. For level 8, the frequency range is centred at

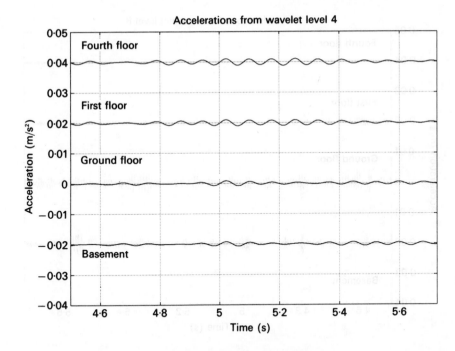

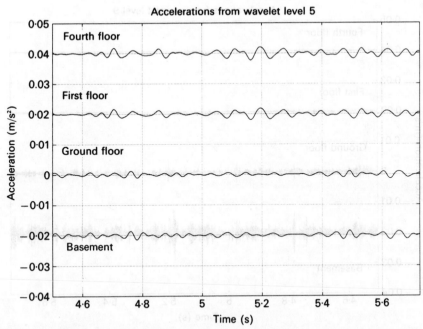

Fig. 17.16 Reconstructed signals from wavelet levels 4 and 5 for the vibration records shown in Fig. 17.15. Records are displaced vertically by arbitrary amounts

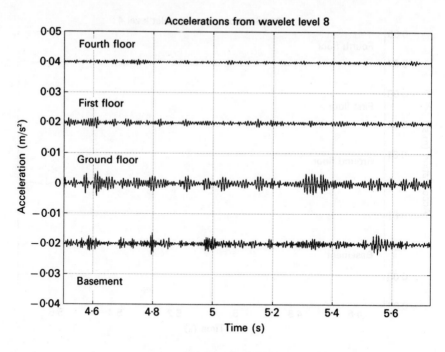

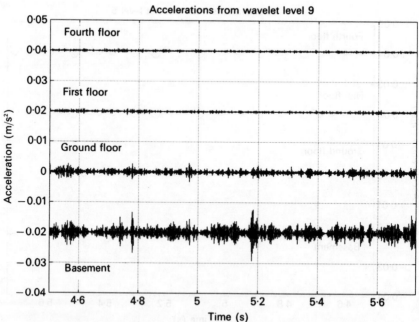

Fig. 17.17 Reconstructed signals from wavelet levels 8 and 9 for the vibration records shown in Fig. 17.15

$256(1667/2048) = 208$ Hz approximately; for level 9 it is 416 Hz approximately. The explanation for this behaviour is that the sprung building tends to move as a unit at low frequencies, as does the ground floor and basement, which is really a separate lower building. At higher frequencies the structure becomes multi-modal and specific vibration records are very sensitive to their position of measurement.

Lastly in this section, we mention a final property of the circular DWT. This concerns the contribution to $f(x)$ from level -1 of its wavelet transform. From (17.97) this is

$$f^\phi(x) = a_0\phi(x). \tag{17.156}$$

That suggests that if $f(x)$ is reconstructed by the IDWT using only the first element $a_0 = a(1)$ of its wavelet transform a, the reconstructed f^ϕ should have the shape of the scaling function. However, by taking examples using the program *iwavedn(a,N)*, we soon see that this is not the case and that the reconstructed $f^\phi(x)$ is always a constant. This is because of the properties of the circular transform. For all cases except $N = 2$ (when the scaling function is a box function and therefore constant), the scaling function occupies more than one interval (if the wavelet has N coefficients the scaling function occupies $N - 1$ intervals). It therefore wraps around the interval $0 \leqslant x < 1$. Take the D4 scaling function (Fig. 17.5) as an example. This extends over three intervals. When wrap-around takes place the circular scaling function in $0 \leqslant x < 1$ is given by

$$\phi(x) + \phi(x + 1) + \phi(x + 2). \tag{17.157}$$

Because these functions may be constructed by iteration from adjacent unit boxes, they are effectively constructed by iteration from a starting function which is $\phi_0(x) = 1$ for all x. Because the wavelet coefficients have to satisfy the accuracy condition (17.30) with $m = 0$, they satisfy

$$\sum_k (-1)^k c_k = 0 \tag{17.158}$$

so that, using also the condition that

$$\sum_k c_k = 2, \tag{17.15}$$

we have seen already (17.78) that

$$\sum_{k \text{ even}} c_k = \sum_{k \text{ odd}} c_k = 1. \tag{17.159}$$

With this result we find that every stage of the iteration generates $\phi_j(x) = 1$ for all x. This may be seen by considering the matrix calculations (17.106) and (17.109) for the first two iterations of the D4 scaling function. The upshot is that, for the circular DWT, the first term in the expansion (17.97) is always

$$f^\phi(x) = a_0 \qquad 0 \leqslant x < 1. \tag{17.160}$$

Mean-square maps

The wavelet expansion (17.97) can be written in the following equivalent form:

$$f(x) = a_0 + \sum_j \sum_k a_{2^j+k} W(2^j x - k) \qquad 0 \leqslant x < 1. \tag{17.161}$$

The wavelet functions $W(2^j x - k)$ are wrapped functions; we have seen that the wrapped scaling function $\phi(x)$ is always unity, so this no longer appears in (17.161). The integer j describes different levels of wavelets, starting with $j = 0$; integer k covers the number of wavelets in each level, so that it covers the range $k = 0$ to $2^j - 1$.

We can calculate the mean-square value of $f(x)$ by squaring both sides of (17.161) and integrating over the interval $0 \leqslant x < 1$ to get

$$\int_0^1 f^2(x)\, dx = a_0^2 + 2a_0 \sum_j \sum_k a_{2^j+k} \int_0^1 W(2^j x - k)\, dx +$$

$$+ \sum_j \sum_k \sum_r \sum_s a_{2^j+k} a_{2^r+s} \int_0^1 W(2^j x - k) W(2^r x - s)\, dx. \tag{17.162}$$

We have seen already that the orthogonality and area properties of wavelets are retained when the wavelet functions are wrapped round the interval $0 \leqslant x < 1$ (see problem 17.7). The results (17.71), (17.73) and (17.76) therefore apply for the wrapped functions when their limits of integration are altered to run from 0 to 1 instead of from $-\infty$ to $+\infty$. With that change, from (17.76)

$$\int_0^1 W(2^j x - k)\, dx = 0 \tag{17.163}$$

and, from (17.71),

$$\int_0^1 W(2^j x - k) W(2^r x - s)\, dx = 0 \quad \text{unless } r = j \text{ and } s = k. \tag{17.164}$$

Hence, from (17.162),

$$\int_0^1 f^2(x)\, dx = a_0^2 + \sum_j \sum_k a_{2^j+k}^2 \int_0^1 W^2(2^j x)\, dx$$

$$= a_0^2 + \sum_j \sum_k a_{2^j+k}^2 \left(\frac{1}{2^j}\right) \tag{17.165}$$

on account of (17.73). Writing out the first few terms of the double summation, we have

$$\int_0^1 f^2(x)\, dx = a_0^2 + a_1^2 + \frac{1}{2}(a_2^2 + a_3^2) + \frac{1}{4}(a_4^2 + a_5^2 + a_6^2 + a_7^2) +$$

$$+ \frac{1}{8}(a_8^2 + a_9^2 + \cdots + a_{15}^2) + \frac{1}{16}(a_{16}^2 + \cdots + a_{31}^2) + \cdots. \tag{17.166}$$

This is a very important result because it shows how the mean square of $f(x)$ is distributed between different wavelet levels and between different wavelets within each level (and therefore different positions within the interval $0 \leqslant x < 1$). When the series is truncated so that there are 2^n terms on the r.h.s. of (17.166), the integral on its l.h.s. can be replaced exactly by

$$\frac{1}{2^n} \sum_{r=0}^{2^n - 1} f_r^2$$

where $f_r \equiv f(x = r/2^n)$ (problem 17.9).

Equation (17.166) can be illustrated graphically by a three-dimensional plot in which the values of the wavelet amplitudes squared are plotted on a grid framework as shown in Fig. 17.18. If a surface is erected on this base, the surface consisting of horizontal planes joined together by vertical walls at the broken lines and at the edges in Fig. 17.18, the volume under this surface represents the mean-square value $\int_0^1 f^2(x)\,dx$. A contour plot of the surface therefore maps the distribution of the mean square of the function $f(x)$ between wavelets of different level and different position.

At level j there are 2^j wavelets in the unit interval. For a train of unit wavelets, all at the same scale, the wavelets are spaced $1/2^j$ apart so that their fundamental frequency is 2^j cycles/unit interval (see problem 17.6). Therefore, wavelet level is a measure of frequency with the fundamental frequency doubling as we move upwards from one level to the next. Since, for vibration records, the independent variable x is usually a measure of time, the mean-square map can be interpreted as a plot of frequency (increasing in octaves) against time.

In order to achieve a diagram that can be interpreted easily, it is helpful to locate each wavelet amplitude (squared) at a position which corresponds to the centre of the corresponding wavelet. Take for example an analysis using the D6 wavelet. This has six coefficients and is five intervals long. Consider level 3 in Fig. 17.18. Element a_8 represents the amplitude of a wavelet which extends over five intervals (each interval at level 3 being $0 \leqslant x < \frac{1}{8}$). This wavelet's centre is in the middle of block 3. To locate element a_8 where the centre of its wavelet occurs, a_8 has to be moved $\frac{1}{2}N - 1 = 2$ intervals to the right in Fig. 17.18.

It turns out that the same result applies to all the elements at levels 1 and above: they all have to be moved $\frac{1}{2}N - 1$ places to the right from the arrangement shown in Fig. 17.18. Elements moved off the right edge of the diagram are wrapped round to return over the left edge so that the whole squence of elements at each level is merely translated $\frac{1}{2}N - 1$ steps to the right before the mean-square surface is plotted.

Figure 17.19 shows two vibration records with the same spectral density. The upper diagram shows the computed response of a two-degree-of-freedom system to white noise excitation; the lower diagram is the impulse response function of the same system. Both records are normalized to have the same mean-square value. Figure 17.20 shows mean-square maps of these two signals

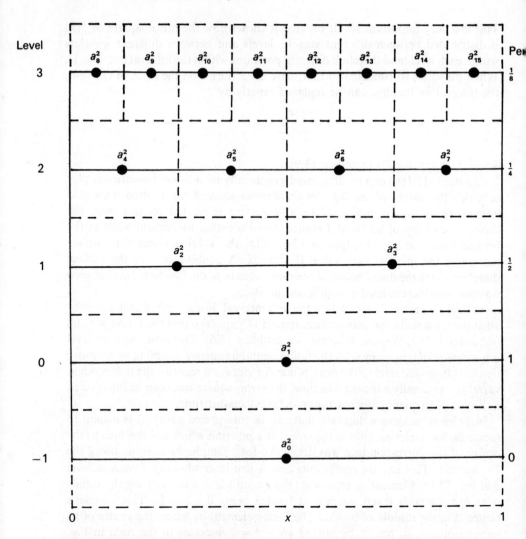

Fig. 17.18 Grid base for plotting wavelet amplitudes (squared) in order to ensure that the volume they enclose is equal to the mean-square value of $f(x)$

calculated using the D4 wavelet. In this example the logarithm of the wavelet coefficients squared has been plotted to accentuate the fine detail at low values of a^2.

By comparing Figs 17.19 and 17.20, the relationship between each mean-square map and the function it represents can be seen. The horizontal axes cover the same interval in all four diagrams. In Fig. 17.19, the integer j runs from 1 to 8192 as that number of points are plotted. The wavelet transform of a sequence of 8192 points has 12 levels (together with an additional zero level

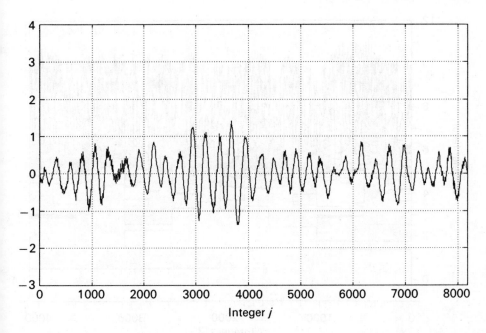

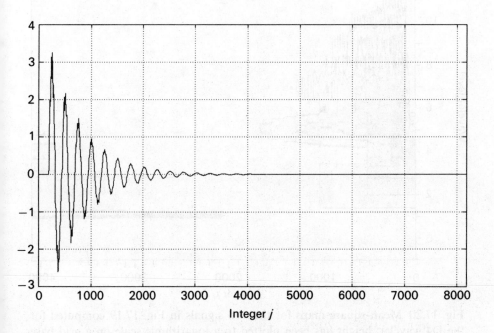

Fig. 17.19 Two signals with the same spectral density: the upper is a realization of a stationary random process, the lower is a non-stationary impulse response

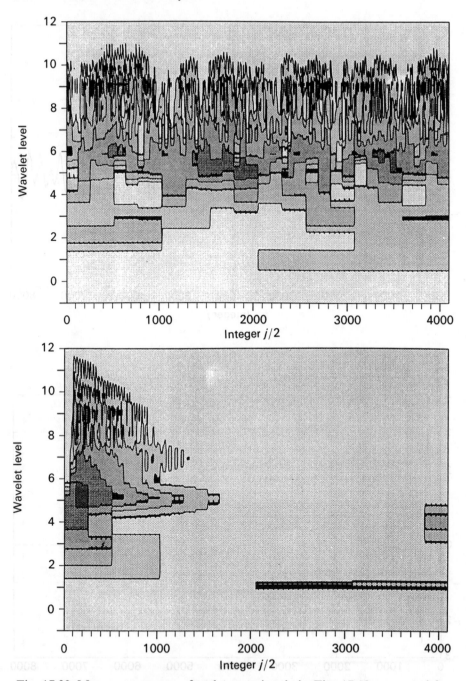

Fig. 17.20 Mean-square maps for the two signals in Fig. 17.19 computed for the D4 wavelet; height has been plotted to a logarithmic scale on a grid base similar to Fig. 17.18 (with the positions of the coefficients adjusted to allow for the differing lengths of the wavelets they define)

and a constant level). The highest level has one wavelet every two positions, so there are 4096 elements of the wavelet transform at level 12. Therefore the two-dimensional array of numbers that defines the surface whose base is like that shown in Fig. 17.18 has $12 + 2 = 14$ rows and 4096 columns. Row 1 has 4096 elements all equal to a_0^2, row 2 has the same number all a_1^2, row 3 has 2048 elements equal to a_2^2 and 2048 equal to a_3^2, and so on.

In Fig. 17.20, the contours are plotted with relatively thick lines and the black regions generated by close contour lines should be distinguished from the different grey scales. The highest ground is in the lower figure at level 5 and horizontal integer about 200, and this level is not reached anywhere in the upper figure. Although the transient response in Fig. 17.19 (lower) has been extinguished at the r.h.s. of the graph, the mean-square map in Fig. 17.20 (lower) shows two promontories of low ground extending from the right-hand edge of the diagram. These arise because of the circular nature of the transform and contribute to the signal in Fig. 17.19 at its left-hand edge. To illustrate the high-frequency behaviour, five contours are plotted, the lowest contour being at a height of approximately 10^{-3} times the highest contour.

A variety of different examples of the application of the wavelet map in vibration analysis and a more detailed examination of the properties of mean-square wavelet maps are given in the author's papers [112, 113].

To be able to appreciate the information mean-square maps provide, it is desirable to plot them in colour. The increasing availability of colour graphics and colour plotters now makes this practicable. As a result, coloured mean-square maps generated by the discrete wavelet transform are likely to become an important new tool in random vibration analysis. They offer considerable scope for studying the time-dependent properties of non-stationary signals.

Convolution by wavelets

It is possible to make use of the orthogonality properties of wavelets in an algorithm for computing the convolution integral

$$y(t) = \int_0^\infty h(\theta)x(t - \theta)\,d\theta. \tag{6.26}$$

In Chapter 6 we saw that this integral allows the response $y(t)$ of a passive linear system to be calculated when $x(t)$ is its excitation and $h(\theta)$, $\theta > 0$, is its impulse response function ($h(\theta) = 0$ for $\theta < 0$).

Consider using (6.26) to calculate $y(t)$ for one specific time only. For this value of $t = t_0$ define a new variable

$$z(\theta) = x(t_0 - \theta) \tag{17.167}$$

so that (6.27) becomes

$$y(t_0) = \int_0^1 h(\theta)z(\theta)\,d\theta. \qquad (17.168)$$

The limits of integration are altered to be from 0 to 1 because we shall assume that $h(\theta)$ decays fast enough that $h(\theta = 1) \to 0$ closely enough.

Now write $h(\theta)$ and $z(\theta)$ in terms of their circular wavelet expansions so that, from (17.161)

$$h(\theta) = h_0 + \sum_j \sum_k h_{2^j+k}W(2^j\theta - k) \qquad (17.169)$$

$$z(\theta) = z_0 + \sum_r \sum_s z_{2^r+s}W(2^r\theta - s) \qquad (17.170)$$

and substitute into (17.168) to get

$$y(t_0) = h_0 z_0 + h_0 \sum_r \sum_s z_{2^r+s} \int_0^1 W(2^r\theta - s)\,d\theta +$$

$$+ z_0 \sum_j \sum_k h_{2^j+k} \int_0^1 W(2^j\theta - k)\,d\theta +$$

$$+ \sum_j \sum_k \sum_r \sum_s h_{2^j+k}z_{2^r+s} \int_0^1 W(2^j\theta - k)W(2^r\theta - s)\,d\theta. \qquad (17.171)$$

Using the conditions (17.163) and (17.164), we then find that

$$y(t_0) = h_0 z_0 + \sum_j \sum_k (1/2^j)h_{2^j+k}z_{2^j+k}$$

$$= [h_0 \quad h_1 \quad h_2/2 \quad h_3/2 \quad h_4/4 \quad h_5/4 \quad \ldots]\begin{bmatrix} z_0 \\ z_1 \\ z_2 \\ \vdots \end{bmatrix} \qquad (17.172)$$

where $[h_0 \quad h_1 \quad h_2 \quad \ldots]$ is the wavelet transform of $h(\theta)$ and $[z_0 \quad z_1 \quad z_2 \quad \ldots]$ is the wavelet transform of $z(\theta)$. Obviously the transforms must be in terms of the same analysing wavelet in both cases. The factor $(1/2^j)$ in (17.172) arises because, from (17.73), we have

$$\int_{-\infty}^{\infty} W^2(2^j\theta)\,d\theta = 1/2^j. \qquad (17.173)$$

The interval over which the calculation is made must be long enough that $h(\theta) \to 0$ sufficiently closely outside it. Within this interval $h(\theta)$ and $z(\theta)$ are sampled at equally-spaced distances along the θ-axis to give 2^n data points where n is a sufficiently large integer to ensure that all the fine detail is properly represented.

The outcome of a calculation using (17.172) leads to a single point on the response graph. By translating the input function horizontally, and repeating the calculation (17.172), a second point is obtained. Unfortunately the wavelet transform is not invariant to translation. A new transform of z has to be calculated for however many times t_0 the response is required. That is a disadvantage for practical calculations. The consequence is that although wavelet convolution provides an interesting example of the properties of the discrete wavelet transform (see problem 17.10), it is not a computationally efficient procedure. Frequency–domain calculations, using the FFT, provide a faster route for computing system response.

Two-dimensional wavelet transforms

The discrete wavelet transform can be extended to two dimensions in the same way that the one-dimensional DFT was extended to two dimensions in Chapter 15. The two-dimensional expansion has terms in $\phi(x)\phi(y)$, $\phi(x)W(y)$, $W(x)\phi(y)$, $W(x)W(y)$, $\phi(x)W(2y)$, $\phi(x)W(2y-1)$ and so on. Instead of the one-dimensional expansion (17.97), we have

$$f(x,y) = W(x)AW^t(y) \tag{17.174}$$

where

$$W(x) = [\phi(x) \quad W(x) \quad W(2x) \quad W(2x-1) \quad W(4x) \quad \ldots \quad W(2^jx - k) \quad \ldots] \tag{17.175}$$

$$W(y) = [\phi(y) \quad W(y) \quad W(2y) \quad W(2y-1) \quad W(4y) \quad \ldots \quad W(2^jy - k) \quad \ldots] \tag{17.176}$$

and

$$A = \begin{bmatrix} a_{00} & a_{01} & a_{02} & a_{03} & a_{04} & \ldots \\ a_{10} & a_{11} & a_{12} & a_{13} & & \\ a_{20} & a_{21} & & & & \\ \vdots & & & & & \end{bmatrix}. \tag{17.177}$$

If the one-dimensional sequence lengths are 2^{n_1} and 2^{n_2}, then the order of the A-matrix in (17.175) is $2^{n_1} \times 2^{n_2}$.

If the function $f(x, y)$ is represented by a two-dimensional array $F(2^{n_1} \times 2^{n_2})$, its two-dimensional wavelet transform $A(2^{n_1} \times 2^{n_2})$ is calculated by repeated application of the one-dimensional discrete wavelet transform just as the two-dimensional discrete Fourier transform is calculated by repeated application of the one-dimensional DFT (see Chapter 15, page 225). The inverse two-dimensional DWT is calculated in the same way, beginning with the A-matrix in (17.177) and making repeated use of the one-dimensional IDWT. The logic for this can be seen as follows.

Let all the elements of A be zero except a_{rs}. This element lies in row $r + 1$ and column $s + 1$ of the matrix, so that

$$A = \begin{bmatrix} & & \\ & & \\ & a_{rs} & \\ & & \\ & & \end{bmatrix} \quad \text{row } r + 1. \tag{17.178}$$

$$\text{column } s + 1$$

Calculate the IDWT of the sequence

$$[0 \quad 0 \quad 0 \quad \ldots \quad a_{rs} \quad \ldots]$$

which is row $r + 1$ of A. This generates a wavelet of amplitude a_{rs} that occupies the whole of row $r + 1$ of an intermediate matrix

$$B = \begin{bmatrix} & & \\ & & \\ \hline \rule{0pt}{2ex} & & \\ & & \end{bmatrix} \quad \text{row } r + 1. \tag{17.179}$$

Each of its coefficients provides the amplitude of a wavelet in the other dimension. These wavelets are calculated by finding the IDWT of each of the columns of B separately. This means that 2^{n_2} IDWTs have to be calculated for sequences like

$$[0 \quad 0 \quad 0 \quad \ldots \quad x \quad \ldots]$$

where the x is the only non-zero element, which is in position $r + 1$. These 2^{n_2} wavelets occupy all the columns of the function matrix $F(2^{n_1} \times 2^{n_2})$ which replaces B and becomes

$$F = \begin{bmatrix} | & | & | & | & | & | & | & | \\ | & | & | & | & | & | & | & | \\ | & | & | & | & | & | & | & | \\ | & | & | & | & | & | & | & | \\ | & | & | & | & | & | & | & | \\ | & | & | & | & | & | & | & | \\ | & | & | & | & | & | & | & | \\ | & | & | & | & | & | & | & | \end{bmatrix} \tag{17.180}$$

Because superposition applies, so that the transform of two elements together is the same as the addition of the transforms of each element alone, when row $r + 1$ of A is a full row the same calculation applies. Similarly, if all the rows of A are full, all of the IDWTs are calculated row by row, to give a full matrix B. Then each of the columns of this new matrix have their IDWTs calculated to determine the matrix F.

This calculation can be programmed quite simply and is the basis of the programs *wave2dn(F,N)* and *iwave2dn(A,N)* in Appendix 7.

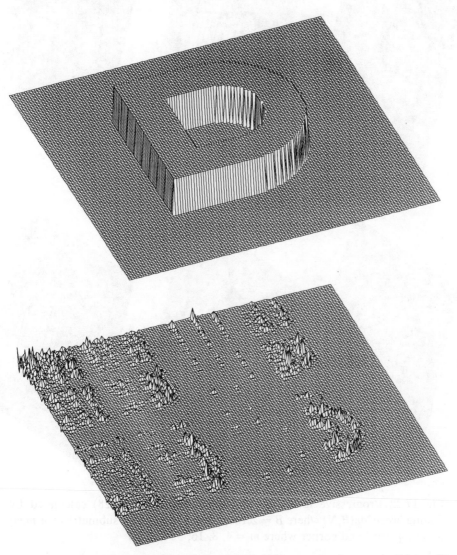

Fig. 17.21 Two-dimensional function $F(128,128)$ representing a raised letter D and a mesh graph of its wavelet transform A calculated by *wave2dn(F,N)* where $N = 20$. The layout of the mesh of A corresponds to the order of the elements in (17.177)

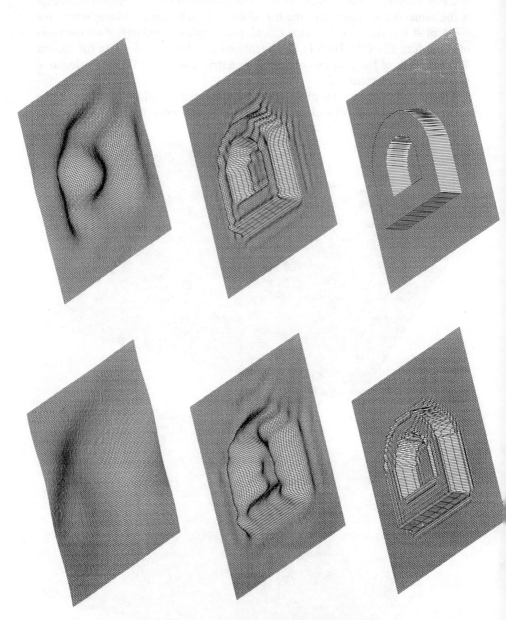

Fig. 17.22 Progressive levels of reconstruction of $F(128,128)$ calculated by finding *iwave2dn(B,N)* where $B = \text{zeros}(128,128)$ except for submatrices $A(m,m)$ in its top left-hand corner where $m = 4, 8, 16, 32$ and 64

One application of these programs is to the analysis and modelling of surface topography. Figure 17.21 shows a raised letter D modelled by a two-dimensional array of numbers $F(128 \times 128)$. Using a D20 wavelet, the discrete wavelet transform of this F has been calculated by implementing $A = wave2dn(F,N)$ with $N = 20$; a mesh plot of the result is shown below F in Fig. 17.21. This A is used in Fig. 17.22 to show successively higher levels of reconstruction of F. In each case all the elements of A are set equal to zero except for a submatrix in the top left-hand corner in (17.177). The order of this non-zero submatrix of the elements of A is progressively increased to cover the range 4×4, 8×8, $16 \times 16, 32 \times 32, 64 \times 64$ as increasing definition is built back into the model.

There appear to be many other possible applications (of both single- and multi-dimensional wavelet transforms). These cover a wide variety of different fields (see, for example, [102] and [106]). At present, progress on the mathematical front (there are reviews in [109] and [115]) runs ahead of applications in science and engineering but this is likely to change as software for the discrete wavelet transform becomes more widely available. Some applications of the DWT in vibration analysis are discussed by the author in [112] and [113].

Harmonic wavelets

The wavelets studied so far in this chapter have all been derived from dilation equations with real coefficients. As the number N of coefficients (which must be even) is increased, the wavelet's Fourier transform becomes increasingly compact. Only values of N up to 20 are included in the accompanying program *dcoeffs*(N), but the numerical values of wavelet coefficients for $N > 20$ can, if required, be computed by solving the N nonlinear algebraic equations that define them. When this is done, it turns out that the underlying spectrum of a wavelet with N coefficients becomes more box-like as N increases. This fact led the author to seek [114] a wavelet $w(x)$ whose spectrum is exactly like a box so that the magnitude of its Fourier transform $W(\omega)$ is zero except for an octave band of frequencies. A convenient way to define $|W(\omega)|$ for different level wavelets is then as shown in Fig. 17.23. All the Fourier transforms are identically zero for $\omega < 0$, and for level zero (for example) the definition of $|W(\omega)|$ is

$$|W(\omega)| = 1/2\pi \quad \text{for} \quad 2\pi \leqslant \omega < 4\pi$$

$$= 0 \text{ elsewhere.} \tag{17.181}$$

If we delete the modulus signs around $W(\omega)$ in (17.181) and put

$$W(\omega) = 1/2\pi \quad \text{for} \quad 2\pi \leqslant \omega < 4\pi$$

$$= 0 \text{ elsewhere} \tag{17.182}$$

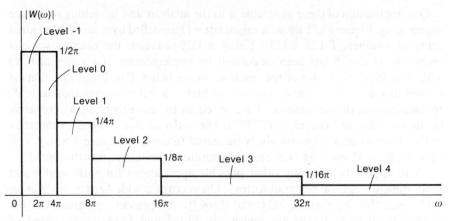

Fig. 17.23 Magnitudes of the Fourier transforms of harmonic wavelets of different level

then, by calculating the inverse Fourier transform of $W(\omega)$, we see that the corresponding complex wavelet is

$$w(x) = (e^{i4\pi x} - e^{i2\pi x})/i2\pi x \tag{17.183}$$

whose real and imaginary parts are shown in Fig. 17.24.

The introduction of a complex function allows two real wavelets to be represented by a single expression. In Fourier analysis the same thing is done when the complex exponential form $e^{i2\pi kt/T}$ is used to represent both an even function $\cos 2\pi kt/T$ and an odd function $\sin 2\pi kt/T$ in the Fourier expansion formula (15.1)

$$x(t) = \sum_{k=-\infty}^{\infty} X_k e^{i2\pi kt/T}. \tag{17.184}$$

In the present case, it can be seen from Fig. 17.24 that the real part of $w(x)$ in (17.183) represents an even wavelet (a wavelet for which $w(x) = w(-x)$) and the imaginary part represents an odd wavelet (for which $w(-x) = -w(x)$).

For the general complex wavelet, at level j, and translated by k steps of size $1/2^j$, we define

$$W(\omega) = (1/2\pi)2^{-j}e^{-i\omega k/2^j} \quad \text{for} \quad 2\pi 2^j \leqslant \omega < 4\pi 2^j$$

$$= \text{elsewhere.} \tag{17.185}$$

On calculating its inverse Fourier transform, this gives

$$w(2^j x - k) = (e^{i4\pi(2^j x - k)} - e^{i2\pi(2^j x - k)})/i2\pi(2^j x - k) \tag{17.186}$$

in which $j = 0$ to ∞, $k = -\infty$ to ∞. For level -1, with the frequency band $0 \leqslant \omega < 2\pi$ in Fig. 17.23, calculating the inverse Fourier transform of

$$W(\omega) = (1/2\pi)e^{-i\omega k} \quad \text{for} \quad 0 \leqslant \omega < 2\pi$$

$$= 0 \text{ elsewhere} \tag{17.187}$$

(a)

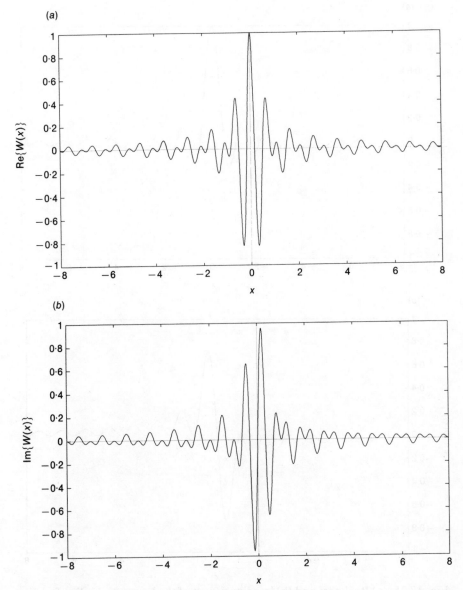

Fig. 17.24 (a) Real part and (b) imaginary part of the harmonic wavelet defined by (17.183)

gives a different function that, by analogy with previous results, we shall call the scaling function

$$\phi(x - k) = (e^{i2\pi(x-k)} - 1)/i2\pi(x - k), \qquad (17.188)$$

$k = -\infty$ to ∞. The real and imaginary parts of $\phi(x)$ are shown in Fig. 17.25.

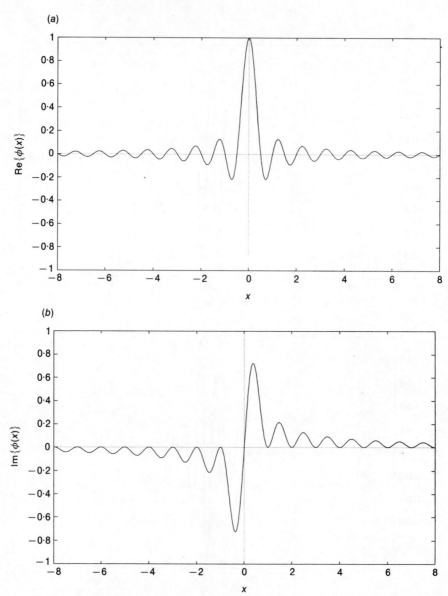

Fig. 17.25 (a) Real part and (b) imaginary part of the harmonic scaling function defined by (17.188)

The rationale behind this choice of wavelet and scaling functions is that they form an orthogonal set and have the properties that

$$\int_{-\infty}^{\infty} w(2^j x - k)w(2^r x - s)\,dx = 0 \quad \text{for all} \quad j, k, r, s$$

$$(j, r \geqslant 0) \quad (17.189)$$

$$\int_{-\infty}^{\infty} w(2^j x - k)w^*(2^r x - s)\, dx = 0 \quad \text{for all } j, k, r, s$$

$$(j, r \geqslant 0) \text{ except for} \\ \text{the case when } r = j \\ \text{and } s = k, \text{ when} \qquad (17.190)$$

$$\int_{-\infty}^{\infty} |w(2^j x - k)|^2\, dx = 1/2^j. \qquad (17.191)$$

Also

$$\int_{-\infty}^{\infty} \phi(x - k)\phi(x - s)\, dx = 0 \quad \text{for all } k, s \qquad (17.192)$$

$$\int_{-\infty}^{\infty} \phi(x - k)\phi^*(x - s)\, dx = 0 \quad \text{for all } k, s \text{ except} \qquad (17.193)$$

$$s = k, \text{ when}$$

$$\int_{-\infty}^{\infty} |\phi(x - k)|^2\, dx = 1 \qquad (17.194)$$

and, finally,

$$\int_{-\infty}^{\infty} w(2^j x - k)\phi(x - s)\, dx = 0 \quad \text{for all } j, k, s \qquad (17.195)$$

$$(j \geqslant 0)$$

$$\int_{-\infty}^{\infty} w(2^j x - k)\phi^*(x - s)\, dx = 0 \quad \text{for all } j, k, s \qquad (17.196)$$

$$(j \geqslant 0).$$

The proof of these results depends on the properties of their Fourier transforms and derives from two results from Fourier transform theory (Problem 17.11) that, if $W(\omega)$ and $V(\omega)$ are the Fourier transforms of $w(x)$ and $v(x)$ respectively, then

$$\int_{-\infty}^{\infty} w(x)v(x)\, dx = \int_{-\infty}^{\infty} W(\omega)V(-\omega)\, d\omega \qquad (17.197)$$

$$\int_{-\infty}^{\infty} w(x)v^*(x)\, dx = \int_{-\infty}^{\infty} W(\omega)V^*(\omega)\, d\omega. \qquad (17.198)$$

Most of the orthogonality properties can be seen immediately from these two equations.

First consider (17.189) and (17.192). Since the $W(\omega)$ are only defined in Fig. 17.23 for $\omega \geqslant 0$ and are identically zero for $\omega < 0$, $W(\omega)$ is always zero for negative ω. Both results, therefore, follow immediately from (17.197).

Next consider (17.190), (17.195) and (17.196). Wavelets of different levels have Fourier transforms that occupy different frequency bands. Hence (17.195) and (17.196) are true and so is (17.190) when $r \neq j$.

For wavelets in the same frequency band, each wavelet is only orthogonal to the complex conjugate of another wavelet if the wavelets are translated by steps of a specific size. For wavelets at level j, the required spacing is $k/2^j$ units (k any integer); for the scaling function, the required spacing is k units. This can be proved by substituting the appropriate expressions for the Fourier transforms from (17.185) and (17.187) into (17.198) and making the integration. It will be found (Problem 17.12) that the result is only zero when the wavelets and scaling function are correctly spaced, which they are when k has any integer value in (17.186) and (17.188).

The result of all this is that the $w(2^j x - k)$ defined by (17.186) together with the $\phi(x - k)$ defined by (17.188) constitute an orthogonal family of wavelets that offer an alternative to wavelets derived from dilation equations. As for dilation wavelets, a series using harmonic wavelets as the basis functions can be shown to have mean-square convergence as the number of its terms increases, provided that $\int_{-\infty}^{\infty} |f(x)|^2 \, dx$ exists [114]. The principal differences are these. Harmonic wavelets can be described by a simple analytical formula, they are compact in the frequency domain, and are described by a complex function so that there are two real wavelets for each j, k pair. Dilation wavelets cannot be expressed in functional form, they are compact in the x-domain, and there is one real wavelet for each j, k pair. The last point is subject to qualification because research is in progress in which mixed families of wavelets are used, when there may be more than one wavelet for each combination of order j and position k. However, the computational algorithms presented in Appendix 7 are for single-family dilation wavelets. A disadvantage of harmonic wavelets is that, although they have convenient band-limited Fourier transforms, the penalty for this is that, in the x-domain, they decay only in proportion to $1/x$. In comparison a wavelet derived from a dilation equation with N coefficients occupies $(N - 1)$ unit intervals and is identically zero outside this range. Conversely, as we shall see, an advantage of harmonic wavelets is their computational simplicity. It is shown in the next section that there is a simple algorithm for harmonic wavelet analysis which uses the FFT and which is generally faster than the algorithm in Appendix 7 for dilation wavelet analysis.

Discrete harmonic wavelet transform

For complex wavelets we have to define two amplitude coefficients

$$a_{j,k} = 2^j \int_{-\infty}^{\infty} f(x)w^*(2^j x - k) \, dx \tag{17.199}$$

$$\tilde{a}_{j,k} = 2^j \int_{-\infty}^{\infty} f(x)w(2^j x - k) \, dx. \tag{17.200}$$

If $f(x)$ is real, $\tilde{a}_{j,k}$ is the complex conjugate of $a_{j,k}$ so that $\tilde{a}_{j,k} = a^*_{j,k}$ but to allow for the case when $f(x)$ is complex we regard $a_{j,k}$ and $\tilde{a}_{j,k}$ as two different amplitudes. Similarly we define

$$a_{\phi,k} = \int_{-\infty}^{\infty} f(x)\phi^*(x-k)\,dx \tag{17.201}$$

$$\tilde{a}_{\phi,k} = \int_{-\infty}^{\infty} f(x)\phi(x-k)\,dx. \tag{17.202}$$

In terms of these coefficients, the wavelet expansion of a general function $f(x)$ for which

$$\int_{-\infty}^{\infty} |f(x)|^2\,dx < \infty \tag{17.203}$$

is given by

$$f(x) = \sum_{k=-\infty}^{\infty} (a_{\phi,k}\phi(x-k) + \tilde{a}_{\phi,k}\phi^*(x-k))$$

$$+ \sum_{j=0}^{\infty} \sum_{k=-\infty}^{\infty} (a_{j,k}w(2^jx-k) + \tilde{a}_{j,k}w^*(2^jx-k)) \tag{17.204}$$

We want a discrete algorithm to compute the coefficients $a_{\phi,k}$, $\tilde{a}_{\phi,k}$, $a_{j,k}$ and $\tilde{a}_{j,k}$ in this expansion. As an example, consider $a_{j,k}$ defined by (17.199). First, substitute for $w^*(2^jx - k)$ in terms of its Fourier transform where, from (17.185),

$$w^*(2^jx-k) = \int_{2\pi 2^j}^{4\pi 2^j} (1/2\pi)2^{-j}\,e^{i\omega k/2^j}\,e^{-i\omega x}\,d\omega \tag{17.205}$$

so that (17.199) becomes

$$a_{j,k} = 1/2\pi \int_{2\pi 2^j}^{4\pi 2^j} d\omega\, e^{i\omega k/2^j} \int_{-\infty}^{\infty} dx\, f(x)\,e^{-i\omega x}\,dx. \tag{17.206}$$

Since the Fourier transform of $w^*(2^jx - k)$ is identically zero outside $2\pi 2^j \leqslant \omega < 4\pi 2^j$, it is legitimate to reverse the order of the integrals in this way. The second integral over x is just the Fourier transform of $f(x)$ multiplied by 2π, so that (17.206) becomes

$$a_{j,k} = \int_{2\pi 2^j}^{4\pi 2^j} F(\omega)\,e^{i\omega k/2^j}\,d\omega. \tag{17.207}$$

Now we replace the integral by a summation. This involves the same sort of approximation that was made in Chapter 10 when deriving the discrete Fourier transform as an approximation for the continuous Fourier transform. In this case we step through the frequency band in increments of 2π (so that there are 2^j steps in level j). Then $F(\omega)$ is replaced by the discrete coefficient F_{2^j+s} where

$$F_{2^j+s} = 2\pi F(\omega = 2\pi(2^j + s)) \tag{17.208}$$

and the integral in (17.207) becomes the summation

$$a_{2^j+k} = \sum_{s=0}^{2^j-1} F_{2^j+s} e^{i2\pi(2^j+s)k/2^j} \qquad (17.209)$$

because $\Delta\omega = 2\pi$ cancels with $1/2\pi$ from (17.208). Also since $e^{i2\pi k} = 1$ for all integers k, we are left with the formula

$$a_{2^j+k} = \sum_{s=0}^{2^j-1} F_{2^j+s} e^{i2\pi sk/2^j}, \quad k = 0 \text{ to } 2^j - 1. \qquad (17.210)$$

This is the inverse discrete Fourier transform for the sequence of frequency coefficients F_{2^j+s}, $s = 0$ to $2^j - 1$.

The conclusion is as follows. To compute the wavelet amplitude coefficients by a discrete algorithm, we begin by representing $f(x)$ by the discrete sequence f_r, $r = 0$ to $N - 1$ (where N is a power of 2). Next the FFT is used to compute the set of (complex) frequency coefficients F_t, $t = 0$ to $N - 1$. Then octave blocks of F_t are processed by the IFFT to generate the amplitudes of the harmonic wavelet expansion of f_r.

For coefficients $\tilde{a}_{2^j+k}$, we begin from (17.200) rather than (17.199) and instead of (17.210) this leads to the result that

$$\tilde{a}_{2^j+k} = \sum_{s=0}^{2^j-1} F_{-(2^j+s)} e^{-i2\pi sk/2^j}, \quad k = 0 \text{ to } 2^j - 1. \qquad (17.211)$$

The negative index of the frequency coefficient $F_{-(2^j+s)}$ is due to the fact that this is the discrete equivalent of $F(-\omega)$. Since the discrete Fourier transform does not have negative indices, but instead replaces F_{-s} by F_{N-s} (which is possible because of the circular property of the DFT, see Chapter 10), (17.211) becomes

$$\tilde{a}_{2^j+k} = \sum_{s=0}^{2^j-1} F_{N-(2^j+s)} e^{-i2\pi sk/2^j}, \quad k = 0 \text{ to } 2^j - 1. \qquad (17.212)$$

For $N = 16$, Fig. 17.26 shows the algorithm diagrammatically, for the case when f_r is a real sequence in (a) and for the case when it is complex in (b). In case (a) it is not necessary to compute the coefficients $\tilde{a}_{2^j+k}$ because these are just the complex conjugates of a_{2^j+k} and so can be written across from the left half of the transform, as shown.

Computation of the amplitudes a_0 and $a_{N/2}$ in the algorithm involve special cases.

First consider a_0. This arises from the discrete equivalent of equations (17.201) and (17.202). Since

$$\phi(x - k) = \int_{-\infty}^{\infty} \Phi(\omega) e^{-i\omega k} e^{i\omega x} d\omega, \qquad (17.213)$$

(17.201) gives

$$a_{\phi,k} = \int_{-\infty}^{\infty} dx \int_{-\infty}^{\infty} d\omega \, \Phi^*(\omega) e^{i\omega k} f(x) e^{-i\omega x} = \int_0^{2\pi} d\omega \, F(\omega) e^{i\omega k} \qquad (17.214)$$

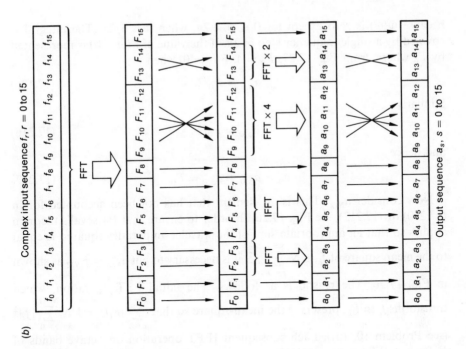

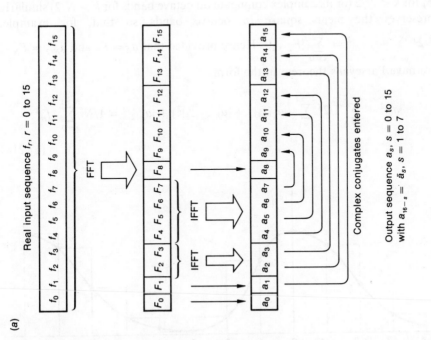

Fig. 17.26 FFT algorithm to compute the harmonic wavelet transform: (a) for a sequence of 16 real elements; (b) for a sequence of 16 complex elements (from Newland, *Proc. R. Soc. London* A, 1993 [114a])

because $\Phi(\omega)$ is zero except for $0 \leqslant \omega < 2\pi$, when it is $1/2\pi$. The integral is covered by a single frequency step of 2π. If the value of $F(\omega = 0)$ is represented by F_0 where, from Fig. 17.27, $F_0 = 4\pi F(\omega = 0)$, then

$$a_{\phi,0} = F_0/2. \qquad (17.125)$$

Similarly,

$$\tilde{a}_{\phi,0} = F_0/2 \qquad (17.216)$$

and hence

$$a_0 = a_{\phi,0} + \tilde{a}_{\phi,0} = F_0. \qquad (17.217)$$

Now consider $a_{N/2}$. This is an element that has not been accounted for in (17.210) or (17.212), see Fig. 17.26. In order to ensure that Parseval's theorem applies so that an appropriate sum of the wavelet amplitudes squared is equal to the mean square of f_r, $1/N \sum_{r=0}^{N-1} f_r^2$, it is necessary to put $a_{N/2} = F_{N/2}$ as shown in Fig. 17.26. The reason is as follows. The initial FFT operation, which transforms f_r to F_k, preserves the mean-square so that $\sum_{k=0}^{N-1} |F_k|^2 = 1/N \sum_{r=0}^{N-1} |f_r|^2$ (see Problem 10.6(ii)). Each subsequent IFFT operation on octave bands of F_k for $k < N/2$ (or its complex conjugate on octave bands for $k > N/2$) similarly preserves the mean square in octave bands so that, for example, $1/2^j \sum_{k=0}^{2^j-1} |a_{2^j+k}|^2 = \sum_{k=0}^{2^j-1} |F_{2^j+k}|^2$. Hence provided that $a_0 = F_0$ and $a_{N/2} = F_{N/2}$, we have Parseval's theorem in the form

$$|a_0|^2 + \sum_{j=0}^{n-2} 1/2^j \sum_{k=0}^{2^j-1} (|a_{2^j+k}|^2 + |a_{N-2^j-k}|^2) + |a_{N/2}|^2 = 1/N \sum_{r=0}^{N-1} |f_r|^2$$

$$(17.218)$$

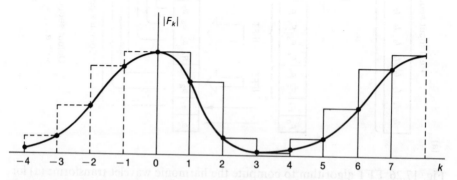

Fig. 17.27 Arrangement of the discrete approximations F_k of the continuous Fourier transform $F(\omega)$ where $\omega_k = 2\pi k$, drawn for a sequence length $N = 8$

where $N = 2^n$. The summation is made over all levels $j = 0$ to $n - 2$, the highest level $n - 2$ being the maximum that can be reached with an initial sequence of length 2^n. For example in Fig. 17.26 an initial sequence length of $2^4 = 16$ can provide harmonic wavelet amplitudes at levels $-1, 0, 1$ and $n - 2 = 2$ only. In comparison, had a wavelet derived from a dilation equation been used, levels up to level $n - 1 = 3$ would be included. This is because, for each j, k pair, there is one (real) dilation wavelet whereas there are two real harmonic wavelets; these are the even and odd wavelets that are given by the real and imaginary parts of the complex harmonic wavelet.

The wavelet amplitude $a_{N/2} = F_{N/2}$ defines a sequence f_r which, from the inverse discrete Fourier transform, is given by

$$f = [1 \quad -1 \quad 1 \quad -1 \quad 1 \quad -1 \quad ...].\qquad(17.219)$$

This can be shown (Problem 17.13) to be orthogonal to any arbitrary discrete harmonic $e^{i2\pi kr/N}$, $r = 0$ to $N - 1$, $k = 0$ to $N/2 - 1$. Since all discrete harmonic wavelets involve summations of discrete harmonics, it follows that the sequence (17.219) is orthogonal to all the discrete harmonic wavelets. Therefore, (17.219) may be thought of as an additional wavelet at the Nyquist frequency which extends over the complete sequence length $r = 0$ to $N - 1$.

With this interpretation, the algorithm illustrated in Fig. 17.26 is a fast way of computing the discrete harmonic wavelet transform. For the inverse transform, the same algorithm works in reverse, starting at the bottom of the diagrams in Fig. 17.26 and working upwards instead of downwards.

Mallat's pyramid algorithm (17.151), for wavelets derived from dilation equations, is a sequential algorithm in the sense that successive wavelets levels are computed in sequence (starting from the highest level and working downwards). In contrast, the FFT algorithm for harmonic wavelets is a parallel algorithm. The FFT is implemented once for the full sequence and then a second time, in octave blocks, all of which may be computed simultaneously. For applications which involve very long sequence lengths, or when real-time processing is required, that may be a significant advantage.

Programs in MATLAB for the harmonic wavelet transform are given in Appendix 7. The discrete harmonic wavelet transform a of a sequence f of length 2^n (whose elements may be complex) is computed by the function $hwtdn(f)$ and its inverse is regained by $ihwtdn(a)$. The defining matrix A for the harmonic wavelet map is computed by the function $hmapdn(f)$. In this case A has $n + 1$ rows and 2^{n-2} columns (compared with $n + 1$ rows and 2^{n-1} columns for the comparable array generated by $mapdn(f)$). This is because A is an array whose typical element is $|a_{2^j+k}|^2 + |a_{N-2^j-k}|^2$ and at level $y = n - 2$ the range of k is $k = 0$ to 2^{n-2}. For level $n - 1$, all the elements in row $n + 1$ are just $|a_{N/2}|^2$. Two-dimensional transforms can be set up to use the one-dimensional transforms twice following the same strategy as that for dilation wavelets in $wave2dn(f,N)$ and $iwave2dn(f,N)$.

Concluding comments

The application of wavelets to vibration analysis is still new. The subject is developing fast and many questions remain to be answered. For example, what is the best choice of wavelet to use for a particular problem? How should wavelet maps be interpreted? How far does the harmonic wavelet's computational simplicity compensate for its slow rate of decay in the x-domain (proportional to x^{-1})? For condition monitoring, the discrete wavelet transform (using families of orthogonal wavelets) will be competing with time–frequency methods using the short-time Fourier transform and the Wigner–Ville distribution (see Chapter 14). Orthogonal wavelets give fast algorithms and there is no redundancy: N data points give N wavelet amplitudes. Instead of a signal's mean-square being given by the area under its spectral density curve, mean-square is given by the volume under a two-dimensional wavelet surface with time (or distance) as one axis and wavelet level (a measure of frequency) as the other axis. In contrast, the STFT and Wigner–Ville methods provide redundant information in the sense that they provide more information than would be needed to reconstruct the signal being analysed and the computations take longer to complete.

In selecting the new material presented in this chapter, the object has been to explain wavelet theory in sufficient detail to allow practical calculations to be made. The harmonic wavelet transform has been worked out only recently by the author [114a]; it has been included because it appears to offer many of the advantages of dilation wavelets but at reduced complexity. A further development allows harmonic wavelets to be defined whose frequency composition is not restricted to octave bands only. For example, one octave may be divided into twelve contiguous narrow bands so that the octave gives rise to twelve different wavelets instead of just one wavelet. This allows greater discrimination in frequency (at the expense of loss of precision in the physical domain) and is helpful in some problems [114b]. Time will tell how far these new methods become part of the vibration engineer's toolbox.

Appendix 1

Table of integrals used in the calculation of mean square response (Chapter 7)

The following list of integrals of the form

$$I_n = \int_{-\infty}^{\infty} |H_n(\omega)|^2 \, d\omega$$

where

$$H_n(\omega) = \left\{ \frac{B_0 + (i\omega)B_1 + (i\omega)^2 B_2 + \cdots + (i\omega)^{n-1} B_{n-1}}{A_0 + (i\omega)A_1 + (i\omega)^2 A_2 + \cdots + (i\omega)^n A_n} \right\}$$

has been derived from previously published results.† The system whose mean square response is being calculated must be a stable system, and this will only be the case if the roots λ of its characteristic equation

$$(A_0 + \lambda A_1 + \lambda^2 A_2 + \cdots + \lambda^n A_n) = 0$$

all have negative real parts, i.e. they lie in the left-hand half of the λ-plane. An alternative equivalent condition is that the roots ω of

$$(A_0 + (i\omega)A_1 + (i\omega)^2 A_2 + \cdots + (i\omega)^n A_n) = 0$$

must all lie in the upper half of the ω-plane. Provided that this condition is satisfied, then the integrals have the following values.

For $n = 1$, $\qquad H_1(\omega) = \dfrac{B_0}{A_0 + i\omega A_1}$,

$$I_1 = \frac{\pi B_0^2}{A_0 A_1}.$$

For $n = 2$, $\qquad H_2(\omega) = \dfrac{B_0 + i\omega B_1}{A_0 + i\omega A_1 - \omega^2 A_2}$,

$$I_2 = \frac{\pi \{A_0 B_1^2 + A_2 B_0^2\}}{A_0 A_1 A_2}.$$

† James *et al.* [38] p. 369. Similar results for the cases $n = 1$ to $n = 4$ are given in Crandall *et al.* [18] p. 72.

For $n = 3$, $H_3(\omega) = \dfrac{B_0 + i\omega B_1 - \omega^2 B_2}{A_0 + i\omega A_1 - \omega^2 A_2 - i\omega^3 A_3}$,

$$I_3 = \frac{\pi\{A_0 A_3(2B_0 B_2 - B_1^2) - A_0 A_1 B_2^2 - A_2 A_3 B_0^2\}}{A_0 A_3(A_0 A_3 - A_1 A_2)}$$

For $n = 4$, $H_4(\omega) = \dfrac{B_0 + i\omega B_1 - \omega^2 B_2 - i\omega^3 B_3}{A_0 + i\omega A_1 - \omega^2 A_2 - i\omega^3 A_3 + \omega^4 A_4}$,

$$I_4 = \frac{\pi\{A_0 B_3^2(A_0 A_3 - A_1 A_2) + A_0 A_1 A_4(2B_1 B_3 - B_2^2) - \\ - A_0 A_3 A_4(B_1^2 - 2B_0 B_2) + A_4 B_0^2(A_1 A_4 - A_2 A_3)\}}{A_0 A_4(A_0 A_3^2 + A_1^2 A_4 - A_1 A_2 A_3)}.$$

For $n = 5$, $H_5(\omega) = \dfrac{B_0 + i\omega B_1 - \omega^2 B_2 - i\omega^3 B_3 + \omega^4 B_4}{A_0 + i\omega A_1 - \omega^2 A_2 - i\omega^3 A_3 + \omega^4 A_4 + i\omega^5 A_5}$,

$$I_5 = \frac{\begin{aligned}\pi\{&A_0 B_4^2(A_0 A_3^2 + A_1^2 A_4 - A_0 A_1 A_5 - A_1 A_2 A_3) + \\ &+ A_0 A_5(2B_2 B_4 - B_3^2)(A_1 A_2 - A_0 A_3) + \\ &+ A_0 A_5(2B_0 B_4 - 2B_1 B_3 + B_2^2)(A_0 A_5 - A_1 A_4) + \\ &+ A_0 A_5(2B_0 B_2 - B_1^2)(A_3 A_4 - A_2 A_5) + \\ &+ A_5 B_0^2(A_1 A_4^2 + A_2^2 A_5 - A_0 A_4 A_5 - A_2 A_3 A_4)\}\end{aligned}}{\begin{aligned}A_0 A_5(&A_0^2 A_5^2 - 2A_0 A_1 A_4 A_5 - A_0 A_2 A_3 A_5 + \\ &+ A_1 A_2^2 A_5 + A_1^2 A_4^2 + A_0 A_3^2 A_4 - A_1 A_2 A_3 A_4)\end{aligned}}.$$

Results for two further cases, $n = 6$ and $n = 7$, can be obtained from the reference quoted.

Appendix 2
Computer programs

Fortran computer subprogram† to calculate the DFT of a sequence A(1), A(2), A(3), . . . , A(NB), where NB = 2ᴺ, by the FFT method

```
      SUBROUTINE FFT (A,N,NB)
      COMPLEX A(NB),U,W,T
C     DIVIDE ALL ELEMENTS BY NB
      DO 1 J = 1,NB
1     A(J) = A(J)/NB
C     REORDER SEQUENCE ACCORDING TO FIG 12.8
      NBD2 = NB/2
      NBM1 = NB − 1
      J = 1
      DO 4 L = 1,NBM1
      IF (L.GE.J) GO TO 2
      T = A(J)
      A(J) = A(L)
      A(L) = T
2     K = NBD2
3     IF (K.GE.J) GO TO 4
      J = J − K
      K = K/2
      GO TO 3
4     J = J + K
C     CALCULATE FFT ACCORDING TO FIG 12.5
      PI = 3.141592653589793
      DO 6 M = 1,N
      U = (1.0,0.0)
      ME = 2**M
      K = ME/2
      W = CMPLX(COS(PI/K),−SIN(PI/K))
      DO 6 J = 1,K
      DO 5 L = J,NB,ME
      LPK = L + K
      T = A(LPK)*U
      A(LPK) = A(L) − T
5     A(L) = A(L) + T
6     U = U*W
      RETURN
      END
```

†After Cooley *et al.* [13]. (Courtesy of the Institute of Electrical and Electronics Engineers, Inc.)

374   *Appendices*

The same program in Basic‡, using real arithmetic only, to calculate the DFT of a sequence Ar(1)+iAi(1), Ar(2)+iAi(2), Ar(3)+iAi(3), . . . , Ar(Nb)+iAi(Nb), where Nb=2ᴺ, by the FFT method

```
1    SUB Fft(Ar(*),Ai(*),N,Nb)
2    ! Divide all elements by Nb
3    FOR J=1 TO Nb
4    Ar(J)=Ar(J)/Nb
5    Ai(J)=Ai(J)/Nb
6    NEXT J
7    ! Reorder sequence according to Fig. 12.8
8    Nbd2=Nb/2
9    Nbm1=Nb-1
10   J=1
11   FOR L=1 TO Nbm1
12   IF (L>J) OR (L=J) THEN GOTO 19
13   Tr=Ar(J)
14   Ti=Ai(J)
15   Ar(J)=Ar(L)
16   Ai(J)=Ai(L)
17   Ar(L)=Tr
18   Ai(L)=Ti
19   K=Nbd2
20   IF (K>J) OR (K=J) THEN GOTO 24
21   J=J-K
22   K=K/2
23   GOTO 20
24   J=J+K
25   NEXT L
26   ! Calculate FFT according to Fig. 12.5
27   FOR M=1 TO N
28   Ur=1
29   Ui=0
30   Me=2∧M
31   K=Me/2
32   Wr=COS(PI/K)
33   Wi=-SIN(PI/K)
34   FOR J=1 TO K
35   FOR L=J TO Nb STEP Me
36   Lpk=L+K
37   Tr=Ar(Lpk)*Ur-Ai(Lpk)*Ui
38   Ti=Ai(Lpk)*Ur+Ar(Lpk)*Ui
39   Ar(Lpk)=Ar(L)-Tr
40   Ai(Lpk)=Ai(L)-Ti
41   Ar(L)=Ar(L)+Tr
42   Ai(L)=Ai(L)+Ti
43   NEXT L
44   Sr=Ur
45   Ur=Ur*Wr-Ui*Wi
46   Ui=Ui*Wr+Sr*Wi
47   NEXT J
48   NEXT M
49   SUBEND
```

Notes

1. The programs calculate

$$X_k = \frac{1}{N}\sum_{r=0}^{N-1} x_r e^{-i2\pi(kr/N)}, \qquad k = 0, 1, 2, ..., N-1, \tag{10.8}$$

using the FFT algorithm described in Chapter 12.

‡The version used is HP-enhanced Basic.

2. The names of some program variables differ from the mathematical symbols adopted in the text. They are related as follows:

Fortran variable name	Basic variable name	Mathematical symbol used in the text
N	N	n
NB	Nb	$N = 2^n$
Program input		
	$Ar(J)$	Real part of x_r
$A(J)$		x_r
	$Ai(J)$	Imaginary part of x_r
Program output		
	$Ar(J)$	Real part of X_k
$A(J)$		X_k
	$Ai(J)$	Imaginary part of X_k
$J = 1, 2, ..., NB$	$J = 1, 2, ..., Nb$	$r, k = 0, 1, ..., N - 1$

3. The programs will accept an input sequence, $y_r, r = 0, 1, 2, ..., N - 1$, whose elements are complex. For vibration analysis, when y_r is derived by sampling a record of a continuous random variable, the elements of y_r are all real and the elements of the input array have zero imaginary parts. Then the input array is as follows:

Fortran program	Basic program
$A(1) = CMPLX(x_0, 0)$	$Ar(1) = x_0$
	$Ai(1) = 0$
$A(2) = CMPLX(x_1, 0)$	$Ar(2) = x_1$
	$Ai(2) = 0$
....	....
$A(NB) = CMPLX(x_{N-1}, 0)$	$Ar(Nb) = x_{N-1}$
	$Ai(Nb) = 0$

4. In the Fortran program, the output array $A(1), A(2), A(3), ..., A(NB)$ is the series of complex numbers $X_0, X_1, X_2, ..., X_{N-1}$ which is the required DFT.

   In the Basic program, the output array $Ar(1), Ar(2), Ar(3), ..., Ar(Nb)$ gives the real parts and the output array $Ai(1), Ai(2), Ai(3), ..., Ai(Nb)$ gives the imaginary parts of the required DFT, $X_0, X_1, X_2, ..., X_{N-1}$.

5. To convert the programs to calculate the inverse DFT according to equation (10.9), the following changes are needed.

   For the Fortran program, lines, 3, 4 and 5 of the program have to be deleted (so that the division of all elements by NB is eliminated), and the program line which is 10 from the bottom has to be changed to

W = CMPLX(COS(PI/K),SIN(PI/K))

(so that the negative sign in front of the SIN term is eliminated).

For the Basic program, lines 2–6 have to be eliminated (so that the division of all elements by Nb is eliminated) and line 33 has to be altered to read

Wi = SIN(PI/K)

(so that the negative sign is eliminated).

Fortran computer subprogram to calculate the DFT of a two-dimensional complex array C(J,K) of order NB1 × NB2 where NB1 = 2^{N1} and NB2 = 2^{N2} by using the subprogram FFT listed above

```
        SUBROUTINE D12FFT(A,B,C,N1,N2,NB1,NB2)
        COMPLEX A(NB1),B(NB2),C(NB1,NB2)
        DO 2 K=1,NB1
        DO 1 J=1,NB2
   1    B(J)=C(K,J)
        CALL FFT(B,N2,NB2)
        DO 2 J=1,NB2
   2    C(K,J)=B(J)
        DO 4 K=1,NB2
        DO 3 J=1,NB1
   3    A(J)=C(J,K)
        CALL FFT(A,N1,NB1)
        DO 4 J=1,NB1
   4    C(J,K)=A(J)
        RETURN
        END
```

The same program in Basic, using real arithmetic only, to calculate the DFT of a two-dimensional array Cr(J,K) + i Ci(J,K) of order Nb1 × Nb2 where Nb1 = 2^{N1} and Nb2 = 2^{N2} by using the subprogram Fft listed above

```
   1    SUB D12fft(Ar(*),Ai(*),Br(*),Bi(*),
                   Cr(*),Ci(*),N1,N2,Nb1,Nb2)
   2    FOR K=1 TO Nb1
   3    FOR J=1 TO Nb2
   4    Br(J)=Cr(K,J)
   5    Bi(J)=Ci(K,J)
   6    NEXT J
   7    CALL Fft(Br(*), Bi(*),N2,Nb2)
   8    FOR J=1 TO Nb2
   9    Cr(K,J)=Br(J)
  10    Ci(K,J)=Bi(J)
  11    NEXT J
  12    NEXT K
  13    FOR K=1 TO Nb2
  14    FOR J=1 TO Nb1
  15    Ar(J)=Cr(J,K)
  16    Ai(J)=Ci(J,K)
  17    NEXT J
  18    CALL Fft(Ar(*),Ai(*),N1,Nb1)
  19    FOR J=1 TO Nb1
  20    Cr(J,K)=Ar(J)
  21    Ci(J,K)=Ai(J)
  22    NEXT J
  23    NEXT K
  24    SUBEND
```

Notes

6. These programs calculate

$$Y_{km} = \frac{1}{N_1 N_2} \sum_{r=0}^{N_1-1} \sum_{s=0}^{N_2-1} y_{rs} \, e^{-i2\pi(kr/N_1 + ms/N_2)} \qquad (15.11)$$

$$k = 0, 1, 2, \ldots, N_1 - 1$$
$$m = 0, 1, 2, \ldots, N_2 - 1$$

by calling the appropriate FFT subprogram given above.

7. The names of the program variables are related to the mathematical symbols as follows:

Fortran variable name	Basic variable name	Mathematical symbol used in the text
N1	N1	n_1
N2	N2	n_2
NB1	Nb1	$N_1 = 2^{n_1}$
NB2	Nb2	$N_2 = 2^{n_2}$
Program input		
	$Cr(J,K)$	Real part of y_{rs}
$C(J,K)$		y_{rs}
	$Ci(J,K)$	Imaginary part of y_{rs}
Program output		
	$Cr(J,K)$	Real part of Y_{km}
$C(J,K)$		Y_{km}
	$Ci(J,K)$	Imaginary part of Y_{km}
$J = 1, 2, \ldots, NB1$	$J = 1, 2, \ldots, Nb1$	$r, k = 0, 1, \ldots, N_1 - 1$
$K = 1, 2, \ldots, NB2$	$K = 1, 2, \ldots, Nb2$	$s, m = 0, 1, \ldots, N_2 - 1$

8. For the Fortran program, the dimensions of the complex arrays are $A(NB1)$, $B(NB2)$ and $C(NB1, NB2)$ where $NB1 = 2^{N1}$ and $NB2 = 2^{N2}$. The input sequence y_{rs}, $r = 0, 1, 2, \ldots, N_1 - 1, s = 0, 1, 2, \ldots, N_2 - 1$, is entered as the complex array $C(J,K)$, where J ranges from 1 to $NB1$ and K ranges from 1 to $NB2$. Usually the elements of y_{rs} will be real so that the imaginary parts of the input array $C(J,K)$ will be zero. The one-dimensional array $A(J)$ is used to store each column and $B(K)$ each row of $C(J,K)$ during the calculation. These must be dimensioned in the calling program but have no data input to or output from the subprogram.

For the Basic program, the dimensions of the arrays are $Ar(Nb1)$, $Ai(Nb1)$, $Br(Nb2)$, $Bi(Nb2)$, $Cr(Nb1, Nb2)$ and $Ci(Nb1, Nb2)$ where $Nb1 = 2^{N1}$ and $Nb2 = 2^{N2}$. The real parts of the input sequence y_{rs}, $r = 0, 1, 2, \ldots, N_1 - 1$,

$s = 0, 1, 2, \ldots, N_2 - 1$, are stored in the array $Cr(J, K)$ and the imaginary parts of the input sequence in the array $Ci(J, K)$, where J ranges from 1 to $Nb1$ and K ranges from 1 to $Nb2$. Usually the imaginary parts of the input sequence will be zero. The one-dimensional arrays $Ar(J)$ and $Ai(J)$ are used to store each column and $Br(K)$ and $Bi(K)$ each row of $Cr(J, K)$ and $Ci(J, K)$ during the calculation. They must be dimensioned in the calling program but have no data input to or output from the subprogram.

9. In the Fortran program, the two-dimensional array Y_{km}, $k = 0, 1, 2, \ldots$, $N_1 - 1, m = 0, 1, 2, \ldots, N_2 - 1$, which is the required two-dimensional DFT, is returned in the complex array $C(NB1, NB2)$.

   In Basic program, the two-dimensional output array Y_{km}, $k = 0, 1, 2, \ldots$, $N_1 - 1, m = 0, 1, 2, \ldots, N_2 - 1$, which is the required two-dimensional DFT, is returned with its real parts in $Cr(Nb1, Nb2)$ and its imaginary parts in $Ci(Nb1, Nb2)$.

10. In order to convert these programs to calculate the two-dimensional inverse DFT, it is necessary to change lines 6 and 12 of the Fortran program and lines 7 and 18 of the Basic program to call a one-dimensional inverse FFT subprogram, which may be obtained from the appropriate one-dimensional FFT listed above by making the alterations given in Note 5.

11. The programs listed in this appendix have been tested on a range of problems for which they have worked correctly. However, their accuracy is not guaranteed and no responsibility can be accepted by the author or the publishers in the event of errors being discovered in the programs.

Appendix 3

The Gaussian probability integral

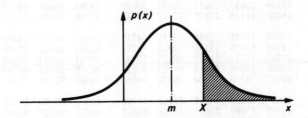

$$\text{Prob}\,(x > X) = (\text{Shaded Area}) = \int_X^\infty p(x)\,dx$$

$$= \int_X^\infty \frac{1}{\sqrt{2\pi}\,\sigma}\, e^{-(x-m)^2/2\sigma^2}\,dx$$

$$= \int_Z^\infty \frac{1}{\sqrt{2\pi}}\, e^{-z^2/2}\,dz = I(Z) \quad (\text{say})$$

$$\text{where } Z = \frac{X-m}{\sigma}.$$

The integral $I(Z)$ is tabulated opposite for values of Z from 0 to 4·99 (corresponding to values of X from $X = m$ to $X = m + 4\cdot99\sigma$). Values of the integral for $Z < 0$ can be deduced from the symmetry of the Gaussian curve, as follows:

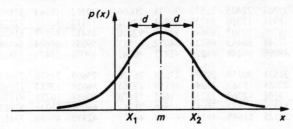

$$\text{Prob}\,(x > X_1) = 1 - \text{Prob}\,(x < X_1)$$

$$= 1 - \text{Prob}\,(x > X_2)$$

$$= 1 - I(Z)$$

$$\text{where} \qquad Z = \frac{X_2 - m}{\sigma} = \frac{m - X_1}{\sigma}.$$

Z		0	1	2	3	4	5	6	7	8	9
0·0	0·	50000	49601	49202	48803	48405	48006	47608	47210	46812	46414
0·1		46017	45620	45224	44828	44433	44038	43644	43251	42858	42465
0·2		42074	41683	41294	40905	40517	40129	39743	39358	38974	38591
0·3		38209	37828	37448	37070	36693	36317	35942	35569	35197	34827
0·4		34458	34090	33724	33360	32997	32636	32276	31918	31561	31207
0·5		30854	30503	30153	29806	29460	29116	28774	28434	28096	27760
0·6		27425	27093	26763	26435	26109	25785	25463	25143	24825	24510
0·7		24196	23885	23576	23270	22965	22663	22363	22065	21770	21476
0·8		21186	20897	20611	20327	20045	19766	19489	19215	18943	18673
0·9		18406	18141	17879	17619	17361	17106	16853	16602	16354	16109
1·0		15866	15625	15386	15151	14917	14686	14457	14231	14007	13786
1·1		13567	13350	13136	12924	12714	12507	12302	12100	11900	11702
1·2		11507	11314	11123	10935	10749	10565	10383	10204	10027	98525
1·3	0·0	96800	95098	93418	91759	90123	88508	86915	85343	83793	82264
1·4		80757	79270	77804	76359	74934	73529	72145	70781	69437	68112
1·5		66807	65522	64255	63008	61780	60571	59380	58208	57053	55917
1·6		54799	53699	52616	51551	50503	49471	48457	47460	46479	45514
1·7		44565	43633	42716	41815	40930	40059	39204	38364	37538	36727
1·8		35930	35148	34380	33625	32884	32157	31443	30742	30054	29379
1·9		28717	28067	27429	26803	26190	25588	24998	24419	23852	23295
2·0		22750	22216	21692	21178	20675	20182	19699	19226	18763	18309
2·1		17864	17429	17003	16586	16177	15778	15386	15003	14629	14262
2·2		13903	13553	13209	12874	12545	12224	11911	11604	11304	11011
2·3		10724	10444	10170	99031	96419	93867	91375	88940	86563	84242
2·4	0·0²	81975	79763	77603	75494	73436	71428	69469	67557	65691	63872
2·5		62097	60366	58677	57031	55426	53861	52336	50849	49400	47988
2·6		46612	45271	43965	42692	41453	40246	39070	37926	36811	35726
2·7		34670	33642	32641	31667	30720	29798	28901	28028	27179	26354
2·8		25551	24771	24012	23274	22557	21860	21182	20524	19884	19262
2·9		18658	18071	17502	16948	16411	15889	15382	14890	14412	13949
3·0		13499	13062	12639	12228	11829	11442	11067	10703	10350	10008
3·1	0·0³	96760	93544	90426	87403	84474	81635	78885	76219	73638	71136
3·2		68714	66367	64095	61895	59765	57703	55706	53774	51904	50094
3·3		48342	46648	45009	43423	41889	40406	38971	37584	36243	34946
3·4		33693	32481	31311	30179	29086	28029	27009	26023	25071	24151
3·5		23263	22405	21577	20778	20006	19262	18543	17849	17180	16534
3·6		15911	15310	14730	14171	13632	13112	12611	12128	11662	11213
3·7		10780	10363	99611	95740	92010	88417	84957	81624	78414	75324
3·8	0·0⁴	72348	69483	66726	64072	61517	59059	56694	54418	52228	50122
3·9		48096	46148	44274	42473	40741	39076	37475	35936	34458	33037
4·0		31671	30359	29099	27888	26726	25609	24536	23507	22518	21569
4·1		20658	19783	18944	18138	17365	16624	15912	15230	14575	13948
4·2		13346	12769	12215	11685	11176	10689	10221	97736	93447	89337
4·3	0·0⁵	85399	81627	78015	74555	71241	68069	65031	62123	59340	56675
4·4		54125	51685	49350	47117	44979	42935	40980	39110	37322	35612
4·5		33977	32414	30920	29492	28127	26823	25577	24386	23249	22162
4·6		21125	20133	19187	18283	17420	16597	15810	15060	14344	13660
4·7		13008	12386	11792	11226	10686	10171	96796	92113	87648	83391
4·8	0·0⁶	79333	75465	71779	68267	64920	61731	58693	55799	53043	50418
4·9		47918	45538	43272	41115	39061	37107	35247	33476	31792	30190

Reproduced by permission from Fisher *et al.* [28].

Appendix 4

Distribution of χ^2_κ

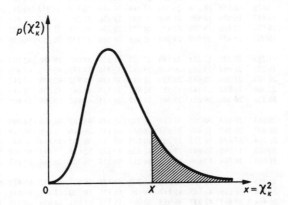

$$\text{Prob}\,(\chi^2_\kappa > X) = (\text{Shaded Area}) = \int_X^\infty \frac{1}{2^{\kappa/2}\,\Gamma(\kappa/2)}\,x^{(\kappa/2-1)}\,e^{-x/2}\,dx$$

(see problem 9.3)

$$= I(X, \kappa).$$

The table gives values of X when the number of degrees-of-freedom κ (left-hand side) and the value of the integral I (along the top) are specified.

					Prob ($\chi_\kappa^2 > X$)								
	0·99	0·98	0·95	0·90	0·80	0·70	0·50	0·30	0·20	0·10	0·05	0·02	0·01	0·0
1	0·0³157	0·0³628	0·00393	0·0158	0·0642	0·148	0·455	1·074	1·642	2·706	3·841	5·412	6·635	10·8
2	0·0201	0·0404	0·103	0·211	0·446	0·713	1·386	2·408	3·219	4·605	5·991	7·824	9·210	13·8
3	0·115	0·185	0·352	0·584	1·005	1·424	2·366	3·665	4·642	6·251	7·815	9·837	11·345	16·2
4	0·297	0·429	0·711	1·064	1·649	2·195	3·357	4·878	5·989	7·779	9·488	11·668	13·277	18·4
5	0·554	0·752	1·145	1·610	2·343	3·000	4·351	6·064	7·289	9·236	11·070	13·388	15·086	20·5
6	0·872	1·134	1·635	2·204	3·070	3·828	5·348	7·231	8·558	10·645	12·592	15·033	16·812	22·4
7	1·239	1·564	2·167	2·833	3·822	4·671	6·346	8·383	9·803	12·017	14·067	16·622	18·475	24·3
8	1·646	2·032	2·733	3·490	4·594	5·527	7·344	9·524	11·030	13·362	15·507	18·168	20·090	26·1
9	2·088	2·532	3·325	4·168	5·380	6·393	8·343	10·656	12·242	14·684	16·919	19·679	21·666	27·8
10	2·558	3·059	3·940	4·865	6·179	7·267	9·342	11·781	13·442	15·987	18·307	21·161	23·209	29·5
11	3·053	3·609	4·575	5·578	6·989	8·148	10·341	12·899	14·631	17·275	19·675	22·618	24·725	31·2
12	3·571	4·178	5·226	6·304	7·807	9·034	11·340	14·011	15·812	18·549	21·026	24·054	26·217	32·9
13	4·107	4·765	5·892	7·042	8·634	9·926	12·340	15·119	16·985	19·812	22·362	25·472	27·688	34·5
14	4·660	5·368	6·571	7·790	9·467	10·821	13·339	16·222	18·151	21·064	23·685	26·873	29·141	36·1
15	5·229	5·985	7·261	8·547	10·307	11·721	14·339	17·322	19·311	22·307	24·996	28·259	30·578	37·6
16	5·812	6·614	7·962	9·312	11·152	12·624	15·338	18·418	20·465	23·542	26·296	29·633	32·000	39·2
17	6·408	7·255	8·672	10·085	12·002	13·531	16·338	19·511	21·615	24·769	27·587	30·995	33·409	40·7
18	7·015	7·906	9·390	10·865	12·857	14·440	17·338	20·601	22·760	25·989	28·869	32·346	34·805	42·3
19	7·633	8·567	10·117	11·651	13·716	15·352	18·338	21·689	23·900	27·204	30·144	33·687	36·191	43·8
20	8·260	9·237	10·851	12·443	14·578	16·266	19·337	22·775	25·038	28·412	31·410	35·020	37·566	45·3
21	8·897	9·915	11·591	13·240	15·445	17·182	20·337	23·858	26·171	29·615	32·671	36·343	38·932	46·7
22	9·542	10·600	12·338	14·041	16·314	18·101	21·337	24·939	27·301	30·813	33·924	37·659	40·289	48·2
23	10·196	11·293	13·091	14·848	17·187	19·021	22·337	26·018	28·429	32·007	35·172	38·968	41·638	49·7
24	10·856	11·992	13·848	15·659	18·062	19·943	23·337	27·096	29·553	33·196	36·415	40·270	42·980	51·1
25	11·524	12·697	14·611	16·473	18·940	20·867	24·337	28·172	30·675	34·382	37·652	41·566	44·314	52·6
26	12·198	13·409	15·379	17·292	19·820	21·792	25·336	29·246	31·795	35·563	38·885	42·856	45·642	54·0
27	12·879	14·125	16·151	18·114	20·703	22·719	26·336	30·319	32·912	36·741	40·113	44·140	46·963	55·4
28	13·565	14·847	16·928	18·939	21·588	23·647	27·336	31·391	34·027	37·916	41·337	45·419	48·278	56·8
29	14·256	15·574	17·708	19·768	22·475	24·577	28·336	32·461	35·139	39·087	42·557	46·693	49·588	58·3
30	14·953	16·306	18·493	20·599	23·364	25·508	29·336	33·530	36·250	40·256	43·773	47·962	50·892	59·7
32	16·362	17·783	20·072	22·271	25·148	27·373	31·336	35·665	38·466	42·585	46·194	50·487	53·486	62·4
34	17·789	19·275	21·664	23·952	26·938	29·242	33·336	37·795	40·676	44·903	48·602	52·995	56·061	65·2
36	19·233	20·783	23·269	25·643	28·735	31·115	35·336	39·922	42·879	47·212	50·999	55·489	58·619	67·9
38	20·691	22·304	24·884	27·343	30·537	32·992	37·335	42·045	45·076	49·513	53·384	57·969	61·162	70·7
40	22·164	23·838	26·509	29·051	32·345	34·872	39·335	44·165	47·269	51·805	55·759	60·436	63·691	73·4
42	23·650	25·383	28·144	30·765	34·157	36·775	41·335	46·282	49·456	54·090	58·124	62·892	66·206	76·0
44	25·148	26·939	29·787	32·487	35·974	38·641	43·335	48·396	51·639	56·369	60·481	65·337	68·710	78·7
46	26·657	28·504	31·439	34·215	37·795	40·529	45·335	50·507	53·818	58·641	62·830	67·771	71·201	81·4
48	28·177	30·080	33·098	35·949	39·621	42·420	47·335	52·616	55·993	60·907	65·171	70·197	73·683	84·0
50	29·707	31·664	34·764	37·689	41·449	44·313	49·335	54·723	58·164	63·167	67·505	72·613	76·154	86·6
52	31·246	33·256	36·437	39·433	43·281	46·209	51·335	56·827	60·332	65·422	69·832	75·021	78·616	89·2
54	32·793	34·856	38·116	41·183	45·117	48·106	53·335	58·930	62·496	67·673	72·153	77·422	81·069	91·8
56	34·350	36·464	39·801	42·937	46·955	50·005	55·335	61·031	64·658	69·919	74·468	79·815	83·513	94·4
58	35·913	38·078	41·492	44·696	48·797	51·906	57·335	63·129	66·816	72·160	76·778	82·201	85·950	97·0
60	37·485	39·699	43·188	46·459	50·641	53·809	59·335	65·227	68·972	74·397	79·082	84·580	88·379	99·6
62	39·063	41·327	44·889	48·226	52·487	55·714	61·335	67·322	71·125	76·630	81·381	86·953	90·802	102·1
64	40·649	42·960	46·595	49·996	54·336	57·620	63·335	69·416	73·276	78·860	83·675	89·320	93·217	104·7
66	42·240	44·599	48·305	51·770	56·188	59·527	65·335	71·508	75·424	81·085	85·965	91·681	95·626	107·2
68	43·838	46·244	50·020	53·548	58·042	61·436	67·335	73·600	77·571	83·308	88·250	94·037	98·028	109·7
70	45·442	47·893	51·739	55·329	59·898	63·346	69·334	75·689	79·715	85·527	90·531	96·388	100·425	112·3

Number of degrees of freedom κ

For odd values of κ between 30 and 70 the mean of the tabular values for $\kappa - 1$ and $\kappa + 1$ may be taken.

Reproduced by permission from Fisher *et al.* [28].

Appendix 5

Random numbers

The table gives 1250 two-figure random numbers and may be used to generate random sequences of digits 0 to 9. Select any row, column or diagonal (in advance of looking at the numbers) and read the digits as they occur (see problem 13.4).

```
03 47 43 73 86    36 96 47 36 61    46 98 63 71 62    33 26 16 80 45    60 11 14 10 95
97 74 24 67 62    42 81 14 57 20    42 53 32 37 32    27 07 36 07 51    24 51 79 89 73
16 76 62 27 66    56 50 26 71 07    32 90 79 78 53    13 55 38 58 59    88 97 54 14 10
12 56 85 99 26    96 96 68 27 31    05 03 72 93 15    57 12 10 14 21    88 26 49 81 76
55 59 56 35 64    38 54 82 46 22    31 62 43 09 90    06 18 44 32 53    23 83 01 30 30

16 22 77 94 39    49 54 43 54 82    17 37 93 23 78    87 35 20 96 43    84 26 34 91 64
84 42 17 53 31    57 24 55 06 88    77 04 74 47 67    21 76 33 50 25    83 92 12 06 76
63 01 63 78 59    16 95 55 67 19    98 10 50 71 75    12 86 73 58 07    44 39 52 38 79
33 21 12 34 29    78 64 56 07 82    52 42 07 44 38    15 51 00 13 42    99 66 02 79 54
57 60 86 32 44    09 47 27 96 54    49 17 46 09 62    90 52 84 77 27    08 02 73 43 28

18 18 07 92 46    44 17 16 58 09    79 83 86 19 62    06 76 50 03 10    55 23 64 05 05
26 62 38 97 75    84 16 07 44 99    83 11 46 32 24    20 14 85 88 45    10 93 72 88 71
23 42 40 64 74    82 97 77 77 81    07 45 32 14 08    32 98 94 07 72    93 85 79 10 75
52 36 28 19 95    50 92 26 11 97    00 56 76 31 38    80 22 02 53 53    86 60 42 04 53
37 85 94 35 12    83 39 50 08 30    42 34 07 96 88    54 42 06 87 98    35 85 29 48 39

70 29 17 12 13    40 33 20 38 26    13 89 51 03 74    17 76 37 13 04    07 74 21 19 30
56 62 18 37 35    96 83 50 87 75    97 12 25 93 47    70 33 24 03 54    97 77 46 44 80
99 49 57 22 77    88 42 95 45 72    16 64 36 16 00    04 43 18 66 79    94 77 24 21 90
16 08 15 04 72    33 27 14 34 09    45 59 34 68 49    12 72 07 34 45    99 27 72 95 14
31 16 93 32 43    50 27 89 87 19    20 15 37 00 49    52 85 66 60 44    38 68 88 11 80

68 34 30 13 70    55 74 30 77 40    44 22 78 84 26    04 33 46 09 52    68 07 97 06 57
74 57 25 65 76    59 29 97 68 60    71 91 38 67 54    13 58 18 24 76    15 54 55 95 52
27 42 37 86 53    48 55 90 65 72    96 57 69 36 10    96 46 92 42 45    97 60 49 04 91
00 39 68 29 61    66 37 32 20 30    77 84 57 03 29    10 45 65 04 26    11 04 96 67 24
29 94 98 94 24    68 49 69 10 82    53 75 91 93 30    34 25 20 57 27    40 48 73 51 92

16 90 82 66 59    83 62 64 11 12    67 19 00 71 74    60 47 21 29 68    02 02 37 03 31
11 27 94 75 06    06 09 19 74 66    02 94 37 34 02    76 70 90 30 86    38 45 94 30 38
35 24 10 16 20    33 32 51 26 38    79 78 45 04 91    16 92 53 56 16    02 75 50 95 98
38 23 16 86 38    42 38 97 01 50    87 75 66 81 41    40 01 74 91 62    48 51 84 08 32
31 96 25 91 47    96 44 33 49 13    34 86 82 53 91    00 52 43 48 85    27 55 26 89 62

66 67 40 67 14    64 05 71 95 86    11 05 65 09 68    76 83 20 37 90    57 16 00 11 66
14 90 84 45 11    75 73 88 05 90    52 27 41 14 86    22 98 12 22 08    07 52 74 95 80
68 05 51 18 00    33 96 02 75 19    07 60 62 93 55    59 33 82 43 90    49 37 38 44 59
20 46 78 73 90    97 51 40 14 02    04 02 33 31 08    39 54 16 49 36    47 95 93 13 30
64 19 58 97 79    15 06 15 93 20    01 90 10 75 06    40 78 78 89 62    02 67 74 17 33

05 26 93 70 60    22 35 85 15 13    92 03 51 59 77    59 56 78 06 83    52 91 05 70 74
07 97 10 88 23    09 98 42 99 64    61 71 62 99 15    06 51 29 16 93    58 05 77 09 51
68 71 86 85 85    54 87 66 47 54    73 32 08 11 12    44 95 92 63 16    29 56 24 29 48
26 99 61 65 53    58 37 78 80 70    42 10 50 67 42    32 17 55 85 74    94 44 67 16 94
14 65 52 68 75    87 59 36 22 41    26 78 63 06 55    13 08 27 01 50    15 29 39 39 43

17 53 77 58 71    71 41 61 50 72    12 41 94 96 26    44 95 27 36 99    02 96 74 30 83
90 26 59 21 19    23 52 23 33 12    96 93 02 18 39    07 02 18 36 07    25 99 32 70 23
41 23 52 55 99    31 04 49 69 96    10 47 48 45 88    13 41 43 89 20    97 17 14 49 17
60 20 50 81 69    31 99 73 68 68    35 81 33 03 76    24 30 12 48 60    18 99 10 72 34
91 25 38 05 90    94 58 28 41 36    45 37 59 03 09    90 35 57 29 12    82 62 54 65 60

34 50 57 74 37    98 80 33 00 91    09 77 93 19 82    74 94 80 04 04    45 07 31 66 49
85 22 04 39 43    73 81 53 94 79    33 62 46 86 28    08 31 54 46 31    53 94 13 38 47
09 79 13 77 48    73 82 97 22 21    05 03 27 24 83    72 89 44 05 60    35 80 39 94 88
88 75 80 18 14    22 95 75 42 49    39 32 82 22 49    02 48 07 70 37    16 04 61 67 87
90 96 23 70 00    39 00 03 06 90    55 85 78 38 36    94 37 30 69 32    90 89 00 76 33
```

Reproduced by permission from Fisher *et al.* [28].

Appendix 6
Distribution of reverse arrangements for a random series

In a series of N terms, all of which are assumed to be derived from a continuous series and therefore be different, a *reverse arrangement* can be defined as occurring:

either (i) when the value of a term in the series is *greater than* a later term in the series,

or (ii) when the value of a term in the series is *less than* a later term in the series.

For a random series of N terms, i.e. a series in which the order of terms is random, the probability P of there being *not more than* c reverse arrangements (defined according to either of the above two definitions) has the following values: (see problem 14.4)

N = 5		N = 6		N = 7		N = 8		N = 9		N = 10		N = 12	
c	P	c	P	c	P	c	P	c	P	c	P	c	P
0	0.008	0	0.001	1	0.001	2	0.001	4	0.001	6	0.001	12	0.002
1	0.042	1	0.008	2	0.005	3	0.003	5	0.003	7	0.002	14	0.004
2	0.117	2	0.028	3	0.015	4	0.007	6	0.006	8	0.005	16	0.010
3	0.242	3	0.068	4	0.035	5	0.016	7	0.012	9	0.008	18	0.022
4	0.408	4	0.136	5	0.068	6	0.031	8	0.022	10	0.014	20	0.043
5	0.592	5	0.235	6	0.119	7	0.054	9	0.038	11	0.023	21	0.058
6	0.758	6	0.360	7	0.191	8	0.089	10	0.060	12	0.036	22	0.076
7	0.883	7	0.500	8	0.281	9	0.138	11	0.090	13	0.054	23	0.098
8	0.958	8	0.640	9	0.386	10	0.199	12	0.130	14	0.078	24	0.125
9	0.992	9	0.765	10	0.500	11	0.274	13	0.179	15	0.108	25	0.155
10	1.000	10	0.864	11	0.614	12	0.360	14	0.238	16	0.146	26	0.190
		11	0.932	12	0.719	13	0.452	15	0.306	17	0.190	27	0.230
		12	0.972	13	0.809	14	0.548	16	0.381	18	0.242	28	0.273
		13	0.992	14	0.881	15	0.640	17	0.460	19	0.300	29	0.319
		14	0.999	15	0.932	16	0.726	18	0.540	20	0.364	30	0.369
		15	1.000	16	0.965	17	0.801	19	0.619	21	0.431	31	0.420
				17	0.985	18	0.862	20	0.694	22	0.500	32	0.473
				18	0.995	19	0.911	21	0.762	23	0.569	33	0.527
				19	0.999	20	0.946	22	0.821	24	0.636	34	0.580
				21	1.000	21	0.969	23	0.870	25	0.700	35	0.631
						22	0.984	24	0.910	26	0.758	36	0.681
						23	0.993	25	0.940	27	0.810	37	0.727
						24	0.997	26	0.962	28	0.854	38	0.770
						25	0.999	27	0.978	29	0.892	39	0.810
						28	1.000	28	0.988	30	0.922	40	0.845
								29	0.994	31	0.946	41	0.875
								30	0.997	32	0.964	42	0.902
								31	0.999	33	0.977	43	0.924
								36	1.000	34	0.986	44	0.942
										35	0.992	45	0.957
										36	0.995	46	0.969
										37	0.998	48	0.984
										38	0.999	50	0.993
										45	1.000	52	0.997
												54	0.999
												66	1.000

The last value of c in each column gives the maximum possible number of reverse arrangements.

N = 14		N = 16		N = 18		N = 21		N = 24		N = 27		N = 30	
c	P	c	P	c	P	c	P	c	P	c	P	c	P
18	0·001	28	0·002	36	0·001	60	0·003	80	0·002	104	0·001	136	0·002
20	0·002	32	0·006	40	0·003	64	0·007	88	0·006	112	0·004	144	0·004
22	0·005	34	0·010	44	0·007	68	0·013	92	0·011	120	0·010	152	0·010
24	0·010	36	0·016	48	0·016	72	0·024	96	0·019	128	0·025	160	0·021
26	0·018	38	0·026	50	0·024	74	0·032	100	0·031	136	0·052	168	0·040
28	0·031	40	0·039	52	0·034	76	0·043	104	0·048	140	0·073	176	0·072
30	0·051	42	0·058	54	0·048	78	0·055	108	0·072	144	0·099	180	0·094
32	0·079	44	0·083	56	0·066	80	0·070	112	0·104	148	0·132	184	0·121
33	0·096	46	0·114	58	0·088	82	0·088	114	0·123	152	0·170	188	0·152
34	0·117	48	0·153	60	0·115	84	0·109	116	0·145	154	0·192	192	0·188
35	0·140	49	0·175	62	0·147	86	0·134	118	0·169	156	0·216	196	0·228
36	0·165	50	0·199	64	0·184	88	0·162	120	0·195	158	0·241	200	0·274
37	0·194	51	0·225	66	0·227	90	0·193	122	0·223	160	0·268	202	0·298
38	0·225	52	0·253	68	0·275	92	0·228	124	0·254	162	0·296	204	0·323
39	0·259	53	0·282	70	0·327	94	0·265	126	0·286	164	0·325	206	0·349
40	0·295	54	0·313	71	0·354	96	0·306	128	0·320	166	0·355	208	0·375
41	0·334	55	0·345	72	0·383	98	0·349	130	0·356	168	0·386	210	0·402
42	0·374	56	0·378	73	0·411	100	0·394	132	0·394	170	0·418	212	0·430
43	0·415	57	0·412	74	0·441	102	0·441	134	0·432	172	0·451	214	0·458
44	0·457	58	0·447	75	0·470	103	0·464	136	0·471	174	0·484	216	0·486
45	0·500	59	0·482	76	0·500	104	0·488	137	0·490	175	0·500	217	0·500
46	0·543	60	0·518	77	0·530	105	0·512	138	0·510	176	0·516	218	0·514
47	0·585	61	0·553	78	0·559	106	0·536	140	0·549	178	0·549	220	0·542
48	0·626	62	0·588	79	0·589	108	0·583	142	0·587	180	0·582	222	0·570
49	0·666	63	0·622	80	0·617	110	0·629	144	0·625	182	0·614	224	0·598
50	0·705	64	0·655	81	0·646	112	0·673	146	0·662	184	0·645	226	0·625
51	0·741	65	0·687	82	0·673	114	0·715	148	0·697	186	0·675	228	0·651
52	0·775	66	0·718	84	0·725	116	0·754	150	0·730	188	0·704	230	0·677
53	0·806	67	0·747	86	0·773	118	0·790	152	0·762	190	0·732	232	0·702
54	0·835	68	0·775	88	0·816	120	0·823	154	0·791	192	0·759	236	0·749
55	0·860	69	0·801	90	0·853	122	0·853	156	0·819	194	0·784	240	0·792
56	0·883	70	0·825	92	0·885	124	0·879	158	0·844	196	0·808	244	0·831
57	0·904	71	0·847	94	0·912	126	0·902	160	0·866	198	0·830	248	0·864
58	0·921	72	0·867	96	0·934	128	0·921	162	0·887	200	0·850	252	0·893
60	0·949	74	0·903	98	0·952	130	0·938	164	0·905	204	0·885	256	0·917
62	0·969	76	0·930	100	0·966	132	0·951	168	0·934	208	0·915	260	0·937
64	0·982	78	0·952	102	0·976	134	0·963	172	0·956	212	0·938	264	0·953
66	0·990	80	0·968	104	0·984	136	0·972	176	0·972	216	0·956	272	0·975
68	0·995	82	0·979	108	0·993	140	0·985	180	0·983	224	0·980	280	0·988
70	0·998	84	0·987	112	0·997	144	0·992	184	0·990	232	0·992	288	0·995
72	0·999	88	0·996	116	0·999	148	0·996	188	0·994	240	0·997	296	0·998
91	1·000	120	1·000	153	1·000	210	1·000	276	1·000	351	1·000	435	1·000

Similar results for values of N up to 20 are given in Kreyszig [45], Table 55 p. 840.

Appendix 7

Wavelet programs

This appendix provides computer programs to calculate and display wavelet transforms using the methods described in Chapter 17.

It has been found convenient to program in the language of MATLAB®*. This is an interactive software package for scientific and engineering numerical computation which uses matrices that do not require dimensioning. Readers who wish can reprogram in Fortran or another basic level language, but the resulting programs are more cumbersome. MATLAB has been developed by and is a registered trade mark of The MathWorks Inc. Details of how to obtain it are given at the end of the book.

An M-file is the name given in MATLAB to a subprogram of normal MATLAB statements which are executed by a single-line command. The M-files here are function files in which arguments are passed from and to the main program; variables that are defined and used inside a file remain local to that file and do not operate globally on the main program. There are ten M-files:

function a = wavedn(f,N)

> to compute the wavelet transform a of an array f of 2^n elements (n an integer) using a wavelet with N coefficients

function f = iwavedn(a,N)

> to compute the inverse wavelet transform f of an array a of 2^n elements (n an integer) using a wavelet with N coefficients

function A = displayn(f,N)

> to compute and display the one-dimensional wavelet transform of an array f of 2^n elements (n an integer); each row of A (of order $n+1$ rows and 2^n columns) gives the contribution to the reconstructed signal of each level of the wavelet transform

function A = mapdn(f,N)

> to compute the two-dimensional array A which defines the mean-square wavelet map of function f; the function f has length 2^n, the wavelet has N coefficients and the order of A is $n+1$ rows and $2^{(n-1)}$ columns

*MATLAB is a registered trademark of The MathWorks Inc.

function A = wave2dn(F,N)

> to compute the wavelet transform A of a two-dimensional array F using a wavelet with N coefficients; the orders of A and F are both $2^{n1} \times 2^{n2}$ (n1 and n2 integers)

function F = iwave2dn(A,N)

> to compute the inverse wavelet transform F of a two-dimensional array A using a wavelet with N coefficients; the orders of A and F are both $2^{n1} \times 2^{n2}$ (n1 and n2 integers)

function c = dcoeffs(N)

> to compute the array of N coefficients (N even) required by the wavelet transform; called by the above functions

function a = hwtdn(f)

> to compute the harmonic wavelet transform a of an array f of 2^n elements (n an integer)

function f = ihwtdn(a)

> to compute the inverse harmonic wavelet transform f of an array a of 2^n elements (n an integer)

function A = hmapdn(f)

> to compute the two-dimensional array A which defines the mean-square harmonic wavelet map of function f; the function f has length 2^n and the order of A is $n + 1$ rows and $2^{(n-2)}$ columns

A description of each M-file, giving its purpose, calling statement, algorithm, source reference and limitations, is given below, followed by the program listing.

Legal disclaimer

These M-files have been tested on a range of problems for which they are believed to have worked correctly. However, their accuracy is not guaranteed and no responsibility can be accepted by the author or the publishers in the event of errors being discovered in the programs. We make no warranties, express or implied, that the programs are free from error, or are consistent with any particular standard of merchantability, or that they meet any particular requirements. They should not be relied upon for solving a problem whose incorrect solution could result in injury to a person or loss of property. The application of these programs is at the user's own risk and the author and publisher disclaim all liability for damages, whether direct, incidental or consequential, arising from such application or any other use of the material in this book.

Copyright

The text of this book, including the computer program listings, is protected by copyright. Purchasers of the book are authorized to make one machine-readable copy of each program for their own use. Distribution of the programs to any other person is not authorized. Information about purchasing additional copies of the programs is given at the back of the book.

wavedn, iwavedn

Purpose:

Wavelet transform and its inverse

Synopsis:

a = wavedn(f,N)
f = iwavedn(a,N)

Description:

wavedn(f,N) returns an array whose elements are the wavelet transform of the sequence of elements in f. The analysing wavelet has N coefficients where N is even. iwavedn(a,N) is the inverse transform.

Algorithm:

Mallat's pyramid algorithm (Mallat [110]; Strang [115]) adapted for a circular transform, see Chapter 17.

Limitations:

It is necessary for the sequence f to have 2^n elements (n an integer); its transform a also has 2^n elements. The wavelet's coefficients are imported from the M-file dcoeffs(N) which is restricted to even values of N in the range 2 to 20.

If f is complex, so that f = fr + i*fi, the wavelet transform a is also complex so that a = ar + i*ai where ar is the transform of fr and ai is the transform of fi, and vice versa.

```
function a = wavedn(f,N)
%
% Copyright (c) D. E. Newland 1992
% All rights reserved
%
M = length(f);
n = round(log(M)/log(2));
c = dcoeffs(N);
clr = fliplr(c);
for j = 1:2:N-1, clr(j) = -clr(j); end
```

```
                          a = f;
                          for k = n:-1:1
                              m = 2^(k-1);
                              x = [0]; y = [0];
                              for i = 1:1:m
                                  for j = 1:1:N
                                      k(j) = 2*i-2 + j;
                                      while k(j) > 2*m, k(j) = k(j)-2*m; end
                                  end
                                  z = a(k);
                                  [mr,nc] = size(z);
                                  if nc > 1, z = z.'; end
                                  x(i) = c*z;
                                  y(i) = clr*z;
                              end
                              x = x/2; y = y/2;
                              a(1:m) = x;
                              a(m + 1:2*m) = y;
                          end


          function f = iwavedn(a,N)
          %
          %  Copyright (c) D. E. Newland 1992
          %  All rights reserved
          %
          M = length(a);
          n = round(log(M)/log(2));
          c = dcoeffs(N);
          f(1) = a(1);
          for j = 1:1:N/2
              c1(1,j) = -c(2*j);
              c1(2,j) = c(2*j-1);
              c2(1,j) = c(N-2*j + 1);
              c2(2,j) = c(N-2*j + 2);
          end
          for k = 1:1:n
              m = 2^(k-1);
              for i = 1:1:m
                  for j = 1:N/2
                      k(j) = m + i-N/2 + j;
                      while k(j) < m + 1, k(j) = k(j) + m; end
                  end
```

```
        z = a(k); [mr,nc] = size(z); if nc > 1, z = z.'; end
        x(2*i-1:2*i) = c1*z;
        zz = f(k-m); [mr, nc] = size(zz); if nc > 1, zz = zz.'; end
        xx(2*i-1:2*i) = c2*zz;
    end
    f(1:2*m) = x + xx;
end
```

displayn

Purpose:

To display the reconstruction of a (real) signal from its wavelet transform.

Synopsis:

A = displayn(f,N)

Description:

Uses a = wavedn(f,N) to compute the wavelet transform of f which is a sequence of length 2^n; then iwavedn(b,N) is used repeatedly to generate successive rows of the matrix A which has $n + 1$ rows and 2^n columns. Row 1 is the level -1 reconstruction (always a constant), row 2 is the level 0 reconstruction, row 3 the level 1 reconstruction and so on. All the rows of A added together regenerate the original signal f. The results are plotted for examination.

Algorithm:

The wavelet transform a is partitioned into $n + 1$ adjacent subarrays of length 1, 1, 2, 4, 8, 16, ..., $2^{(n-1)}$ elements. Each array b of length 2^n consists of a null matrix except for an embedded subarray from a. The corresponding level of the reconstruction of f is obtained by computing iwavedn(b,N). A detailed description is given in Chapter 17.

Limitations:

Restricted to even values of N in the range 2 to 20. The signal f must have only real elements. Plotting instructions are for MATLAB version 3.

Example:

```
f = [-9  -7  4  4  4  12  2  6];
A = displayn(f,2)
```

gives

$$A = \begin{matrix} 2 & 2 & 2 & 2 & 2 & 2 & 2 & 2 \\ -4 & -4 & -4 & -4 & 4 & 4 & 4 & 4 \\ -6 & -6 & 6 & 6 & 2 & 2 & -2 & -2 \\ -1 & 1 & 0 & 0 & -4 & 4 & -2 & 2 \end{matrix}$$

```
function A = displayn(f,N)
%
% Copyright (c)  D. E. Newland 1992
% All rights reserved
%
M = length(f);
n = round(log(M)/log(2));
subplot(211)
plot(f), grid, title('signal f')
a = wavedn(f,N);
plot(a), grid, title(['wavelet transform wavedn(f,',int2str(N),')'])
xlabel('PRESS TO CONTINUE ...  AND WAIT PLEASE')
pause
%
b = zeros(1,M);
b(1) = a(1);
A(1,:) = iwavedn(b,N);
for i = 2:1:n + 1
    b = zeros(1:M);
    b(2^(i-2) + 1:2^(i-1)) = a(2^(i-2) + 1:2^(i-1));
    A(i,:) = iwavedn(b,N);
end
%
j = 0;
for i = 1:1:n + 1
    plot(A(i,:)),grid,title(['wavelet level ',int2str(i-2)])
    j = j + 1; if j == 2, xlabel('PRESS TO CONTINUE'), pause, j = j-2; end
end
plot(sum(A)),grid,title('Reconstructed signal: all levels added')
xlabel('PRESS TO CONTINUE')
pause
clg, subplot(111)
```

mapdn

Purpose:

> Computes a two-dimensional array A which defines the mean-square map for a one-dimensional function f.

Synopsis:

> A = mapdn(f,N)

Description:

> Uses a = wavedn(f,N) to compute the wavelet transform of f which is a sequence of length 2^n; then reorders and arranges the square of the elements $(1:2^n)$ to form the two-dimensional array A which has $(n + 1)$ rows and $2^{(n-1)}$ columns; the volume under this surface is proportional to the mean-square value of f.

Algorithm:

> The elements of the wavelet transform a are reordered according to the strategy described in Chapter 17 so that when the matrix A is formed each element (squared) is located at the centre of the wavelet it represents. The location of the elements (before reordering) is as shown in Fig. 17.18.

Limitations:

> Restricted to even values of N in the range 2 to 20.

Example:

> f = [−9 −7 4 4 4 12 2 6];
> A = mapdn(f,2)
> gives
> A = 4   4   4   4
>   16   16   16   16
>   36   36   4   4
>   1   0   16   4

```
function A = mapdn(f,N)
%
% Copyright (c) D. E. Newland 1992
% All rights reserved
%
M = length(f);
n = round(log(M)/log(2));
```

```
a = wavedn(f,N);
b(1) = a(1); b(2) = a(2);
for j = 1:n-1
  for k = 1:2^j
    index = 2^j + k + N/2-1;
    while index > 2^(j + 1), index = index-2^j; end
    b(index) = a(2^j + k);
  end
end
a = b;
for j = 1:2^(n-1)
  A(1,j) = a(1);
end
for j = 2:n + 1
  for k = 1:2^(j-2)
    for m = 1:2^(n-j + 1)
      A(j,(k-1)*2^(n-j + 1) + m) = a(2^(j-2) + k);
    end
  end
end
A = A.*conj(A);
```

wave2dn, iwave2dn

Purpose:

Two-dimensional wavelet transform and its inverse.

Synopsis:

```
A = wave2dn(F,N)
F = iwave2dn(A,N)
```

Description:

wave2dn(F,N) returns a matrix whose elements are the wavelet transform of the elements in F. The analysing wavelet has N coefficients where N is even. iwave2dn(A,N) is the inverse transform.

Algorithm:

Mallat's pyramid algorithm (Mallat [110]; Strang [115]) adapted for a circular transform as described in Chapter 17.

Limitations:

It is necessary for the matrix F to have $2^{n1} \times 2^{n2}$ elements (*n1* and *n2* integers); its transform A also has $2^{n1} \times 2^{n2}$ elements. The wavelet's coefficients are imported from the M-file dcoeffs(N) which is restricted to even values of N in the range 2 to 20.

If F is complex so that F = Fr + i*Fi, the wavelet transform A is also complex so that A = Ar + i*Ai, where Ar is the transform of Fr and Ai is the transform of Fi, and vice versa.

```
function A = wave2dn(F,N)
%
%  Copyright (c) D. E. Newland 1992
%  All rights reserved
%
[M1,M2] = size(F);
for k = 1:M1
B(k,:) = wavedn(F(k,:),N);
end
for j = 1:M2
A(:,j) = (wavedn(B(:,j).',N)).';
end

function F = iwave2dn(A,N)
%
%  Copyright (c) D. E. Newland 1992
%  All rights reserved
%
[M1,M2] = size(A);
for k = 1:M1
B(k,:) = iwavedn(A(k,:),N);
end
for j = 1:M2
F(:,j) = (iwavedn(B(:,j).',N)).';
end
```

dcoeffs

Purpose:

To generate the N coefficients required by the wavelet transform.

Synopsis:

c = dcoeffs(N)

Description:

Generates an array with the N coefficients required by the wavelet transform.

Algorithm:

Numerical data are derived from results published by Daubechies [104].

Limitations:

Restricted to even values of N in the range 2 to 20.

```
function c = dcoeffs(N)
%
% Copyright (c) D. E. Newland 1992
% All rights reserved
%
if N = = 2,
  c(1) = 1;
  c(2) = 1;
end
if N = = 4,
  c(1) = (1 + sqrt(3))/4;
  c(2) = (3 + sqrt(3))/4;
  c(3) = (3-sqrt(3))/4;
  c(4) = (1-sqrt(3))/4;
end
%
if N = = 6,
  s = sqrt(5 + 2*sqrt(10));
  c(1) = (1 + sqrt(10) + s)/16;
  c(2) = (5 + sqrt(10) + 3*s)/16;
  c(3) = (5-sqrt(10) + s)/8;
  c(4) = (5-sqrt(10)-s)/8;
  c(5) = (5 + sqrt(10)-3*s)/16;
  c(6) = (1 + sqrt(10)-s)/16;
end
%
if N = = 8,
  c(1) = .325803428051;
  c(2) = 1.010945715092;
  c(3) = .892200138246;
  c(4) = -.039575026236;
  c(5) = -.264507167369;
```

```
  c(6) = .043616300475;
  c(7) = .046503601071;
  c(8) = -.014986989330;
end
%
if N = = 10,
  c(1) = .226418982583;
  c(2) = .853943542705;
  c(3) = 1.024326944260;
  c(4) = .195766961347;
  c(5) = -.342656715382;
  c(6) = -.045601131884;
  c(7) = .109702658642;
  c(8) = -.008826800109;
  c(9) = -.017791870102;
  c(10) = .004717427938;
end
%
if N = = 12,
  c(1) = .157742432003;
  c(2) = .699503814075;
  c(3) = 1.062263759882;
  c(4) = .445831322930;
  c(5) = -.319986598891;
  c(6) = -.183518064060;
  c(7) = .137888092974;
  c(8) = .038923209708;
  c(9) = -.044663748331;
  c(10) = .000783251152;
  c(11) = .006756062363;
  c(12) = -.001523533805;
end
%
if N = = 14,
  c(1) = .110099430746;
  c(2) = .560791283626;
  c(3) = 1.031148491636;
  c(4) = .664372482211;
  c(5) = -.203513822463;
  c(6) = -.316835011281;
  c(7) = .100846465010;
  c(8) = .114003445160;
  c(9) = -.053782452590;
```

```
  c(10) = -.023439941565;
  c(11) = .017749792379;
  c(12) = .000607514996;
  c(13) = -.002547904718;
  c(14) = .000500226853;
end
%
if N = = 16,
  c(1) = .076955622108;
  c(2) = .442467247152;
  c(3) = .955486150427;
  c(4) = .827816532422;
  c(5) = -.022385735333;
  c(6) = -.401658632782;
  c(7) = .000668194093;
  c(8) = .182076356847;
  c(9) = -.024563901046;
  c(10) = -.062350206651;
  c(11) = .019772159296;
  c(12) = .012368844819;
  c(13) = -.006887719256;
  c(14) = -.000554004548;
  c(15) = .000955229711;
  c(16) = -.000166137261;
end
%
if N = = 18,
  c(1) = .053850349589;
  c(2) = .344834303815;
  c(3) = .855349064359;
  c(4) = .929545714366;
  c(5) = .188369549506;
  c(6) = -.414751761802;
  c(7) = -.136953549025;
  c(8) = .210068342279;
  c(9) = .043452675461;
  c(10) = -.095647264120;
  c(11) = .000354892813;
  c(12) = .031624165853;
  c(13) = -.006679620227;
  c(14) = -.006054960574;
  c(15) = .002612967280;
  c(16) = .000325814672;
```

```
c(17) = -.000356329759;
c(18) = .000055645514;
end
%
if N = = 20,
  c(1) = .037717157593;
  c(2) = .266122182794;
  c(3) = .745575071487;
  c(4) = .973628110734;
  c(5) = .397637741770;
  c(6) = -.353336201794;
  c(7) = -.277109878720;
  c(8) = .180127448534;
```

```
c(9) = .131602987102;
c(10) = -.100966571196;
c(11) = -.041659248088;
c(12) = .046969814097;
c(13) = .005100436968;
c(14) = -.015179002335;
c(15) = .001973325365;
c(16) = .002817686590;
c(17) = -.000969947840;
c(18) = -.000164709006;
c(19) = .000132354366;
c(20) = -.000018758416;
end
```

hwtdn, ihwtdn

Purpose:

Harmonic wavelet transform and its inverse.

Synopsis:

```
a = hwtdn(f)
f = ihwtdn(a)
```

Description:

hwtdn(f) returns an array whose elements are the harmonic wavelet transform of the sequence of elements in f. ihwtdn(a) is the inverse transform.

Algorithm:

The harmonic wavelet transform algorithm using the fft described in Chapter 17 and illustrated in Fig. 17.26.

Limitations:

It is necessary for the sequence f to have 2^n elements (n an integer); its transform a also has 2^n elements. The elements of f may be real or complex; those of a are complex.

```
function a = hwtdn(f)
%
% Copyright (c) D. E. Newland 1992
% All rights reserved
%
N = length(f);
n = round(log(N)/log(2));
F = fft(f)/N;
a(1) = F(1); a(2) = F(2);
for j = 1:n-2
    a(2^j + 1:2^(j + 1)) = ifft(F(2^j + 1:2^(j + 1)))*2^j;
    a(N-2^(j + 1) + 2:N-2^j + 1) = fliplr(fft(fliplr(F(N-2^(j + 1) + 2:N-2^j + 1))));
end
a(N/2 + 1) = F(N/2 + 1);
a(N) = F(N);
```

```
function f = ihwtdn(a)
%
% Copyright (c) D. E. Newland 1992
% All rights reserved
%
N = length(a);
n = round(log(N)/log(2));
F(1) = a(1); F(2) = a(2);
for j = 1:n-2
    F(2^j + 1:2^(j + 1)) = fft(a(2^j + 1:2^(j + 1)))/2^j;
    F(N-2^(j + 1) + 2:N-2^j + 1) = fliplr(ifft(fliplr(a(N-2^(j + 1) + 2:N-2^j + 1))));
end
F(N/2 + 1) = a(N/2 + 1);
F(N) = a(N);
f = ifft(F)*N;
```

hmapdn

Purpose:

Computes a two-dimensional array A which defines the mean-square harmonic wavelet map for a one-dimensional function f.

Synopsis:

A = hmapdn(f)

Description:

Uses a = hwtdn(f) to compute the harmonic wavelet transform of f which is a sequence of length 2^n; then arranges the squares of the elements $(1:2^n)$ to form the two-dimensional array A which as $(n + 1)$ rows and $2^{(n-2)}$ columns; the volume under this surface is proportional to the mean-square of f.

Algorithm:

Each element of A is the sum of the squares of the moduli of a(r) and $a(N + 2 - r), r = 2$ to $N/2$; a(1) and $a(N/2 + 1)$ are special cases, see Chapter 17, Fig. 17.26.

Limitations:

The length of sequence f must be 2^n (n an integer). The elements of f may be real or complex.

```
function A = hmapdn(f)
%
%  Copyright (c) D. E. Newland 1992
%  All rights reserved
%
N = length(f);
n = round(log(N)/log(2));
a = hwtdn(f);
%
for j = 1:2^(n-2)
  A(1,j) = abs(a(1))^2;
  A(n + 1,j) = abs(a(2^(n-1) + 1))^2;
end
for j = 2:n
  for k = 1:2^(j-2)
    for m = 1:2^(n-j)
      A(j,(k-1)*2^(n-j) + m) = abs(a(2^(j-2) + k))^2 + abs(a(N + 2-2^(j-2)-k))^2;
    end
  end
end
```

Problems

Note: Problem numbers are prefixed by the number of the chapter to which they relate.

1.1. A random variable x is distributed between 0 and 1 so that

$$p(x) = 1 \qquad \text{for} \qquad 0 \leqslant x \leqslant 1$$

and

$$p(x) = 0 \qquad \begin{array}{ll} \text{for} & x < 0 \\ \text{and} & x > 1. \end{array}$$

Determine $E[x]$, $E[x^2]$ and σ_x.

1.2. A random variable x is obtained by sampling a regular saw-tooth waveform $x(t)$ at randomly chosen times, all times being equally likely. Construct the probability density function $p(x)$ and the probability distribution function $P(x)$ for x when the amplitude of the saw-tooth wave is $\pm a$.

1.3. A random variable y is defined by

$$y = x^2$$

where x is another random variable which satisfies a Gaussian distribution with zero mean and unit variance ($\sigma_x^2 = 1$). Find $E[y]$, $E[y^2]$ and σ_y.

Hint. Equations (1.20) may be extended by adding the results

$$\int_0^\infty y^3 e^{-y^2/2\sigma^2}\, dy = 2\sigma^4$$

$$\int_0^\infty y^4 e^{-y^2/2\sigma^2}\, dy = \tfrac{3}{2}\sqrt{2\pi}\,\sigma^5.$$

Show also that the probability distribution function $P(y)$ for y may be written as

$$P(y = a^2) = \int_{-a}^{a} p(x)\, dx$$

and differentiate $P(y)$ to verify that

$$p(y) = \begin{cases} \dfrac{1}{\sqrt{2\pi y}}e^{-y/2} & \text{for} \quad y \geqslant 0 \\ 0 & \text{for} \quad y < 0. \end{cases}$$

1.4. Two random variables x and y have mean values $E[x]$ and $E[y]$ respectively. If x and y are *statistically independent*, the average value of the product of x and y, $E[xy]$, is equal to the product of the separate mean values, i.e.

$$E[xy] = E[x] \cdot E[y].$$

If $z = x + y$, where x and y are statistically independent, show that

$$E[z^2] = E[x^2] + E[y^2] + 2E[x]E[y].$$

If, in addition,

$$p(x) = \begin{cases} 0 \cdot 5 & \text{for} \quad 0 \leqslant x \leqslant 2 \\ 0 & \text{for} \quad x < 0 \quad \text{and} \quad x > 2 \end{cases}$$

$$p(y) = \begin{cases} 1 & \text{for} \quad 0 \leqslant y \leqslant 1 \\ 0 & \text{for} \quad y < 0 \quad \text{and} \quad y > 1 \end{cases}$$

find $E[z^2]$ and σ_z.

1.5. A random variable z is defined by

$$z = x + y$$

where x and y are two separate, statistically independent random variables. If $p(x)$ is the probability density function for x and $P(y)$ and $P(z)$ are the probability distribution functions for y and z, verify that

$$P(z) = \int_{-\infty}^{z} P(y = z - x)p(x)\,\mathrm{d}x.$$

Hence determine $P(z)$ and, by differentiation, $p(z)$ when $p(x)$ and $p(y)$ are as given in problem 1.4.

Hint. Notice that the three cases (i) $0 \leqslant z \leqslant 1$, (ii) $1 < z \leqslant 2$ and (iii) $2 < z \leqslant 3$ must be considered separately because of the discontinuities in $p(x)$ and $P(y)$.

2.1. Two random variables x and y have the joint probability density function

$$p(x, y) = \frac{c}{1 + x^2 + y^2 + x^2 y^2}\,,$$

where c is a constant. Verify that x and y are statistically independent and find the value of c for $p(x, y)$ to be correctly normalized. Check that $E[x] = E[y] = 0$ and that $E[x^2]$ and $E[y^2]$ are both infinite.

2.2. Two Gaussian random variables x and y both have zero means and the same variance a^2. If their normalized covariance $\rho_{xy} = 1/2$, write down the second-order probability density function $p(x, y)$ and check that the first-order probability density functions $p(x)$ and $p(y)$ can be obtained from $p(x, y)$ by applying (2.6). Hence determine the conditional probability density function $p(x|y)$ and find the mean value of x when y is fixed.

2.3. The joint probability distribution function $P(x, y)$ for two continuous random variables $x(t)$ and $y(t)$ is defined so that

$$P(x, y) = \text{Prob}\,(x(t_0) < x \text{ and } y(t_0) < y)$$

where $x(t)$ and $y(t)$ are sampled at arbitrary time t_0. Verify that

$$p(x, y) = \frac{\partial^2}{\partial x\, \partial y} P(x, y)$$

and check also that

$$P(x|y) = \frac{P(x, y)}{P(y)}$$

where $P(x|y)$ is the probability that $x(t_0) < x$ when $y(t_0) < y$.

For the case when

$$p(x, y) = \begin{cases} cxy & \begin{cases} \text{for} & 0 \leqslant x \leqslant 1 \\ \text{and} & 0 \leqslant y \leqslant 1 \end{cases} \\ 0 & \text{elsewhere} \end{cases}$$

find the value of the constant c if $p(x, y)$ is to be correctly normalized, and determine
(i) $\text{Prob}\,(x > 1/2)$, (ii) $\text{Prob}\,(x > y)$ and (iii) $\text{Prob}\,(x > 1/2|y < 1/2)$.

Hint. For (ii) and (iii) integrate $p(x, y)$ over the appropriate regions of x and y.

2.4. A random process is composed of an ensemble of sample functions, each of which is a sine wave of constant amplitude x_0 and frequency ω. The frequency is the same for all samples but the amplitude varies randomly from sample to sample. Each sample also has a different (constant) phase angle ϕ which also varies randomly from sample to sample. A typical sample is therefore represented by the equation

$$x(t) = x_0 \sin(\omega t - \phi)$$

where x_0, ω and ϕ are all constant for the sample. If the joint probability density function for x_0 and ϕ is

$$p(x_0, \phi) = \begin{cases} \dfrac{1}{2\pi X}\left\{1 + \left(2\dfrac{x_0}{X} - 1\right)\cos\phi\right\} & \\ & \text{for} \begin{cases} 0 \leqslant \phi \leqslant 2\pi \\ 0 \leqslant x_0 \leqslant X \end{cases} \\ 0 & \text{elsewhere} \end{cases}$$

Problems

determine the ensemble averages $E[x^2]$ and $E[x^2|\phi_0]$ where the latter denotes $E[x^2]$ for only those samples for which $\phi = \phi_0$ (i.e. $\phi_0 < \phi < \phi_0 + d\phi_0$). Is $x(t)$ an ergodic process? Is it a stationary process?

3.1. (i) If x is a random variable and c is an arbitrary constant, show that the minimum value of $E[(x - c)^2]$ is σ_x^2.

(ii) If x and y are two correlated random variables with mean values m_x and m_y, and variances σ_x^2 and σ_y^2, verify that (3.6) gives the line of regression of y on x and (3.7) gives the line of regression of x on y.

(iii) Consider the case when x and y are both continuous Gaussian random variables with zero means, and roughly sketch their joint probability density function (2.12) against rectangular axes representing x/σ_x and y/σ_y. Show that the probability surface is a circular bell when $\rho = 0$ but becomes distorted when $\rho \neq 0$ and planes of symmetry through the axis of the bell make angles of 45° and 135° with the x/σ_x axis where they cut the x–y plane. Hence show that, on a graph of x/σ_x against y/σ_y, the lines of regression of y on x and of x on y are bisected by the planes of symmetry of the probability surface.

3.2. A random process $x(t)$ consists of an ensemble of sample functions, each of which is a square wave of amplitude $\pm a$ and period T, Fig. 3.8(a). The "phase" of each sample is defined as the time $t = \phi$ at which the sample first switches from $+a$ to $-a$ (for $t > 0$). ϕ varies randomly from sample to sample with a uniform probability distribution of the form shown in Fig. 3.8(b).

Show that the time history of a single sample function may be represented by the Fourier series expansion

$$x(t) = \frac{4a}{\pi} \sum_{n=1,3,5,\ldots} \frac{1}{n} \sin \frac{2\pi n}{T}(\phi - t)$$

and hence calculate the ensemble average $E[x(t)x(t + \tau)]$ to find the autocorrelation function for the $x(t)$ process. Verify that this agrees with the result shown in Fig. 3.9.

Hint. The Fourier series "recipe" is given by equations (4.1) and (4.2).

3.3. Each sample function $x(t)$ of a random process is given by

$$x(t) = a_1 \sin(\omega_1 t + \phi_1) + a_2 \sin(\omega_2 t + \phi_2)$$

where a_1, a_2, ω_1 and ω_2 are constants but ϕ_1 and ϕ_2, although constant for each sample, vary randomly from one sample to the next. If ϕ_1 and ϕ_2 are statistically independent and each is uniformly distributed between 0 and 2π, determine the autocorrelation function $R_x(\tau)$.

The corresponding sample functions of a separate random process are given by

$$y(t) = b_1 \sin(\omega_1 t + \omega_1 t_0 + \phi_1) + b_2 \sin(\omega_2 t + \phi_3)$$

where b_1, b_2 and t_0 are constants and ϕ_3 varies randomly from one sample to the next. If ϕ_3 is statistically independent of ϕ_1 and ϕ_2 and is also uniformly distributed between 0 and 2π, find the cross-correlation functions $R_{xy}(\tau)$ and $R_{yx}(\tau)$.

3.4. A forging hammer operates at regular intervals and generates in its foundation a stress $x(t)$ which has the approximate time history shown in Fig. P3.4(a). The duration of each impact is short so that $b \ll T$.

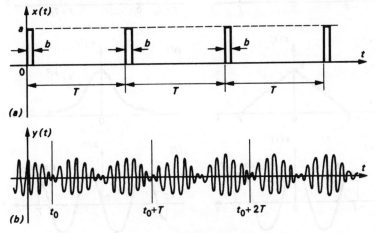

Fig. P3.4

As a result, the ground acceleration at a location some distance away from the hammer has the time history $y(t)$, Fig. P3.4(b), which is represented approximately by

$$y(t) = c \sin \Omega(t - t_0) \sin \frac{2\pi}{T}(t - t_0)$$

where c, Ω, T and t_0 are all constants.

Determine the cross-correlation functions $R_{xy}(\tau)$ and $R_{yx}(\tau)$.

Hint. Calculate a sample average by integrating $x(t)y(t + \tau)$ over one period.

How will the results change if $y(t)$ becomes

$$y(t) = c \sin \Omega(t - t_0) \sin \frac{2\pi}{T}(t - t_0) + z(t)$$

where $z(t)$ is the response to an independent source of high-frequency noise with zero mean?

4.1. Use equation (4.10) to determine the Fourier transforms of the six different functions of time shown in Fig. P4.1. Approximately sketch

$X(\omega)$ against ω in each case and compare particularly cases (a) and (c) and also cases (e) and (f) when $T \gg 2\pi/\omega_0$.

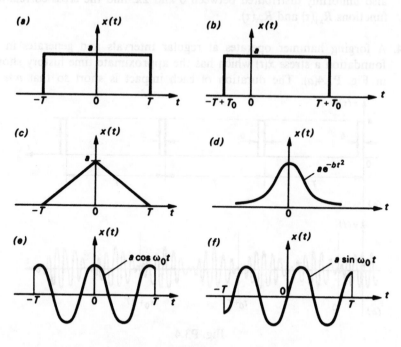

Fig. P4.1

Hint. Make use of the standard result (1.20) to evaluate case (d).

4.2. Determine the Fourier transform of the function $x(t)$ shown in Fig. P4.2 and compare this with the discrete Fourier series representation of a continuous square wave of amplitude a and period $2T$. Show that the values of $\mathrm{Im}\{X(\omega)\}$ at $\omega = \pi/T$, $3\pi/T$, $5\pi/T$, etc., are in proportion to the amplitudes of corresponding terms in the discrete Fourier series.

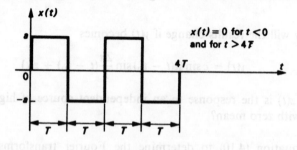

Fig. P4.2

4.3. Prove *Parseval's theorem* which states that, if $x(t)$ is any arbitrary periodic function of time represented by (4.1), then

$$\frac{1}{T}\int_0^T x^2(t)\,dt = a_0^2 + \sum_{k=1}^\infty \tfrac{1}{2}(a_k^2 + b_k^2).$$

Show also that, if $x_1(t)$ and $x_2(t)$ are aperiodic functions of t which satisfy (4.7) and whose Fourier transforms are $X_1(\omega)$ and $X_2(\omega)$, then

$$\int_{-\infty}^\infty x_1(t)x_2(t)\,dt = \int_{-\infty}^\infty dt\, x_1(t) \int_{-\infty}^\infty d\omega\, X_2(\omega)e^{i\omega t}$$

and, by reversing the order of integration,

$$\int_{-\infty}^\infty x_1(t)x_2(t)\,dt = 2\pi \int_{-\infty}^\infty d\omega\, X_1^*(\omega)X_2(\omega)$$

where $X_1^*(\omega)$ denotes the complex conjugate of $X_1(\omega)$. Hence prove that Parseval's theorem for an aperiodic function $x(t)$ has the form

$$\frac{1}{2\pi}\int_{-\infty}^\infty x^2(t)\,dt = \int_{-\infty}^\infty |X(\omega)|^2\,d\omega.$$

4.4. (i) Prove the *duality theorem* of Fourier analysis which states that, if $X(\omega)$ is the Fourier transform of $x(t)$, then $(1/2\pi)x(-\omega)$ is the Fourier transform of $X(t)$.

(ii) Consider the Fourier transform of the rectangular pulse shown in Fig. P4.1(a) and take the limiting case when $T \to 0$ and $a \to \infty$ but the product aT remains finite so that $2aT \to c$. Hence find the Fourier transform of an impulse of magnitude c at time $t = 0$.

(iii) Beginning with the Fourier transform of a rectangular pulse centred at time T_0, Fig. P4.1(b), find also the Fourier transform of an impulse of magnitude c at $t = T_0$.

(iv) Hence find the Fourier transforms of the pairs of impulse functions shown in Figs. P4.4(a) and (b).

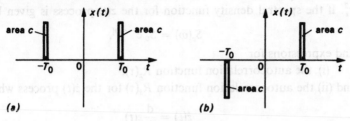

Fig. P4.4

(v) Now use the duality theorem to show that the Fourier transform of the function $x(t) = a\cos\omega_0 t$ is a pair of "impulse functions" of area $a/2$ each at frequencies $+\omega_0$ and $-\omega_0$. Similarly show that the Fourier transform of $x(t) = a \sin \omega_0 t$ is an imaginary quantity represented by a pair of equal and opposite impulse functions of area $a/2$ at $-\omega_0$ and $-a/2$ at $+\omega_0$.

Note. An impulse function is an example of a *generalized function* by which the scope of Fourier analysis can be greatly extended. Notwithstanding the condition (4.7), every generalized function has a Fourier transform. A constant and sine and cosine waves are other examples of generalized functions and so, as shown above, we can calculate their Fourier transforms even though (4.7) is not satisfied. The formal definition of a generalized function is quite complicated (see Lighthill [46]), but, for practical purposes, if we can find the Fourier transform of a function then it must be a generalized function. Impulse functions (also called delta functions) are applied in Chapters 5 and 6 and used frequently throughout the later chapters of this book.

5.1. The single-sided spectral density of the deflection $x(t)$ of a machine component is

$$W_0 = 0.01 \text{ cm}^2/\text{Hz} \quad \text{(constant)}$$

over the frequency band 0 to 200 Hz and is zero for frequencies above 200 Hz. Determine the mean square deflection $E[x^2]$ and obtain an expression for the autocorrelation function $R_x(\tau)$ for the $x(t)$ process, where the units of τ are seconds.

5.2. The autocorrelation function for a stationary random process $x(t)$ with zero mean and variance σ_x^2 is given approximately by the expression

$$R_x(\tau) = \sigma_x^2 e^{-(\omega_1 \tau)^2} \cos \omega_2 \tau$$

where ω_1 and ω_2 are constants. Calculate and sketch the corresponding spectral density.

Hint. Put $\cos \omega_2 \tau = \frac{1}{2}(e^{i\omega_2\tau} + e^{-i\omega_2\tau})$ and use the standard integrals (1.20).

5.3. A stationary Gaussian random process $z(t)$ has zero mean and variance σ_z^2. If the spectral density function for the $z(t)$ process is given by

$$S_z(\omega) = S_0 e^{-c|\omega|},$$

find expressions for
    (i) the autocorrelation function $R_z(\tau)$,
and (ii) the autocorrelation function $R_{\dot{z}}(\tau)$ for the $\dot{z}(t)$ process where

$$\dot{z}(t) = \frac{d}{dt} z(t).$$

Also (iii) show that, for any stationary process $z(t)$,

$$\frac{d}{d\tau} R_z(\tau) \bigg|_{\tau=0} = E[z(t)\dot{z}(t)] = -E[\dot{z}(t)z(t)]$$

and hence prove that $E[z\dot{z}] = 0$, i.e. that $z(t)$ and its derivative $\dot{z}(t)$ (provided that this exists) are always uncorrelated, and that the slope of $R_z(\tau)$ at $\tau = 0$ must be zero.

Hence find

(iv) the joint probability distribution for $z(t)$ and $\dot{z}(t)$.

Hint. If $z(t)$ is a Gaussian process then so is $\dot{z}(t)$ and so is any other process which describes the response of a linear system to $z(t)$ (see Chapter 7).

5.4. Each sample function $x(t)$ of a stationary random process is defined by the expression

$$x(t) = a\cos(\omega t + \phi)$$

where a, ω and ϕ are all constant. The amplitude a is the same for all samples, but ω and ϕ vary randomly from one sample to the next. Consider the case when ω and ϕ are independent and

$$p(\phi) = \begin{cases} 1/2\pi & \text{for} \quad 0 \leqslant \phi \leqslant 2\pi \\ 0 & \text{elsewhere} \end{cases}$$

$$p(\omega) = \begin{cases} 1/2\omega_0 & \text{for} \quad -\omega_0 \leqslant \omega \leqslant \omega_0 \\ 0 & \text{elsewhere}. \end{cases}$$

Find the autocorrelation function $R_x(\tau)$ and hence show that the spectral density of the $x(t)$ process is constant for $|\omega| \leqslant \omega_0$ and is zero for all higher frequencies.

5.5. Figure 3.9 shows the autocorrelation function $R_x(\tau)$ for a random process whose sample functions $x(t)$ are each periodic square waves, Fig. 3.8. Find the discrete Fourier series representation for $R_x(\tau)$ and for one sample of $x(t)$ when the time origin of the latter is adjusted so that $x(t)$ switches from $-a$ to $+a$ at $t = 0$.

Use the result of part (v) of problem 4.4 to draw a graph of $S_x(\omega)$ against ω and of $X(\omega)$ (the Fourier transform of the periodic function $x(t)$) against ω.

If S_k is the area under the "impulse" at $\omega = \omega_k$ and X_k is the area under the corresponding impulse at $\omega = \omega_k$ on these two graphs, show that

$$S_k = X_k^* X_k.$$

This is a special case of an important result which will be used later in Chapter 10 when we are considering digital spectral analysis.

5.6. Determine the cross-spectral densities $S_{xy}(\omega)$ and $S_{yx}(\omega)$ for the correlated functions $x(t)$ and $y(t)$ described in problem 3.4.

Hint. Make use of the result of part (v) of problem 4.4.

6.1. Find the frequency response function $H(\omega)$ and the impulse response function $h(t)$ for the four different systems shown in Fig. P6.1. In each case $x(t)$ is the input and $y(t)$ the output.

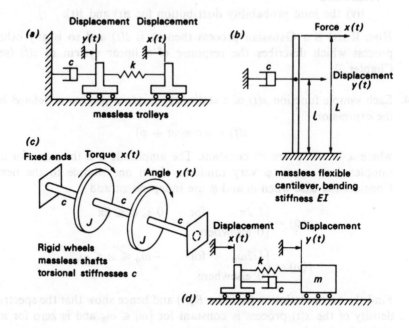

Fig. P6.1

Notes:

All parts. The physical units of $H(\omega)$ are those of (output/input) while those of $h(t)$ are (output)/(unit impulse of input) = (output)/(input × time).

Parts (a) and (d). The impulsive displacement input $x(t) = I\delta(t)$ requires $x(t)$ to move rapidly forward and then immediately backward to its initial position in such a way that the area under the displacement–time curve during this brief period is unity.

Part (b). The deflection at distance x from its fixed end for a cantilever of length a loaded at its free end by a lateral force P is

$$(P/6EI)(3x^2a - x^3).$$

Part (c). Assume that the system has very small damping and that it is at rest before the impulsive input is applied.

Part (d). The initial conditions for $h(t)$ are

$$h(0+) = c/m \quad \text{and} \quad \dot{h}(0+) = k/m - c^2/m^2.$$

To find these, note that, since there are discontinuities in both y and $\dot{y}$ at $t = 0$, we can put, for $t \simeq 0$,

$$\dot{y}(t) = \begin{cases} a\delta(t) + b & \text{for} \quad t > 0 \\ a\delta(t) & \text{for} \quad t < 0 \end{cases}$$

$$\ddot{y}(t) = a\frac{d}{dt}\delta(t) + b\delta(t)$$

where a and b are constant.

Assume that $c^2 < 4mk$, i.e. that damping is less than critical.

6.2. A vehicle moving in a straight line carries a mass m which is attached by a spring k and a viscous dashpot c, Fig. P6.2. The distance travelled by the vehicle is described by the coordinate $x(t)$ and the displacement of the mass m relative to the vehicle by the coordinate $y(t)$. Assuming that the

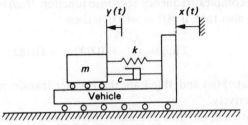

Fig. P6.2

paths of motion of the vehicle and the mass and the axes of the spring and the dashpot are all parallel, determine:

(i) the frequency response function $H(\omega)$ relating the output $y(t)$ to the input *acceleration* $\ddot{x}(t)$;

(ii) the impulse response function $h(t)$ in response to a delta function input acceleration of unit magnitude (dimensions of velocity) at $t = 0$ (assuming $c^2 < 4mk$, i.e. less than critical damping);

and check

(iii) that $H(\omega)$ can be determined from $h(t)$ by applying the Fourier transform relation (6.21).

6.3. The system shown in Fig. P6.1(a) is subjected to a step input $x(t) = a$, suddenly applied at $t = 0$ and lasting for time T. At $t = T$, $x(t)$ suddenly returns to zero. Assuming that $x(t) = 0$ for $t < 0$ and $t > T$, determine the system's response $y(t)$ to this input:

(i) by classical methods of solving the governing differential equation of motion;

(ii) by applying the convolution integral (6.24);

and

(iii) by determining the Fourier transform of the input, applying (6.20), and then calculating the inverse transform of $Y(\omega)$.

Hint for part (iii). The definite integral in the final step of part (iii) should be determined by the method of residues. See, for instance, Kreyszig [45], Ch. 15. Knowledge of this method is not required elsewhere and, if desired, it is sufficient to accept the result stated in part (iii) without proof.

6.4. (i) Take Fourier transforms of both sides of equation (6.27) to show that

$$Y(\omega) = \frac{1}{2\pi} \int_{-\infty}^{\infty} dt \int_{-\infty}^{\infty} d\theta \, e^{-i\omega t} h(\theta) x(t - \theta).$$

Now (i) integrate with respect to $\tau = (t - \theta)$, keeping θ constant, and (ii) integrate with respect to θ, to prove that

$$Y(\omega) = H(\omega)X(\omega)$$

where the complex frequency response function $H(\omega)$ is defined by (6.21).
(ii) Prove also that, if $y(t) = w(t)x(t)$, then

$$Y(\omega) = \int_{-\infty}^{\infty} W(\Omega)X(\omega - \Omega)\,d\Omega$$

where $X(\omega)$, $Y(\omega)$ and $W(\omega)$ are the Fourier transforms of $x(t)$, $y(t)$ and $w(t)$ respectively.

Convolution in the time domain is therefore equivalent to *multiplication* in the frequency domain (part i), and vice versa (part ii).

7.1. For the system described in problem 6.2, determine the mean square displacement of the mass m, $E[y^2]$, when the spectrum of the input acceleration $\ddot{x}$ is white with (constant) spectral density S_0.

Calculate also the r.m.s. force in the spring when the input acceleration is a white noise process with spectral density S_0 superimposed on a constant acceleration of magnitude a, i.e. when

$$\ddot{x}(t) = a + b(t)$$

where $b(t)$ is a white noise process with spectral density S_0. Assume that the initial transients excited by first application of the steady input acceleration have disappeared.

7.2. For the system shown in Fig. 6.3(a), determine the mean output displacement $E[y]$ and the mean square value $E[y^2]$ when:
(i) the spectral density of the excitation $S_x(\omega) = S_0$ (constant), i.e. when there is white noise excitation,
and when
(ii) $S_x(\omega) = S_0$ for $-\Omega < \omega < \Omega$ but $S_x(\omega) = 0$ for other values of ω, i.e. when the excitation is "band-limited" white noise.

For case (i) only, determine also the autocorrelation function for the input and use (7.10) to calculate the output autocorrelation function

$R_y(\tau)$. Check that the correct output spectral density may be obtained from this $R_y(\tau)$ by applying (5.1).

7.3. The dynamic characteristics of a block of visco-elastic material may be modelled approximately by two springs and a viscous dashpot, arranged as shown in Fig. P7.3. If the block is subjected to a randomly varying force $f(t)$ and responds by moving distance $x(t)$, determine the mean square response $E[x^2]$ when $f(t)$ is a member function of a stationary random process whose mean square spectral density is given by

$$S_f(\omega) = \frac{S_0}{(1 + \omega^2/\omega_0^2)}.$$

Hints. Introduce an additional coordinate $y(t)$ (say) between the spring k_2 and the dashpot and obtain two equations of motion relating $x(t)$, $y(t)$ and $f(t)$. Then put $f(t) = e^{i\omega t}$ and eliminate $y(t)$ to find $H(\omega)$ where $x(t) = H(\omega) e^{i\omega t}$. Write

$$S_f(\omega) = \frac{S_0}{|1 + i\omega/\omega_0|^2}$$

in order to obtain $E[x^2]$ in the standard form given in Appendix 1.

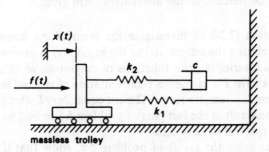

Fig. P7.3

7.4. For the two degree-of-freedom system shown in Fig. P7.4, determine the frequency response function $H(\omega)$ relating the output displacement $x_2(t)$ to the input force $F(t)$ and hence find the average kinetic energy of the mass m_2 when the spectral density of $F(t)$ is $S_F(\omega) = S_0$ (constant).

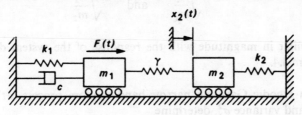

Fig. P7.4

7.5. A linear system has two inputs $x_1(t)$ and $x_2(t)$ and two outputs $y_1(t)$ and $y_2(t)$. The frequency response functions relating output $y_1(t)$ to inputs $x_1(t)$ and $x_2(t)$ are $H_{11}(\omega)$ and $H_{12}(\omega)$ respectively and those relating $y_2(t)$ to $x_1(t)$ and $x_2(t)$ are $H_{21}(\omega)$ and $H_{22}(\omega)$ respectively. Show that the cross-spectral densities between the two outputs, $S_{y_1y_2}(\omega)$ and $S_{y_2y_1}(\omega)$, are related to the input spectra by the equations

$$S_{y_1y_2}(\omega) = H_{11}^*(\omega)H_{21}(\omega)S_{x_1}(\omega) + H_{11}^*(\omega)H_{22}(\omega)S_{x_1x_2}(\omega) +$$
$$+ H_{12}^*(\omega)H_{21}(\omega)S_{x_2x_1}(\omega) + H_{12}^*(\omega)H_{22}(\omega)S_{x_2}(\omega)$$

$$S_{y_2y_1}(\omega) = H_{21}^*(\omega)H_{11}(\omega)S_{x_1}(\omega) + H_{21}^*(\omega)H_{12}(\omega)S_{x_1x_2}(\omega) + \quad (7.30)$$
$$+ H_{22}^*(\omega)H_{11}(\omega)S_{x_2x_1}(\omega) + H_{22}^*(\omega)H_{12}(\omega)S_{x_2}(\omega)$$

and to the cross-spectra between inputs and outputs by the equations

$$S_{y_1y_2}(\omega) = H_{11}^*(\omega)S_{x_1y_2}(\omega) + H_{12}^*(\omega)S_{x_2y_2}(\omega)$$
$$S_{y_2y_1}(\omega) = H_{21}^*(\omega)S_{x_1y_1}(\omega) + H_{22}^*(\omega)S_{x_2y_1}(\omega).$$

Hint. Derive the cross-correlation function $R_{y_1y_2}(\tau)$ in a form similar to (7.10) and then take its Fourier transform to find $S_{y_1y_2}(\omega)$. The result, equation (7.30), is needed in Chs 14 and 16. Make use of (7.22) to show that (7.30) can be reduced to the alternative form given.

7.6. Use equation (7.24) to investigate the form of the frequency response function relating the output $y(t)$ to the input $x(t)$ in problem 3.4. Assume that $x(t)$ is a series of delta functions of magnitude ab at spacing T.
(i) Calculate the Fourier series representation of $x(t)$ and hence show that, if the system is linear, frequency Ω must equal $2\pi n/T$ where n is an integer.
(ii) Use the result of the last part of problem 5.5 to find an expression for the input spectral density $S_x(\omega)$.
(iii) Hence, using the result of problem 5.6, show that $|H(\omega)| = cT/4ab$ at $\omega = \pm(n - 1)2\pi$ and $\pm(n + 1)2\pi/T$, and is zero at all other frequencies for which $S_x(\omega)$ and $S_{xy}(\omega)$ are defined.
(v) Lastly sketch $|H(\omega)|$ versus ω for the system of problem 7.4 for the case when c and γ are both small. Compare the response of this system when

$$\sqrt{\frac{k_1}{m_1}} \quad \text{and} \quad \sqrt{\frac{k_2}{m_2}}$$

are similar in magnitude with the response of the system described in problem 3.4.

8.1. For an ergodic Gaussian narrow-band random process $x(t)$ with zero mean and variance σ_x^2, determine
(i) the proportion of time which $x(t)$ spends outside the levels $x = \pm\sigma_x$;

(ii) the probability that any peak, selected at random, has a magnitude greater than σ_x;

(iii) the average level of all peaks;

and

(iv) the level which is exceeded by only 1 per cent of all peaks.

Hint. For part (i) refer to Appendix 3 for values of the Gaussian probability distribution function.

8.2. Consider the response $y(t)$ of the single degree-of-freedom oscillator shown in Fig. 7.2(a) when it is excited by a stationary Gaussian process $x(t)$ with spectral density

$$S_x(\omega) = \frac{S_0}{1 + \omega^2/\omega_0^2}.$$

Find expressions for the following statistics of the oscillator's response:

(i) the average frequency of positive slope crossings of the level $y = a$;

(ii) the probability density function for the distribution of peaks;

(iii) the average frequency of occurrence of maxima in $y(t)$.

Compare the frequency of positive slope zero crossings and the frequency of maxima. Explain why these are different and verify that, when there is vanishingly small damping, they become the same.

Hint. See the note following problem 7.3.

8.3. Calculate the statistical average frequency v_a^+ and the peak probability density function $p_p(a)$ for a stationary random process $y(t)$ whose joint probability density function is

$$p(y, \dot{y}) = \begin{cases} \dfrac{1}{4y_0\dot{y}_0} \text{ (constant)} & \begin{array}{l} \text{for} \quad -y_0 \leqslant y \leqslant y_0 \\ \text{and} \quad -\dot{y}_0 \leqslant \dot{y} \leqslant \dot{y}_0 \end{array} \\ 0 \text{ elsewhere.} \end{cases}$$

Use the result of problem 1.2 to describe the time history of sample functions of such a process.

Hint. Remember that $p(y)$ and $p(\dot{y})$ can be found from $p(y, \dot{y})$ by using (2.6) and (2.7).

8.4. Calculate the statistical average frequency v_0^+ and the frequency of maxima μ for the response $x_2(t)$ of mass m_2 of the two degree-of-freedom oscillator shown in Fig. P7.4, when its excitation $F(t)$ is a stationary, Gaussian, white noise process.

Check that, when $\gamma \to \infty$,

$$v_0^+ \to \frac{1}{2\pi} \sqrt{\left(\frac{k_1 + k_2}{m_2} \right)}.$$

Should this value not be

$$\frac{1}{2\pi} \sqrt{\left| \left(\frac{k_1 + k_2}{m_1 + m_2} \right) \right|} \ ?$$

Explain the discrepancy.

Hint. Remember that, to find v_0^+ and μ, it is only necessary to calculate the *ratios* $\sigma_{\dot{x}}/\sigma_x$ and $\sigma_{\ddot{x}}/\sigma_{\dot{x}}$ so that the evaluation of integral I_4 from Appendix 1 is greatly simplified; the denominator of I_4 does not need to be found.

8.5. Verify that the time history $x(t)$ of the simple harmonic motion

$$x(t) = r \sin \omega t$$

(where r and ω are constant) can be represented as a circle of radius r on a graph of $\dot{x}/\omega$ versus x, Fig. P8.5. If point P denotes the combined values of $x(t)$ and $\dot{x}(t)/\omega$ at time t, P progresses round the circle in a clockwise direction as t increases.

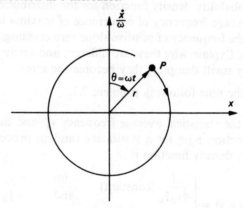

Fig. P8.5

Now consider the trajectory of point P when $x(t)$ is a sample function of a smooth narrow band random process. P spirals slowly inwards and outwards as the magnitude r of individual cycles falls and rises. Verify that the probability that, at any arbitrary time, the radius of the trajectory is less than r will be given by

$$\int dx \int \frac{d\dot{x}}{\omega} p\left(x, \frac{\dot{x}}{\omega}\right)$$

integrated over a circle of radius r and that, if $p(r)$ is the probability density function for r, then

$$p(r)\,dr = \int dx \int \frac{d\dot{x}}{\omega} p\left(x, \frac{\dot{x}}{\omega}\right)$$

integrated over an annular ring of width dr.

Hence show that, for a stationary Gaussian process $x(t)$, $p(r)$ is given by

$$p(r) = \frac{r}{\sigma_x^2} e^{-(r^2/2\sigma_x^2)}$$

which is the density function for the Rayleigh distribution (8.25).

Although, in this case, the *envelope* distribution $p(r)$ and the *peak* distribution $p_p(r)$ are the same, in general this is not so. Why? Consider, for example, the joint distribution given in problem 8.3.

Hint. Express x and $\dot{x}/\omega$ in terms of polar coordinates (r, θ) and put $\sigma_{\dot{x}}/\sigma_x = \omega$ from (8.21).

9.1. An analogue spectrum analyser has a filter bandwidth of 1/10th octave and a 10 s averaging time. Calculate the standard deviation of a measurement as a percentage of its mean value when the filter centre frequency is set at (i) 10 Hz, (ii) 100 Hz and (iii) 1000 Hz.

In order to improve the accuracy obtained, measurements are taken at 10 s intervals and the results averaged. Assuming that the input is a sample from a stationary random process, calculate the time for which readings would have to be taken in the three cases to reduce the standard deviation of the measurement to 1 per cent of the true mean spectral density at the appropriate frequency.

9.2. A numerical procedure for calculating estimates of spectral density introduces a spectral window whose shape is given by

$$W(\Omega) = \begin{cases} c \cos \frac{\pi}{2} \left| \frac{\Omega}{\Omega_0} \right| & \text{for} \quad -\Omega_0 \leqslant \Omega \leqslant \Omega_0 \\ 0 \text{ elsewhere.} \end{cases}$$

Determine the constant c if $W(\Omega)$ is to be normalized correctly, and find the window's effective bandwidth, B_e.

If the record being analysed is a sine wave of constant amplitude a and constant angular frequency Ω_1 and the record length T is very long (assume that T is infinite), find an expression for the smoothed spectrum $\tilde{S}(\omega)$ obtained by using this window.

9.3. The chi-square probability density function for κ degrees of freedom is given by

$$p(y) = \frac{1}{2^{\kappa/2} \, \Gamma(\kappa/2)} y^{(\kappa/2 - 1)} e^{-y/2} \qquad \text{for} \qquad y > 0$$

where $y = \chi_\kappa^2$ and Γ denotes the gamma function defined by

$$\Gamma(n) = \int_0^\infty y^{n-1} e^{-y} dy \qquad \text{for} \qquad n > 0.$$

(i) Verify that the expression for $\Gamma(n)$ can be integrated by parts to give

$$\Gamma(n) = (n - 1)\Gamma(n - 1).$$

(ii) Show that

$$\Gamma(1) = 1$$

and

$$\Gamma(\tfrac{1}{2}) = \sqrt{(\pi)}.$$

Hint. Put $y = \tfrac{1}{2}z^2$ and use (1.20).

(iii) Check that, for one degree-of-freedom, $p(y)$ agrees with the result derived in problem 1.3.

(iv) For the case of two degrees-of-freedom, find the limiting values within which there is an 80 per cent chance of finding χ_2^2.

(v) Lastly, for the case of 10 degrees-of-freedom, calculate the probability that χ_{10}^2 exceeds $(m + \sigma)$, where m and σ are the mean and standard deviation of χ_{10}^2, and compare this with the corresponding probability for a quantity with a Gaussian distribution with the same mean and standard deviation.

9.4. When a measurement of spectral density is taken for long enough that $\sigma/m = 1$ per cent (see problem 9.1), what are the 80 per cent and the 99 per cent confidence limits for the true mean value m in terms of the measured value S_0?

Hint. Assume that the distribution of measured values may be approximated by a Gaussian distribution (why?) and refer to Appendix 3 for values of the Gaussian probability distribution function.

10.1. Calculate the DFT's of the discrete time series:

(i) $(0, 1/\sqrt{2}, 1, 1/\sqrt{2}, 0, -1/\sqrt{2}, -1, -1\sqrt{2})$

and

(ii) $(1, 1/\sqrt{2}, 0, -1/\sqrt{2}, -1, -1/\sqrt{2}, 0, 1/\sqrt{2})$

and compare the results with the (continuous) Fourier transforms of the periodic functions $\sin 2\pi t/T$ and $\cos 2\pi t/T$. Then generalize the conclusion to find the DFT of the series:

(iii) $\{x_r\}$, $r = 0, 1, 2, \ldots, (N - 1)$, where $x_r = \sin 2\pi r/N$.

10.2. Check that equation (10.18) is only true when the terms in the $\{x_r\}$ series are all real and use (10.17) and (10.18) to show that, in this case, X_0 and $X_{N/2}$ (assuming N even) are real.

For an eight-term DFT of a series of real terms, verify that

$$X_4 = X_4^*$$
$$X_5 = X_3^*$$
$$X_6 = X_2^*$$
$$X_7 = X_1^*$$

and use this information to help find the DFT of the series

$$(1, 1, 1, 1, -1, -1, -1, -1).$$

Compare the result with the (continuous) Fourier transform of a periodic square wave (see problem 3.2) and comment on the difference.

10.3. Use the computer program listed in Appendix 2 to calculate the DFT of the 64-term series

$$x_r = \begin{cases} 1 & \text{for} & r = 0, 1, 2, \ldots, 31 \\ -1 & \text{for} & r = 32, 33, \ldots, 63 \end{cases}$$

and compare this with the (continuous) Fourier transform of a square wave (found in problem 10.2).

10:4. Use the IDFT to find the discrete time series $\{x_r\}$, $r = 0, 1, 2, \ldots, (N - 1)$, whose DFT's are:

(i) $X_k = 0$ for all k except $k = 0$
$X_0 = 1$

(ii) $X_k = 0$ for all k except $k = N/2$ (N even)
$X_{N/2} = 1$

(iii) $X_k = 0$ for all k except $k = N/4$ and $3N/4$ ($N/2$ even)

$$X_{N/4} = \frac{1}{2} - i\frac{1}{2}$$

$$X_{3N/4} = \frac{1}{2} + i\frac{1}{2}$$

(iv) $X_k = 0$ for all k except $k = 1$ and $k = (N - 1)$

$$X_1 = -i\frac{1}{2}$$

$$X_{N-1} = i\frac{1}{2}.$$

10.5. Find the discrete time series $\{x_r\}$, $r = 0, 1, 2, \ldots, (N - 1)$, whose DFT is

$$X_k = 0 \text{ for } k = 0, 1, 2, \ldots, (N - 1) \text{ except } k = m \text{ and } k = (N - m)$$

$$X_m = a_m - ib_m$$

$$X_{N-m} = a_m + ib_m.$$

Hence show that, if

$$\{X_k\} = (a_0, a_1 - ib_1, a_2 - ib_2, \ldots, a_{N/2}, \ldots, a_2 + ib_2, a_1 + ib_1)$$

where N is even, then

$$x_r = a_0 + \sum_{m=1}^{N/2-1} \left(2a_m \cos\frac{2\pi m}{N}r + 2b_m \sin\frac{2\pi m}{N}r\right) + a_{N/2} \cos \pi r$$

for $r = 0, 1, 2, \ldots, (N - 1)$.

Compare this result with equation (10.1) and confirm that harmonics with frequencies up to the Nyquist frequency $1/2\Delta$ can be detected in data sampled at intervals Δ. Note that the harmonic whose frequency is exactly $1/2\Delta$ cannot be properly detected. Hence show that, if $x(t)$ can be represented by a continuous Fourier series of period $T = N\Delta$ in which the frequency of the highest harmonic is less than $1/2\Delta$, then $x(t)$ can be exactly reconstructed (for all t) from the DFT of the discrete series $\{x_r\}$. This is a special case of *Shannon's Sampling Theorem* (Shannon [61], p. 10).

10.6. (i) Use the result of the first part of problem 4.3 to show that, if $x(t)$ is given by equation (10.1), then

$$\frac{1}{T}\int_0^T x^2(t)\,dt = a_0^2 + 2\sum_{k=1}^{\infty}(a_k^2 + b_k^2)$$

where a_k and b_k are as defined in (10.1) (instead of by equation (4.1)).

(ii) Prove also that Parseval's theorem for discrete transforms becomes, if the terms of $\{x_r\}$ are all real,

$$\frac{1}{N}\sum_{r=0}^{N-1} x_r^2 = \sum_{k=0}^{N-1} |X_k|^2.$$

Hint. Begin by writing $|X_k|^2 = X_k X_k^*$ and substitute for X_k and X_k^* from (10.8) using index r for X_k and s for X_k^*.

(iii) Hence show that if, as required by (10.17) and (10.18),

$$X_0 = a_0$$

$$X_k = a_k - ib_k \quad \text{for} \quad k = 1, 2, \ldots, (N/2 - 1)$$

$$X_{N/2} = a_{N/2}$$

$$X_{N-k} = a_k + ib_k \quad \text{for} \quad (N - k) = (N/2 + 1), \ldots, (N - 1)$$

(N is assumed even) then

$$\frac{1}{N}\sum_{r=0}^{N-1} x_r^2 = a_0^2 + 2\sum_{k=1}^{N/2-1}(a_k^2 + b_k^2) + a_{N/2}^2$$

and compare this result with the result of part (i).

10.7. Consider a discrete time series $\{x_r\}$, $r = 0, 1, 2, \ldots, (N - 1)$, whose terms are given by

$$x_r = 2a_m \cos\frac{2\pi m}{N}r + 2b_m \sin\frac{2\pi m}{N}r$$

where a_m, b_m, and the integer m are constants and $0 < m < N/2$. Determine the series $\{R_r\}$ whose terms are the discrete autocorrelation function

$$R_r = \frac{1}{N} \sum_{s=0}^{N-1} x_s x_{s+r} \qquad r = 0, 1, 2, \ldots, (N-1)$$

and find $\{S_k\}$, the DFT of $\{R_r\}$.

Hence show that, if $\{X_k\}$ is the DFT of $\{x_r\}$, then

$$S_k = X_k^* X_k$$

in agreement with (10.24).

11.1. Consider the discrete series

$$\{x_s\} = (x_0, x_1, x_2, x_3, 0, 0, 0, 0, 0, 0,)$$

and

$$\{y_s\} = (y_0, y_1, y_2, y_3, 0, 0, 0, 0, 0, 0,).$$

Determine
(i) the series $\{R_r\}$ where R_r is defined by equation (10.20);
(ii) the series $\{\hat{R}_r\}$ and $\{\hat{R}_{-r}\}$ where $\hat{R}_r$ and $\hat{R}_{-r}$ are defined by (11.8) and (11.9);
and verify that
(iii) corresponding terms of these series satisfy (11.10).

Now assume that the six zeros have been artificially added to the $\{x_s\}$ and $\{y_s\}$ series (so that $N = 4$, $L = 6$) and determine:
(iv) the corrected series $\{\hat{R}_r\}$ and $\{\hat{R}_{-r}\}$ from (11.42) (why are these different from (11.8) and (11.9)?);
and verify that
(v) corresponding terms of $\{R_r\}$ may be interpreted as arising from the summation of a train of functions each of the form

$$\left(\frac{N - |r|}{N + L}\right)\hat{R}_r, \qquad -N \leqslant r \leqslant N,$$

and spaced $(N + L)$ points apart.

11.2. Consider the triangular wave $v(t)$ shown in Fig. P11.2(a). Check that

$$v(t) = \frac{a}{2} + \frac{4a}{\pi^2}\left\{\cos\frac{\pi t}{T} + \frac{1}{9}\cos\frac{3\pi t}{T} + \frac{1}{25}\cos\frac{5\pi t}{T} + \cdots\right\}$$

and hence obtain the coefficients V_k in the expression

$$V(\omega) = \sum_{k=-\infty}^{\infty} V_k \delta\left(\omega - \frac{\pi k}{T}\right)$$

where $V(\omega)$ is the (continuous) Fourier transform of $v(t)$.

If $U(\omega)$ is the Fourier transform of the triangular function $u(t)$ shown in Fig. P11.2(b) (see problem 4.1), show that each of the coefficients V_k is given by

$$V_k = \frac{\pi}{T} U\left(\omega = \frac{\pi k}{T}\right)$$

which is in agreement with (11.18) allowing for the fact that the period here is $2T$ (rather than T).

If a new function $v'(t)$ is made up of a train of overlapping functions similar to $u(t)$ and spaced at regular intervals T (not $2T$ as before) find by inspection the coefficients V_k' and verify that these also satisfy (11.18).

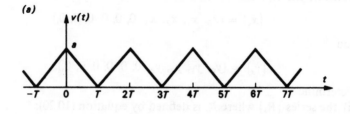

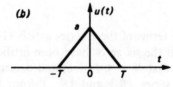

Fig. P11.2

11.3. The spectral density of an ergodic random process $x(t)$ is

$$S_x(\omega) = \begin{cases} S_0 & \text{for} & -\omega_0 \leqslant \omega \leqslant \omega_0 \\ 0 & \text{elsewhere.} \end{cases}$$

(i) What is the value of S_0 when $\omega_0 = 200\pi$ rad/s and $E[x^2] = 1.0$ cm²?
(ii) Member functions of the ensemble $x(t)$ are sampled at interval $\Delta = 0.0025$ s to form an ensemble of discrete time series $\{x_r\}$. Each of these series has 2^{12} terms. The DFT $\{X_k\}$ of each series $\{x_r\}$ is calculated and a new series $\{S_k\}$ formed where

$$S_k = X_k^* X_k.$$

There is then an ensemble of discrete frequency series $\{S_k\}$ and corresponding terms are averaged. What will be the value of each term in the final (ensemble averaged) series $\{S_k\}$? Assume that the inherent spectral window, Fig. 11.4(b), may be approximated by a rectangular function of width $2\pi/T$ where T is the record length.
(iii) Determine how the result of part (ii) will be altered by the presence

of a deterministic sine wave of constant amplitude 0·1 cm and frequency 350 Hz superimposed on the random process.

11.4. Use equation (11.18) and the result of problem 4.1(e) to find the (continuous) Fourier transform $X(\omega)$ of the periodic function $x(t)$ shown in Fig. P11.4(a).

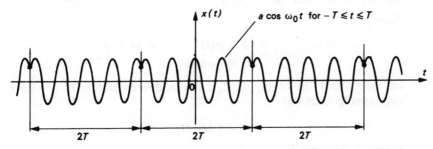

Fig. P11.4(a)

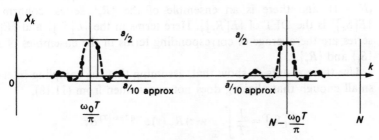

Fig. P11.4(b)

Check that, when $\omega_0 T = \pi n$ where n is an integer, $X(\omega)$ correctly reduces to a pair of delta functions of area $a/2$ at $\omega = \pm\omega_0$.

Now consider sampling $x(t)$ at interval Δ (where $\pi/\Delta \gg \omega_0$) between time $t = 0$ to $t = 2T$ to generate the discrete time series $\{x_r\}$. Verify that the DFT of $\{x_r\}$ is zero except in the regions $k \simeq (\omega_0 T/\pi)$ and $(N - \omega_0 T/\pi)$. Check that unless $\omega_0 T = n\pi$, there is more than one non-zero value of X_k in each region, Fig. P11.4(b), but that not more than two of these values of $|X_k|$ can be greater than $(a/10)$ approx. Check also that, when $\omega_0 T$ is large, the sum of squares of X_k is *approximately* equal to $a^2/2$. Why is this sum not *exactly* equal to $a^2/2$?

Would it make any difference to $\{X_k\}$ if $\{x_r\}$ were derived by sampling $x(t)$ from $t = -T$ to $t = T$ (instead of from $t = 0$ to $t = 2T$)? Assume that N is even so that the same data points are used in both cases.

11.5. (i) Consider an ensemble of series $\{R_r\}$, $r = 0, 1, 2, ..., (N - 1)$, and take the ensemble average of both sides of equation (11.4) to show that

$$E[R_r] = \frac{N-r}{N}R_{xy}(\tau = r\Delta) + \frac{r}{N}R_{xy}(\tau = -(N-r)\Delta) \qquad 0 \leqslant r \leqslant N$$

where $R_N = R_0$, because of the circular property of R_r arising from (10.21). Hence prove that

$$E[R_r] = \frac{T - \tau}{T} R_{xy}(\tau) + \frac{\tau}{T} R_{xy}(\tau - T) \qquad (11.51)$$

where $\tau = r\Delta$ and $0 \leqslant \tau \leqslant T$.

Now put

$$w(\tau) = \begin{cases} 1 - |\tau|/T & 0 \leqslant |\tau| \leqslant T \\ 0 & \text{elsewhere} \end{cases}$$

and show that the terms of $\{E[R_r]\}$ are discrete values of a train of overlapping functions

$$w(\tau) R_{xy}(\tau)$$

spaced at intervals T.

If $\{S_k\}$, $k = 0, 1, 2, \ldots, (N - 1)$, is the DFT of $\{R_r\}$, $r = 0, 1, 2, \ldots, (N - 1)$, and there is an ensemble of the $\{R_r\}$ series, confirm that $\{E[S_k]\}$ is the DFT of $\{E[R_r]\}$. Here terms in the $\{E[S_k]\}$ and $\{E[R_r]\}$ series are the average of corresponding terms in the ensembles of series $\{S_k\}$ and $\{R_r\}$.

Use these results to show that, assuming that the sampling interval is small enough that aliasing does not occur, then from (11.18),

$$E[S_k] = \frac{1}{T} \int_{-\infty}^{\infty} w(\tau) R_{xy}(\tau) e^{-i(2\pi k\tau/T)} \, d\tau$$

$$= \frac{1}{T} \int_{-T}^{T} \left\{ \frac{T - |\tau|}{T} \right\} R_{xy}(\tau) e^{-i(2\pi k\tau/T)} \, d\tau \qquad (11.52)$$

in agreement with (11.21) and (11.22).

(ii) Now consider the case when the series $\{x_r\}$ is modified by a symmetrical weighting function $\{d_r\}$ (see Fig. 11.8) before analysis, to form the new series $\{x_r d_r\}$. The DFT of $\{x_r d_r\}$ is calculated and a modified spectral estimate S_k found by the usual procedure. Following the approach used in part (i), show that the modified lag window $w(\tau)$ is now given by

$$w(\tau = r\Delta) = \frac{1}{N} \sum_{s=0}^{N-1-r} d_s d_{s+r} \qquad (11.53)$$

and that the resulting spectral estimates must be divided by a correction factor

$$\frac{1}{N} \sum_{r=0}^{N-1} d_r^2$$

to account for the loss of amplitude caused by the weighting $\{d_r\}$.

Hint. $w(\tau)$ must satisfy the second of (11.23). Why?

(iii) Investigate the effectiveness of a *cosine taper data window* as follows. Using the computer program listed in Appendix 2, determine the DFT's of

(a) $\{x_r\}$, $r = 0, 1, \ldots, 127$, where

$$x_r = \sin \frac{31}{256}\pi r,$$

(b) $\{y_r\}$, $r = 0, 1, \ldots, 127$, where

$$y_r = \frac{d_r x_r}{D} \text{ with } x_r \text{ as in part } (a)$$

$$d_r = \begin{cases} \frac{1}{2}\left(1 - \cos\frac{\pi}{16}r\right), & r = 0, 1, \ldots, 15 \\ 1 \cdot 0, & r = 16, 17, \ldots, 112 \\ \frac{1}{2}\left(1 - \cos\frac{\pi}{16}(128 - r)\right), & r = 113, 114, \ldots, 127 \end{cases}$$

and

$$D = \frac{1}{128}\sum_{r=0}^{127} d_r^2,$$

(c) $\{x_r\}$, $r = 0, 1, 2, \ldots, 1023$, where

$$x_r = \sin \frac{255}{2048}\pi r,$$

and

(d) y_r, $r = 0, 1, 2, \ldots, 1023$, where

$$y_r = \frac{d_r x_r}{D} \text{ with } x_r \text{ as in part } (c)$$

$$d_r = \begin{cases} \frac{1}{2}\left(1 - \cos\frac{\pi}{128}r\right), & r = 0, 1, \ldots, 127 \\ 1 \cdot 0, & r = 128, 129, \ldots, 896 \\ \frac{1}{2}\left(1 - \cos\frac{\pi}{128}(1024 - r)\right), & r = 897, 898, \ldots, 1023 \end{cases}$$

and

$$D = \frac{1}{1024}\sum_{r=0}^{1023} d_r^2.$$

Cases (a) and (b) involve a sine wave of length $7\frac{3}{4}$ cycles and cases (c) and (d) a sine wave of length $63\frac{3}{4}$ cycles. The step length is

$$\Delta = \pi\left(\frac{1}{8} - \frac{1}{256}\right)$$

in the first two cases and

$$\Delta = \pi\left(\frac{1}{8} - \frac{1}{2048}\right)$$

in the last two cases. Compare the results obtained and so form a judgement about the effectiveness of tapering data before analysis.

11.6. Show that, by putting $R(\tau) = R_{xy}(\tau)$ and

$$w(\tau) = \begin{cases} 1 - \dfrac{|\tau|}{T} & \text{for} \quad 0 \leqslant |\tau| \leqslant T \\ 0 \text{ elsewhere,} \end{cases} \tag{11.31}$$

equation (11.51) of problem 11.5 may be written in the form

$$E[R_r] = w(\tau)R(\tau) + w(T - \tau)R(T - \tau)$$

where $\tau = r\Delta$.

Hence show that

$$E[S_k] = \frac{1}{N}\sum_{r=0}^{N-1} E[R_r]e^{-i(2\pi kr/N)}$$

$$= \frac{1}{N}\sum_{r=0}^{N-1} w(r\Delta)R(r\Delta)e^{-i(2\pi kr/N)} + \frac{1}{N}\sum_{s=-N}^{-1} w(s\Delta)R(s\Delta)e^{-i(2\pi ks/N)}$$

$$= \frac{1}{N}\sum_{r=-N}^{N-1} w(r\Delta)R(r\Delta)e^{-i(2\pi kr/N)}$$

Now introduce the generalized function

$$\sum_{r=-\infty}^{\infty} \delta(\tau - r\Delta)$$

to show that, assuming $w(\tau)$ and $R(\tau)$ are continuous, an alternative form of this result is

$$E[S_k] = \frac{1}{N}\sum_{r=-\infty}^{\infty}\int_{-\infty}^{\infty} \delta(\tau - r\Delta)w(\tau)R(\tau)e^{-i(2\pi k\tau/T)}\,d\tau \tag{11.54}$$

where $T = N\Delta$. Lastly, use equation (11.16) to show that

$$E[S_k] = \frac{1}{T}\sum_{j=-\infty}^{\infty}\int_{-\infty}^{\infty} w(\tau)R(\tau)e^{-i(\omega_k + j(2\pi/\Delta))\tau}\,d\tau$$

and hence prove that

$$E[S_k] = \frac{2\pi}{T}\sum_{j=-\infty}^{\infty} \tilde{S}\left(\omega_k + j\frac{2\pi}{\Delta}\right) \tag{11.55}$$

where $\tilde{S}(\omega)$ is the weighted spectrum defined by (11.24).

Notice that S_k is now the "aliased" version of the result given in the text (equation (11.21)). Check that, provided $\tilde{S}(\omega)$ is zero for $\omega \geqslant \pi/\Delta$, then all terms in the summation except the $j = 0$ term are zero, and so

$$E[S_k] = \frac{2\pi}{T}\tilde{S}(\omega_k)$$

in agreement with (11.21).

11.7. Beginning with equation (11.55) of problem 11.6, show that, if S_k is related to the true spectrum $S(\omega)$ by the equation

$$E[S_k] = \frac{2\pi}{T}\int_{-\infty}^{\infty} W_a(\omega - \omega_k)S(\omega)\,d\omega \qquad (11.56)$$

(compare equations (11.21) and (11.28)), then the *aliased spectral window* function $W_a(\omega)$ is given by

$$W_a(\omega) = \sum_{j=-\infty}^{\infty} W\left(\omega - j\frac{2\pi}{\Delta}\right) \qquad (11.57)$$

where $W(\omega)$ is defined, as before, by equation (11.25).

Now find an expression for the aliased spectral window corresponding to the basic window

$$W(\omega) = \frac{T}{2\pi}\left(\frac{\sin \omega T/2}{\omega T/2}\right)^2 \qquad (11.32)$$

and show that, since

$$\sum_{j=-\infty}^{\infty} \frac{1}{(\theta - j)^2} = \frac{\pi^2}{\sin^2 \pi\theta}$$

(see, for instance, Jolley [40] p. 152), this may be written as (Cooley *et al.* [10])

$$W_a(\omega) = \frac{T}{2\pi N^2}\left(\frac{\sin \omega T/2}{\sin \omega T/2N}\right)^2. \qquad (11.58)$$

Sketch $W_a(\omega)$ against ω and compare this with a similar graph of $W(\omega)$ against ω.

11.8. A random process is made up of an ensemble of member functions

$$x(t) = a\sin(\omega_0 t + \phi) \qquad ,$$

where a and ω_0 are constants and ϕ is constant for each $x(t)$ but varies from one member function to the next. The distribution of ϕ is defined by

$$p(\phi) = \begin{cases} \dfrac{1}{2\pi} & 0 \leqslant \phi \leqslant 2\pi \\ 0 & \text{elsewhere.} \end{cases}$$

(i) A discrete time series with four terms $\{x_r\}$, $r = 0, 1, 2, 3$ is constructed from a member function $x(t)$ by putting

$$x_r = x(t = r\Delta)$$

where Δ is the sampling interval. Consider calculating the DFT $\{X_k\}$, $k = 0, 1, 2, 3$ of $\{x_r\}$ and find the *ensemble averaged* values of the four terms of $\{S_k\}$ where

$$S_k = X_k^* X_k.$$

(ii) Show that the true spectral density of the $x(t)$ process is

$$S_x(\omega) = \tfrac{1}{4}a^2\{\delta(\omega - \omega_0) + \delta(\omega + \omega_0)\}$$

and use equation (11.56) of problem 11.7 to calculate an expression for the terms S_k using the aliased spectral window (11.58).

(iii) Verify that values of S_k obtained from part (ii) are in agreement with those found in part (i).

11.9. The method of digital spectral analysis described in Chapter 11 involves calculating a set of coefficients $\{S_k\}$ which are related to the DFT $\{X_k\}$ of a sample record $\{x_r\}$ by the equation

$$S_k = X_k^* X_k.$$

We have seen that S_k is an estimate for a weighted spectral density value $\tilde{S}(\omega = \omega_k)$ where, from (11.21),

$$E[S_k] = \frac{2\pi}{T}\tilde{S}(\omega = \omega_k) \tag{11.59}$$

when T is the record length analysed. According to (11.28), the weighted value $\tilde{S}(\omega)$ is related to the true spectrum $S(\omega)$ by the convolution

$$\tilde{S}(\omega) = \int_{-\infty}^{\infty} W(\omega - \Omega)S(\Omega)\,d\Omega \tag{11.28}$$

where $W(\omega)$ is the spectral window function associated with the calculation. $W(\omega)$ is defined as the (continuous) Fourier transform of the lag window $w(\tau)$, which for the basic calculation using unweighted data values is given by (11.31).

The theoretical development in Chapter 11 assumes that aliasing is not a problem, but in problem 11.7 we showed that equation (11.59) still holds when aliasing occurs, provided that an aliased spectral window

$$W_a(\omega) = \sum_{j=-\infty}^{\infty} W\left(\omega - j\frac{2\pi}{\Delta}\right) \tag{11.57}$$

replaces $W(\omega)$ in (11.28) to give

$$\tilde{S}(\omega) = \int_{-\infty}^{\infty} W_a(\omega - \Omega)S(\Omega)\,d\Omega. \tag{11.60}$$

The weighted spectral density $\tilde{S}(\omega)$ calculated from the latter equation is often referred to as the *periodogram*.

(i) Beginning from (11.59) and (11.60) show that the average values of the coefficients S_k are given by the following equation (Cooley *et al.* [10]):

$$E[S_k] = \frac{1}{N} \sum_{m=-N}^{N} w(\tau = m\Delta)R(\tau = m\Delta)\cos(\omega_k m\Delta). \quad (11.61)$$

Hints. (a) First substitute for $W_a(\omega - \Omega)$ in (11.60) from (11.57).

    (b) Express $W(\omega)$ as the Fourier transform of $w(\tau)$ using (11.25).

    (c) Introduce $R(\tau)$ as the Fourier transform of $S(\omega)$ remembering that $R(-\tau) = R(\tau)$.

    (d) Make use of (11.16) after replacing ω by τ and T by $2\pi/\Delta$.

(ii) Beginning from (11.57), show also that the aliased spectral window $W_a(\omega)$ is given by (Cooley *et al.* [10])

$$W_a(\omega) = \frac{\Delta}{2\pi} \sum_{m=-N}^{N} w(\tau = m\Delta) \cos \omega m\Delta \quad (11.62)$$

and use this expression to calculate $W_a(\omega)$ when, according to (11.31),

$$w(\tau) = \begin{cases} 1 - \dfrac{|\tau|}{T} & \text{for} \quad 0 \leqslant |\tau| \leqslant T \\ 0 & \text{elsewhere.} \end{cases} \quad (11.31)$$

Check that the result agrees with that given in equation (11.58) of problem 11.7.

Hints. To derive (11.62) use steps (b) and (d) of part (i). To obtain (11.58) from (11.62) the following two summations are needed together with some determined trigonometry:

$$\sum_{m=1}^{N-1} \cos m\theta = \frac{\cos \frac{1}{2}(N+1)\theta \sin N\theta/2}{\sin \theta/2} - \cos N\theta$$

$$\sum_{m=1}^{N-1} m \cos m\theta = \frac{N}{2} \frac{\sin(N - \frac{1}{2})\theta}{\sin \theta/2} - \frac{(1 - \cos N\theta)}{4 \sin^2 \theta/2}$$

(see Jolley [40] pp. 78 and 80).

12.1. (i) Follow the logical steps illustrated in Figs. 12.2 and 12.4 to calculate the IDFT of the series

$$(0,\ 0,\ 0,\ 0,\ 1,\ 0,\ 0,\ 0,)$$

by the FFT method.

(ii) Then modify the FFT computer program of Appendix 2 so that it calculates the IDFT (instead of the DFT) of the input array and use it to calculate the IDFT of $\{X_k\}$, $k = 0, 1, 2, \ldots, 63$, where

$$X_k = \begin{cases} 1 & \text{for } k = 32 \\ 0 & \text{for all other values of } k. \end{cases}$$

12.2. (i) The DFT of the series $\{x_r\}$, $r = 0, 1, 2, \ldots, (n-1)$, is $\{X_k\}$, $k = 0, 1, 2, \ldots, (n-1)$. Consider a new series $\{y_s\}$ of length $N = mn$ terms constructed by taking adjacent values of $\{x_r\}$ and separating them by $(m-1)$ zeros, so that

$$y_s = \begin{cases} x_{r=s/m} & \text{for } s = 0, m, 2m, \ldots, (n-1)m \\ 0 & \text{for all other values of } s. \end{cases}$$

If the DFT of $\{y_s\}$, $s = 0, 1, 2, \ldots, (N-1)$, is $\{Y_l\}$, $l = 0, 1, 2, \ldots, (N-1)$, show that

$$Y_l = \frac{1}{m}X_{k=l}. \tag{12.15}$$

(ii) Consider also another new series $\{z_r\}$, $r = 0, 1, 2, \ldots, (n-1)$, constructed from the x_r series by putting

$$z_r = x_{(r-t)}$$

where t is a constant integer and show that, if $\{Z_k\}$ is the DFT of $\{z_r\}$, then

$$Z_k = X_k W^{kt} \tag{12.16}$$

where $W = e^{-i(2\pi/N)}$.

(iii) Now apply these results to the following problem. A series $\{y_s\}$, $s = 0, 1, 2, \ldots, (mn-1)$, is partitioned into m sub-series each of length n which are called $\{x_{0_r}\}$, $\{x_{1_r}\}$, $\{x_{2_r}\}$, etc., where $r = 0, 1, 2, \ldots, (n-1)$. Successive terms of the sub-series $\{x_{j_r}\}$ are the j-th, $m + j$-th, $2m + j$-th, etc., terms of the series $\{y_s\}$.

Show that the DFT $\{Y_l\}$ of the original series can be expressed in terms of the DFT's $\{X_{j_k}\}$ of the sub-series by the equation

$$Y_l = \frac{1}{m}\sum_{j=0}^{m-1} X_{j_{k=l}} W^{jl} \tag{12.17}$$

where $W = e^{-i(2\pi/nm)}$.

(iv) Check that the computational "butterfly" equation (12.9) is a special case of this result when $mn = N$ and $m = 2$.

12.3. Calculate the DFT of the following series of 64 terms

$$(1, 1, 1, 1, -1, -1, -1, -1, 1, 1, 1, 1, -1, -1, -1, -1, \ldots, 1, 1,$$
$$-1, -1, -1, -1)$$

by partitioning this series into eight sub-series, and then combining the DFT's of the sub-series by using equation (12.17) of part (iii) of problem 12.2.

12.4. (i) The series $\{x_r\}$, $r = 0, 1, 2, \ldots, (N - 1)$, where N is even, has DFT $\{X_k\}$, $k = 0, 1, 2, \ldots, (N - 1)$. Consider two new series $\{y_s\}$ and $\{z_s\}$, $s = 0, 1, 2, \ldots, (N/2 - 1)$, constructed from the $\{x_r\}$ series in such a way that

$$y_s = (x_{r=s} + x_{r=s+N/2})$$

and

$$z_s = (x_{r=s} - x_{r=s+N/2})W^s \qquad (12.18)$$

where $W = e^{-i(2\pi/N)}$. Determine the DFT's $\{Y_l\}$ and $\{Z_l\}$ of the two new series and hence show that

$$\left.\begin{array}{l} Y_l = 2X_{k=2l} \\ \\ Z_l = 2X_{k=(2l+1)} \end{array}\right\} \quad l = 0, 1, 2, \ldots, (N/2 - 1). \qquad (12.19)$$

(ii) Construct a signal flow graph similar to Fig. 12.4 to show how repeated application of equations (12.18) and (12.19) can serve as the basis for an alternative version of the FFT algorithm. Take a sequence of eight terms, $\{x_r\}$, $r = 0, 1, 2, \ldots, 7$. Draw the butterfly graphs by which two 4-term sequences $\{y_s\}$ and $\{z_s\}$, $s = 0, 1, 2, 3$ are generated by equation (12.18) and verify that the terms of their DFT's are, in order, $2X_0, 2X_2, 2X_4, 2X_6$, and $2X_1, 2X_3, 2X_5, 2X_7$. Now draw another line of butterflies to generate four 2-term sequences according to (12.18) and verify that the DFT's of these sub-series are, in order, $4X_0, 4X_4$; $4X_2$, $4X_6$; $4X_1, 4X_5$; and $4X_3, 4X_7$. Finally draw the last line of butterflies to generate eight single-term sequences which (since the DFT of a single-term sequence is the term itself) are, in order, $8X_0, 8X_4, 8X_2, 8X_6, 8X_1, 8X_5, 8X_3, 8X_7$.

This is called a *decimation in frequency* algorithm because alternate terms of the DFT are calculated from each sub-series. Notice that computation begins with the terms of $\{x_r\}$ in their original order but the terms of $\{X_k\}$ are generated in "bit-reversed" order.

13.1. Each sample of a random binary process $x(t)$ is the result of "tossing coins" at regular intervals Δt, and its value is either $+a$ or $-a$ for each interval. The phase ϕ of each sample function $x(t)$ is defined as shown in Fig. 13.1(a) and has the distribution shown in Fig. 13.1(c). In this case, however, the coin is biased so that $+a$ (for "heads") occurs more frequently than $-a$ (for "tails"). Consider the case when for every n heads there are only $n - 1$ tails, and find the autocorrelation function $R_x(\tau)$ and the spectral density $S_x(\omega)$ of the $x(t)$ random process.

How would these results differ if the bias was reversed and there were $n - 1$ heads for every n tails?

13.2. Each sample of a random binary process $x(t)$ is the result of "tossing unbiased coins" at regular intervals Δt, and its value is either $+a$ or $-a$ for each interval. The phase ϕ of each sample function is again uniform as shown in Fig. 13.1.

A four-level random process $y(t)$ is constructed from the $x(t)$ process by summing three adjacent values of $x(t)$ on the same sample so that

$$y(t) = x(t) + x(t - \Delta t) + x(t - 2\Delta t).$$

(i) Verify that

$$E[x(t)x(t + \tau)] = \begin{cases} a^2\left(1 - \dfrac{|\tau|}{\Delta t}\right) & \text{for} \quad 0 \leqslant |\tau| \leqslant \Delta t \\[2mm] 0 & \text{elsewhere.} \end{cases}$$

(ii) Show that $R_y(\tau)$ can be expressed as the sum of nine terms of the form $E[x(t_1)x(t_2)]$, each of which satisfies (i).

(iii) Hence prove that

$$R_y(\tau) = \begin{cases} a^2\left(3 - \dfrac{|\tau|}{\Delta t}\right) & \text{for} \quad 0 \leqslant |\tau| \leqslant 3\Delta t \\[2mm] 0 & \text{elsewhere.} \end{cases}$$

13.3. (i) Each sample function $x(t)$ of a random process is periodic, and is the same except for its "phase" ϕ which is defined in Fig. P13.3. The distribution of ϕ from sample to sample is given by

$$p(\phi) = \begin{cases} \dfrac{1}{T} & \text{for} \quad 0 \leqslant \phi \leqslant T \\[2mm] 0 & \text{elsewhere} \end{cases}$$

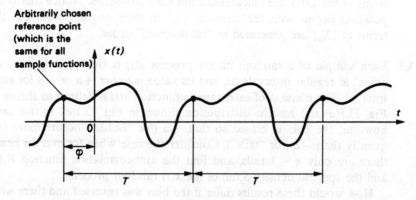

Fig. P13.3

where T is the period. Show that the ensemble averages $E[x]$ and $R_x(\tau)$ are given by

$$E[x] = \frac{1}{T} \int_0^T x(t)\,dt$$

and

$$R_x(\tau) = \frac{1}{T} \int_0^T x(t)x(t + \tau)\,dt. \tag{13.27}$$

(ii) Consider the generation of pseudo random binary numbers by a feedback shift register which performs the operation

$$x_n = (x_{n-1} + x_{n-m}) \text{ modulo 2.}$$

Starting from an initial state in which the content of each stage of the register is 1, determine the sequence of states of x_n for the cases $m = 3$, $m = 4$ and $m = 6$. Verify that the length of the sequence in each case is $N = (2^m - 1)$ and that each sequence has one more state 1 than it has state 0.

This result is only true for certain values of m, when the resulting sequence is called a *full length sequence*. Check that a full length sequence is not generated for $m = 5$, but that by changing the feedback connections so that

$$x_n = (x_{n-2} + x_{n-5}) \text{ modulo 2}$$

a full length sequence can be obtained.

(iii) Now consider pseudo random binary signals $x(t)$, each generated by the full length sequences obtained in part (ii). The clock interval is Δt and $x(t) = +a$ for the interval when $x_n = 1$ and $-a$ when $x_n = 0$. If $x_r = x(t = r\Delta t)$, check, by taking several test cases, that

$$\sum_{r=1}^{N} x_r = a$$

and

$$\sum_{r=1}^{N} x_r x_{r+s} = \begin{cases} Na^2 & \text{for } s = 0, N, 2N, \text{ etc.} \\ -a^2 & \text{for other values of } s. \end{cases}$$

(iv) A random process $\{x(t)\}$ consists of an ensemble of sample functions, each of which is generated from the same full-length pseudo random binary sequence as described in part (iii). Each sample is the same but has a different randomly chosen phase as defined in part (i). Use the above results to confirm that the autocorrelation function for this process is

$$R_x(\tau) = \begin{cases} \left(1 - \frac{|\tau|}{\Delta t}\right)a^2 - \frac{|\tau|}{\Delta t}\frac{a^2}{N} & \text{for} \quad 0 \leqslant |\tau| \leqslant \Delta t \\ & \text{repeating at intervals } N\Delta t \\ -\frac{a^2}{N} & \text{elsewhere} \end{cases}$$

and find the corresponding spectral density $S_x(\omega)$. Compare your results with those from problem 13.1.

13.4. (i) Consider the discrete time series $\{x_r\}$, $r = 0, 1, 2, \ldots, (N-1)$ where

$$x_r = 2 \sum_{k=1}^{n} (S_k)^{1/2} \cos\left(\frac{2\pi kr}{N} - \phi_k\right)$$

and $n < N/2$, and use the result of problem 10.5 to show that, if $\{X_k\}$ is the DFT of $\{x_r\}$, then

$$X_k^* X_k = \begin{cases} S_k & \text{for} \quad k = 1, 2, \ldots, n \quad \text{and for} \\ & \quad k = (N-n), \ldots, (N-1) \\ 0 & \text{for other values of } k. \end{cases}$$

(ii) Then write a computer program to generate a typical series $\{x_r\}$ for the case when the phase angles ϕ_k are randomly distributed between 0 and 2π. Take $n = 100$ and $N = 1024$ and use equation (13.26) with $m = 6$ to generate the random phase angles. The initial six random numbers required in (13.26) should themselves be randomly distributed between 0 and 1. Suppose that each number is to have five digits. It can be generated by devising an experiment which has ten equally likely possible results and allotting the numbers 0 to 9 to each different result. The experiment is then carried out five times to generate the first random number, five more times to generate the second number, and so on. However a quicker method is to refer to a table of random digits, Appendix 5, which have already been generated by such a process and to use 30 digits in order from this table. Put $S_k = 0.005$ for $k = 1, 2, 3, \ldots,$ 100, so that $\sigma_x^2 = 1$ (see the result of problem 10.6).

Calculate values of the probability distribution function $P(\phi)$ for the distribution of phase angles by including in the program a sub-routine to count the number of points for which $\phi_k < \phi$ where ϕ has a series of values between 0 and 2π and compare this with the uniform distribution intended.

Calculate also the probability distribution function $P(x)$ by counting the number of points for which $x_r < x$ where x has a series of values between $-3\sigma_x$ and $+3\sigma_x$. Compare this with the Gaussian distribution predicted by the central limit theorem when $n \to \infty$.

(iii) Use the FFT program in Appendix 2 to generate the DFT of the $\{x_r\}$ series and verify that the original values of S_k can be regained correctly.

14.1. The height z of a point on a road surface above a horizontal reference plane is a function of the position of the point (x, y), Fig. P14.1(a). It has been found (Dodds *et al.* [24]) that many typical roads have surface irregularities which may be described by a correlation function

$$E[z(x_1, y_1)z(x_2, y_2)]$$

which depends only on the horizontal distance

$$\rho = \sqrt{\{(x_2 - x_1)^2 + (y_2 - y_1)^2\}}$$

between the points (x_1, y_1) and (x_2, y_2). The surface is then said to have properties which are homogeneous (the same at all points) and isotropic (the same in all directions). Let this correlation function be $R(\rho)$.

(i) Consider the displacement excitation which such a surface imparts to the four wheels of a vehicle when it is driven along a straight path across the surface at steady speed V. Let the wheelbase of the vehicle be l and the wheel spacing b, and number the wheels 1, 2, 3, 4, as shown in Fig. P14.1(b). Find expressions for the time spectral densities $S_1(\omega)$,

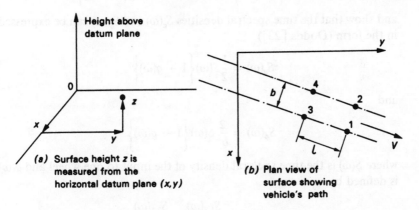

(a) Surface height z is measured from the horizontal datum plane (x, y)

(b) Plan view of surface showing vehicle's path

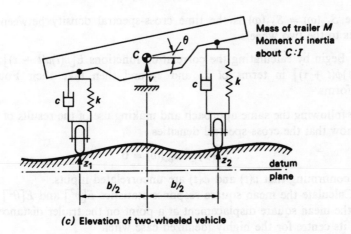

(c) Elevation of two-wheeled vehicle

Fig. P14.1

$S_2(\omega)$, etc., and the cross-spectral densities $S_{12}(\omega)$, $S_{13}(\omega)$, $S_{14}(\omega)$, etc., in terms of $R(\rho)$, V, l and b.

(ii) Now let the "vehicle" be a two-wheeled trailer, Fig. P14.1(c). The only vibration inputs to the trailer come through its wheels which are assumed to be small enough to follow the surface profile exactly. Obtain the two differential equations for vertical and roll motion of the trailer, in terms of the (small) vertical displacement v of the centre of mass of the trailer and the (small) roll angle θ. Assume that, when the trailer is in static equilibrium with $z_1 = z_2 = 0$, then $v = \theta = 0$. Verify that these equations have the form of (14.29) where the corresponding input is $\frac{1}{2}(z_1 + z_2)$ for the vertical mode and $(1/b)(z_2 - z_1)$ for the roll mode.

(iii) Let

$$u = \frac{1}{2}\left(z_1 + z_2\right) \qquad \text{and} \qquad \phi = \frac{1}{b}\left(z_2 - z_1\right)$$

and show that the time spectral densities $S_u(\omega)$ and $S_\phi(\omega)$ can be expressed in the form (Dodds [23])

$$S_u(\omega) = \frac{1}{2}S(\omega)\left\{1 + g(\omega)\right\}$$

and

$$S_\phi(\omega) = \frac{2}{b^2}S(\omega)\left\{1 - g(\omega)\right\}$$

where $S(\omega)$ is the time spectral density of the input to one wheel and $g(\omega)$ is defined by

$$g(\omega) = \frac{S_{12}(\omega)}{S(\omega)} = \frac{S_{21}(\omega)}{S(\omega)}$$

where $S_{12}(\omega) = S_{21}(\omega)$ is the time cross-spectral density between the inputs to the two wheels.

Hint. Begin by calculating the correlation functions $E[u(t)u(t + \tau)]$ and $E[\phi(t)\phi(t + \tau)]$ in terms of z_1 and z_2 and then take their Fourier transforms.

By following the same approach and making use of the results of part (i), show that the cross-spectral densities

$$S_{u\phi}(\omega) = S_{\phi u}(\omega) = 0$$

thus confirming that $u(t)$ and $\phi(t)$ are uncorrelated inputs.

(iv) Calculate the mean square response statistics $E[v^2]$ and $E[\theta^2]$ and also the mean square displacement at a point on the trailer distance $b/2$ from its centre for the highly idealized case when

$$R(\rho) = 2\pi S_0\, \delta(\rho).$$

This correlation function corresponds to a "white" spatial spectral density of level S_0; in practice it has been found (Dodds *et al.* [24]) that the spatial spectral density $S(\gamma)$ of road surfaces depends on the wavenumber γ and typically has the form

$$S(\gamma) = \frac{S_0}{(\gamma/\gamma_0)^n}$$

where n is approximately 2 for values of γ greater than about 0·01 cycle/m (wavelengths less than 100 m).

14.2. A linear system has two uncorrelated inputs $x(t)$ and $z(t)$, Fig. P14.2, and a single output $y(t)$. If the spectral density of the output is $S_{yy}(\omega)$ and the ordinary coherence function between $y(t)$ and $x(t)$ is $\eta_{yx}^2(\omega)$, show that, when the input $x(t)$ is removed, the spectral density of the residual output process $S_{yy|x}(\omega)$ is given by

$$S_{yy|x}(\omega) = S_{yy}(\omega)\{1 - \eta_{yx}^2(\omega)\}.$$

Hint. Since the system is linear, Fig. P14.2 can be re-drawn in the same form as Fig. 14.9 where $H_2(\omega)$ in the latter is replaced by $H(\omega)H_1(\omega)$. Then apply equations (7.22) and (14.60).

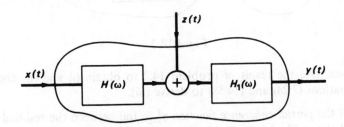

Fig. P14.2

14.3. Fig. 14.9 shows a linear system with two inputs $x_1(t)$ and $x_2(t)$ and a single output $y(t)$. Consider the case when the input process $x_2(t)$ is the addition of (a) the response of another (unknown) linear system to $x_1(t)$ and (b) uncorrelated noise $z(t)$, as illustrated in Fig. P14.3.

If $S_{yy}(\omega)$ = the output spectral density

$S_{yy|x_1}(\omega)$ = the spectral density of the residual output when $x_1(t) = 0$

$S_{yx_2|x_1}(\omega) = S_{yz|x_1}(\omega)$
= the cross-spectral density between the residual output when $x_1(t) = 0$ and the input $z(t)$

$\eta_{yx_1}^2(\omega)$ = the ordinary coherence function between $y(t)$ and $x_1(t)$

$\eta_{x_2x_1}^2(\omega)$ = the ordinary coherence function between $x_1(t)$ and $x_2(t)$

$\eta^2_{y.x}(\omega)$ = the multiple coherence function between $y(t)$ and the two inputs $x_1(t)$ and $x_2(t)$

show that

(i) $S_{zz}(\omega) = S_{x_2x_2|x_1} = S_{x_2x_2}\{1 - \eta^2_{x_2x_1}(\omega)\}$

(ii) $S_{yy|x_1}(\omega) = S_{yy}\{1 - \eta^2_{yx_1}(\omega)\}$

(iii) $S_{yx_2|x_1}(\omega) = S_{x_2y}\left\{1 - \dfrac{S_{x_1y}S_{x_2x_1}}{S_{x_2y}S_{x_1x_1}}\right\}.$

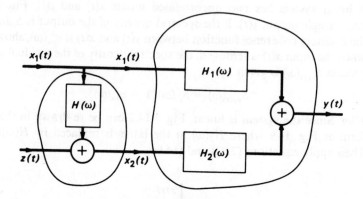

Fig. P14.3

Hints. Use the result of problem 14.2 to obtain (i) and (ii) and apply equations (7.24) and (14.58) to derive (iii).

If the *partial coherence function* $\eta^2_{yx_2|x_1}(\omega)$ between the residual output and the residual input $x_2(t)$ when $x_1(t) = 0$ is defined as

$$\eta^2_{yx_2|x_1} = \frac{|S_{yx_2|x_1}|^2}{S_{x_2x_2|x_1} S_{yy|x_1}}$$

show that (Jenkins *et al.* [39] p. 489)

(iv) $1 - \eta^2_{yx_2|x_1} = \dfrac{1 - \eta^2_{y.x}}{1 - \eta^2_{yx_1}}.$

Notice that, since $\eta^2_{y.x}$ is unity for a linear system, the partial coherence function $\eta^2_{yx_2|x_1}$ will also be unity. Similarly $\eta^2_{yx_1|x_2}$ will be unity. This is not the case for the ordinary coherence functions between the output and one of the inputs; these will in general be less than unity (see equation (ii) above).

14.4. A possibly nonstationary random process $x(t)$ is being studied. Only one sample time history exists, so this is divided into N "slices" of equal length and the short-time mean square value for each slice is calculated. These values are then set down in order of their occurrence to form a series of length N.

(i) Show that there are $N!$ different orders in which the N values could have occurred and confirm that, if $x(t)$ is stationary, the probability of any particular order is $1/N!$.

(ii) Consider the first value in the series. If this exceeds all the other values, it will contribute $(N - 1)$ reverse arrangements. If it is less than all the other values, it will contribute no reverse arrangements. Confirm that, if the values are in a random order, the first value is equally likely to be larger or smaller than each succeeding value, and so the average number of reverse arrangements it contributes is $\frac{1}{2}(N - 1)$. Similarly show that the average number of reverse arrangements contributed by the second value in the series is $\frac{1}{2}(N - 2)$, and so on.

Hence prove that, if the series is drawn in a random order, the average total number of reverse arrangements is

$$\tfrac{1}{4}N(N - 1).$$

(iii) Now take the case of a *random* series with $N = 4$ and, by considering all the possible ways in which the terms may be ordered, draw up a table to show the probability that there will be 0, 1, 2, 3, 4, 5 and 6 reverse arrangements. Hence show that the probability of 2 or less reverse arrangements is $3/8$.

(iv) The method of part (iii) may be extended to the case when $N = 5$ and it is not difficult to show that the probability that there will be 0, 1, 2, 3, 4, 5, 6, 7, 8, 9 and 10 reverse arrangements is

$$\frac{1}{5!}, \frac{4}{5!}, \frac{9}{5!}, \frac{15}{5!}, \frac{20}{5!}, \frac{22}{5!}, \frac{20}{5!}, \frac{15}{5!}, \frac{9}{5!}, \frac{4}{5!} \quad \text{and} \quad \frac{1}{5!}$$

respectively. Use this result and the result of part (iii) to check that corresponding results in the general case may be obtained by successive application of the following algorithm:

$$P_N(c) = \frac{1}{N!} \qquad \text{for} \qquad c = 0$$

$$P_N(c) = P_N(c - 1) + \frac{1}{N}P_{N-1}(c) \qquad \text{for} \qquad 1 \leqslant c \leqslant N - 1$$

$$P_N(c) = P_N(c - 1) + \frac{1}{N}P_{N-1}(c) - \frac{1}{N}P_{N-1}(c - N)$$
$$\text{for} \qquad N \leqslant c \leqslant \tfrac{1}{2}(N - 1)(N - 2)$$

$$P_N(c) = P_N(c - 1) - \frac{1}{N}P_{N-1}(c - N)$$
$$\text{for} \qquad \tfrac{1}{2}(N - 1)(N - 2) < c \leqslant \tfrac{1}{2}N(N - 1)$$

in which

$$c = \text{number of reverse arrangements}$$

$P_N(c) = $ the probability of c reverse arrangements in a series of N terms

$P_{N-1}(c) = $ the probability of c reverse arrangements in a series of $(N-1)$ terms.

(v) By using this algorithm, find the probability of there being $0, 1, 2, \ldots, 12$ reverse arrangements when $N = 10$, and hence find the probability of there being not more than 12 reverse arrangements in a series of 10 values.

(vi) Now consider the situation when a reverse arrangement is defined as occurring whenever the value of one term in the series is less than (instead of greater than) the value of a later term in the series. How are the above results changed by this new definition?

Note. Appendix 6 gives a limited tabulation of the probability of there being not more than c reverse arrangements in a random series of N terms for values of N up to 30. These values have been calculated by using the algorithm in part (iv); if required, results for $N > 30$ can be obtained by the same method. The reverse arrangement trend test is discussed briefly in Kreyszig [45] p. 820 and in Otnes *et al.* [52] pp. 399–405.

14.5. (i) For a narrow band process $\{y(t)\}$ with zero mean, show that, if y and $\dot{y}$ are statistically independent,

$$p_p(a) = -\frac{\left(\dfrac{dp(y)}{dy}\right)_{y=a}}{p(y)_{y=0}}$$

where $p_p(a)$ is the probability density function for peaks of height a and $p(y)$ is the probability density function for y.

(ii) Hence show that, if

$$p_p(a) = \frac{d}{da}\{\text{Prob}(\text{Peak} < a)\} = \frac{d}{da}\{1 - e^{-\alpha a^k}\}$$

for a Weibull distribution of peaks, the probability density function $p(y)$ is given by

$$p(y) = p(y)_{y=0}\, e^{-\alpha y^k}, \qquad y > 0.$$

(iii) For $p(y)$ to be normalized correctly, it is necessary for

$$\int_{-\infty}^{\infty} p(y)\, dy = 1.$$

Check that this requires

$$p(y)_{y=0} = \frac{k\alpha^{1/k}}{2\Gamma(1/k)}$$

where Γ is the gamma function defined in problem 9.3.

(iv) Use the above result to show that

$$\sigma_y^2 = 2 \int_0^\infty y^2 p(y)\,dy = \left(\frac{1}{\alpha}\right)^{2/k} \frac{\Gamma(3/k)}{\Gamma(1/k)}.$$

(v) Lastly, verify that the median peak height a_0 is given by

$$a_0 = \left(\frac{\ln 2}{\alpha}\right)^{1/k}$$

and hence show that

$$\frac{\sigma_y^2}{a_0^2} = (\ln 2)^{-2/k} \frac{\Gamma(3/k)}{\Gamma(1/k)}.$$

Check that, for a Rayleigh distribution, when $k = 2$, this result reduces to

$$\frac{\sigma_y^2}{a_0^2} = \frac{1}{2\ln 2}$$

in agreement with (14.98).

15.1. The random processes $\{x(t)\}$ and $\{y(t)\}$ have sample functions $x(t)$ and $y(t)$. For a specified time interval, corresponding lengths of the sample records $x(t)$ and $y(t)$ are represented by the discrete time series $\{x_r\}$ and $\{y_r\}$, $r = 0$, $1, 2, \ldots, (N-1)$. Consider synthesizing $\{x_r\}$ and $\{y_r\}$ using the algorithm

$$x_r = \sum_{k=0}^{N-1} \sqrt{S_{xx_k}}\, e^{i[\theta_k + 2\pi(kr/N)]}$$

$$y_r = \sum_{k=0}^{N-1} \sqrt{S_{yy_k}}\, e^{i[\theta_k + \Psi_k + 2\pi(kr/N)]}$$

where $\{\theta_k\}$ is a set of randomly chosen phase angles (which differ from one pair of sample functions to the next pair), uniformly distributed in the range 0 to 2π, and $\{\Psi_k\}$ is a set of defined phase angles which is the same for each pair of the sample functions.

Find the ensemble-averaged discrete cross-spectral density between the synthesized $\{x(t)\}$ and $\{y(t)\}$, and hence confirm that the (discrete) coherence between the two artificial processes is always unity.

Show also that, if

$$y_r = \left\{\sum_{k=0}^{N-1} \eta_{xy_k} \sqrt{S_{xx_k}}\, e^{i[\theta_k + 2\pi(kr/N)]}\right\} + \sum_{k=0}^{N-1} \sqrt{1 - \eta_{xy_k}^2}\, \sqrt{S_{xx_k}}\, e^{i[\phi_k + 2\pi(kr/N)]}$$

where $\{\phi_k\}$ is a second set of randomly chosen phase angles uniformly distributed between 0 and 2π, then, after ensemble averages have been taken,

$$E[S_{xy_k}] = \eta_{xy_k} S_{xx_k}$$

and the ensemble-averaged coherence between the two processes is then
$\eta_{xy_k}^2$.

15.2. Prove that, for the two-dimensional discrete Fourier transform, (15.12) is the exact inverse of (15.11).

Hint. Begin by rewriting (15.11) with the integers t and u replacing r and s. Then substitute this revised expression for Y_{km} into (15.12) and interchange the order of summation, making use of the fact that all the summations of the complex exponential functions become zero unless the exponent of the exponential is zero (see page 116).

Also prove Parseval's theorem for the continuous two-dimensional case (15.23) and for the discrete two-dimensional case (15.25).

Hints. To prove (15.23), begin by expressing $y^2(x_1, x_2)$ in the form

$$y^2(x_1, x_2) = \left\{ \sum_k \sum_m Y_{km} e^{-i2\pi(kx_1/L_1 + mx_2/L_2)} \right\} \times$$
$$\times \left\{ \sum_n \sum_p Y_{np} e^{-i2\pi(nx_1/L_1 + px_2/L_2)} \right\}.$$

Substitute this expression into the left-hand side of (15.23), and then interchange the order of integration and summation and show that the double integral is zero unless $n = -k$ and $p = -m$.

To prove (15.25), use (15.11) to substitute for Y_{km} and Y_{km}^* in the r.h.s. of (15.25), using subscripts r, s for Y_{km} and t, u for Y_{km}^*. Then make the summations over k and m.

15.3. (i) Use one of the two-dimensional Fourier transform programs in Appendix 2 to calculate the two-dimensional DFT's of the array of numbers given by

(a)
$$y_{rs} = \cos 2\pi\left(\frac{r}{8}\right)\cos 2\pi\left(\frac{3s}{8}\right)$$

$$r, s = 0, 1, 2, ..., 7$$

and by

(b)
$$y_{rs} = \sin 2\pi\left(\frac{r}{8}\right)\cos 2\pi\left(\frac{3s}{8}\right)$$

$$r, s = 0, 1, 2, ..., 7.$$

(ii) Write a program to compute the double summations on both sides of Parseval's equation (15.25) and hence verify that the results in part (i) satisfy Parseval's theorem.

(iii) Calculate the two-dimensional DFT of the array of numbers given by

$$y_{rs} = \cos 2\pi\left(\frac{r}{8}\right)\cos 2\pi\left(\frac{5s}{8}\right)$$

$$r, s = 0, 1, 2, ..., 7$$

and check that, because of aliasing, this is the same as the result for part (i)(a) of this problem.

(iv) Now modify the program to calculate the inverse two-dimensional DFT (as described in the notes in Appendix 2) and check that the inverse two-dimensional DFT of the array

$$Y_{km} = 0, \qquad k, m = 0, 1, 2, \ldots, 7$$

except for

$$Y_{1,3} = Y_{1,5} = Y_{7,3} = Y_{7,5} = 1/4$$

yields the starting arrays used in part (i)(a) and part (iii).

15.4. Consider a two-dimensional stationary random process $\{y(X_1, X_2)\}$ which has correlation function $R_{yy}(x_1, x_2)$ and spectral density $S_{yy}(\gamma_1, \gamma_2)$ where

$$R_{yy}(x_1, x_2) = E[y(X_1, X_2)y(X_1 + x_1, X_2 + x_2)] \qquad \text{(i)}$$

and

$$S_{yy}(\gamma_1, \gamma_2) = \int_{-\infty}^{\infty} dx_1 \int_{-\infty}^{\infty} dx_2\, R_{yy}(x_1, x_2)\, e^{-i(\gamma_1 x_1 + \gamma_2 x_2)}. \qquad \text{(ii)}$$

According to the theory of generalized Fourier analysis, if the sample function $y(X_1, X_2)$ is physically realisable, then we can expect it to have a Fourier transform (which will be a generalized function of some sort). Let $Y(\gamma_1, \gamma_2)$ be this transform, defined by

$$Y(\gamma_1, \gamma_2) = \int_{-\infty}^{\infty} dX_1 \int_{-\infty}^{\infty} dX_2\, y(X_1, X_2)\, e^{-i(\gamma_1 X_1 + \gamma_2 X_2)}. \qquad \text{(iii)}$$

Consider the ensemble-averaged value of the product $Y(\gamma_1, \gamma_2)Y(\gamma_1', \gamma_2')$. From (iii), we find that

$$E[Y(\gamma_1, \gamma_2)Y(\gamma_1', \gamma_2')] =$$

$$= \int_{-\infty}^{\infty} dX_1 \int_{-\infty}^{\infty} dX_2 \int_{-\infty}^{\infty} dX_1' \int_{-\infty}^{\infty} dX_2'\, E[y(X_1, X_2)y(X_1', X_2')] \times$$
$$\times\, e^{-i(\gamma_1 X_1 + \gamma_2 X_2 + \gamma_1' X_1' + \gamma_2' X_2')}. \qquad \text{(iv)}$$

Now substitute for $R_{yy}(X_1' - X_1, X_2' - X_2)$ in the r.h.s. of (iv) from (i), change the variables X_1', X_2' to x_1, x_2 defined by

$$X_1' = X_1 + x_1$$
$$X_2' = X_2 + x_2 \qquad \text{(v)}$$

and use the result that

$$2\pi\, \delta(\gamma) = \int_{-\infty}^{\infty} e^{-i\gamma X}\, dX$$

to show that (see Dowling *et al.* [80], p. 222)

$$E[Y(\gamma_1, \gamma_2)Y(\gamma_1', \gamma_2')] = (2\pi)^2 S_{yy}(\gamma_1', \gamma_2')\, \delta(\gamma_1 + \gamma_1')\, \delta(\gamma_2 + \gamma_2'). \qquad \text{(vi)}$$

This means that $E[Y(\gamma_1, \gamma_2)Y(\gamma_1', \gamma_2')]$ is zero everywhere except where $\gamma_1' = -\gamma_1$ and $\gamma_2' = -\gamma_2$, when it is infinite. Use (vi) to demonstrate that the two-dimensional spectral density function $S(\gamma_1, \gamma_2)$ is a non-negative function.

Hint. From (iii), $Y(-\gamma_1, -\gamma_2)$ is the complex conjugate of $Y(\gamma_1, \gamma_2)$, and we know from (15.31) that $S(\gamma_1, \gamma_2)$ is an even function.

15.5. (i) Show that the random variables A_{km} and B_{km} defined by equation (15.111) will have zero means provided that the y_{rs} in (15.111) are derived by sampling a random process $\{y(X, t)\}$ for which $E[y(X, t)] = 0$.

(ii) Consider the mean-square value of A_{km} and show that this may be written as

$$E[A_{km}^2] = \frac{1}{(N_1 N_2)^2} \sum_{r=0}^{N_1-1} \sum_{s=0}^{N_2-1} \sum_{t=0}^{N_1-1} \sum_{u=0}^{N_2-1} R_{yy}\{(t-r)\Delta_1, (u-s)\Delta_2\} \times$$

$$\times \cos 2\pi\left(\frac{kr}{N_1} + \frac{ms}{N_2}\right)\cos 2\pi\left(\frac{kt}{N_1} + \frac{mu}{N_2}\right)$$

where $R_{yy}(x, \tau)$ is the two-dimensional correlation function for $\{y(X, t)\}$ which is assumed to be homogeneous and stationary, and for which the sampling intervals are Δ_1 over X and Δ_2 over t.

(iii) Hence show that, when the correlation function is zero unless $r = t$ and $s = u$, so that

$$R_{yy}\{(t-r)\Delta_1, (u-s)\Delta_2\} = E[y^2] \quad \text{for } r = t \text{ and } s = u$$
$$= 0 \quad \text{elsewhere}$$

then

$$E[A_{km}^2] = \frac{1}{(N_1 N_2)} E[y^2](\tfrac{1}{2})$$

and that, in this limiting case, A_{km} and B_{km} have the same variance.

15.6. (i) Show that the DFT of the circular spectral sequence

$$S_{c_k} = S_k\left(1 - \frac{k}{N}\right) + S_{N-k}\left(\frac{k}{N}\right) \qquad k = 0, 1, 2, \ldots, N-1 \qquad (15.132)$$

may be written in the form

$$R_r = \sum_{k=-N}^{N} S_k\left(1 - \frac{|k|}{N}\right)e^{i2\pi(kr/N)} \qquad (i)$$

when $S_{-k} = S_k$.

(ii) When the ensemble-averaged coefficients S_k are related to the true spectrum by

$$E[S_k] = S\left(\gamma = \frac{k}{N}\gamma_0\right)\left(\frac{\gamma_0}{N}\right) \qquad \text{(ii)}$$

where, in the usual notation, $\gamma_0 = 2\pi/\Delta$, show that by taking the ensemble average of equation (i) we obtain

$$E[R_r] = \frac{\gamma_0}{N}\sum_{k=-N}^{N} W(\gamma)S(\gamma)\,e^{i\Delta\gamma r} \qquad \text{(iii)}$$

where

$$\gamma = \frac{2\pi k}{N\Delta} \qquad \text{(iv)}$$

and

$$W(\gamma) = \begin{cases} 1 - \dfrac{\Delta}{2\pi}|\gamma| & \text{for} \quad 0 \leqslant |\gamma| \leqslant 2\pi/\Delta \\ 0 & \text{elsewhere.} \end{cases} \qquad \text{(v)}$$

(iii) By following the steps used to derive equations (15.82), show that equation (iii) may be written as

$$E[R_r] = \int_{-\infty}^{\infty} dx\left(\sum_{j=-\infty}^{\infty} \frac{\gamma_0\Delta}{(2\pi)^2} w(x + jN\Delta)R(r\Delta - x)\right) \qquad \text{(vi)}$$

where $w(x)$ is the inverse Fourier transform of $W(\gamma)$ and $R(x)$ is the inverse Fourier transform of $S(\gamma)$. Hence prove that

$$E[R_r] = \int_{-\infty}^{\infty} dx\, w_a(r\Delta - x)R(x) \qquad (15.133)$$

where

$$w_a(x) = \sum_{j=-\infty}^{\infty} w(x + jN\Delta) \qquad (15.134)$$

and

$$w(x) = \frac{\gamma_0}{2\pi}\left(\frac{\sin\dfrac{\gamma_0 x}{2}}{\dfrac{\gamma_0 x}{2}}\right)^2. \qquad (15.135)$$

15.7. (i) The height of a surface is defined by ordinates $y_{s,t}$ on a rectangular grid with spacing Δ_1 in the s-direction and spacing Δ_2 in the t-direction. If $Y_j^{(1)}$ is the DFT of the sequence $\{y_{s,0}\}, s = 0, 1, 2, \ldots, N_1 - 1$, and $Y_j^{(2)}$ is the DFT of the sequence $\{y_{r,t}\}, r = 0, 1, 2, \ldots, N_1 - 1$, show that the cross-spectral density coefficients $S_j^{(1,2)}$ for two parallel tracks distance $t\Delta_2$ apart are given by

$$S_j^{(1,2)} = Y_j^{(1)} * Y_j^{(2)} = \frac{1}{N_1^2} \sum_{s=0}^{N_1-1} \sum_{r=0}^{N_1-1} y_{s,0} y_{r,t} e^{-i(2\pi/N_1)j(r-s)}$$

$$j = 0, 1, 2, \ldots, N_1 - 1.$$

(ii) If the ordinates are generated artificially by the algorithm (15.123), show that, when the phase angles $\phi_{k,m}$ are distributed uniformly between 0 and 2π (subject to the conditions (15.124) to (15.127)), then

$$E[y_{s,0} y_{r,t}] = \sum_{k=1}^{N_1-1} \sum_{m=1}^{N_2-1} S_{km} e^{i2\pi[(r-s)/N_1]k} e^{i2\pi(mt/N_2)}.$$

Hint. Because the phase angles are distributed uniformly, $E[e^{i(\phi_{km}+\phi_{np})}]$ is zero unless $\phi_{km} + \phi_{np} = 0$, which is when $n = N_1 - k$ and $p = N_2 - m$.
(iii) Hence show that the ensemble-averaged cross-spectral density coefficients $E[S_j^{(1,2)}]$ are given by

$$E[S_j^{(1,2)}] = \sum_{m=0}^{N_2-1} S_{jm} e^{i2\pi(mt/N_2)}$$

in agreement with (15.151) except that the coefficients S_{jm} do not have to be averaged because the artificial generation of the surface by (15.123) ensures that they are correct.

15.8. (i) For the same surface as problem 15.7, show that the discrete cross-correlation coefficients $R_r^{(1,2)}$ between two parallel tracks distance $t\Delta_2$ apart, which are defined by (15.149), can be written as

$$R_r^{(1,2)} = \left(1 - \frac{t}{N_2}\right)\left\{\left(1 - \frac{r}{N_1}\right)\hat{R}_{r,t} + \frac{r}{N_1}\hat{R}_{-(N_1-r),t}\right\} +$$

$$+ \frac{t}{N_2}\left\{\left(1 - \frac{r}{N_1}\right)\hat{R}_{r,-(N_2-t)} + \frac{r}{N_1}\hat{R}_{-(N_1-r),-(N_2-t)}\right\}$$

where $\hat{R}_{r,t}$ is an unbiased estimate for $R(r\Delta_1, t\Delta_2)$, etc.

(ii) Hence show that when $S_j^{(1,2)}$ is defined by (15.150), we find that

$$E[S_j^{(1,2)}] = \left(1 - \frac{t}{N_2}\right)\frac{2\pi}{L_1} \int_{-\infty}^{\infty} W_a(\gamma_k - \gamma) S^{(1,2)}(\gamma) \, d\gamma +$$

$$+ \frac{t}{N_2} \frac{2\pi}{L_1} \int_{-\infty}^{\infty} W_a(\gamma_k - \gamma) S^{(1,3)}(\gamma) \, d\gamma$$

in agreement with (15.162) where

$S^{(1,2)}(\gamma)$ is the exact cross-spectral density between parallel tracks distance $t\Delta_2$ apart
$S^{(1,3)}(\gamma)$ is the exact cross-spectral density between parallel tracks distance $(t - N_2)\Delta_2$ apart
$W_a(\gamma)$ is the aliased spectral window defined by (15.83) with $L = L_1$.

16.1. Investigate further the response of a taut string to broad-band random excitation applied at one point, Fig. 16.2, as follows:

(i) Study the effect of different levels of damping on the mean-square velocity response. The results shown in Fig. 16.3(c) are for the case when $T = 1$, $L = 1$, $a = L/3$, $m = 1/400$ and $c = 2\pi/400$ in consistent units. Calculate and plot out the results when the damping coefficient c is reduced by a factor of 10 to $c = 2\pi/4000$. Confirm that both the peaks of mean-square velocity at $L/3$ and $2L/3$ are now approximately the same height which is about 50 per cent greater than the average mean-square velocity.

Calculate also the same results when the damping is increased to $c = 2\pi/100$ and show that the mean-square velocity now begins to lose its essentially uniform distribution. The peaks at $L/3$ and $2L/3$ are still present but there is now a trend for the mean-square velocity response to decrease from the point of excitation, $L/3$, towards the two ends of the string.

(ii) Investigate the distribution of mean-square acceleration along the length of the string. Modify (16.25) to calculate the mean-square acceleration at position x

$$E[\ddot{y}^2(x, t)]$$

and calculate and plot out this parameter for the same numerical data as for Fig. 16.3(c). Verify that the distribution of mean-square acceleration is broadly the same as the distribution of mean-square velocity except that the amplitude of the ripples is increased, although the heights of the peaks at $L/3$ and $2L/3$ remain approximately the same. There is a notable increase in the peakiness of the acceleration response close to the two ends of the string.

16.2. Show that the response $y(\mathbf{r}, t)$ of a continuous, linear system at position $\mathbf{r}$ and time t can be expressed in terms of the distributed excitation $p(\mathbf{s}, t)$ per unit area at position $\mathbf{s}$ and time t in the following form

$$y(\mathbf{r}, t) = \int_R d\mathbf{s} \int_{-\infty}^{\infty} d\theta\, h(\mathbf{r}, \mathbf{s}, \theta) p(\mathbf{s}, t - \theta) \qquad (16.127)$$

where $h(\mathbf{r}, \mathbf{s}, t)$ gives the response of the system at position $\mathbf{r}$ to unit impulse excitation applied at $t = 0$ at position $\mathbf{s}$.

Hence show that the cross-correlation function for the two outputs at $\mathbf{r}_1$ and $\mathbf{r}_2$, $R_{yy}(\mathbf{r}_1, \mathbf{r}_2, \tau)$, can be written in terms of the cross-correlation function for the excitation at positions $\mathbf{s}_1$ and $\mathbf{s}_2$, $R_{pp}(\mathbf{s}_1, \mathbf{s}_2, \tau)$ in the form

$$R_{yy}(\mathbf{r}_1, \mathbf{r}_2, \tau) =$$
$$= \int_R d\mathbf{s}_1 \int_R d\mathbf{s}_2 \int_{-\infty}^{\infty} d\theta_1 \int_{-\infty}^{\infty} d\theta_2 h(\mathbf{r}_1, \mathbf{s}_1, \theta_1) h(\mathbf{r}_2, \mathbf{s}_2, \theta_2) R_{pp}(\mathbf{s}_1, \mathbf{s}_2, \tau + \theta_1 - \theta_2).$$
$$(16.128)$$

Show that this equation leads directly to the result (16.34).

Hint. Before taking Fourier transforms, introduce the multiplying factors

$$e^{i\omega\theta_1} e^{-i\omega\theta_2} e^{-i(\omega\theta_1 - \omega\theta_2)}$$

and then change the variable of integration for the Fourier transform calculation to $\tau + \theta_1 - \theta_2$.

16.3. (i) Show that the normal modes and natural frequencies of a taut string of length L, tension T and mass per unit length m are given by

$$\Psi_j(x) = \sqrt{2} \sin\left(j\frac{\pi x}{L}\right)$$

and

$$\omega_j = j\frac{\pi}{L}\sqrt{\frac{T}{m}}.$$

(ii) Show also that the frequency response function for the displacement of the string as a result of a unit harmonic transverse force $e^{i\omega t}$ applied at $x = a$ can be written as the summation

$$H(x,a,\omega) = \sum_{j=1}^{\infty} \frac{2\sin\left(j\frac{\pi x}{L}\right)\sin\left(j\frac{\pi a}{L}\right)}{mL\left\{-\omega^2 + \left(\frac{c}{m}\right)i\omega + \omega_j^2\right\}}.$$

where c is defined in equation (16.4).

(iii) Write a computer program to calculate $H(x, a, \omega)$ when $T = 1, L = 1$, $a = 1/3, m = 1/400$ and $c = 2\pi/400$ in consistent units. Compute $H(x, a, \omega)$ for $x = 1/10, x = 1/3$ and $x = 2/3$ at 400 discrete frequencies equally spaced between 0 and 400π rad/s. Make one set of calculations with the summation covering 20 modes and a second set of calculations with the summation covering 30 modes. Investigate the effect of including more than 20 modes in the summation.

Use the results for the 30-mode summation to calculate the mean-square velocity response of the string at $x = 1/10, 1/3$ and $2/3$ for broad-band stationary random excitation with a constant spectral density to the cut-off frequency of 400π rad/s, and hence check the relative heights of three points on the response curve shown in Fig. 16.3(c).

16.4. Consider again the random vibration of a taut string of length L when its excitation $f(x, t)$ per unit length is independent of position x, so that

$$f(x, t) = f(t).$$

(i) Confirm that the two-dimensional spectral density of the distributed excitation can be written

$$S_{ff}(\gamma, \omega) = S_{ff}(\omega)\delta(\gamma)$$

and that the cross-spectral density of the excitation is

$$S_{ff}(x_1, x_2, \omega) = S_{ff}(\omega).$$

(ii) When the excitation is uncorrelated in time, so that $S_{ff}(\omega) = S_0$, show that the cross-spectral density of the response can be expressed as

$$S_{yy}(x_1, x_2, \omega) = S_0 \left\{ \int_0^L ds_1\, H^*(x_1, s_1, \omega) \right\} \left\{ \int_0^L ds_2\, H(x_2, s_2, \omega) \right\}$$

where $H(x, s, \omega)$ is given by the expression in problem 16.3.

(iii) On the assumption of small damping, so that the modal bandwidths are small compared with the spacing between adjacent natural frequencies, show that $S_{yy}(x_1, x_2, \omega)$ may be written approximately as

$$S_{yy}(x_1, x_2, \omega) = \frac{16 S_0}{m^2 \pi^2} \sum_{j \text{ odd}} \frac{\sin^2\left(j \dfrac{\pi x}{L} \right)}{j^2(-\omega^2 + \beta_j i\omega + \omega_j^2)}.$$

(iv) Hence show that, for this excitation, the mean-square displacement response of the string at position x is given approximately by

$$E[y^2(x, t)] = \frac{S_0 L^2}{Tc} \left(\frac{2}{3} \pi^3 \right) \left\{ 3\left(\frac{x}{L} \right)^2 - 4\left(\frac{x}{L} \right)^3 \right\}, \quad 0 \leqslant \frac{x}{L} \leqslant \frac{1}{2}$$

and the mean-square velocity response is given approximately by

$$E[\dot{y}^2(x, t)] = \frac{4\pi S_0}{cm} \left(\frac{x}{L} \right), \quad 0 \leqslant \frac{x}{L} \leqslant \frac{1}{2}$$

(Crandall [76], equations (23) and (24)).

Hint. The following results will be needed

$$\left(1 + \frac{1}{3^2} + \frac{1}{5^2} + \cdots \right) = \frac{\pi^2}{8}$$

$$\left(1 + \frac{1}{3^4} + \frac{1}{5^4} + \cdots \right) = \frac{\pi^4}{96}$$

$$\left(\cos \alpha + \frac{\cos 3\alpha}{3^2} + \frac{\cos 5\alpha}{5^2} + \cdots \right) = \frac{\pi}{8}(\pi - 2\alpha), \quad 0 \leqslant \alpha \leqslant \pi$$

$$\left(\cos \alpha + \frac{\cos 3\alpha}{3^4} + \frac{\cos 5\alpha}{5^4} + \cdots \right) = \frac{\pi}{96}(4\alpha^3 - 6\pi\alpha^2 + \pi^3), 0 \leqslant \alpha \leqslant \pi$$

(Jolley [40], series 307, 518 and 531).

16.5. (i) For a one-dimensional structure of length L subjected to uncorrelated excitation for which $S_{pp}(\gamma, \omega) = S_0$, show that the cross-correlation between the excitation at position s_1 and position s_2 is given by

$$R_{pp}(s_1, s_2, \tau) = (2\pi)^2 S_0\, \delta(\tau)\, \delta(s_1 - s_2)$$

and use equation (16.128) of problem 16.2 to show that the autocorrelation of the response at position r may be expressed as

$$R_{yy}(r, r, \tau) = \int_0^L ds \int_{-\infty}^{\infty} d\theta_1 \int_{-\infty}^{\infty} d\theta_2\, h(r, s, \theta_1) h(r, s, \theta_2)(2\pi)^2 S_0\delta(\tau).$$

By following the hint in problem 16.2, check that this result reduces correctly to equation (16.80) for the case when $r_1 = r_2 = r$.

(ii) Show, by starting from equation (16.87), that the mean-square displacement of a uniform beam of length L at distance x from one end is given by

$$E[y^2(x, t)] = \frac{2\pi^2 S_0 L^3}{3EIc} \left(\frac{x}{L}\right)^2 \left(1 - \frac{x}{L}\right)^2, \qquad 0 \leqslant \frac{x}{L} \leqslant 1.$$

by using the results that

$$\left(1 + \frac{1}{2^4} + \frac{1}{3^4} + \cdots\right) = \frac{\pi^4}{90}$$

$$\left(\cos\alpha + \frac{\cos 2\alpha}{2^4} + \frac{\cos 3\alpha}{3^4} + \cdots\right) =$$

$$= \frac{1}{48}\left\{2\pi^2(\alpha - \pi)^2 - (\alpha - \pi)^4\right\} - \frac{7\pi^4}{720} \qquad 0 \leqslant \alpha \leqslant 2\pi$$

(Jolley [40], series 305 and 529).

17.1. Find a solution of (17.1) for the case when (i) $c_0 = c_1 = 1$, $c_2 = c_3 = 0$; (ii) $c_0 = c_2 = \frac{1}{2}$, $c_1 = 1$, $c_3 = 0$; (iii) $c_0 = 2$, $c_1 = c_2 = c_3 = 0$; (iv) $c_0 = c_2 = 1$, $c_1 = c_3 = 0$.

Hint. Explore the result of making the iteration (17.2) starting from the box function $\phi_0(x) = 1$, $0 \leqslant x < 1$.

Generate wavelets from the scaling functions found above by applying (17.12) where N is the number of non-zero wavelet coefficients (but put $N = 3$ for case (iv)).

The Haar wavelet (i) is the simplest useful wavelet. Note that

$$\int_0^1 W(x)W(2x)\,dx = \int_0^1 W(x)W(2x - 1)\,dx = 0$$

so that this is an orthogonal wavelet. The inverted Mexican hat wavelet (ii) is not orthogonal because, for example,

$$\int_0^2 W(x)W(2x)\,dx \neq 0.$$

The results from (iii) and (iv) are curiosities. The delta function from (iii) and the comb function from (iv) do not provide useful basis functions for signal decomposition.

17.2. A recursive method is one in which one or more terms of a sequence can be determined *exactly* if other terms are known. Instead of finding the scaling function $\phi(x)$ by iterating until convergence is achieved, an alternative approach is to find particular values of $\phi(x)$ exactly and design a recursive scheme to find all the other values progressively from these known terms.

Consider the definition (17.1) written for compactness as

$$\phi(x) = \sum_{k=0}^{3} c_k \phi(2x - k)$$

which, on replacing x by $x/2$ becomes

$$\phi(x/2) = \sum_{k=0}^{3} c_k \phi(x - k). \tag{a}$$

If $\phi(x)$, $\phi(x-1)$, $\phi(x-2)$, $\phi(x-3)$ are all known, then we can find $\phi(x/2)$ exactly. For four non-zero wavelet coefficients, we see from Fig. 17.4 that $\phi(x)$ is zero for $x \leqslant 0$ and $x \geqslant 3$. It can be shown that this is a general result. Suppose that we can find $\phi(0)$, $\phi(1)$, $\phi(2)$, $\phi(3)$. We know that $\phi(-1)$, $\phi(4)$, etc. are all zero. Hence from (a) we can calculate

$$\phi(1/2), \; \phi(3/2), \; \phi(5/2).$$

Then, by again using (a) and these new values we can calculate

$$\phi(1/4), \; \phi(3/4), \; \phi(5/4), \; \phi(7/4), \; \phi(9/4), \; \phi(11/4)$$

and so on, until all the required ordinates of $\phi(x)$ have been found exactly.

In order to make the recursion work, we must find the starting values $\phi(0)$, $\phi(1)$, $\phi(2)$, $\phi(3)$. For the D4 coefficients in (17.3), we know from Fig. 17.4 that $\phi(0) = \phi(3) = 0$. Use (a) to express $\phi(1)$ in terms of $\phi(2)$ to $\phi(-1)$ to prove that

$$\frac{\phi(1)}{\phi(2)} = \frac{1 + \sqrt{3}}{1 - \sqrt{3}}.$$

Show that the same result is obtained if (a) is used to express $\phi(2)$ in terms of $\phi(4)$ to $\phi(1)$.

The scale of $\phi(x)$ is arbitrary, but set

$$\phi(1) = (1 + \sqrt{3})/2 \qquad \phi(2) = (1 - \sqrt{3})/2$$

and then use recursion to find by hand

$$\phi(1/4), \; \phi(1/2), \; \phi(3/4), \; \ldots, \; \phi(11/4).$$

Set up a computer program to make this calculation and use it to check Fig. 17.5.

Find by hand the values of $W(1)$, $W(3/2)$, $W(2)$ by using (17.5). Then set up a computer program to use this equation to calculate $W(x)$ from $\phi(x)$ and hence check Fig. 17.6.

17.3. Verify that the matrix equation (17.7) correctly generates the 10 wavelet ordinates given in the text above (17.7) and check that (17.8) produces the same result and so is identical with (17.7). Similarly, verify that (17.9) and (17.10) give identical results.

17.4. According to (17.25), if $P(\xi)$ is the Fourier transform of a scaling function $\phi(x)$, then it is always possible to write $P(\xi)$ as the infinite product

$$P(\xi) = p(\xi/2)p(\xi/4)p(\xi/8) \ldots (1/2\pi) \tag{a}$$

where

$$p(\xi) = \frac{1}{2}\sum_k c_k e^{-i\xi k} \tag{b}$$

and c_k are the wavelet coefficients which appear in the dilation equation (17.11).

(i) Examine the shape of the function $p(\xi)$ for the D4 wavelet coefficients defined by (17.3) by preparing a computer program to graph the real and imaginary parts of $p(\xi)$ against ξ for the range $0 \leqslant \xi \leqslant 9$.

(ii) Check from (b) that $p(\xi)$ and $(d/d\xi)p(\xi)$ are both zero at $\xi = \pi$ and confirm that this agrees with the graphs from part (i).

(iii) Hence check that $p(\xi/2)$ and its derivative are zero when $\xi = 2\pi, 6\pi, 10\pi$, etc; also that $p(\xi/4)$ and its first derivative are zero for $\xi = 4\pi, 12\pi, 20\pi$, etc. and that $p(\xi/8)$ and its first derivative are zero for $\xi = 8\pi, 24\pi, 40\pi$, etc.

(iv) Modify the computer program from part (i) to calculate the product $p(\xi/2)p(\xi/4)p(\xi/8)$ and plot its real and imaginary parts over the range $0 \leqslant \xi \leqslant 35$. Hence verify that, for the D4 scaling function, its Fourier transform $P(\xi)$ and its first derivative $dP(\xi)/d\xi$ are zero for $\xi = 2\pi, 4\pi, 6\pi, \ldots$.

(v) Because of the result in (iv), the D4 wavelet is the basis for an expansion of the function $f(x)$ which can reproduce $f(x)$ exactly provided that $f(x)$ consists only of terms like $\alpha + \beta x$ where α and β are constants. Examine this proposition by calculating the Fourier transforms $F(\xi)$ of $f(x)$ for the cases when

(a) $f(x) = 1$ for $-\frac{1}{2} \leqslant x \leqslant \frac{1}{2}$
      $= 0$ elsewhere

(b) $f(x) = x$ for $-\frac{1}{2} \leqslant x \leqslant \frac{1}{2}$
      $= 0$ elsewhere.

Show that, for case (a), $F(\xi)$ has zeros for $\xi = 2\pi, 4\pi, 6\pi, \ldots$ and that, for case (b), $F(\xi)$ has zeros at intervals of approximately 2π and $dF/d\xi$ becomes progressively closer to zero when $F(\zeta) = 0$ as $\xi \to \infty$.
(vi) Verify that the D4 scaling function can represent $f(x) = \alpha + \beta x$ exactly by computing $\phi(x) + \phi(x - 1) + \phi(x - 2)$ for $0 \leqslant x \leqslant 5$ and $\phi(x) + 2\phi(x - 1)$ for $0 \leqslant x \leqslant 4$. Use the computer program prepared for calculating $\phi(x)$ by recursion (problem 17.2) and compute $2^8 = 256$ points in the unit interval $0 \leqslant x < 1$.

17.5. Use the program $a = wavedn(f, N)$ to compute the wavelet transforms for $N = 2, 4, 6, 20$ of each of the following sequences:
  (i) $f(j) = 1$,
  (ii) $f(j) = (j/512)^2 - 2(j/512)^3 + (j/512)^4$,
  (iii) $f(j) = \sin(2\pi(j - 1)/64)$,
  (iv) $f(j) = \sin(2\pi(j - 1)/50)$,
when j covers the range $j = 1$ to 512. In each case plot f and a and examine how the elements of a are altered as the number of wavelet coefficients N is increased.
  For case (ii), plot $\log_{10}|a(r)|$ for $N = 2, 4, 6$ and 20 to compare higher-order terms in the wavelet transform. Why does the result for $N = 20$ show peaks for the amplitudes of wavelets positioned around $x = 0$ and $x = 1$?
  Compute $f = iwavedn(a, 4)$ when $a(r) = 0$, $r = 1$ to 512, except for $a(r) = 1$ when $r = 5, 9, 17, 33, 65, 129$ and 257.
  Compare $f(r)$, $r = 1$ to 16, with the scaling function ϕ calculated for a sequence length of 48 using the program prepared for problem 17.2. Confirm that your conclusion agrees closely with the result suggested by (17.94) with $k = 0$. Explore this result when $a(r), r = 5, 9$, etc. have different values.
  Now use (17.12) to show that $W(x/2)$ can be expressed as a sum of terms in $\phi(x)$ and its translates. Show also that $W(x/4)$ is the sum of terms in $\phi(x/2)$ and its translates and that, because of (17.11), $W(x/4)$ may also be expressed as a sum of terms in $\phi(x)$ and its translates.
  Use these results to prove that the sum of a sequence of wavelets at levels $-1, -2, -3, \ldots, -\infty$ can always be expressed in terms of the scaling function $\phi(x)$ and hence that (17.94) is true.

17.6. Use the program $f = iwavedn(a, N)$ to compute the inverse wavelet transforms of the following cases:
(i) $a(r) = 0$, $r = 1$ to 512, except $a(24) = 1$ for $N = 2, 4, 6, 10, 20$. Note that the relative position of the centre of the wavelet changes as N increases. The scale of the wavelets at this level 4 is such that there are $2^4 = 16$ adjacent wavelets in the unit interval. The most leftward of these is that for which $a(17) = 1$.
(ii) Compute the wavelet for which $a(r) = 0, r = 1$ to 512, except $a(17) = 1$

when $N = 20$. For $N = 20$, the length of the wavelet is 19 intervals, which in this case is $\Delta x = 19/16$ so that the wavelet wraps round the interval $0 \leqslant x \leqslant 1$. The "centre" of the wavelet lies in the middle of interval 9 which is at $x = 19/32$. For 512 ordinates, the wavelet's centre is at ordinate $512 \, (19/32) = 304$.

(iii) Mark ordinate 304 on the graph obtained in part (ii).

(iv) Use the $f(j)$, $j = 1$ to 512, computed from part (ii) with the function file *mapdn(f,N)* to compute $A = mapdn(f,20)$. Use MATLAB's *contour* function to draw a contour plot of this A. Check that the centre of the non-zero block of A is at $304/2 = 152$, and that it lies in row 6 corresponding to level 4 of the wavelet transform.

(v) Compute the inverse wavelet transform $f(j)$ of $a(r) = 0$ for $r = 1$ to 512 except $a(r) = 1$ for $r = 17$ to 32. Examine this function, compute $A = mapdn(f,20)$ and draw its contour plot.

(vi) Compute the discrete Fourier transform F of the functions f from part (iv) and part (v), and compare these by plotting $|F(j)|$ against j in the two cases. Check that the maximum value of $|F(j)|$ from part (v) occurs at $j = 17$ (where the constant component occurs at $j = 1$, not $j = 0$).

(vii) Compare the result of part (v) with a sine wave of the same magnitude and phase. Calculate its DFT and plot this for comparison with the result of part (vi).

17.7. (i) Check that, subject to the condition (17.47), the wrapped matrices M_1, M_2, M_3 defined by (17.113), (17.114), (17.115) are orthogonal matrices and so satisfy.

$$\alpha M^t M = I$$

where $\alpha = 1/\underset{k}{\Sigma} c_k^2$.

(ii) Using the terminology that, in the discrete wavelet transform, $f^{(0)}(1:8)$ represents eight equally-spaced samples of $W(x)$ in the interval $0 \leqslant x < 1$ and that $f^{(1,1)}(1:8)$ represents the same for $W(2x)$, where $W(x)$ and $W(2x)$ are the wrapped functions, use the result of part (i) with (17.120) and (17.124) to show that

$$f^{(0)}f^{(1,1)t} = 0.$$

Hence confirm that

$$\int_0^1 W(x)\,W(2x)\,dx = 0.$$

(iii) Examine the first step of the construction of the wrapped scaling function $\phi(x)$ and wrapped wavelet $W(x)$ by iteration from a unit box. Show that, instead of (17.51)

$$\phi_1(x) = (c_0 + c_2)\phi_0(2x) + (c_1 + c_3)\phi_0(2x - 1)$$

and find a corresponding expression for $W_1(x)$. Hence show that

$$\int_0^1 \phi_1(x)W_1(x)\,dx = 0$$

whatever the values of the wavelet coefficients.

(iv) Also examine the generation of the wrapped wavelet functions $W(2x)$ and $W(2x - 1)$ by iteration from the boxes

$$\phi_0(x) = 1 \qquad 0 \leqslant x < \tfrac{1}{2}$$

and $\phi_0(x - \tfrac{1}{2})$. After one iteration show that

$$\int_0^1 W_1(2x)W_1(2x - 1)\,dx = 0$$

provided that

$$c_0c_2 + c_1c_3 = 0.$$

17.8. Compute $f(j)$ the inverse wavelet transform $iwavedn(a,20)$ where $a(r) = 0$ for $r = 1$ to 512 except $a(33) = 1$. Compare this function with the Gaussian windowed sine wave

$$p(j) = \exp(-(j - 152)^2/1024)\sin(2\pi(j - 1)/32), \qquad j = 1 \text{ to } 512.$$

Compute the DFTs $F(k)$ of $f(j)$ and $P(k)$ of $p(j)$ and compare these by plotting $\log|F(k)|$ and $\log|P(k)|$ against k for $k = 1$ to 256. Use $mapdn(f,20)$ to compute the mean-square maps of both the above functions and compare the results.

17.9. Use Mallat's tree algorithm (17.134) for the IDWT to write $f(1:4)$ in terms of its DWT $a(1:4)$. Hence show that

$$\frac{1}{4}\sum_{r=0}^3 f_r^2 = a_0^2 + a_1^2 + \frac{1}{2}(a_2^2 + a_3^2)$$

and confirm that, when the expansion is truncated to stop after 2^n terms, the left-hand side of (17.166) can be replaced exactly by

$$\frac{1}{2^n}\sum_{r=0}^{2^n-1} f_r^2$$

where $f_r \equiv f(x = r/2^n)$.

17.10. (i) Use the wavelet convolution formula (17.172) to compute the response of a lightly damped oscillator to a periodic square wave. Let $x(r) = 1$ for $r = 1$ to 64 and 129 to 192 and $x(r) = -1$ for $r = 65$ to 128 and 193 to 256. Calculate the impulse response function $h(t)$ for the second-order system

$$\ddot{y}(t) + 0{\cdot}2\dot{y}(t) + y(t) = x(t).$$

Hence form a sequence $h(r) \equiv h(t = r\Delta)$ such that the detail in $h(t)$ is shown in a graph of $h(r)$ against r. Obtain $z(r) = x(R - r)$ from $x(r)$ according to (17.167) where $t_0 = R\Delta$ when Δ is the time step (a suitable value for Δ is $\frac{1}{4}$). Compute the wavelet transforms of z and h and then execute (17.172) to find a single-response point $y(t_0 = R\Delta)$. Calculate new sequences z by changing R, and hence compute and plot the response $y(r)$, $r = 1$ to 256.

17.11. Prove that, if the functions $w(x)$ and $v(x)$ have Fourier transforms $W(\omega)$ and $V(\omega)$, then (17.197) and (17.198) are true. *Hint*: Begin by using their inverse Fourier transforms to substitute for $w(x)$ and $v(x)$ and then use the identity for an infinite integral of a complex exponential given in Problem 15.4.

17.12. Show that the complex conjugate of one harmonic wavelet will be orthogonal to another harmonic wavelet of the same level j provided that their spacing is a multiple of $1/2^j$. *Hint*: Substitute from (17.185) in (17.198) and evaluate the integral over frequency.

17.13. Show that the sequence f defined by (17.219) and of length N is orthogonal to all discrete harmonic wavelets $e^{i2\pi kr/N}$, $r = 0$ to $N - 1$, $k = 0$ to $N/2 - 1$.

Answers to problems

1.1. $\frac{1}{2}$, $\frac{1}{3}$, 0·29 approx.

1.2. $1/2a$ for $|x| \leqslant a$; 0 for $|x| > a$
$\frac{1}{2}(x/a + 1)$ for $|x| \leqslant a$; 0 for $x < -a$; 1 for $x > a$.

1.3. 1, 3, $\sqrt{2}$

1.4. 2·66, 0·64 approx.

1.5. Case (i) $\frac{1}{4}z^2$; $\frac{1}{2}z$
Case (ii) $\frac{1}{2}z - \frac{1}{4}$; $\frac{1}{2}$
Case (iii) $-\frac{1}{4}z^2 + \frac{3}{2}z - \frac{5}{4}$; $-\frac{1}{2}z + \frac{3}{2}$

2.1. $1/\pi^2$

2.2. $\dfrac{1}{\sqrt{3\pi a^2}} e^{-(2/3a^2)(x^2 + y^2 - xy)}$

$\dfrac{1}{\sqrt{2\pi a}} e^{-x^2/2a^2}$; $\quad \dfrac{1}{\sqrt{2\pi a}} e^{-y^2/2a^2}$; $\quad \dfrac{\sqrt{2}}{\sqrt{3\pi a}} e^{-(2/3a^2)(x - y/2)^2}$

$y/2$

2.3. $c = 4$; $\frac{3}{4}, \frac{1}{2}, \frac{3}{4}$

2.4. $\frac{1}{6}X^2$; $\quad \frac{1}{6}X^2 \sin^2(\omega t - \phi_0)(2 + \cos\phi_0)$

Non-ergodic since the sample means and the ensemble mean are different; non-stationary since some ensemble averages are dependent on absolute time t.

3.2. $R_x(\tau) = \displaystyle\sum_{n=1,3,5,\cdots} \frac{8a^2}{\pi^2 n^2} \cos\frac{2\pi n\tau}{T}$

3.3. $\dfrac{a_1^2}{2} \cos\omega_1\tau + \dfrac{a_2^2}{2} \cos\omega_2\tau$; $\quad \dfrac{a_1 b_1}{2} \cos\omega_1(\tau + t_0)$; $\quad \dfrac{a_1 b_1}{2} \cos\omega_1(\tau - t_0)$

3.4. $\dfrac{abc}{T} \sin\Omega(\tau - t_0)\sin\dfrac{2\pi}{T}(\tau - t_0)$; $\quad \dfrac{abc}{T} \sin\Omega(\tau + t_0)\sin\dfrac{2\pi}{T}(\tau + t_0)$

The ensemble averages for $R_{xy}(\tau)$ and $R_{yx}(\tau)$ will not change although sample averages will show (small) random differences (unless averaged over an infinite time span).

4.1. (a) $\dfrac{aT}{\pi}\left(\dfrac{\sin \omega T}{\omega T}\right)$

(b) $e^{-i\omega T_0}\left(\dfrac{aT}{\pi}\right)\left(\dfrac{\sin \omega T}{\omega T}\right)$

(c) $\dfrac{aT}{2\pi}\left(\dfrac{\sin \omega T/2}{\omega T/2}\right)^2$

(d) $\dfrac{a}{2\sqrt{(\pi b)}}e^{-\omega^2/4b}$

(e) $\dfrac{aT}{2\pi}\left\{\dfrac{\sin(\omega_0 - \omega)T}{(\omega_0 - \omega)T} + \dfrac{\sin(\omega_0 + \omega)T}{(\omega_0 + \omega)T}\right\}$

(f) $-\dfrac{iaT}{2\pi}\left\{\dfrac{\sin(\omega_0 - \omega)T}{(\omega_0 - \omega)T} - \dfrac{\sin(\omega_0 + \omega)T}{(\omega_0 + \omega)T}\right\}$

4.2. $X(\omega) = \dfrac{4aT}{\pi}\dfrac{\sin^2(\omega T/2)\cos \omega T}{\omega T}(\sin 2\omega t + i\cos 2\omega t)$

$x(t) = \displaystyle\sum_{n=1,3,5,\cdots}\dfrac{4a}{n\pi}\sin\dfrac{n\pi t}{T}$

4.4. (ii) $c/2\pi$ (iii) $e^{-i\omega T_0}(c/2\pi)$ (iv) $(c/\pi)\cos \omega T_0$ and $-(ic/\pi)\sin \omega T_0$

5.1. $2\ \text{cm}^2$; $2\left(\dfrac{\sin 400\pi\tau}{400\pi\tau}\right)\text{cm}^2$

5.2. $S(\omega) = \dfrac{\sigma_x^2}{4\sqrt{\pi}\omega_1}(e^{-\{(\omega - \omega_2)/2\omega_1\}^2} + e^{-\{(\omega + \omega_2)/2\omega_1\}^2})$

5.3. (i) $\dfrac{2S_0 c}{c^2 + \tau^2}$

(ii) $\dfrac{4S_0 c(c^2 - 3\tau^2)}{(c^2 + \tau^2)^3}$

(iv) $\dfrac{c^2}{4\sqrt{2\pi S_0}}e^{-(z^2 c/4S_0 + \dot{z}^2 c^3/8S_0)}$

5.4. $\dfrac{a^2}{2}\left(\dfrac{\sin \omega_0 \tau}{\omega_0 \tau}\right)$; $\dfrac{a^2}{4\omega_0}$ for $-\omega_0 \leqslant \omega \leqslant \omega_0$

5.5. $R(\tau) = \dfrac{8a^2}{\pi^2}\left\{\displaystyle\sum_{k=1,3,5,\cdots}\dfrac{1}{k^2}\cos\dfrac{2\pi k\tau}{T}\right\}$; $x(t) = \dfrac{4a}{\pi}\left\{\displaystyle\sum_{k=1,3,5,\cdots}\dfrac{1}{k}\sin\dfrac{2\pi k t}{T}\right\}$

5.6. $S_{xy}(\omega) = S_{yx}^*(\omega) = \dfrac{abc\,e^{-i\omega t_0}}{4T}\left\{\delta\left(\omega - \Omega + \dfrac{2\pi}{T}\right) + \delta\left(\omega + \Omega - \dfrac{2\pi}{T}\right) - \right.$

$\left. - \delta\left(\omega - \Omega - \dfrac{2\pi}{T}\right) - \delta\left(\omega + \Omega + \dfrac{2\pi}{T}\right)\right\}$

where $\delta(\omega - \Omega + 2\pi/T)$ denotes an impulse (delta function) of unit magnitude at $\omega = \Omega - 2\pi/T$, etc.

6.1. (a) $\dfrac{k}{k + ic\omega}$; $\quad \dfrac{k}{c}e^{-(k/c)t}$ for $t \geqslant 0$

(b) $\dfrac{\frac{1}{2}(3L/l - 1)}{3EI/l^3 + ic\omega}$; $\quad \dfrac{(3L/l - 1)}{2c}e^{-(3EI/cl^3)t}$ for $t \geqslant 0$

(c) $\dfrac{c}{(J\omega^2 - 3c)(J\omega^2 - c)}$; $\quad \dfrac{1}{2\sqrt{(cJ)}}\left\{ \sin\sqrt{\left(\dfrac{c}{J}\right)}\, t - \dfrac{1}{\sqrt{3}}.\sin\sqrt{\left(\dfrac{3c}{J}\right)}\, t \right\}$

$$\text{for}\quad t \geqslant 0$$

(d) $\dfrac{ic\omega + k}{-m\omega^2 + ic\omega + k}$; $\quad e^{-(c/2m)t}\left\{ \dfrac{c}{m}\cos\sqrt{\left(\dfrac{k}{m} - \dfrac{c^2}{4m^2}\right)}\, t + \right.$

$$\left. + \dfrac{k/m - c^2/2m^2}{\sqrt{(k/m - c^2/4m^2)}}\sin\sqrt{\left(\dfrac{k}{m} - \dfrac{c^2}{4m^2}\right)}\, t \right\}$$

$$\text{for}\quad t \geqslant 0$$

6.2. $\dfrac{-m}{-m\omega^2 + ic\omega + k}$; $\quad \dfrac{-e^{-(c/2m)t}\sin\sqrt{(k/m - c^2/4m^2)}\, t}{\sqrt{(k/m - c^2/4m^2)}}$

6.3. $y(t) = 0$ for $t \leqslant 0$

$\quad y(t) = a(1 - e^{-(k/c)t})$ for $0 \leqslant t \leqslant T$

$\quad y(t) = a(e^{-(k/c)(t - T)} - e^{-(k/c)t})$ for $t \geqslant T$

7.1. $\dfrac{\pi S_0 m^2}{kc}$; $\quad m\sqrt{\left(\dfrac{\pi S_0 k}{c} + a^2\right)}$

7.2. (i) 0, $\dfrac{\pi S_0}{kc}$

Note: $E[x]$ must be zero; otherwise $S_x(\omega)$ would have to include a delta function at $\omega = 0$ to account for the finite contribution to $E[x^2]$ at $\omega = 0$.

(ii) 0, $\dfrac{2S_0}{kc}\tan^{-1}\dfrac{\Omega c}{k}$

$\quad R_x(\tau) = 2\pi S_0\delta(\tau)$; $\quad R_y(\tau) = \dfrac{\pi S_0}{kc}e^{-(k/c)|\tau|}$; $\quad S_y(\omega) = \dfrac{S_0}{k^2 + c^2\omega^2}$

7.3. $\pi S_0\left[\dfrac{ck_2(k_1 + k_2)/\omega_0 k_1 + c^2}{\{c(k_1 + k_2)/\omega_0\}\{c(k_1 + k_2) + k_1 k_2/\omega_0\}}\right]$

7.4. $H(\omega) = \dfrac{\gamma}{\{m_1 m_2\omega^4 - im_2 c\omega^3 - \{(k_1 + \gamma)m_2 + (k_2 + \gamma)m_1\}\omega^2 + i(k_2 + \gamma)c\omega + k_1 k_2 + \gamma(k_1 + k_2)\}}$

$\quad \dfrac{\pi S_0}{2c}$

7.6. $x(t) = \dfrac{ab}{T}\left\{1 + 2\cos\dfrac{2\pi t}{T} + 2\cos\dfrac{4\pi t}{T} + \cdots\right\}$

$S_x(\omega) =$ an infinite "comb" of delta functions of magnitude $a^2 b^2/T^2$ at spacing $2\pi/T$ along the frequency axis $(-\infty < \omega < \infty)$.

8.1. 32 per cent, 61 per cent, $1.25\sigma_x$, $3.03\sigma_x$

8.2. (i) $v_a^+ = \dfrac{1}{2\pi}\sqrt{\left(\dfrac{k}{m + c/\omega_0}\right)}\,e^{-a^2/2\sigma^2}$

(ii) $p_p(a) = \dfrac{a}{\sigma^2}e^{-a^2/2\sigma^2}$

where $\sigma^2 = \dfrac{\pi S_0/kc}{\left(1 + \dfrac{k/m}{\omega_0^2 + c\omega_0/m}\right)}$

(iii) $\mu = \dfrac{1}{2\pi}\sqrt{\left(\dfrac{k}{m} + \dfrac{c\omega_0}{m}\right)}$

8.3. $v_a^+ = \dfrac{1}{8}\dfrac{\dot{y}_0}{y_0} = \dfrac{1}{8}\dfrac{\sigma_{\dot{y}}}{\sigma_y}$ for $0 \leqslant a \leqslant y_0$; $\quad \delta(a - y_0)$

Sample functions could have a saw-tooth waveform, each with the same amplitude y_0, but with different periods and phasing (uniformly distributed).

8.4. $v_0^+ = \dfrac{1}{2\pi}\sqrt{\left\{\dfrac{k_1 k_2 + \gamma(k_1 + k_2)}{m_2(k_1 + \gamma)}\right\}}$; $\quad \mu = \dfrac{1}{2\pi}\sqrt{\left(\dfrac{k_2 + \gamma}{m_2}\right)}$

v_0^+ and μ are only the same for $\gamma \to 0$. Otherwise the response of $x_2(t)$ is not a smooth narrow band process. When γ is large, the higher frequency mode contributes falsely to v_0^+ (as individual "cycles" are superimposed on the lower frequency mode and do not all cross the zero level). Therefore v_0^+ does not approach the value $(1/2\pi)\sqrt{\{(k_1 + k_2)/(m_1 + m_2)\}}$ for a single degree-of-freedom system when $\gamma \to \infty$ unless the spectral density of the input process is limited at high frequencies. If $S_F(\omega) = S_0/(1 + \omega^2/\omega_0^2)$ (say) then v_0^+ does converge to its correct value when $\gamma \to \infty$.

9.1. 38 per cent, 12 per cent, 3.8 per cent; 4 h, 24 min, 2.4 min.

9.2. $\dfrac{\pi}{4\Omega_0}$; $\left(\dfrac{4}{\pi}\right)^2 \Omega_0$;

$\tilde{S}(\omega) = \begin{cases} \dfrac{\pi a^2}{16\Omega_0}\cos\dfrac{\pi(\omega - \Omega_1)}{2\Omega_0} & \text{for } \Omega_1 - \Omega_0 \leqslant \omega \leqslant \Omega_1 + \Omega_0 \\[3mm] \dfrac{\pi a^2}{16\Omega_0}\cos\dfrac{\pi(\omega + \Omega_1)}{2\Omega_0} & \text{for } -\Omega_1 - \Omega_0 \leqslant \omega \leqslant -\Omega_1 + \Omega_0 \\[3mm] 0 \text{ elsewhere} \end{cases}$

9.3. (iv) $0 \cdot 21 < \chi_2^2 < 4 \cdot 6$

(v) $\text{Prob}(\chi_{10}^2 > y) = e^{-y/2}\{1 + \frac{1}{2}y + \frac{1}{8}y^2 + \frac{1}{48}y^3 + \frac{1}{384}y^4\}$

$0 \cdot 15$ for chi-square cf. $0 \cdot 16$ for Gaussian.

9.4. $0 \cdot 987 S_0 < m < 1 \cdot 013 S_0$; $0 \cdot 974 S_0 < m < 1 \cdot 026 S_0$

10.1. (i) $(0, -\frac{1}{2}i, 0, 0, 0, 0, 0, \frac{1}{2}i)$

(ii) $(0, \frac{1}{2}, 0, 0, 0, 0, 0, \frac{1}{2})$

(iii) $(0, -\frac{1}{2}i, 0, 0, 0, ..., 0, 0, 0, \frac{1}{2}i)$

10.2. $(0, \frac{1}{4} - i\frac{1}{4}(\sqrt{2} + 1), 0, \frac{1}{4} - i\frac{1}{4}(\sqrt{2} - 1), 0, \frac{1}{4} + i\frac{1}{4}(\sqrt{2} - 1), 0, \frac{1}{4} + i\frac{1}{4}(\sqrt{2} + 1))$

If only the first two harmonics of the square wave were present, the DFT would be

$$\left(0, -\frac{2}{\pi}i, 0, -\frac{2}{3\pi}i, 0, \frac{2}{3\pi}i, 0, \frac{2}{\pi}i\right)$$

(this result is obtained by extending the result of problem 10.1 (iii) – see also problem 10.5) but the higher harmonics distort the calculated spectrum at lower frequencies due to aliasing.

10.3. $X_k = \begin{cases} 0 \text{ for } k \text{ even} \\ \dfrac{1}{32}\left(1 - i\cot\dfrac{\pi k}{64}\right) \text{ for } k \text{ odd} \end{cases}$

10.4. (i) $x_r = 1$ for all r

(ii) $x_r = \begin{cases} 1 \text{ for } r \text{ even} \\ -1 \text{ for } r \text{ odd} \end{cases}$

(iii) $\{x_r\} = (1, 1, -1, -1, 1, 1, -1, -1, ..., -1, -1)$

(iv) $x_r = \sin 2\pi r/N$ for all r

where $0 \leqslant r \leqslant (N - 1)$ in each case.

10.5. $2a_m \cos\dfrac{2\pi m}{N}r + 2b_m \sin\dfrac{2\pi m}{N}r$

10.7. $R_r = 2(a_m^2 + b_m^2)\cos\dfrac{2\pi m}{N}r$ for $r = 0, 1, 2, ..., (N - 1)$

$S_k = 0$ for all k except $k = m$ and $k = N - m$, for which $S_m = S_{N-m} = (a_m^2 + b_m^2)$.

11.1. $\{R_r\} = (\frac{1}{10}(x_0 y_0 + x_1 y_1 + x_2 y_2 + x_3 y_3), \frac{1}{10}(x_0 y_1 + x_1 y_2 + x_2 y_3),$
$\frac{1}{10}(x_0 y_2 + x_1 y_3), \frac{1}{10}(x_0 y_3), 0, 0, 0, \frac{1}{10}(x_3 y_0), \frac{1}{10}(x_2 y_0 + x_3 y_1),$
$\frac{1}{10}(x_1 y_0 + x_2 y_1 + x_3 y_2))$

$\left. \begin{array}{l} \{\hat{R}_r\} = (R_0, \frac{10}{9}R_1, \frac{10}{8}R_2, \frac{10}{7}R_3, 0, 0, 0, 0, 0, 0) \\ \{\hat{R}_{-r}\} = (R_0, \frac{10}{9}R_9, \frac{10}{8}R_8, \frac{10}{7}R_7, 0, 0, 0, 0, 0, 0) \end{array} \right\}$ for part (ii)

$$\{\hat{R}_r\} = (\tfrac{10}{4}R_0, \tfrac{10}{3}R_1, \tfrac{10}{2}R_2, \tfrac{10}{1}R_3, 0, 0, 0, 0, 0, 0) \biggr\}$$
$$\{\hat{R}_{-r}\} = (\tfrac{10}{4}R_0, \tfrac{10}{3}R_9, \tfrac{10}{2}R_8, \tfrac{10}{1}R_7, 0, 0, 0, 0, 0, 0) \biggr\} \quad \text{for part (iv)}$$

11.2. $\{V_k\} = \left(\dfrac{a}{2}, \dfrac{2a}{\pi^2}, 0, \dfrac{2a}{9\pi^2}, 0, \dfrac{2a}{25\pi^2}, \, \cdots\right)$

$\{V_k'\} = (a, 0, 0, 0, 0, 0, \ldots)$

11.3. (i) $S_0 = \dfrac{1}{400\pi}$ cm^2 s

(ii) $S_k = \begin{cases} \dfrac{1}{2048} \text{ cm}^2 & \text{for } 0 \leqslant k \leqslant 1023 \quad \text{and} \quad 3073 \leqslant k \leqslant 4095 \\[2mm] \dfrac{1}{4096} \text{ cm}^2 & \text{for } k = 1024 \quad \text{and} \quad k = 3072 \\[2mm] 0 \text{ for } 1025 \leqslant k \leqslant 3071 \end{cases}$

(iii) S_k = as above except

$$S_{512} = S_{3584} = \left(\dfrac{1}{2048} + 0{\cdot}0025\right) \text{cm}^2$$

due to aliasing.

Values of S_k for $k > 2048$ (i.e. for frequencies above the Nyquist frequency) have no practical value.

11.4. $X(\omega) = \displaystyle\sum_{k=-\infty}^{\infty} \dfrac{a}{2}\left[\dfrac{\sin(\omega_0 - \omega)T}{(\omega_0 - \omega)T} + \dfrac{\sin(\omega_0 + \omega)T}{(\omega_0 + \omega)T}\right]\delta\!\left(\omega - \dfrac{\pi k}{T}\right)$

Since there is not a complete number of cycles of the cosine function on the record, $E[x^2] \neq \tfrac{1}{2}a^2$.

$|X_k|$ would be the same, although alternate terms of $\{X_k\}$ have opposite sign.

11.8. $\left(\dfrac{a^2}{2}\cos^2\omega_0\Delta\cos^2\dfrac{\omega_0\Delta}{2}, \ \dfrac{a^2}{4}\sin^2\omega_0\Delta, \ \dfrac{a^2}{2}\cos^2\omega_0\Delta\sin^2\dfrac{\omega_0\Delta}{2}, \ \dfrac{a^2}{4}\sin^2\omega_0\Delta\right)$

$S_k = \dfrac{a^2}{64}\left[\left\{\dfrac{\sin(2\omega_0\Delta - \pi k)}{\sin\left(\dfrac{2\omega_0\Delta - \pi k}{4}\right)}\right\}^2 + \left\{\dfrac{\sin(2\omega_0\Delta + \pi k)}{\sin\left(\dfrac{2\omega_0\Delta + \pi k}{4}\right)}\right\}^2\right]$

12.1. $x_r = \begin{cases} 1 \text{ for } r \text{ even} \\ -1 \text{ for } r \text{ odd} \end{cases}$

12.3. $\{Y_l\}$ where $Y_l = 0$ for all l except

$$Y_8 = \dfrac{1}{4} - i\dfrac{(\sqrt{2} + 1)}{4} = Y_{56}^*$$

$$Y_{24} = \dfrac{1}{4} - i\dfrac{(\sqrt{2} - 1)}{4} = Y_{40}^*$$

13.1. $R_x(\tau) = \begin{cases} a^2\left(1 - \dfrac{1}{N^2}\right)\left(1 - \dfrac{|\tau|}{\Delta t}\right) + \dfrac{a^2}{N^2} & \text{for } 0 \le |\tau| \le \Delta t \\ a^2/N^2 & \text{elsewhere} \end{cases}$

$\left.\begin{array}{l} \\ \\ \\ \\ \end{array}\right\}$ where $N = (2n - 1)$

$S_x(\omega) = \dfrac{a^2}{N^2}\delta(\omega) + a^2\left(1 - \dfrac{1}{N^2}\right)\dfrac{\Delta t}{2\pi}\left(\dfrac{\sin \omega\Delta t/2}{\omega\Delta t/2}\right)^2$

Reversing the bias would not alter these results.

13.3. $S_x(\omega) = \dfrac{a^2}{N^2}\delta(\omega) + a^2\dfrac{(N+1)}{N^2}\left(\dfrac{\sin \omega\Delta t/2}{\omega\Delta t/2}\right)^2 \displaystyle\sum_{\substack{k=-\infty \\ k \ne 0}}^{\infty} \delta\left(\omega - \dfrac{2\pi k}{N\Delta t}\right)$

The difference between this answer and that to problem 13.1 is that the latter is for genuinely random noise, whereas the peculiar shape of $R_x(\tau)$ (and therefore of $S_x(\omega)$) in 13.3 arises from the repetitive properties of pseudo random sequences.

14.1. (i) $S_1(\omega) = S_2(\omega) = S_3(\omega) = S_4(\omega) = \dfrac{1}{2\pi V}\displaystyle\int_{-\infty}^{\infty} R(\rho)e^{-i(\omega\rho/V)}\,d\rho$

$S_{13}(\omega) = S_{31}^*(\omega) = S_{24}(\omega) = S_{42}^*(\omega) = \dfrac{1}{2\pi V}\displaystyle\int_{-\infty}^{\infty} R(l - \rho)e^{-i(\omega\rho/V)}\,d\rho$

$S_{12}(\omega) = S_{21}(\omega) = S_{34}(\omega) = S_{43}(\omega)$

$= \dfrac{1}{2\pi V}\displaystyle\int_{-\infty}^{\infty} R(\sqrt{(b^2 + \rho^2)})e^{-i(\omega\rho/V)}\,d\rho$

$S_{14}(\omega) = S_{41}^*(\omega) = S_{23}(\omega) = S_{32}^*(\omega)$

$= \dfrac{1}{2\pi V}\displaystyle\int_{-\infty}^{\infty} R(\sqrt{\{b^2 + (l - \rho)^2\}})e^{-i(\omega\rho/V)}\,d\rho$

(ii) $M\ddot{v} + 2c\dot{v} + 2kv = 2c\dot{u} + 2ku$

$I\ddot{\theta} + \tfrac{1}{2}cb^2\dot{\theta} + \tfrac{1}{2}kb^2\theta = \tfrac{1}{2}cb^2\dot{\phi} + \tfrac{1}{2}kb^2\phi$

(iii) $S_u(\omega) = \dfrac{S_0}{2V}, \quad S_\phi(\omega) = \dfrac{2S_0}{b^2 V}$

$E[v^2] = \dfrac{\pi S_0\omega_1(1 + 4\zeta_1^2)}{4\zeta_1 V}, \quad E[\theta^2] = \dfrac{\pi S_0\omega_2(1 + 4\zeta_2^2)}{b^2\zeta_2 V}$

$\dfrac{\pi S_0}{4V}\left\{\dfrac{\omega_1}{\zeta_1}(1 + 4\zeta_1^2) + \dfrac{\omega_2}{\zeta_2}(1 + 4\zeta_2^2)\right\}$

where $\omega_1^2 = \dfrac{2k}{M}, \quad 2\zeta_1\omega_1 = \dfrac{2c}{M}, \quad \omega_2^2 = \dfrac{kb^2}{2I}, \quad 2\zeta_2\omega_2 = \dfrac{cb^2}{2I}$

14.4. (iv) $\dfrac{1}{4!}, \dfrac{3}{4!}, \dfrac{5}{4!}, \dfrac{6}{4!}, \dfrac{5}{4!}, \dfrac{3}{4!}$ and $\dfrac{1}{4!}$ respectively

(v) 3·6 per cent approx.

(vi) There is no change

Note. There are no separate answers for problems relating to Chapters 15 and 16.

17.1. (i) $\phi(x) = 1, 0 \leqslant x < 1$
(ii) $\phi(x) = x, 0 \leqslant x < 1$
   $\phi(x) = (2 - x), 1 \leqslant x < 2$
(iii) $\phi(x) = \delta(x)$
(iv) $\phi(x)$ is a comb of closely packed infinitesimal pencil functions covering the range $0 \leqslant x < 2$ with the property that $\int_0^2 \phi(x)\,dx = 1$. The corresponding wavelets are as follows:

(i) Haar: $W(x) = -1, 0 \leqslant x < \frac{1}{2}, W(x) = +1, \frac{1}{2} \leqslant x < 1$.
(ii) Inverted Mexican hat: $W(x) = x, 0 \leqslant x < \frac{1}{2}, W(x) = 2 - 3x, \frac{1}{2} \leqslant x < 1$ and its mirror image about $x = 1$ for $1 \leqslant x < 2$.
(iii) $W(x) = \delta(x)$
(iv) $W(x) = \phi(x)$.

17.2. $\phi(1/2) = (2 + \sqrt{3})/4$
$\phi(3/2) = 0$
$\phi(5/2) = (2 - \sqrt{3})/4$
$\phi(1/4) = (5 + 3\sqrt{3})/16, \phi(7/4) = (1 - \sqrt{3})/8$
$\phi(3/4) = (9 + 5\sqrt{3})/16, \phi(9/4) = (9 - 5\sqrt{3})/16$
$\phi(5/4) = (1 + \sqrt{3})/8, \phi(11/4) = (5 - 3\sqrt{3})/16$

$W(1) = (-1 + \sqrt{3})/2, W(3/2) = -\sqrt{3}, W(2) = (1 + \sqrt{3})/2$

17.4. $\phi(x) + \phi(x - 1) + \phi(x - 2) = 1$ for $2 \leqslant x < 3$

$\dfrac{d}{dx}\{\phi(x) + 2\phi(x - 1)\} = 1$ for $1 \leqslant x < 2$.

17.5. Consider $f(j)$ to be the result of sampling $f(x) = x^2 - 2x^3 + x^4$. At $x = 0$ and $x = 1$, $f(x)$, $f'(x)$ and $f''(x)$ are the same but $f'''(0) = -12$, $f'''(1) = +12$. Therefore there is a discontinuity when the record is wrapped which gives rise to the local peaks in a graph of $\log_{10}|a(r)|$ versus r.

17.6. (vii) A set of 16 D20 wavelets at level 4 is almost indistinguishable from a sine wave. The DFT has small odd harmonics which can be identified on a graph of $\log|F(j)|$ against j.

References

1. BEAUCHAMP, K. G.
   Signal Processing Using Analog and Digital Techniques, George Allen & Unwin, London, 1973.
2. BENDAT, J. S. and PIERSOL, A. G.
   Measurement and Analysis of Random Data, John Wiley, New York, 1966.
3. BENDAT, J. S. and PIERSOL, A. G.
   Random Data: Analysis and Measurement Procedures, John Wiley, New York, 1971.
4. BINGHAM, C., GODFREY, M. D. and TUKEY, J. W.
   'Modern Techniques of Power Spectrum Estimation', *IEEE Trans. Audio and Electroacoustics*, Vol. AU-15, 1967, No. 2, 56–66.
5. BLACKMAN, R. B. and TUKEY, J. W.
   The Measurement of Power Spectra, Dover, New York, 1959.
6. BRIGGS, P. A. N., HAMMOND, P. H., HUGHES, M. T. G. and PLUMB, G. O.
   'Correlation Analysis of Process Dynamics using Pseudo-Random Binary Test Perturbations', *Proc. I. Mech. E.*, Vol. 179, 1964–65, Part 3H, 53–67.
7. BRIGHAM, E. O.
   The Fast Fourier Transform, Prentice-Hall, Englewood Cliffs, New Jersey, 1974.
8. CHURCHILL, R. V.
   Fourier Series and Boundary Value Problems, McGraw-Hill, New York, 1941.
9. COOLEY, J. W. and TUKEY, J. W.
   'An Algorithm for the Machine Calculation of Complex Fourier Series', *Mathematics of Computation*, Vol. 19, 1965, 297–301. (Reprinted in A. V. Oppenheim (ed.) *Papers on Digital Signal Processing*, MIT Press, Cambridge, Mass., 1969.)
10. COOLEY, J. W., LEWIS, P. A. W. and WELCH, P. D.
    The Application of the Fast Fourier Transform Algorithm to the Estimation of Spectra and Cross Spectra. IBM Research Paper, IBM Watson Research Center, Yorktown Heights, N.Y., 1967. This work is included in a chapter entitled 'The Fast Fourier Transform and its Applications to Time Series Analysis' in Enslein et al. (26).
11. COOLEY, J. W., LEWIS, P. A. W. and WELCH, P. D.
    'Application of the Fast Fourier Transform to Computation of Fourier Integrals, Fourier Series and Convolution Integrals', *IEEE Trans. Audio and Electroacoustics*, Vol. AU-15, 1967, No. 2, 79–84.
12. COOLEY, J. W., LEWIS, P. A. W. and WELCH, P. D.
    'Historical Notes on the Fast Fourier Transform', *IEEE Trans. Audio and Electroacoustics*, Vol. AU-15, 1967, No. 2, 76–9.
13. COOLEY, J. W., LEWIS, P. A. W. and WELCH, P. D.
    'The Fast Fourier Transform and its Applications', *IEEE Transactions on Education*, Vol. 12, 1969, No. 1, 27–34.

14. COOLEY, J. W., LEWIS, P. A. W. and WELCH, P. D.
    'The Finite Fourier Transform', *IEEE Trans. Audio and Electroacoustics*, Vol. AU-17, 1969, No. 2, 77–85.
15. COOLEY, J. W., LEWIS, P. A. W. and WELCH, P. D.
    'The Fast Fourier Transform Algorithm: Programming Considerations in the Calculation of Sine, Cosine and Laplace Transforms', *J. Sound Vib.*, Vol. 12, 1970, 315–37.
16. CRANDALL, S. H. (ed.)
    Random Vibration, M.I.T. Press, Cambridge, Mass. and John Wiley, New York, 1958.
17. CRANDALL, S. H. (ed.)
    Random Vibration – Vol. 2, M.I.T. Press, Cambridge, Mass., 1963.
18. CRANDALL, S. H. and MARK, W. D.
    Random Vibration in Mechanical Systems, Academic Press, New York, 1963.
19. CRANDALL, S. H., CHANDIRAMANI, K. L. and COOK, R. G.
    'Some First-Passage Problems in Random Vibration', *J. Appl. Mech., Trans. ASME.*, Vol. 33, 1966, 532–8.
20. CRANDALL, S. H.
    'First Crossing Probabilities of the Linear Oscillator', *J. Sound Vib.*, Vol. 12, 1970, No. 3, 285–99.
21. CRANDALL, S. H.
    'Distribution of Maxima in the Response of an Oscillator to Random Excitation', *J. Acoust. Soc. Am.*, Vol. 47, 1970, No. 3 (Part 2), 838–45.
22. DAVENPORT, W. B. Jr. and ROOT, W. L.
    An Introduction to the Theory of Random Signals and Noise, McGraw-Hill, New York, 1958.
23. DODDS, C. J.
    'The Laboratory Simulation of Vehicle Service Stress', *J. Engng. Ind., Trans. ASME*, Vol. 96, 1974, No. 3, 391–8.
24. DODDS, C. J. and ROBSON, J. D.
    'The Description of Road Surface Roughness', *J. Sound Vib.*, Vol. 31, 1973, No. 22, 175–83.
25. DURRANI, T. S. and NIGHTINGALE, J. M.
    'Data Windows for Digital Spectral Analysis', *Proc. IEE*, Vol. 119, 1972, No. 3, 343–52.
26. ENSLEIN, K., RALSTON, A. and WILF, H. S. (eds)
    Statistical Methods for Digital Computers (Vol. 3 of *Mathematical Methods for Digital Computers*), John Wiley, New York, 1975.
27. FELLER, W.
    Probability Theory and its Applications, John Wiley, New York, 1950.
28. FISHER, R. A. and YATES, F.
    Statistical Tables for Biological, Agricultural and Medical Research, Sixth edition, Oliver and Boyd, Edinburgh, 1963.
29. GOLD, B. and RADER, C. M.
    Digital Processing of Signals, McGraw-Hill, New York, 1969.
30. GOLOMB, S. W.
    Shift Register Sequences, Holden-Day, San Francisco, 1967.
31. GREEN, B. F., SMITH, J. E. K. and KLEM, L.
    'Empirical Tests of an Additive Random Number Generator', *J. Assoc. Computer Machinery*, Vol. 6, 1959, No. 4, 527–37.
32. HALFMAN, R. L.
    Dynamics (2 Vols.), Addison-Wesley, Reading, Mass., 1962.
33. HARTLEY, M. G.
    'Development, Design and Test Procedures for Random Generators using Chaincodes', *Proc. IEE.*, Vol. 116, 1969, 22–6.

34. HARTLEY, M. G.
    'Evaluation of Performance of Random Generators Employing Chaincodes',
    Proc. IEE., Vol. 116, 1969, 27–34.
35. HEATH, F. G. and GRIBBLE, M. W.
    'Chaincodes and their Electronic Applications', *Proc. IEE.*, Vol. 108C, 1961,
    50–7.
36. HOESCHELE, D. F. Jr.
    Analog-to-Digital/Digital-to-Analog Conversion Techniques, John Wiley, New
    York, 1968.
37. HUFFMAN, P. A.
    'The Synthesis of Linear Sequential Coding Networks', in C. Cherry (ed.),
    Information Theory, Butterworth, 1956.
38. JAMES, H. M., NICHOLS, N. B. and PHILLIPS, R. S.
    Theory of Servomechanisms, MIT Radiation Laboratory Series, Vol. 25, McGraw-
    Hill, New York, 1947.
39. JENKINS, G. M. and WATTS, D. G.
    Spectral Analysis and its Applications, Holden-Day, San Francisco, 1968.
40. JOLLEY,.L. B. W.
    Summation of Series, Second edition, Dover, New York, 1961.
41. KENDALL, M. G. and STUART, A.
    The Advanced Theory of Statistics, Charles Griffin, London, (a) Vol. 1 (2nd
    edition) 1963, (b) Vol. 2 1961, (c) Vol. 3 1966.
42. KENDALL, M. G.
    Time-Series, Charles Griffin, London, 1973.
43. KLINE, MORRIS
    Mathematical Thought from Ancient to Modern Times, Oxford University Press,
    New York, 1972.
44. KOOPMANS, L. H.
    The Spectral Analysis of Time Series, Academic Press, New York, 1974.
45. KREYSZIG, E.
    Advanced Engineering Mathematics, Second edition, John Wiley, New York,
    1967.
46. LIGHTHILL, M. J.
    Introduction to Fourier Analysis and Generalised Functions, Cambridge University
    Press, 1962.
47. MACLAREN, M. D. and MARSAGLIA, G.
    'Uniform Random Number Generators', *J. Assoc. Computer Machinery*, Vol. 12,
    1965, 83–9.
48. MEYER, P. L.
    Introductory Probability and Statistical Applications, Addison-Wesley, Reading,
    Mass., 1965.
49. MIDDLETON, D.
    An Introduction to Statistical Communication Theory, McGraw-Hill, New York,
    1960.
50. MINER, M. A.
    'Cumulative Damage in Fatigue', *J. Appl. Mech., Trans. ASME*, Vol. 12, 1945,
    A159–64.
51. MUSTER, D. and CRENWELGE, O. E. Jr.
    'Simulation of Complex Excitation of Structures in Random Vibration by One-
    Point Excitation', Proc. Soc. of Environmental Engineers Symposium – *Vibra-
    tions in Environmental Engineering*, London, 1973, M1–M13.
52. OTNES, R. K. and ENOCHSON, L.
    Digital Time Series Analysis, John Wiley, New York, 1972.
53. PALMGREN, A.
    'Die Lebensdauer von Kugellagem', *Ver. Deut. Ingr.*, Vol. 68, 1924, 339–41.

54. PARZAN, E.
   Modern Probability Theory and its Applications, John Wiley, New York, 1960.
55. PIERSOL, A. G.
   'Power Spectra Measurements for Spacecraft Vibration Data', *J. Spacecraft and Rockets*, Vol. 4, 1967, 1613.
56. POWELL, ALAN
   'On the Fatigue Failure of Structures due to Vibrations Excited by Random Pressure Fields', *J. Acoust. Soc. Am.*, Vol. 30, 1958, No. 12, 1130–5.
57. PRICE, W. G. and BISHOP, R. E. D.
   Probabilistic Theory of Ship Dynamics, Chapman & Hall, London, 1974.
58. RICE, S. O.
   'Mathematical Analysis of Random Noise', *Bell System Tech. J.*, Vol. 23, 1944, 282–332, and Vol. 24, 1945, 46–156, both of which are reprinted in N. Wax (ed.), *Selected Papers on Noise and Stochastic Processes*, Dover, New York, 1954.
59. ROBSON, J. D.
   An Introduction to Random Vibration, Edinburgh University Press, 1963.
60. ROBSON, J. D. and ROBERTS, J. W.
   'A Theoretical Basis for the Practical Simulation of Random Motions', *J. Mech. Engng. Sci.*, Vol. 7, 1965, No. 3, 246–51.
61. SHANNON, C. E.
   'Communication in the Presence of Noise', *Proc. IRE.*, Vol. 37, 1949, 10–21.
62. SHINOZUKA, M. and JAN, C. M.
   'Digital Simulation of Random Processes and its Applications', *J. Sound. Vib.*, Vol. 25, 1972, 111–28.
63. SHINOZUKA, M.
   'Simulation of Multivariate and Multidimensional Random Processes', *J. Acoust. Soc. Am.*, Vol. 49, 1971, 357–67.
64. SINGLETON, R. C.
   'A Method for Computing the Fast Fourier Transform with Auxiliary Memory and Limited High-Speed Storage', *IEEE Trans. Audio and Electroacoustics*, Vol. AU-15, June, 1967, No. 2, 91–8.
65. SLOANE, E. A.
   'Comparison of Linearly and Quadratically Modified Spectral Estimates of Gaussian Signals', *IEEE Trans. Audio and Electroacoustics*, Vol. AU-17, 1969, No. 2, 133–7.
66. VIRCHIS, V. J. and ROBSON, J. D.
   'The Response of an Accelerating Vehicle to Random Road Undulation', *J. Sound Vib.*, Vol. 18, 1971, 423–7.
67. WATTS, D. G.
   'A General Theory of Amplitude Quantization with Application to Correlation Determination', *Proc. IEE.*, Vol. 109C, 1962, 209–18.
68. WELCH, P. D.
   'The Use of Fast Fourier Transform for the Estimation of Power Spectra: A Method Based on Time Averaging Over Short, Modified Periodograms', *IEEE Trans. Audio and Electroacoustics*, Vol. AU-15, June 1967, No. 2, 70–3.
69. WIENER, N.
   The Fourier Integral and Certain of Its Applications, Cambridge University Press, 1933.
70. WIENER, N.
   Extrapolation, Interpolation and Smoothing of Stationary Time Series, John Wiley, New York, 1949.

References for the second edition

71. BATCHELOR, G. K.
    The Theory of Homogeneous Turbulence, Cambridge University Press, 1960.
72. BEAUCHAMP, K. G. and YUEN, C. K.
    Digital Methods for Signal Analysis, George Allen & Unwin, London, 1979.
73. BRACEWELL, RONALD N.
    The Fourier Transform and its Applications (2nd edn), McGraw-Hill, New York, 1978.
74. COURANT, R. and HILBERT, D.
    Methods of Mathematical Physics, Vol. 1, Interscience, New York, 1953.
75. CRANDALL, S. H. and WITTIG, L. E.
    'Chladni's Patterns for Random Vibration of a Plate', *Dynamic Response of Structures*, G. Herrmann and N. Perrone (eds.), Pergamon Press, New York, 1972, 55–71.
76. CRANDALL, S. H.
    'Wide-Band Random Vibration of Structures', *Proc. 7th US National Congress of Applied Mechanics*, ASME, New York, 1974, 131–8.
77. CRANDALL, S. H.
    'Structured Response Patterns due to Wide-band Random Excitation', *Stochastic Problems in Dynamics*, B. L. Clarkson (ed.), Pitman, London, 1977, 366–87.
78. CRANDALL, S. H.
    'Random Vibration of One- and Two-Dimensional Structures', *Developments in Statistics*, P. R. Krishnaiah (ed.), Vol. 2, Academic Press, New York, chapter 1, 1979, 1–82.
79. DAVENPORT, W. B.
    Probability and Random Processes, McGraw-Hill, New York, 1970.
80. DOWLING, A. P. and FFOWCS WILLIAMS, J. E.
    Sound and Sources of Sound, Ellis Horwood, Chichester, 1983.
81. ELISHAKOFF, ISAAC
    Probabilistic Methods in the Theory of Structures, John Wiley, New York, 1983.
82. ERINGEN, A. C.
    'Response of Beams and Plates to Random Loads', *J. Appl. Mechs.*, Vol. 24, 1957, 46–52.
83. HARRIS, FREDRIC J.
    'On the Use of Windows for Harmonic Analysis with the Discrete Fourier Transform', *Proc. IEEE*, Vol. 66, 1978, 51–83.
84. LIN, Y. K.
    Probabilistic Theory of Structural Dynamics, McGraw-Hill, New York, 1967.
85. LONGUET-HIGGINS, M. J.
    'The Statistical Analysis of a Randomly Moving Surface', *Trans. Roy. Soc. Lond.* A 249, 1957, 321–87.
86. LYON, R. H.
    Statistical Energy Analysis of Dynamical Systems: Theory and Applications, MIT Press, Cambridge, Massachusetts, 1975.
87. MANSOUR, A.
    'Methods of Computing and Probability of Failure under Extreme Values of Bending Moment', *J. Ship Res.*, Vol. 16, 1972, 113–23.
88. MELBOURNE, W. H.
    'Probability Distributions Associated with the Wind Loading of Structures', *Civil Engng Trans., Instn Engrs Australia*, Vol. 19, 1977, 58–67.

89. NIGAM, N. C.
  Introduction to Random Vibrations, MIT Press, Cambridge, Massachusetts, 1983.
90. OHTA, M., HATAKEYAMA, K., HIROMITSU, S. and YAMAGUCHI, S.
  'A Unified Study of the Output Probability Distribution of Arbitrary Linear Vibratory Systems with Arbitrary Random Excitation', *J. Sound Vib.*, Vol. 43, 1975, 693–711.
91. OPPENHEIM, A. V. and SCHAFER, R. W.
  Digital Signal Processing, Prentice-Hall, Englewood Cliffs, New Jersey, 1975.
92. PAPOULIS, ATHANASIOS
  The Fourier Integral and its Applications, McGraw-Hill, New York, 1962.
93. PRIESTLEY, M. B.
  Spectral Analysis and Time Series, Vols. 1 and 2, Academic Press, New York, 1981.
94. RABINER, L. R. and GOLD, B.
  Theory and Applications of Digital Signal Processing, Prentice-Hall, Englewood Cliffs, New Jersey, 1975.
95. SAXTON, W. O.
  Computer Techniques for Image Processing in Electron Microscopy, Academic Press, New York, 1978.
96. VANMARCKE, E. H.
  'Method of Spectral Moments to Estimate Structural Damping', *Stochastic Problems in Dynamics*, B. L. Clarkson (ed.), Pitman, London, 1977, 515–24.
97. VANMARCKE, ERIK
  Random Fields, MIT Press, Cambridge, Massachusetts, 1983.
98. VINCENT, C. H.
  'The Generation of Truly Random Binary Numbers', *J. Physics E: Scientific Instruments*, Vol. 3, 1970, 594–8.
99. VINCENT, C. H.
  Random Pulse Trains – their Measurement and Statistical Properties, IEE Monograph Series 13, Peter Pereginus for the IEE, London, 1973.
100. YAGLOM, A. M.
  An Introduction to the Theory of Stationary Random Functions, Prentice-Hall, Englewood Cliffs, New Jersey, 1962.

References for the third edition

101. CHUI, C. K.
  An Introduction to Wavelets, Academic Press, San Diego, 1992.
102. CHUI, C. K. (ed.)
  Wavelets: A Tutorial in Theory and Applications, Academic Press, San Diego, 1992.
103. DAUBECHIES, I.
  'Orthonormal Bases of Wavelets with Finite Support – Connection with Discrete Filters', *Proc. Int. Conf. on Wavelets*, Marseille, 1987, published as pages 38–66 of *Wavelets, Time-Frequency Methods and Phase Space*, J. M. Combes, A. Grossmann and Ph. Tchamitchian (eds), Springer-Verlag, Berlin–Heidelberg, 1989.
104. DAUBECHIES, I.
  'Orthonormal Bases of Compactly Supported Wavelets', *Comm. Pure & Appl. Maths*, Vol. XLI, 1988, 909–96.

105. DAUBECHIES, I.
   'The Wavelet Transform, Time-Frequency Localization and Signal Analysis',
   IEEE Trans. on Information Theory, Vol. 36, 1990, 961–1005.
106. DAUBECHIES, I., MALLAT, S. and WILLSKY, A. S. (eds)
   'Wavelet Transforms and Multiresolution Signal Analysis', *Special Issue of
   IEEE Trans. Information Theory*, Vol. 38, No. 2, March 1992.
107. GOUPILLAUD, P., GROSSMANN, A. and MORLET, J.
   'Cycle-Octave and Related Transforms in Seismic Signal Analysis',
   Geoexploration, Vol. 23, 1984, 85–102.
108. HAAR, A.
   'Zur Theorie der orthogonalen Funktionensysteme', *Math. Ann.*, Vol. 69, 1910,
   331–71.
109. HEIL, C. E. AND WALNUT, D. F.
   'Continuous and Discrete Wavelet Transforms', *SIAM Review* (Soc. for
   Industrial and Appl. Maths), Vol. 31, 1989, 628–66.
110. MALLAT, S.
   'A Theory for Multiresolution Signal Decomposition: The Wavelet
   Representation', *IEEE Trans. Pattern Anal. and Machine Intell.*, Vol. 11, 1989,
   674–93.
111. MALLAT, S. and ZHONG, S.
   'Signal Characterization from Multiscale Edges', *Proc. 10th Int. Conf. on
   Pattern Recognition*, IEEE Comput. Soc. Press, Atlantic City, 1990, 891–6.
112. NEWLAND, D. E.
   'Wavelet Analysis of Vibration', *Proc. Structural Dynamics and Vibration
   Symposium*, ASME Energy-Sources Technology Conf., Houston, 1993,
   PD-Vol. 52, 1–12.
113. NEWLAND, D. E.
   'Some Properties of Discrete Wavelet Maps', *Probabilistic Engineering
   Mechanics*, Vol. 8, 1993.
114a. NEWLAND, D. E.
   'Harmonic Wavelet Analysis', *Proc. R. Soc. Lond.* A, 1993 (to be published).
114b. NEWLAND, D. E.
   'Musical Wavelets', 1993 (to be published).
115. STRANG, G.
   'Wavelets and Dilation Equations: A Brief Introduction', *SIAM Review*, Vol.
   31, 1989, No. 4, 614–27.
116. STRANG, G. and FIX, G.
   'A Fourier Analysis of the Finite Element Variational Method', in *Constructive
   Aspects of Functional Analysis*, Edizioni Cremonese, Rome, 1973.
117. VILLE, J.
   'Théorie et Applications de la Notion de Signal Analytique', *Câbles et
   Transmission*, Vol. 2, 1948, 61–74.
118. WIGNER, E. P.
   'On the Quantum Correction for Thermodynamic Equilibrium', *Physical
   Review*, Vol. 40, 1932, 749–59.

Index

Accuracy of
  measurements, 95
  spectral analysis, example calculation, 102
  spectral estimates, 138
Addition of zeros to sample record, 139
Aliased spectral window, 251, 427
Aliasing, 118
  distortion, 120
Alias spectra, 120
Analogue
  digital conversion, 149
    error analysis, 149
  spectrum analysis, 95
  spectrum analyzer, 95
Analysing wavelet, 298
Aperiodic functions, Fourier transform of a
    train of, 129
Artificial generation of
  isotropic surface, 258, 263
  random process, 167
  two-dimensional random process, 256
Autocorrelation, 25
  function
    definition of, 25
    for a pseudo random binary signal, 173
    for a random binary process, 168
    for white noise, 46
    properties of, 26, 27
    spatial, 198
  input–output relation for a linear system, 69
Average frequency of crossings, 86
Averages, second order, 14
Average value of a function of random
    variables, 14
Averaging
  spectral estimates, 138
  time of a spectrum analyzer, 95

Bandwidth
  half-power, 191
  mean square, 191
  of analogue spectrum analyzer, 96
  of spectral window, 108
Beam, response to random excitation, 281, 285

Bias error of spectral measurement, 103
Binary
  random process, 168
  sequence, full length, 433
Bingham window, 146
Binomial probability distribution, 175
Broad band random process, 44

Calculation
  of averages, 7
  of spectral estimates from discrete data, 120
Central limit theorem, 80
Chain code, 183
Chi-square
  random variable, 109
    distribution of, 110, 417
  probability density function, 417
  probability distribution
    application to spectral analysis, 138
    dependence on degrees-of-freedom, 110
  tabulated values of distribution, 381
Circular correlation function, 122
  interpretation in terms of linear correlation
      function estimate, 129
  relationship with linear correlation function,
      128, 129
  two-dimensional, 241, 243, 261
Circular wavelet transforms, 322
Clock frequency of random binary process,
    170
Coherence functions, 207
  digital calculation of, 210
  multiple, 207, 209
  ordinary, 209
  partial, 438
Complex frequency
  response function, 56
    measurement of, 206
Computer programs
  FFT in Basic, 374
  FFT in Fortran, 373
  Two-dimensional DFT, 376
  Wavelet, 387
Conditional probability, 17

Confidence limits
of spectral measurements, 108
sample calculation, 112
Continuous time series, 113
Convolution, 64, 412
by wavelets, 353
integral, 64
Correction for slow trend in spectral analysis, 147
Correlated random noise, synthesis of, 185
Correlation, 21
coefficient, 23, 241
function (*see also* autocorrelation, cross-correlation and circular correlation)
calculation from data extended by additional zeros, 139
circular, 122
consistent estimate for, 127
error due to "wrap around", 125
sample average, 126
separation of overlapping parts by added zeros, 142
Cosine data taper function, 147
Covariance, 23
Cross-correlation, 29
function
definition of, 29
properties of, 30
input–output relation for a linear system, 77
Crossing analysis, 85
Cross-spectral density
between parallel tracks on a random surface, 263
definition of, 48
from multi-dimensional spectral density, 279
from two-dimensional spectral density, 237
input–output relations for a linear system, 79, 414
properties of, 49

Data
taper function, 147
window, 147
cosine taper, 146, 425
Degrees-of-freedom, for chi-square distribution, 110
Delta function, units of, 46
Derived process, 46
spectral density of, 46
Digital filtering, 182
Digital simulation of random processes, 186
Digital spectral analysis, 113
correction for slow trend, 147
data weighting, 146
advantage of, 148
example showing choice of parameters, 145
practical considerations, 146
quantization error, 148
summary of method, 142
windows and smoothing, 125

Dilation equations, 300
Dilation wavelets, 303
Dirac's delta function, 46
Direct method of spectral analysis, 135
Discrete Fourier transform, 114
definition, 116
Parseval's theorem for,
one-dimensional, 420
two-dimensional, 233
properties of, 118
properties of two-dimensional, 229
two-dimensional, 226
Discrete harmonic wavelet transform, 364, 398
Discrete spectral density, 122
two-dimensional, 239, 262
Discrete time series, 113
Discrete wavelet analysis, 295 et seq.
Discrete wavelet transform, 326
properties of, 339
Distribution of peaks, 90
Distribution of reverse arrangements for a random series, table of values, 385
Double-sided function of frequency, 52, 199
Duality theorem of Fourier analysis, 407
DWT (*see* discrete wavelet transform), 326
Dynamic characteristics, measurement of, 167, 206

Effective bandwidth of spectral window, 108, 141
Ensemble averaging, 19
concept for a linear system subjected to random excitation, 67
Envelope distribution for a narrow band random process, 417
Environmental vibration simulation, 201
number of separate inputs required, 205
Ergodic random process, 20, 211
Ergodicity, discussion of, 211
Excitation by random surface irregularities, 196
Excitation–response relations, 53
Extending record length by adding zeros, 139
to separate overlapping parts of linear correlation function, 142

Failure due to random vibration, 191
mechanisms of, 192, 193
Fast Fourier transform, 114, 150
algorithm, 155
alternative algorithms, 165
basic theory, 150
bit-reversed counter, 162
butterfly calculation, 152
computer program, 373
decimation in frequency algorithm, 166, 431
decimation in time algorithm, 166
flow chart for bit-reversed counter, 163
flow chart for re-ordering data sequence, 164
practical value, 162
programming flow chart, 155, 156, 159
sample calculation, 152

Fast Fourier transform, *continued*
  signal flow graphs, 155, 157
  twiddle factor, 155
Fatigue failure
  fractional damage, 192
  Palmgren–Miner hypothesis, 192
  prediction of mean lifetime, 193
Fatigue law of a material, 192
FFT (*see* fast Fourier transform), 150
FFT computer program, 373
Finite length records, analysis of, 103
First crossing problem, 194
  mean time to failure, 196
  standard deviation of time to failure, 196
First passage probability density, 196
Folding frequency, 120
Fourier analysis, 33
  duality theorem, 407
Fourier integral
  and Fourier series, relationship between, 36
Fourier series, 33
  calculation of coefficients, 34
  graphical representation of coefficients, 34
  multi-dimensional, 226
  two-dimensional, 225
Fourier transform, 35
  complex form of, 38
  fast, 150
  of a train of aperiodic functions, 129
  of periodic function, 116
  pair, 39
  short-time, 218
Fourth-order average for a Gaussian process, 98
Frequency
  negative, 39
  of crossings, 86
    for a Gaussian process, 88
  of maxima, 92
    comparison with frequency of zero
      crossings, 93
  of zero crossings, 89
    comparison with frequency of maxima, 93
  response function, 55
    measurement of, 206
    relation to impulse response function, 60
  response method, 55
Full-length binary sequence, 433

Gamma function, definition of, 417
Gaussian
  distribution, tabulated values, 379
  probability distribution, 7
    second-order, 18
  probability integral, table of, 379
  random process, 20
Generalized function, 46, 408
Generation of random numbers, 183

Haar wavelet, 321, 322, 450
Half-power bandwidth, 191

Harmonic wavelet, 359, 398
Harmonic wavelet transform, FFT algorithm
  for, 367

Impulse response
  function, 58
  method, 58
Inverse discrete Fourier transform, 116
Inverse Fourier transform, 35
Isotropic random process, 239, 258

Joint probability, 12
  density function, 12

Lag window, 131
  basic form of, for digital spectral analysis, 136
  relationship with spectral window, 136
Line of regression, 22
Linear superposition, principle of, 53
Linear system
  equations of, 53
  excitation-response relations, 53
  principle of superposition, 53
  response to an arbitrary input, 62
  transmission of random vibration, 67
Locally stationary process, 217

Mallat's algorithm,
  tree (pyramid), 326, 389
Matlab information, 477
Mean level, input–output relation for a linear
  system, 68
Mean square bandwidth, 191
Mean square response
  approximate calculation of, 189
  exact calculation of, 73, 371
  input–output relation for a linear system, 73
  list of integrals, 371
Mean square spectral density, 42
Mean square wavelet maps, 348, 393, 399
Measurement accuracy, 95
Modal mean square bandwidth, 191
Modulo, 187
Multiple coherence function, 207, 209

Narrow band random process, 44
  crossing analysis, 85
  description of, 82
  distribution of peaks, 90
  envelope distribution, 417
  frequency of maxima, 92
  peak distribution, 90
  statistics of, 82
Negative frequency, 39
Non-stationary random process, 211
  analysis of locally stationary process, 217
  short-sample average, 212
  spectral density of, 214
  testing for, 212
Non-stationary trend, testing for, 212

Normal mode analysis, 280
Normalized covariance, 23
Nyquist frequency, 120

Ordinary coherence function, 209

Parseval's theorem, 233, 407, 420, 442
  for a two-dimensional series, 233
Partial coherence function, 438
Pascal's triangle, 177
Peak distribution,
  Rayleigh, 91
  Weibull, 219
Periodic function, Fourier transform of, 116
Periodogram, 429
Plate, response to random excitation, 283,
  290, 292
P.r.b.s., *see* pseudo-random binary signal, 172
Pre-whitening before spectral analysis, 148
Principle of linear superposition, 53
Probability analyzer, 6
Probability density function
  calculation of first-order from second-order,
    13
  chi-square, 417
  conditional, 17
  first-order, 2, 4
  Gaussian (normal), 7, 18
  second-order, 12, 17
Probability distribution
  binomial, 175
  chi-square, 110, 381, 417
  function, 10
  Gaussian (normal), 7, 18, 379
Probability integral, tabulated values of
  Gaussian, 379
Probability of a first crossing in time T,
  195
Pseudo random binary sequence, 172, 183, 433
Pseudo random binary signal, 172, 183, 433
  autocorrelation function of, 173, 433
  clock frequency, 173, 182
  sequence lengths, 172, 182, 433
  spectral density of, 174, 434
Pseudo random process (*see also* pseudo
  random binary signal), 167, 432
  application of, 167
  multi-level, 175, 432
    spectral density of, 179
Pyramid algorithm, 326, 389

Quantization error, 148

Random binary process, 168
  autocorrelation function of, 168
  clock frequency, 170
  derived from a binary sequence, 168
  pseudo, *see* pseudo random binary process
    spectral density of, 171
Random multi-level process, 175

Random numbers,
  generation of, 183
  table of, 383
Random process
  binary, 168
  broad band, 44
  digital simulation of, 185
  Gaussian, 20
  isotropic, 239, 258, 263
  locally stationary, 217
  multi-dimensional, 233
  multi-level, 175
  narrow band, 44, 82
  non-stationary, 211
  simulation of, 168, 185
Random surface irregularities
  excitation by, 196
  response to, 197
Random variables, statistical independence of,
  18, 311
Random vibration environment, simulation by
  single point excitation, 201, 203, 205
Rayleigh distribution of peaks, 91
Resonant mode, response to broad band
  excitation, 189
Response of a continuous system
  by normal mode analysis, 280
  to distributed excitation, 277
  to point excitation, 268
Response of a linear system, 53
  to random excitation
    autocorrelation, 69
    cross-correlation, 77
    cross-spectral density, 78, 414
    mean level, 68
    mean square response, 73
    spectral density, 71
    probability distributions, 79
Response of a resonant mode to broad band
  excitation, 189
Reverse arrangements
  distribution for a random series, 385
  tabulated values of distribution, 385
  trend test, 213
Road surface roughness, 197, 437

Sampling theorem, 420
Scaling function, 301
Separation of overlapping linear correlation
  functions by added zeros, 142
Second-order averages, 14
Shannon's sampling theorem, 420
Shift register, 183
  operation of, 184
Signal flow graphs, 155
Simulation of random environments, 201
Simulation of random processes, 185
Smoothed spectral density, 135
  estimate of, 135

Smoothing spectral estimates, 137
  two-dimensional, 254
Smoothing with a spectral window, 107, 136
S–N fatigue curve, 192
Space lag, 198
Spatial spectral density of road surface, 198, 258
Spectral analysis, 41
  analogue, 95
  bias error, 103
  digital, 113
    correction for added zeros, 142
    correction for slow trend, 147
    data window, 146, 425
    estimate of smoothed spectrum, 135
    final smoothing, 144
    periodogram, 429
    practical considerations for, 146
    pre-whitening, 148
    quantization error, 148
    summary of method, 142
  direct method, 135
  of finite length records, 103
  summary of digital method, 142
  variance of measurement, 97
Spectral density, 42
  double frequency, 214
  for a multi-dimensional process, 233, 237
  for a non-stationary process, 214
  input–output relation for a linear system, 71
  of derived process, 46
  of multi-level random process, 179
  of pseudo random binary process, 173, 174
  of random binary process, 171
  properties of, 43
  relation between single-sided and
      double-sided, 52
  smoothed, 107, 135
  spatial for road surfaces, 197, 437
  units of, 42, 51
Spectral estimates
  accuracy of, 138
  averaging, 138
  calculation of, 120
  smoothing, 137
  two-dimensional smoothing, 254
Spectral measurements, confidence limits, 108
Spectral window, 104, 131, 138
  aliased, 251, 427
  bandwidth of, 108, 138, 141
  basic form of, for digital spectral analysis, 136
  relation with lag window, 136
  shape of, 107, 136, 138, 141
  smoothing with, 107
  two-dimensional, 245
  two-dimensional aliased, 247, 251
Spectrum, *see* spectral density
Spectrum analysis, *see* spectral analysis
Spectrum analyzer
  averaging time of, 95
  variance of measurement, 97, 101

Standard deviation, 8
Stationarity, discussion of, 212
Stationary random process, 20, 212
Statistical energy analysis, 292
STFT (*see* short-time Fourier transform), 218
Strictly stationary process, 20
String, response to random excitation, 269, 275
Superposition integral, 64
Synthesis of correlated noise sources, 185

Time lag, 25, 198
Time series analysis, 113
Transmission of random vibration, 67
Tree algorithm, 326, 389
Trend test, 213
  reverse arrangement, 213

Uncertainty principle, 218
Unit impulse, 58
  dimensions of, 58

Variance, 8
Variance of spectral analyzer measurement,
      97, 101
Vehicle response to road irregularities, 197,
      199
  example calculation, 199
Ville, J., 218

Wavelet
  analysing, 298
  compactly supported, 308
  definition of D4, 321
    of D6, 322
  dilation, 300
  Haar, 321, 322, 450
  harmonic, 359, 398
    real and imaginary parts of, 361
Wavelet analysis, 295 et seq.
Wavelet coefficients,
  conditions,
    collected, 320
    for accuracy, 309
    for orthogonality, 313
  properties of, 307 et seq.
  summary of orthogonal properties, 318
Wavelet computer programs, 387
Wavelet convolution, 353
Wavelet generation,
  by iteration, 301 et seq.
  by recursion, 451
Wavelet level, 295
Wavelet maps,
  grid base for, 350
  mean square, 348, 393, 399
  re-ordering elements of, 349
Wavelet scaling function, 301
Wavelet transforms,
  circular, 322
  discrete, 326

Wavelet transforms, *continued*
  discrete harmonic, 364, 398
  FFT algorithm for, 367
  two-dimensional, 355, 394
Wavelets, 219, 295 et seq.
Wavenumber, 196
Weakly stationary process, 20
Weibull distribution of peaks, 219
Weighted spectral density, 135
Wigner, E. P., 218

Wigner distribution, 218
Wigner-Ville method, 218
Window function, 133, 135, 137
  two-dimensional, 245
White noise, 45

Zero crossings, frequency of, 89
Zeros, addition to sample record before
  analysis, 139

INFORMATION ABOUT COMPUTER SOFTWARE

The computer programs in Appendix 7 use the interactive software package MATLAB®. MATLAB is available for MS-DOS® based personal computers, Macintosh® computers, and most workstations and larger computers.

For information on MATLAB, contact

> The MathWorks, Inc.
> Cochituate Place
> 24 Prime Park Way
> Natick
> MA 01760
> USA
>
> Tel: +1 508 653 1415
> Fax: +1 508 653 2997
> email: info@mathworks.com

A wavelet toolbox of MATLAB M-files based on the programs described in Appendix 7 of *Random Vibrations, Spectral and Wavelet Analysis* is planned. Details can be obtained from the author:

> Prof. D. E. Newland
> Cambridge University Engineering Department
> Trumpington Street
> Cambridge
> CB2 1PZ, England
>
> Fax: +44 223 359153*
> email: den@eng.cam.ac.uk

*The area code is expected to change from 223 to 1223 in April 1995.
MATLAB is a registered trademark of The MathWorks, Inc.
MS-DOS is a registered trademark of Microsoft Corporation
Macintosh is a registered trademark of Apple Computer, Inc.